T0118600

Science in Denmark
A Thousand-Year History

Science in Denmark
A Thousand-Year History

Aarhus University Press |

Science in Denmark

Designed and typeset by Jørgen Sparre
Cover design by Jørgen Sparre
Cover illustration: S.P.L. Sørensen in the Carlsberg Laboratory.
Painted by Niels Vinding Dorph, 1922.
Printed by Narayana Press, Denmark
Printed in Denmark 2008
ISBN 978 87 7934 317 7

Aarhus University Press
Langelandsgade 177
DK-8200 Aarhus N
www.unipress.dk

White Cross Mills
Hightown, Lancaster, LA1 4XS
United Kingdom
www.gazellebookservices.co.uk

PO Box 511
Oakville, CT 06779
www.oxbowbooks.com

Published with the financial support of
The Carlsberg Foundation

Contents

Preface

Science in Denmark: A Thousand-Year History is based on *Dansk Naturvidenskabs Historie*, its four-volume predecessor written in Danish and spanning 2,000 pages. Those volumes, initiated in 2001 and published successively during 2005 and 2006, were the result of a research project conceived by the History of Science Department, now the Department of Science Studies, at the University of Aarhus. Well aware that Danish is a language read by relatively few people, we, the editors, decided early on that the original four volumes deserved to be condensed into an English-language version that would allow an international readership to learn about the development of science in the Kingdom of Denmark. The fruits of our labour lie before you.

Had it not been for the coordinated efforts of a considerable number of people, *Science in Denmark: A Thousand-Year History* would never have existed. We take this opportunity to sincerely thank the more than 30 different authors who have contributed chapters to one or multiple volumes of the original *Dansk Naturvidenskabs Historie.* Drawing on their studies of a variety of public and private archives, libraries and other records, each and every author has delivered an important contribution to researching and expanding our knowledge of science. What is more, they have unflinchingly taken part in editorial dialogues to determine the best possible outline for their contributions to the Danish four-volume work. In this work, however, the four editors are fully responsible for the cropping, selection and wording that has resulted in this abridged English version, except in certain sections marked by an explanatory footnote.

We also thank Aarhus University Press, who has been an excellent partner throughout the project, and whose staff has consistently strived to achieve a handsome and thoroughly professional end product. Furthermore, many grateful thanks are due to the Carlsberg Foundation for so generously supporting the underlying research and pre-press work that went into all five of the project's publications. We have done our utmost to prove ourselves worthy of the trust placed in us, and it is no overstatement that without the financial backing from the foundation, this project could never have been brought to fruition.

Last but not least we would like to thank our translator, Heidi Flegal, for her warm enthusiasm and her unfailing interest in making the English text as faithful and accurate as possible.

Aarhus, May 2008

Helge Kragh, Peter C. Kjærgaard,
Henry Nielsen and Kristian H. Nielsen

Prologue

Science has played an important role in Denmark's history ever since astronomer Tycho Brahe's day. Even so, the history of Danish science has never been described as a contextualized whole, as an integrated part of its cultural and social development – which is what this work intends to do. Preparing comprehensive national accounts of areas such as literature, the visual arts, ballet, architecture, sports and philosophy has been considered quite natural, but science has never been included in this tradition. The closest thing to such a history would have to be the impressive book project associated with the University of Copenhagen's 500th anniversary in 1979, three volumes of which chronicle the academic history of the scientific and medical subjects taught there.[1] While this is an extremely valuable source, its scope is limited to the history of university subjects in Copenhagen, in addition to which it suffers from several weaknesses arising, not least, from the rather narrow subject-based or discipline-oriented approach that typifies it. What is more, the work was written in Danish and has not since been abridged or converted into an English-language publication. One does occasionally find historical accounts of Danish science written in English, with notable contributions made by Valdemar Meisen, Jole Shackelford and Andrew Jamison. These are brief, however, and in no way constitute the sort of modern, comprehensive coverage that is called for.[2]

There is a long-standing tradition – one that is also prevalent in Denmark – of scientists studying the historical development of their own fields. The products of this tradition have usually been field-specific, internalistic descriptions that often focus on the nation's great scientists and their innovative efforts. The body of history-of-science literature in this genre is very large. It is also very scattered and, on the whole, lacks a broader perspective, often displaying a certain historiographic naivety as well. This literature is usually directed towards the researchers themselves rather than being aimed at a wider, historically interested audience. The present history of science in Denmark seeks to remedy the situation by giving a comprehensive account that integrates scientific developments with general history wherever this is feasible. Our goal has been to weave science into Denmark's historical tapestry – to bestow upon science a status comparable to that of other cultural factors – and to integrate it into the larger fabric of international development.

The approach in this work is contextual, in that the authors have sought to place the scientists and their activities - their opinions, theories and experiments - within a wider social, cultural, ideological, or other similar context. This is not tantamount to writing a history of Danish science that neglects the scientific content. Rather, it is an attempt to show equal appreciation for both aspects. Even though scientific disciplines such as botany, geology and physiology are realities and have provided essential support for the structure of scientific development, it is important to point out that this history of science is more than a sum of discipline-based historical accounts. Not only have the various disciplines changed significantly over the years. They have also, at times, formed alliances that now seem foreign to us, and for many modern-day disciplines it would be positively anachronistic to trace them back to the 1800s or earlier. It would not be unreasonable to claim that it is also anachronistic to speak of "scientists" (in the sense "scientific researchers") in early history, when conditions were very different from those prevailing today. And yet we do just that, partly because there is a tradition of doing so, and partly because at the end of the day scientific development embodies a continuity. The "savants" and "naturalists" and "natural philosophers" of old belonged to a breed not entirely unlike the "scientists" of later eras.

Denmark, being a small country, can only lay claim to a very modest share of the important breakthroughs that have been internationally decisive to humankind's progress in understanding the physical world. Nevertheless, the country has made some remarkable contributions, and of course this book describes these highlights, exemplified by such luminaries as Tycho Brahe, Hans Christian Ørsted and Niels Bohr. Yet in striking resemblance to the topography of Denmark itself, be the high points ever so prominent, they are surrounded by a flat landscape that is far more typical of the Danish history of science. That is why this level landscape, though not always as intellectually stimulating as the rolling hills, must be thoroughly described. Most particularly, in a small country like Denmark natural science (like other sciences, and like culture in general) has largely developed by means of receiving, and adapting to, science from abroad. Reception history therefore holds a prominent place in this work. The history of institutions and organizations is another area we have given a high priority, in contrast to the place it holds in most traditional accounts. In general, we put more emphasis on the physical, financial and cultural confines within which scientific research was obliged to operate. Such circumstances alone do not determine the results that are achieved, but they do set the pace locally and give certain areas precedence over others.

Finally, the history of science is obviously not identical with the history of the research that the scientists carried out. It encompasses, or is linked to, many other areas as well, including the various fields of higher education and the wider propagation of scientific knowledge. Popular science in Denmark goes back to about 1600, and it has since continued to play a significant role – which is why this oft-neglected aspect receives considerable attention in our accounts. Undoubtedly, one reason why the history of the natural sciences does not enjoy the same status as the history of other scientific, scholarly and cultural fields is that the exact sciences are perceived as "difficult". And so they are, although hardly more so than many other branches of learning. Even with the contextual approach, we try to give readers additional insight into the content of the scientific contributions originating in Denmark. Indeed, here context and content are inseparable. The level of technical-scientific complexity is modest, however, and our intended target group will hardly find it daunting. At any rate, this book does not aim to instruct its readers in the technical details underlying the scientific exploits of the past, but those with a particular interest in such details should find the secondary literature in the reference lists helpful.

Given that development in Denmark has always been highly dependent upon events in other countries, it is not possible to understand Danish science without also considering the wider international perspectives. The text therefore includes brief accounts of international developments, as well as references to other useful sources, and generally it embeds the country's national history of science in an international history. One could, of course, ask whether it is even remotely meaningful to portray the history of science within a national framework, since natural science, perhaps more so than any other branch of science, pays no heed to national borders – at least as far as its methods and findings are concerned. Science is, in essence, transnational, and very often international. We do not discuss that issue here, however, except to point out that from a historiographic point of view there is no decisive difference between the history of science and the history of, say, literature or art. Electromagnetism is admittedly a universal phenomenon, independent of time and place, but the phenomenon itself was proved to exist in a specific historical and local setting by a Danish scientist working in Copenhagen, and the circumstances that led to the discovery and subsequent events were part of a particular Danish history that unfolded around 1820. It is equally clear that every scientific institution in Denmark has been influenced by the conditions that reigned in Denmark and by the country's general history.

In short, it is in many ways quite meaningful to describe science, like other spheres of activity, within a national framework, although clearly such a history should not be restricted to that framework alone. Certainly our book is not the first modern example of a national history of science written with an international audience in mind. The genre has earned considerable interest over the last decade or so and been cultivated in several small countries. Excellent examples are the comprehensive history of science published in the Netherlands in 1999, and the Belgian project that was completed in 2001 and resulted in two exquisite volumes that appeared in French and Flemish.[3]

On the other hand, writing a national history of science is no easy task, and at all events the attempt is bound to result in a different sort of historiography than one that is internationally oriented – the latter normally being unconcerned with the scientists' nationalities. One of the obvious problems is striking a balance between, or rather determining the difference between, national and nationalistic historiography. Many traditional scientific histories (and other sorts of histories) written from a national perspective have belonged to the nationalistic or patriotic genre. They have been barely disguised tributes to the authors' own native country and its great sons – or, so much more rarely, its great daughters. There is no reason why a history resting within a national framework must necessarily be a part of this dubious genre, however; no reason why the history of science in Denmark cannot be written adhering to the same critical and objective standards that generally apply to historiography.

We have made a conscious effort to offer a critical and reflective account that pays no less attention to weak points and setbacks than it does to strong points and progress. Science, in terms of its institutions and its cognitive advances, is strongly typified by progress, although this is mainly true when it is regarded over long periods of time and from an international perspective. If one descends to a national level, delving into the finer structural details, it becomes absolutely clear that scientific development is not the result of an unbroken sequence of progressive steps forward. A nationally oriented history cannot, and should not, be merely a fragment of the internationally accepted history of science. It is legitimate, indeed unavoidable, that some individuals and institutions are magnified, simply because they *were* important in a local context, however insignificant they may have been internationally. This kind of "legitimate magnification" seems justified in the case of figures who influenced the organizational, political and educational aspects of science, or who contributed to the dissemination and popularization of knowledge. When it comes to the cognitive level, however – the factual scientific contributions – such magnification becomes much less legitimate.

As for the phrase "Danish history of science", it contains an implicit claim that words such as "Danish" and "science" denote historically meaningful unitary concepts; that these words have maintained some sort of invariance over a period of more than 500 years. At the very least, a work like this one must take a position on the corresponding issues of limitation or demarcation, for truly the question of how one is to construe these phenomena in a historical sense is far from simple. Dealing first with the designation "Danish", it mainly refers to the scientific activities that took place within the varying boundaries of the Danish (or Danish-Norwegian) kingdom, such as they were at any given time. One consequence of this is that science relating to Norway is dealt with up until 1814 – a pivotal year in Norway's struggle for independence – whereas developments in Holstein – the disputed duchy which, to all intents and purposes, remained German – and its university in Kiel are only dealt with sporadically. Several of the scientists mentioned in this book worked in Denmark without being Danish citizens or otherwise having any national affiliation with the country. Generally we have not been very interested in each scientist's nationality, and this work does not deal primarily with science done by Danish scientists, but with research that took place in, and was relevant to, the natural and cultural entity that was Denmark.

One recurring question is whether there are any particular characteristics that apply to the Danish history of science, and which are, so to speak, contingent upon Danishness. Yes, science is international, but might there still be some particularly Danish approach, or certain attitudes and priorities that typify Danish scientists throughout the entire period? The issue has been raised, and some have claimed that Danish scientists, more so than their colleagues in other countries, are notable for their moderation, eclecticism and pragmatism.[4] According to the Danish physician and historian of medicine Julius Petersen, such personal qualities were specifically characteristic for Danes, and he detected their presence in Danish researchers working within the fields of medicine and natural science. As he concluded in his book from 1893: "One might venture to say that a certain sobriety, a certain critical awareness, and hence an overall lack of inflammability in the face of explosive movements abroad, both in former and more recent times, is a fairly prominent and not undesirable national characteristic of the Danes."[5] The soundness of this character description is, of course, open to question. But what is surely beyond question is that this characteristic has been, and may perhaps still be, a significant element in the Danes' self-understanding – as witnessed by the following description appearing in a publication from 1937 that portrayed the Danish 16th-century physician and professor Christian Therkelsen Morsing: He was "a true Danish scientist, seldom extreme, preferring to remain sober-minded."[6]

Moving on to the concept of a "national style", the question of whether this concept is applicable to science is both interesting and problematic. Although we shall refrain from discussing the issue here, it is worth noting that what is purportedly the "Danish national scientific style" is almost identical to the research style that several historians and scientists from the Netherlands have found to be characteristic of their own compatriots.[7] According to this perception, Dutch scientific research is thought to have been particularly practical and professional, while being foreign, or even hostile, to the more metaphysical speculations that typified scientific activity in many other countries. The very fact that one can apparently observe the same "national style" in the Netherlands and Denmark should, perhaps, give rise to some measure of scepticism regarding the concept's historiographic value.

Chronologically, *Science in Denmark: A Thousand-Year History* begins in time immemorial, which for our purposes means around the year 1200. The historical descriptions progress from that point onwards to the 1960s, alternating between sections that are chronological, thematic and subject-oriented. Developments over the last several decades, approaching the new millennium, are only outlined in the broadest of contours. Like all cut-off dates, the ones we use to delimit the individual chapters and parts are, to some degree, defined by conventional usage and practical circumstances. However, they are by no means arbitrary: The years around 1540, 1730, 1850 and 1920 marked significant shifts in Danish science and in the conditions under which it existed.

Finally, as far as the concept of "science" is concerned we have chosen quite a liberal interpretation, since it is necessary to consider what was perceived as "natural science" or "scientific research" in the past, and what was relevant to that category of knowledge. This not only means that our account includes areas that are seen today as non-scientific or even pseudoscientific (such as alchemy, electrotherapy and phrenology). It also means that we mention some of the trends and ideas that in various ways place the development of Danish science in perspective. Although we pay special attention to discoveries and factual advances achieved through basic research, it is impossible to separate this from the applied research benefitting such fields as medicine, agriculture and technology. A number of these applied sciences are obvious candidates for inclusion here, whereas the clinical and social aspects of medical science are rarely mentioned. The same goes for traditional, practice-based technology. This work is quite comprehensive, but needless to say it is not a "complete" history of science in Denmark. Finally it is worth noting that although they have been omitted here, many events and figures that are mainly, if not solely, of local interest are covered in the larger Danish version.

The book is based, in part, on new research, but naturally it also relies on the large body of literature in the field – which is mainly written in Danish and typically deals with specialized subject matter. This makes *Science in Denmark: A Thousand-Year History*, with its numerous bibliographical references, uniquely useful as a guide to the extensive secondary literature that already exists.

PART I

From Medieval Scholarship to New Science

Circa 1000–1730

Helge Kragh

From Viking Age to Absolute Monarchy

I

Science, in a form that would be more or less recognizable today, became established in Denmark during the 1500s, but its roots stretch much farther back. There were interesting examples of what could be called protoscience as far back as the Viking Age. It has even been claimed that the runic inscriptions on the Golden Horns – two national heirlooms dating from the 400s, unearthed in 1639 and 1734 – show evidence of astronomical and astrological learning. The historians of science Willy Hartner and Aage Drachmann believe that the Golden Horns demonstrate an influence from southern European culture and show that there were people in south Jutland who harboured interests that were scientific in nature.[8] This interpretation is controversial, however, and not widely accepted. Actually it is not until Denmark reaches the Catholic medieval period that one can really detect a development containing any sort of sustained natural philosophy or – to use the term somewhat anachronistically – any sort of science. Danish historiography often marks the beginning of the Middle Ages around 1050 with king Sven Estridsen's establishment of bishoprics, letting it end with the dawn of the Reformation in Denmark in 1536. Throughout this period of almost 500 years, the Catholic Church was the supreme institution in Denmark, and in most of the rest of Europe as well. Indeed, the earliest seats of learning in the region revolved around the cathedrals in Roskilde and Lund, and the first Danish contributions to natural philosophy originated within these bishoprics. Later, in the 1200s, a small but remarkable circle of Danish scholars appeared at the university in Paris and elsewhere in the countries south of Denmark.

After 1300 there was a prolonged lull in the advancement of Danish knowledge and learning, and the country was little influenced by the ideological currents of nascent Renaissance humanism that were stirring in southern Europe. Even so, there is good reason to mention Claudius Clavus, who was born in a village on the island of Funen but came to hold an important place in the history of cartography. Around 1420, probably while in Rome, Clavus drew some of the earliest maps of the Nordic countries, thereby significantly influencing later cartographers working in the 1500s. He was definitely an exception, however, being the only Dane meriting international notice on the scientific stage.

Then, at the end of the fifteenth century, two events occurred that were momentous for Danish science and scholarship – events that would profoundly affect their development in later years. One was the founding of the university in Copenhagen in 1479, and the other was the introduction of printed books into the country followed, from 1482 onwards, by the printing of books in Denmark itself. Despite its importance, the short-lived medieval university did not bring about a blossoming of Danish academic or intellectual life, and only a modest number of books were printed in Denmark prior to the Reformation.

As far as Danish scholarship and higher education were concerned, other nations, notably Germany, were far more important to the Danes than their own country. Between 1451 and 1535 more than 1,400 Danish students were matriculated at the universities in Rostock, Greifswald and Cologne. Generally in the late Middle Ages and subsequent periods the history of Danish scholarship and science was at least as European as it was Danish, if not more so.

During the long and crucial period stretching from about 1540 to 1730, the kingdom of Denmark-Norway underwent a number of drastic changes. After thriving in a favourable economic climate throughout the 1500s, the country was weakened by a series of financial and military setbacks and ravaged by disease under the long reign of King Christain IV, from 1596 to 1648. The Danish army suffered defeat in the Thirty Years War and was forced to sign a humiliating peace treaty in 1629. At roughly the same time, a sustained economic recession began, followed by a string of devastating epidemics and no less devastating wars against a Sweden that was in the process of successfully establishing itself as one of the great powers of Northern Europe. Denmark's decline became evident in demographic terms as well. In the mid-1600s, Denmark proper (without Norway) had a population of about 825,000, but after relinquishing the Swedish territories this figure was reduced to about 610,000. Norway at the time counted a mere 110,000. In 1645 Copenhagen, the capital of the kingdom and heart of the nation's scholarly life, was home to about 30,000 souls, but by 1660 that number had shrunk to 25,000, the same as in 1625.[9] After Denmark lost the regions of Scania in southern Sweden, its inhabitants constituted less than 1% of the total European population, which naturally meant a correspondingly small number of scientists and scholars. It is worth noting, however, that no correlation can be traced between the country's dire politico-economic situation and its scientific development. On the contrary, it is surprising that 1645–1665 – incontestably the country's direst period – was in fact a golden age for science in Denmark.

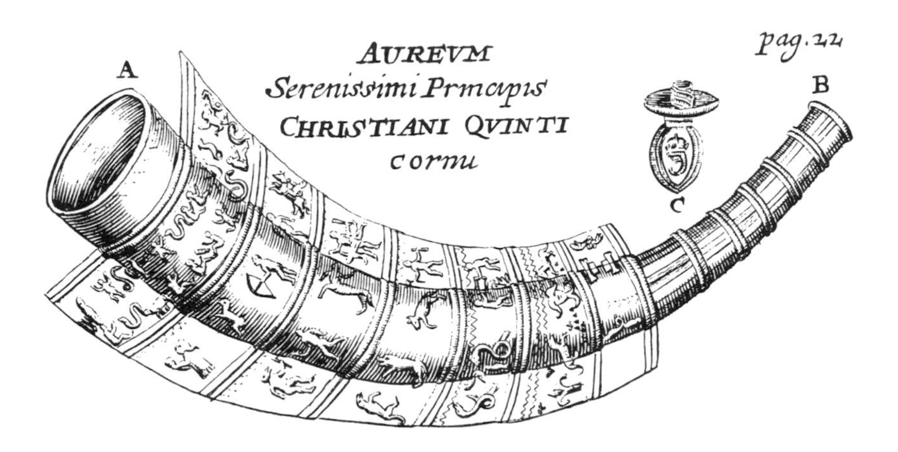

1.1 *The larger of the two ancient Golden Horns as depicted in 1678 in the French Academy of Sciences'* Journal de sçavans, *based on the description Ole Worm gave in his* De aureo cornu *from 1641. This richly ornamented horn, presumably ceremonial, caused quite a stir among the scholars of Europe.*

It is typical that in terms of science, the Dual Monarchy in Denmark-Norway was all but monopolized by Denmark, yet although very little scientific activity took place in Norway there was some interest in the area. In the latter half of the 1600s, for instance, a number of Norwegian physicians cultivated various forms of alchemy and hermetic chemistry, and several priests had botanical collections and did topographical studies. Throughout the entire period, particularly in the 1600s, the significance of Copenhagen grew, not only as the political, administrative and military powerhouse of the Dual Monarchy, but also as the realm's unrivalled cultural and academic capital. Some speak of the metropolization and the aristocratization of the nation's scholarship and intelligence, referring to the way in which the prominent professorial families formed alliances in the 1600s with Copenhagen's most powerful merchants, becoming established as a special group that enjoyed equally high status in the academic, financial and political spheres.

From a political point of view, the most significant changes took place in 1536 and 1660, and both years heralded momentous changes in the conditions that prevailed over science in Denmark, which was still in its initial stages. The political revolution in 1536, associated with the Reformation, had transformed Denmark into a society determined by rank. Made up of noblemen, the *rigsraad*, or Council of the Realm, was the political authority that governed the four social classes – nobility, clergy, commoners and peasants – and alongside the Crown it ruled the country. The decisive influence lay with the nobility, most particularly with the members of the Council, who looked after the interests of the other three classes in name only. As the highest class in the realm, the nobility was also crucial to the existence of science and scholarship, though only indirectly – Tycho Brahe being one monumental exception to the general rule. Noblemen with scholarly interests could cultivate a penchant for alchemy or other sciences, but it would have been beneath the dignity of their class to contribute to the scientific literature. The kingdom's mightiest men played a vital indirect role in upholding the scholarly system, however, by acting as patrons and protectors of science. Anyone without a personal benefactor or other support within the highest circles of the royal court might just as well give up any hope of a prosperous scientific career. Of course the attitude of the reigning monarch was also extremely important, and in this respect conditions during the century of 1570–1670 were favourable. Frederik II and his successors Christian IV and Frederik III were greatly interested in the new ideas of natural science, whereas Christian V and Frederik IV were not.

Throughout the entire period stretching from 1536 to 1730, the single factor that was, without doubt, the most significant to scholarly life in Denmark was the Lutheran religion and its Church, whose purpose was to instruct in, and to supervise the practise of, the true faith. According to its charter from 1539, the university in Copenhagen was essentially a seminary whose basis of existence was ultimately linked to the interests of the Crown's Church. Science, too, existed under conditions defined by the country's religion, although these conditions did not control it. The relationship between the Church and secular scholarship, including the natural sciences, changed throughout the period, but at no point up to 1730 was there any doubt as to where the centre of gravity lay.

The secondary role that science in many ways played does not mean that Danish scientists were forced to submit to the spiritual yoke of the prevailing religion, or that they stood in opposition to that religion. The idea of a "war" between religion and science is a fabrication from a later era. At the time, the difference was simply a non-issue, and the idea would hardly have held any meaning for Danish scientists from that period. What is more, those very scientists could have been theologians as well, or vice versa, and many scholars were both. Eventually the distinction between scientists and theologians would become a reality, but in seventeenth-century Denmark that was not yet the case.

As will become evident, quite a few of the scientists operating in Denmark were foreigners, and many of them were from Germany. Conversely, there were also Danish scientists who for various reasons went abroad to work – and those who left Denmark were not the least talented. Steno is surely the most well known, but he was not the only one to pursue a career outside his homeland. The mathematician Georg Mohr spent most of his time in the Netherlands, France and Germany and earned international respect for his works on geometry. Nicolaus Mercator, a native of Holstein who contributed to mathematics and theoretical astronomy, lived mainly in England, where in 1666 he was honoured with a membership of the Royal Society. The physician Jacob Winsløw, who converted to Catholicism as a young man, worked in France under the name of Jacques-Bénignes Winslow and became one of the foremost physicians of his day.

With Denmark's economic and political conditions gradually improving from the 1670s, one would have expected to see Danish science continue to advance. On the contrary, however, a dark period of severe decline began during which Ole Rømer was really the only spark of light. It is fitting to set 1730 as the unprecedented low point. Two years before, a great fire had ravaged Copenhagen, devastating the city and destroying its anatomical theatre and virtually all of the books in the university's library. Soon after, in 1732, a new university charter was drawn up that further weakened the natural sciences. Only then could science once again begin to advance in small, slow steps. The deterioration of science can largely be seen as a delayed consequence of the Absolute Monarchy's indifference to scientific activity as a field that was valuable in its own right. The new system held military and administrative capabilities in high esteem and also valued entertainment. Not so with scientific research, which was therefore marginalized and came to be seen as inconsequential. Scientists experienced a rapid loss of prestige, identity and power, and the only way they could recoup this loss was by applying their talents as administrators for the royal court and the all-powerful Crown. A royal academy of learning like those found in other countries would have changed the situation, but such an academy was not established in Denmark until 1742.

The Middle Ages 2

To the extent that one can even speak of Danish science in the Middle Ages, it existed not so much as science in Denmark, but rather as scientific and academic contributions from Danes who studied at the universities spread across southern Europe. More than a few Danish scholars played a role in shaping the European tradition of learning, and some, like Boethius de Dacia and Petrus de Dacia, played roles that were quite significant. They were Danish scholars, but since no Danish tradition of learning as yet existed, it is hardly reasonable to talk about Danish scholarship and science within a strictly national framework.

The Danish scholars studying and working in Paris and Bologna during the second half of the 1200s were part of a larger tradition focused on, and inspired by, a new awareness of Aristotle, Hippocrates, Galen, Ptolemy and other great writers of Antiquity. During this period many people regarded Aristotelianism as a progressive intellectual force, and several of the Danish scholars helped to make the Greek philosopher's thinking more widely known. On the contrary, no Danes seem to have contributed to what could be termed a bit simplistically as the next phase: the critique of Aristotle that stimulated new ways of thinking about philosophy and physics, and which heralded the new understanding of nature that arose around 1600.

Denmark's budding book culture

The beginning of a culture of learning was contingent upon peaceful contact with other countries, and upon a literary tradition based on books written either in Latin or in the local language. These conditions did not arise in Denmark until the late 1100s. Books were chiefly found in the monasteries and the chapter houses of the cathedrals, and these ecclesiastical institutions were therefore of seminal importance to the emergence of a tradition of learning in the medieval Danish region. The first monasteries in Denmark were founded by Benedictine monks arriving in the late 1000s. The Dominican order came to the Nordic region roughly a century later, and in 1223, at the request of

Archbishop Anders Sunesen, they built their first Nordic monastery in the Scanian city of Lund, in what is today southern Sweden. The Dominicans were particularly dedicated to learning and to the scholarship of theology and canon law, as we know it was practised in Lund – which for most of the Middle Ages was the intellectual and spiritual nucleus of the Danish kingdom. Only Roskilde, situated centrally on the island of Zealand and home to the great cathedral completed around 1275, was at times able to rival Lund in power and splendour. The Cistercians arrived in the mid-1100s, and the first Cistercian monastery in the realm, built near the Scanian town of Herrevad, was established in 1144 with the encouragement of Archbishop Eskil. Unlike the Dominicans, the Cistercians took little interest in book-learning. On the other hand, they energetically applied themselves to solving practical and technological tasks.[10]

Throughout the 1200s Denmark gradually evolved from a peripheral and rather barbaric country into an integrated part of European culture, based on Christianity and the Latin literature. The German chronicler Arnold of Lübeck gave an account of the Danish people around 1200 in his *Chronica Slavorum*, lauding the Danes for their insights into secular and theological knowledge alike. The Danes, he wrote, had mastered the literature and advanced greatly in the field of learned studies. In reality, however, the number of books in the country was extremely limited, and a modicum of fairness would preclude the use of the word "learned", except when referring to a few select Danes. The library that Anders Sunesen left when he died in 1228, and which his contemporaries regarded as extensive, boasted somewhere between 30 and 40 works. Estimates say that in the late 1200s, Denmark would have had no more than a few hundred people with a genuine command of Latin, and who in that sense could be termed "learned".

The literature in the book collections at the monasteries were mainly theological works, whereas the number of books dealing with natural philosophy and medical issues was infinitesimal. This literature existed in the form of books written on vellum (made of calfskin or sheepskin), as was also the case elsewhere in Europe. The use of paper for making books was known in Europe from the early 1300s, but the technique did not catch on in Denmark until the following century. The paper must have been imported, as no paper mills were known to exist in Denmark in the Middle Ages. The first paper mill in the kingdom was set up in 1567, and that mill was located in Scania.

Even though the crafting and distribution of books was clearly very limited, interesting works did appear, and these occasionally contained descriptions of nature or reflections on natural philosophy. One of these was the *Hexaëme-*

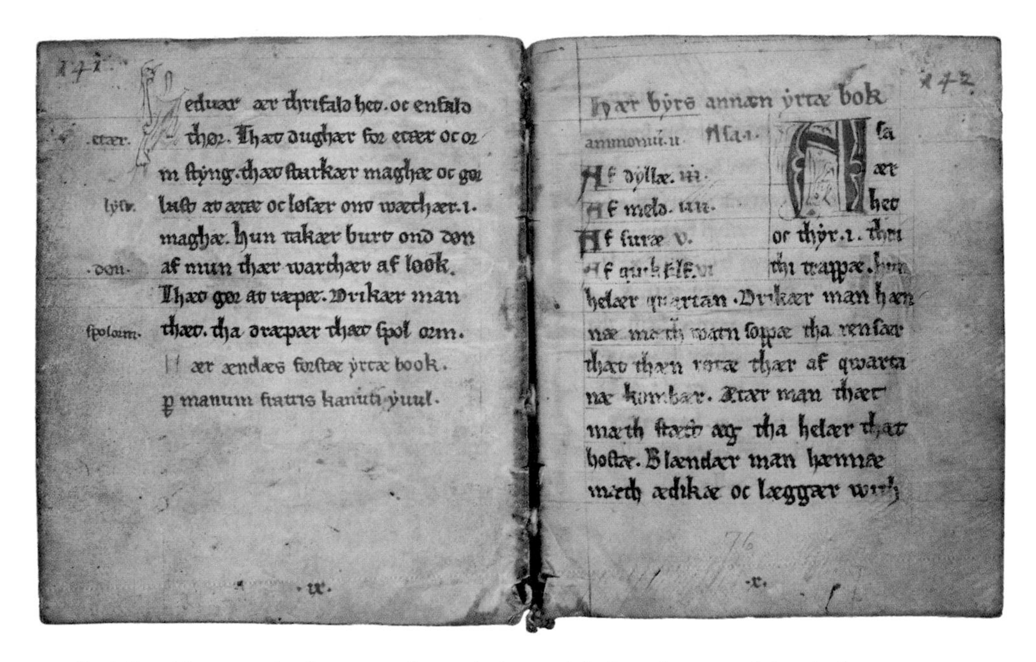

[H]eduær ær thrifald het. oc enfald
thor. Thæt dughær for etter oc or
m styng. thæt starkær maghæ oc gø
... oc løsær ond wærthær. i.
maghæ. hun takær burt ond døn
af mun thær warthær af look.
Thæt gør at ræpæ. Drikær man
thæt. tha drapær thæt spol orm.
[H]ær ændæs førstæ yrtæ book.
p manum fratris kanuti yuul.

Hær byrs annæn yrtæ bok

2.1 *Early Danish manuscript from around 1300 dealing with herbs and stones and their healing properties. Such "herbals" were popular in the Middle Ages. One well-known author of this genre was the Danish physician Henrik Harpstring. Sometimes – and not entirely without reason – he is referred to as Denmark's first scientific researcher.*

ron, Anders Sunesen's great poetic account of Creation written around 1200. Only a single contemporary specimen of this work exists today, copied some time in the late 1200s.[11] Although Sunesen's monumental poem was essentially a presentation of the dogmas of the Church, it also provided information on the prevailing world view at the time, which formed a natural part of the detailed description of God's creation of the Earth. This view states that the Sun is eight times the size of the Earth, which in turn is larger than the Moon – a numerical ratio often seen in medieval literature. Sunesen cites theological reasons for his rejecting a deterministic brand of astrology, which was also in accordance with the recognized view in the Middle Ages. Actual passages describing nature occur in Sunesen's depictions of meteorological phenomena, which are treated in keeping with the Aristotelian ideas, for instance in his explanation of how the light of the Sun heats land and sea. The *Hexaëmeron* ends with Judgement Day, in which the Earth is utterly destroyed, but like other writers Sunesen believes (based to some degree on the Bible) that out of Armageddon a new and better world will arise. This world

Dette er troheb vden alt vetzele Siwen-
de er evindeligh glede Ther giffue off ihs
ius cristus hwilkẽ som er wel signet ewĩ
delighe e for vten ende A M E N

Trucke i Køpenhafn hooff
Godfred aff ghemen Anno
domini Tusent.d.x.invigilia
palmarum.

2.2 *The Danish version of the* Lucidarius. *At right, a page of the manuscript. At left, a page of the printed edition from 1510, a project undertaken by Gotfred of Ghemen, a native of the Netherlands who became the first book printer in Copenhagen. The text beneath his printer's mark states that the book was printed in Copenhagen on the eve of Palm Sunday.*

> ... shall rise once again, fresh and new, for the better changed;
> Heav'n and Earth shall be not as before, tho' the matter remains;
> as matter unbounded has been created, so it lasts eternal.

The popular book *Lucidarius* belongs to a completely different literary genre than the *Hexaëmeron*, but nevertheless the two works share certain features.[12] *Lucidarius* is known as a Danish manuscript from around 1460, and as a printed book from around 1510, but it is modelled on German and Latin originals dating from the 1100s. The first Danish adaptation is from around 1360. The work, which unfolds as a dialogue between a pupil and a schoolmaster (a *magister*), includes a good bit of popular science presented as information on such topics as geography, natural history and astronomy, which is naturally kept at an elementary level. Explaining the occasional eclipses of the Sun, for instance, the learned schoolmaster explains: "It comes of nothing but the passage of the

Moon, which is such that he moves to stand right between us and the Sun. And because the Moon is so very thick and large, we cannot see the bright light of the Sun through the Moon ere she passes round him." When describing the Earth – which is "plump as an egg" – and its situation in space, the schoolmaster largely follows Aristotle: The Earth is surrounded by 12 crystal spheres that hold the celestial bodies of the Moon, the Sun, the planets and the fixed stars. When explaining why the oceans are salty, he refers to the heat of the Sun, outlining a circulation process that separates the water into its salt-water and fresh-water components, all in accordance with Aristotle's *Meteorologica.* Finally it is worth mentioning that *Lucidarius* contains imaginative descriptions of the exotic country of India and of the strange peoples believed to live there. The curious customs and wondrous creatures it mentions were also found in the well-known work *Mandeville's Travels,* which appeared in Danish in the 1440s.[13]

The art of printing, based on Johann Gutenberg's invention from around 1450, was quite late in coming to Denmark, arriving with an itinerant German printer named Johann Snell, who in 1482 printed a prayer book (*Breviarum Ottoniense*) in Odense on the island of Funen. An actual Danish printing tradition arose even later, mainly thanks to Gotfred of Ghemen, a Dutchman who settled in Copenhagen in 1490. His many publications included the printed edition of the above-mentioned Danish *Lucidarius* from 1510. The number of Danish books long remained very small, and until 1550 or so the country's annual book production (judging from the books we know today) was limited to three or four titles. As for Norway, no books were printed there until the 1600s, and the first book-printing shop on Norwegian soil was established as late as 1643.

The art of medicine and pharmacy

Of those who were knowledgeable in medicine in the Nordic region, the greatest medieval figure – and the only one whose name was known abroad – was Henrik Harpstring (or Harpestreng to the Danes), author of *The Danish Book of Herbs* as well as the more scholarly Latin work *Liber herbarum.* Little is known of Harpstring's life and deeds, except that he was canon of the cathedral chapter in Roskilde and died there in 1244. He undoubtedly studied abroad, probably at the famous school of medicine in Salerno, which disseminated to Europe's physicians the medical ideas of Antiquity, based on Galen's humoral pathology.[14] Although *Liber herbarum* was largely a compilation of literature from the school in Salerno, it also contained original contributions on a few Nordic plants and their medicinal properties, which were classified according to Galenic principles and credited with various "virtues". Garlic, the text explained, was good against the bite of an adder, and cress was an anti-aphrodisiac that could "bind a man's desire for women".

Liber herbarum quickly became popular and won renown all across Europe.[15] The herbal was reproduced in a variety of handwritten copies and translations, including versions in Swedish, Norwegian and Icelandic. It is not unreasonable to regard Harpstring's book as the first scientific contribution from a Danish scholar, even though it would not soon be followed by any comparable contributions from other Danish physicians or learned writers.

We know that the Danish scholars of the Middle Ages were interested, albeit not intensely so, in the art of medicine. However, no reliable sources mention any interest on their part in chemistry or alchemy, although undoubtedly some among the clergy were acquainted with such pursuits from their travels abroad. It is said that Bishop Bo of Aarhus was known for his mastery of "a peculiar art", which may have involved alchemical experiments. No Danish book describes distillation processes or equipment until 1534, when the Renaissance humanist Christiern Pedersen published his medical and herbal guide *On Herbal Water.*

Pharmacists, who long continued to function as practical chemists as well, figure in Danish history from about 1500, but even before that time the occasional pedlar would travel the countryside operating as a sort of unauthorized pharmacist. During the 1400s Copenhagen had a handful of permanent establishments that sold medicinal preparations, and the first known licence to run a pharmacy in the city dates from 1514. This document, signed by Christian II, shows that a few pharmacists had worked in the city before that time.[16] Yet theirs was a rare profession, as witnessed by the medical professor Christian Morsing, whose foreword to Henrik Smith's medical guide from 1546 declares that there is but a single apothecary in "all of the Danish Kingdom and Norway". This situation would soon change, however, and pharmacists would come to play a prominent role in Danish science, exploring the fields of botany, chemistry and pharmacy.

Danish scholars in a Latinized Europe

There were a few Danish scholars during the Middle Ages whose contributions were known outside their native country, and who earned notice at the

2.3 *During the Middle Ages and the Renaissance, God was often portrayed as an architect or an engineer. Because the Lord had created the world according to certain scientific principles, it was a divine duty to seek to elucidate these principles. This way of thinking became very significant for the new field of natural science. From the Austrian National Library in Vienna.*

universities that had begun to spread across Europe from about 1200 onwards. In a Nordic context, Danish scholars were mainly attracted to the celebrated university in Paris. The powerful Danish archbishop Jacob Erlandsen had himself studied in Paris, and in 1253 he set up a charitable foundation in Roskilde to send hopeful young scholars abroad, enabling them to pursue further studies. The exact number of Danes studying in Paris and elsewhere is unknown, but we know of at least 75 Danes who in the 1200s had earned the academic title of *magister.* The Danes held the lead in scholarly activities in the Nordic region, whereas few Norwegians chose the path of learning. Only 81 Norwegian students are known to have studied abroad during the period 1200–1350, and none of them seem to have contributed to the scholarly literature.[17]

The intense Danish interest in the great Parisian university waned during the 1300s, and in the century that followed, those with academic aspirations were mainly drawn to the new German universities, especially those in Rostock, Erfurt, Leipzig and Cologne. From around 1400 to 1450, the number of Danish students at these institutions is believed to have been at least 170, 75, 59 and 11, in that order. The university in Rostock, founded in 1419, was particularly popular, and this trend continued well into the 1600s. One reason was undoubtedly the city's location, as Rostock's geographical proximity limited the travel time, and travel costs, for students arriving from Denmark.

The extent to which Danes were pursuing academic studies in the late Middle Ages is illustrated by numbers from the universities in Rostock, Greifswald and Cologne between 1451 and 1535. All in all during this period, the three institutions matriculated more than 1,400 Danish students.[18] Between 1367 and 1536, the total body of Nordic nationals studying at universities in Germany, the Netherlands and Eastern Europe consisted of 2,146 Danes, 724 Swedes, 219 Norwegians and 97 Finns. Very few ever earned a *magister* degree from the faculty of *artes*, however, and fewer still graduated from one of the faculties of higher learning. Records from the university in Cologne for 1451–1535 clearly show the pre-eminent position of the *artes* faculty, as well as the special status that the faculty of law held among the higher studies: While 239 Danes attended the *artes* classes, the three higher faculties had only 32 of these students, 25 of whom studied law. The number of Danish students achieving a degree from the faculty of medicine during that period – a mere 3 – is negligible.

In the field of natural philosophy, or natural science, there are four Danish scholars who merit special mention for influencing the intellectual life of the Middle Ages. The first was Boethius de Dacia, or Bo of Denmark, who taught at the university in Paris around 1270, after which he may have become a Dominican monk, which is all we know about his personal life. His contemporaries regarded him as an eminent representative of radical Aristotelianism, also known as "Averroism" after the Spanish-Arab scholar Ibn Rushd, known throughout Latin Europe as Averroës. Boethius's academic work lay mainly in the fields of philosophy and grammatical analysis, but his most valuable contribution in terms of natural philosophy is the cosmological work *De aeternitate mundi*.[19]

Boethius upheld the Aristotelian views that nothing comes from nothing and hence the world must be eternal – which, needless to say, was at odds with the teachings of Christianity. He nevertheless claimed that this clash only *seemed* to exist, as the theological and the scientific viewpoint were independent from one another. As a philosopher, Boethius believed that the world must be eternal. Good Christian that he was, however, he still concluded that the world had been created in accordance with the Bible. No one who is truly a scientist or a philosopher at heart can accept the idea of a created world, since the idea of actual creation out of nothing (*ex nihilo*) lies beyond the confines of science. So naturally Boethius did not conclude that the world was not created; merely that it was "not [possible to] show, by means of any rational human reasoning, that the first movement and the world are new". There were two truths: one theological and one philosophico-rational, but they did not carry identical weight, for in the event of a conflict, faith would always prevail.

Several of Boethius's arguments were discussed among the more radical circles in Paris, along with other Averroistic ideas that the Church considered extremely controversial, and which were proposed by Siger of Brabant and others. The bishop of Paris actually reacted against this radical Aristotelianism by condemning a wide variety of hypotheses as heretical in 1277, including the claim that "no question is disputable by reason which a philosopher ought not to dispute and determine, since arguments [*rationes*] are based on things."[20] This condemnation was very precisely aimed at a particular wording that Boethius had presented in *De aeternitate mundi.*

While commenting on Aristotle's *Physica*, Boethius also discussed another theologically volatile issue, namely the idea of "the Great Year", a period lasting about 36,000 calendar years, after which all heavenly bodies would have returned to the position they had at the beginning of the cycle. But was the world destined, then, to repeat itself in every detail? The Church rejected the thought of such astral determinism, and the idea was among the heretical hypotheses denounced in 1277. Even though Boethius believed in the Great Year, he still argued that the repeatability only applied to the physical processes and not to those defined by the human soul. This enabled him to claim that an exact deterministic repetition of the world in some distant future would not come to pass after all.

Whereas Boethius de Dacia can be described as a philosopher with an interest in cosmology, the second noteworthy Danish medieval scholar – who also studied and worked in Paris but lived somewhat later – merits the designation of scientist for his work in mathematics and astronomy. Petrus de Dacia, Peter Nightingale (or Nattergal) among the Danes, is known to have had ties to Roskilde, where he held the office of canon in the early 1300s.[21] Before that time he may have been the creator of a calendar, no longer extant but referred to as *Liber daticus Roskildensis*, which for each day of the year indicated the height of the Sun at midday and the length of the day. We know that

2.4 *An astronomical calculating instrument, as described in a fourteenth-century manuscript. This sort of instrument, an "aequatorium", was originally developed by Islamic scholars and was not used in Christian Europe until around 1260, by Campanus of Novara. Peter Nightingale's instruments, constructed some years later, were among the most advanced in their day.*

in the early 1290s Peter Nightingale lived in Bologna, probably studying at the faculty of medicine. There he wrote a commentary on the arithmetic book *Algorismus vulgaris* by the Englishman John of Holywood (or John Halifax), known across Europe as Johannes de Sacrobosco. Among other things, Peter's commentary contained a new arithmetic technique for extracting the cubic root from a number, as well as a theory for arithmetical progressions based on works by Leonardo Fibonacci, the prominent mathematician from Pisa. Some time after his stay in Bologna, we find Peter in Paris, working on a new-moon calendar covering the period 1293–1368. This calendar, which could be extrapolated to cover another period of 76 years (1369–1442, to be precise) became very widely known, especially in England and Germany.

Peter Nightingale played a material role in the efforts going on around 1300 to recreate astronomy as an exact science, prompted by the rediscovery of Ptolemy's *Almagest* and the scholars' gradual digestion of the technically demanding contents in that ancient work. Peter mainly contributed to this process by developing new computation methods and constructing a new variety of calculating devices, *aequatoria*, that could be used for efficiently preparing tables. His most significant works from the 1290s – *Tractatus de semissis* and *Tractatus exlipsorii* (although some doubt his authorship of the former) – both described the construction and use of such astronomical calculating instruments. In these treatises Peter presented methods for quickly determining the longitude of the seven planets in the ecliptica, thereby significantly improving on the existing techniques. So while the distinguished astronomer Campanus de Novara was obliged to use six *aequatoria* to calculate longitudes, Peter was able to solve the task using a single instrument. Several of the devices widely employed in the 1300s were based on Peter's design principles, such as one instrument described in 1391 by Geoffrey Chaucer, the author of the *Canterbury Tales.*

After his stay in Roskilde in the early 1300s, Peter Nightingale may once again have sought intellectual stimulation abroad, and we know neither the time nor the place of his death. Like so many other medieval scholars, Peter and his scientific contributions soon faded into the mists of oblivion. He was not entirely forgotten, however, and as late as 1494 – two centuries after Peter had been active in Paris – the Benedictine monk Johannes Trithemius wrote of him that "as [a] philosopher, mathematician and astronomer [he] was one of the most excellent and famous of his time. ... He is said to have written many works bearing on astronomy and arithmetic which are regarded as being not without use for the ecclesiastical account of time, whereby he has preserved his name for posterity. ... He was at the height of his fame in the days of Emperor Albrecht, anno 1300."[22]

Apparently, after Peter Nightingale many years passed before the appearance of another Danish scholar with an affinity for astronomy and mathematics. Not until the 1400s did such a scholar arise, and we hear of him in Vienne in the Duchy of Savoy – residing, predictably enough, outside his native Denmark. His name, Johannes Simon or John Simonis of Selandia, indicates that he was probably born on Zealand. In 1417 he completed a work, well known in his own lifetime, in which he described various calculating devices used to mechanically determine the positions of the planets.[23] Hans *Speculum planetarum* was an insightful work on planetary theory, and the *aequatorium* it described was one of the most advanced of its day. The work was actually quite well-known into the 1400s. Numerous copies were made, and it was even reproduced by the prominent German astronomer Regiomontanus (alias Johannes Müller).

More or less concurrently with John Simonis, a fourth Danish scholar shone brightly on the European academic firmament. This Claudius Clavus (or Claus Claussøn Swart to his compatriots) was born on the island of Funen in 1388.[24] By drawing some of the world's first maps of the Nordic countries, Clavus presented a new picture of the region that had previously been called "Scandia," but which he labelled "Sielandia", or Zealand. His maps were not drawn in Denmark but probably in Rome, where Clavus arrived around 1423, and where he later completed a paper on the geography of the Nordic region accompanied by a map. This map exists today as a copy, reduced in size and included in a manuscript that was done just a few years after the original. Following his stay in Rome, Clavus continued his travels and returned to the Nordic countries. We do not know for certain what later befell Clavus. One guess is that the king, Eric of Pomerania, also encouraged him to prepare a map of Denmark, but if he did so the map has not survived up to our day.

Clavus's most notable achievements were placing Greenland on the world map and generally giving a better picture of the Far North, of which next to nothing was known. The latitude he indicated for the southern tip of Greenland was remarkably precise, and the 59°46' he notes is just a few minutes of arc away from the correct latitude. Although Clavus was well aware that his maps demonstrated the inadequacy of Ptolemy's portrayal of the Earth, he was by no means attempting to undermine Ptolemy's status as an authority; at most he sought to make the great Greek geographer's work a little more complete. Like other Renaissance scholars acquainted with the Antique literature, Clavus treated the classical works with a reverent deference and harboured no desire to be original. Wherever possible he built his work on Ptolemy's geography, which he employed as a natural, acknowledged model and essentially duplicated. Apart from this predominant source his description and map are also based on his own travels, though to what extent is uncertain. He undoubtedly travelled in Norway, but his claim to have visited Greenland is hardly credible.

At any rate, whether Clavus was in Greenland or not, he provided a description of the ultimate North covering the landscapes and their inhabitants, the Eskimos or "pygmies". He wrote of the "savage Lapps, people living wholly in the wild and entirely covered with hair", and of "the small pygmeyes, two feet in height, whom I have seen after they were captured at sea in a boat made of skins, which now hangs in the cathedral of Trondheim." Clavus's map and geographical descriptions were extremely important for cartographers working in the 1500s, in whose work his influence can be traced. One of these was the Venetian geographer Niccolo Zeno, whose map from 1558 borrowed elements from Clavus, and through Zeno the Dane's depiction of Greenland was conveyed to Gerhard Mercator's famous map from 1569. In Denmark, however, Clavus's work played no role at all, and not until the 1550s do we come across the first printed maps of Denmark, prepared by Marcus Jordan, who was a professor of mathematics.

Copenhagen's Catholic university

Overwhelmingly, medieval scholarly life in Europe, including the pursuit of topics relating to natural philosophy, unfolded either at or in association with the new universities that were established across Europe from around 1200 – the first in Bologna in 1188. The major early universities were those founded in Paris (circa 1200), Oxford (circa 1220), Padua (1222), Cambridge (circa 1225) and Salerno (circa 1250), whereas the Nordic region lagged behind until the late 1400s. In 1419, the very year Rostock founded its own university, Erik of Pomerania – king of Denmark from 1397 to 1439 – applied to the Pope for permission to set up a university in Denmark. His application was granted, though with a proviso that the university be established within a period of two years. According to the papal letter the university was to enjoy the same privileges as the university in Paris and to have a bishop as its chancellor. Another condition was that the Danish university must *not* have a faculty of theology, which seems more than a little odd, but this condition was not specifically targeted at the Danish plans. It was, rather, a symptom of the intense and ongoing power struggle within the Catholic Church during that period, the Pope believing that learned theological debate was inherently detrimental to his own authority. Not that any of these conditions made a difference: Two years after papal permission had been granted Denmark was still without a university, by which time the permit had expired.

It was not until 1479, after a second attempt, that a university was successfully established in Denmark, following a visit in 1474–1475 by King Christian I and Queen Dorothea to Rome, where the royal couple had promoted the cause. On 19 June 1475, Pope Sixtus IV issued a papal "bull" or official document, thereby permitting the undertaking. This document stipulated that the university would be a *studium generale* with all four standard faculties, including theology. As to its location, the Pope merely stated that the university must

2.5 *Wilhelm Marstrand's painting of the inaugural ceremony at the new university in Copenhagen, held on 1 June 1479, as imagined by the painter in 1871. Here the bishop, Ole Mortensen, brings the first professors before the king, among them Peder Albertsen, who served as the university's rector from 1479 until his death in 1517.*

be set up in connection with a cathedral. In earlier times this would have put Roskilde or Lund in the running, but after Erik of Pomerania had declared Copenhagen the capital of the realm in 1417, it was clear that the country would have a university in Copenhagen linked to the cathedral chapter at the Church of Our Lady.[25]

The charter was issued by the king in October 1478, and the new academic institution was inaugurated with fitting pomp and circumstance on 1 June the following year. Remarkably, Sweden – which received its papal bull in February 1477, almost two years after Denmark – was able to inaugurate Uppsala university, the first institution of higher learning in the Nordic region, in September of that year. Around that time Europe already had about 70 universities, so the two Nordic countries were doing no more than following an established tradition.

According to the papal bull, the Danish university was to be modelled on the university in Bologna. In practice, however, its ideals and principles were more after the fashion of the university in Cologne (founded in 1388), and that was because the task of finding teachers for the new university fell to Peder Albertsen, Petrus Alberti, who had studied philosophy and medicine there. Not unpredictably, Albertsen travelled to the university in Cologne and found his first teachers there: a Scotsman called Petrus Scotus and a Dutchman by the name of Tileman Slecht. Right from the outset the university was international, as most European universities were. Albertsen was appointed rector of the university and held that important position until his death in 1517.

So it was that Copenhagen got its university at last, but it was a small institution that brought no significant increase in the level of academic achievement, and it definitely did not lead to any activities that could be termed scientific. The intention behind the university was not to create a place of higher learning and scientific study, but to establish an education for those who would later hold the higher ecclesiastical and legal positions in Danish society. The intake of students was very modest, and the early 1500s even saw years with no new matriculations at all. Nor did foreign students apply at the university in Copenhagen, which around the year 1500 would hardly have had more than 50 students. One of the reasons for the tiny student body was that many Danes still preferred the larger and more alluring universities abroad. A ban was therefore decreed in 1498 that prohibited anyone from studying abroad if they had not first spent two years at the Danish university, but apparently the ban had little or no effect.

During the latter part of the 1400s, new humanist ideas slowly began to well up and trickle through the educated circles of Europe. Initially they had no impact on the Danish university, but around 1520 these currents surged more strongly, developing into something of a humanist revival. The only early university humanist who worked in the field of science was Christiern Therkelsen Morsing, who became the rector of the university in 1522. In 1528, Morsing wrote a textbook of mathematics entitled *Arithmetica breuis*. Not only was this the first textbook on the subject written by a Dane; the book was also used outside Denmark. Morsing spent many years abroad, and he taught mathematics and astronomy at various universities including Cologne, Leipzig, Montpellier and Louvain. Morsing was the only professor from the original Catholic university who was also employed at the reformed university. This was in 1537 after Morsing had earned his doctorate of medicine in Basle. He was a prominent figure at the new university in Copenhagen, not only as a mathematics teacher but also as a professor of medicine and, not least, as the university's rector.

In the late 1520s the Catholic Church in Denmark was heading towards complete dissolution. This resulted in a similarly critical situation for the university, which was intimately linked to the Church and was, in many respects, a bastion of Catholic orthodoxy. Years of turmoil ended with the introduction of the Reformation in 1536, but that was not the only outcome. About five years before, the university for all intents and purposes had ceased to exist, although it was not formally closed. The venerable institution was not revived until 1537, shortly before being re-established as an Evangelical Lutheran university in 1539 by royal decree from His Majesty King Christian III.

Institutions, travel and literature

3

In the new Protestant Denmark, the beginnings of a scientific and scholarly infrastructure began to solidify into a network of emerging institutions and techniques that enabled the fragile scientific system to grow in accordance with the Christian faith. Most of the institutions were founded during the reign of Christian IV and were associated with the university in Copenhagen – notably the university library, the Round Tower, the anatomical theatre and the botanical gardens. But Denmark had other institutions as well, such as Tycho Brahe's scientific centre on the island of Hven, the royal cabinet of curiosities, and the academy at Sorø. The institutions that engaged in research of a scientific nature did not have to be public or linked to the Crown, and a few were even founded by private individuals. This was true of several libraries and natural-history collections, of which prominent physician Ole Worm's museum was the most outstanding, and the only one to earn international renown.

Scholars universally used Latin as their common language, and this made it easier to go on educational journeys, a practice that became widespread during the century covering 1580 to 1680, and which was vital to Danish science and natural philosophy. Another factor crucial to the dissemination of knowledge was the art of printing. During the 1600s, this technique not only gave rise to a steady stream of books and dissertations, but also enabled the earliest scientific journals to appear from the 1660s onwards, originating the mode of communication that even today remains prevalent in scientific circles. These journals grew from printed volumes of professional correspondence, a genre known in Denmark from Tycho Brahe and Thomas Bartholin, who will be dealt with in some detail later. In fact, Bartholin was decisive in shaping this development, as *Acta medica* (1670–1680), the journal he created, was one of the first professional journals known to the history of science. *Acta medica* was one of Bartholin's means of establishing himself as a scientific entrepreneur with an extensive network of valuable associates. This project might very well have led to the formation of a learned academy along the lines of the Royal Society in London, had the political and cultural climate in Denmark been receptive at the time. Unfortunately it was not.

The Sorø Academy and other scholarly institutions

Throughout the entire period stretching from the Reformation to the mid-1900s, the university in Copenhagen stood unrivalled as Denmark's chief institution of science and learning. This university, essentially conservative in nature, enjoyed a monopoly in several areas. Even so, its dominance was not absolute, and occasionally its position was called into question. Although Copenhagen's university had a solid hold on academic education in Denmark, there were some who suggested as early as the 1600s that the university be supplemented by other higher-level schools and educational institutions that were similar in nature to universities. The first written proposal for a new university dates from the 1720s, however, and was prepared by the clergyman and writer Christen Lassen Tychonius, who recommended founding a new university in the historic town of Viborg in central Jutland. Tychonius gave a ruthless (and not unreasonable) portrayal of Copenhagen's university, describing the condition of the old institution's buildings, as well as its moral and scientific state, as "a total shambles". His proposal came to nothing, however, and was ridiculed and resolutely rebuffed by the Copenhagen professors.

From a scientific point of view the most important non-university research institution in Denmark in early Protestant times was, without doubt, Tycho Brahe's scientific centre on Hven, which was truly remarkable in every way, as will become clear in chapter 5. In addition, the founding of the Sorø Academy for young noblemen in 1623 provided a sort of counterweight to the university in Copenhagen, or at least complemented that venerable institution and for a short time expanded the scientific infrastructure of the kingdom.[26] During the Catholic era, Sorø – a small town located in the heart of Zealand, about 80 km from Copenhagen – had been home to a Cistercian monastery, which was later converted into a grammar school catering mainly to the sons of the nobility. Across Europe, particularly in Italy, Germany and France, the early 1600s saw the founding of academies or "colleges" for noblemen's sons, and these are what inspired the Council of the Realm to transform the school at Sorø into an academy for the nobility.

The principal founder and organizer of Sorø Academy was the Danish nobleman Holger Rosenkrantz, who was himself well-educated in the humanities at various German universities and was profoundly interested not only in theology, but also in science and pedagogical reform. Among other things, Rosenkrantz personally appointed the Academy's first teachers, most of whom he found in Germany. In 1623 he even tried to get the renowned physician, chemist and natural philosopher Daniel Sennert to come to Denmark, but

Sennert preferred Wittenberg and his position there as a professor of medicine.

One major motivation for establishing the Sorø Academy was a desire to limit the educational and formative journeys that many sons of the nobility would make abroad – often for long periods of time and at great expense. Estimates say that the travels of young noblemen during the first decade of the 1600s cost Denmark a loss of capital roughly equivalent to 50,000 rixdollars per year – about the same in monetary terms as it would have cost to employ 300 professors! In 1623, Christian IV ordered that young noblemen under the age of 19 were not to study abroad, but must instead use the new academy. Although the order did temporarily put a lid on their travel activities, in the long run Sorø Academy was unable to serve as a replacement for the more interesting and instructive journeys to foreign lands. Throughout its life span of about 40 years the Academy turned out 400 or so graduates of noble blood. By the early 1660s, however, the basis for continuing to prop up an academy already on its knees had disappeared.

The Sorø Academy was conceived as a university for the nobility, while the lower classes would continue to use the university in Copenhagen. When the Academy's university-like status was formalized in 1643 it was granted the right to award academic degrees, which only happened in the rarest of cases, however. The Academy's instruction covered modern languages such as German, Italian and French (which were not taught at the university in Copenhagen), as well as history, theology, political science, mathematics and medicine, the last of which incorporated elements of physics. The natural sciences had no prominent status, but a number of the Academy's students and teachers set their own mark on the Danish scientific community of the 1600s. Among them were Hans Willumsen Lauremberg and Joachim Burser, both Germans by birth.

Besides his professional career as a mathematician and cartographer Lauremberg, who was educated in Rostock, was also a celebrated writer of satirical plays and songs. His book *Logarithmus seu canon numerorum* etc. from 1628 introduced the use of logarithms in Denmark, and he presented them as a practical tool that was especially useful in trigonometry. He was also the only person in Denmark to publish a cartography textbook in the seventeenth century (his *Gromaticae libri tres* from 1640), sharing the knowledge he had acquired during his studies in Rostock. In 1631 Lauremberg was charged, by royal order, with preparing a map of the entire kingdom, but his progress was so slow that he lost the commission in 1647. His compatriot Joachim Burser was the academy's professor of medicine and physics, and he was also the

STELLÆBURGI
REGII
HAUNIENSIS
AUREA IN SCRIPTIO.
Turris fortissima nomen JEHOVÆ: Ad eam curret

3.1 *Engraving of the Round Tower from 1648. This image was first used in* Phosphorus inscriptionis hierosymbolicae *from 1648, a work written by the minister and philologist Thomas Bang, who presented his solution of the rebus that decorates the tower. Written in a mixture of Latin and Hebrew, this puzzling inscription can be deciphered as "Let Jehovah direct the scholarship and justice in the heart of Christian IV".*

licensed apothecary for the town of Sorø. Actually, Burser's favourite pursuit was botany, and his numerous journeys through southern and central Europe had enabled him to assemble an impressive herbarium with 3,500 specimen sheets – a collection that earned him quite a reputation among the scholars of Europe. What is more, Burser was the first to create a separate herbarium of Danish plants, which he entitled *Observationis seu catalogi plantarum specimen.* Although this herbarium was probably meant to be a draft for a herbal guide, it also contained many plants that had no known medicinal use.

Certainly the most remarkable scientific institution in seventeenth-century Denmark, and the most enduring, was the astronomical observatory that was constructed atop the Round Tower, which lay in the heart of Copenhagen and was part of the city's university.[27] This tower was dedicated to astronomical science and was among the first university observatories in the world, predated only by the one in Leiden (the university there completing its observatory in 1633, nine years before Copenhagen). The construction project commenced in 1637 under the leadership of Dutch master builder Hans van Steenwinckel the Younger, who was succeeded after his death by a fellow Dutchman named Leenart Blaszius. The 36-metre-tall astronomical tower housing a unique spiral walkway more than 200 metres long was completed in 1642, whereas the adjacent Church of the Holy Trinity was not ready for consecration until 1656. Despite its close proximity to the church, the Round Tower was not a church tower, and it never served as such. On the contrary, its function as a university institute was underscored by the fact that ascending the spiral walkway was the official way to access the university's library hall, which was located above the nave of the church.

The reason the Round Tower was placed at the very centre of Copenhagen is that Christian IV wanted it to be a prestigious building project whose magnificence would reflect upon the realm's capital and king. From an astronomical point of view the location was a disaster, and the university's esteemed professor of astronomy, Christian Longomontanus, would certainly have preferred an observatory located outside the city, but arguing against the king's wishes was, of course, out of the question. In 1639 Longomontanus, then 77 years old, described the Round Tower as a work in progress in a paper entitled *Introductio in theatrum astronomicum*, declaring on the title page that the new astronomical theatre was being built "in honour of the Creator of the Heavens, the supremely benevolent and omnipotent God, and for the benefit of all". This paper, with its proposal for the outfitting and arrangement of the observatory, was strongly criticized by the French mathematician and astrologer Jean-Baptiste Morin, who, in addition to being a notorious polemicist, was a professor at the Collège Royal.

The Round Tower quickly became one of Copenhagen's major attractions, a must-see for all foreigners arriving in the city. One newcomer was an Englishman, the learned William Oliver, who described his travels through Holland and Denmark in *Philosophical Transactions.* Among other things he wrote of how he had visited the island of Hven and there beheld the sad remains of Tycho Brahe's Renaissance castle, which now lay "demolish'd, and quite raz'd to the Ground". He did, however, have the pleasure of studying Tycho's great celestial globe in the Round Tower. "This Tower was built 1601", states the slightly misinformed Englishman, continuing, "Monsieur Romer, the great Mathematician and Astronomer of the present Age, has converted the upper part of this Tower now to other uses, where in a dark Room he has his Instruments for observation. Here I saw his Machine for observing the Stars by day."[28]

Since a later chapter will look in greater detail at the astronomical research in Denmark, the following merely outlines the highlights in the Round Tower's early history as an observatory. As early as 1643 the tower was fitted with a telescope that had been obtained abroad, but nothing is known of this first Danish astronomical telescope, or of it uses. As for the other observation equipment, this consisted of traditional observation instruments such as sextants and quadrants, similar in type to those Tycho had used on Hven half a century before. Longomontanus, still a professor at the time, had little faith in "the optical tube", which he believed would reveal little more than the physical conditions of the heavenly bodies, and would by no means advance astronomy as such. The first known observation of a lunar eclipse is from 1653, but the first real systematic and sustained measurements were not carried out until Rasmus Bartholin did so in the 1660s. The Round Tower had its heyday under Bartholin's successor, the celebrated astronomer Ole Rømer, who in 1685 was appointed director of the observatory. Under his leadership the Round Tower was equipped with more and better instruments, and for a time it played a significant role in European astronomy.

Museum Wormianum

The golden age of the "cabinet of curiosities" ran throughout the sixteenth and seventeenth centuries. During this period, such rooms with their collections of specimens and rarities were an important and integral part of the burgeoning field of natural science, and they were used for both research and educational purposes. Some of the most famous cabinets, which attracted enormous attention from scholars all over Europe, included Ferrante Imperato's collection in Naples, and Ulisse Aldrovandi's in Bologna. Some of these cabinets – which were, in effect, the earliest museums – focused mainly on natural wonders and were effectively collections for studying natural history. Other cabinets, those of kings or princes, focused more on art and artefacts and were more universal, reflecting the structure of the universe (macrocosmos) in the objects arrayed in the collection (microcosmos).

The first Danish cabinet of curiosities – the Royal Kunstkammer – is believed to have originated in the 1500s, beginning when noblemen, after travels abroad, brought back rarities that were kept at Krogen Castle – predecessor of the present-day Kronborg Castle in Elsinore. Not until 1620 or so did the

3.2 *Engraving from Ole Worm's* Historiarum anatomicarum rariorum, *published by Elsevir in Leiden in 1655. The numerous objects in the picture may seem randomly placed, but the collection was in fact assembled and arranged according to a strictly systematic logic; a practise Worm had adopted after observing colleagues abroad.*

kingdom see its first significant museum collection, however, when the professor Ole Worm began what would become quite an impressive accumulation of natural, scientific, archaeological and ethnographic objects. Worm's Museum, as it became known, was located in Copenhagen. It was, first and foremost, a study collection, and it was one of the first in Europe to be used for practical exercises associated with the university's courses.

Worm was employed as a professor of rhetoric at the university in Copenhagen from 1613 to 1624, after which he advanced to a professorship in medicine. During his extensive travels abroad as a young man, Worm had come to know the famous European collections created by the likes of Imperato, Bernhard Paludanus and Konrad Gesner, which inspired him to begin and cultivate his own collection. His final description of the collection was published in Leiden under the title *Museum Wormianum* in 1655, the year after his death. The great folio work had four parts, beginning with the mineral kingdom and progressing through flora and fauna to *artificialia*, meaning objects and utensils manufactured by human hands. In the sections dealing with natural history – by far the greater part of the work – readers could find stones, minerals, dried plants and fish, wild animals stuffed and mounted, seeds and fruits, conchs and corals.[29] It is worth noting that all this was a natural part of a medico-scientific collection, since at that point in time natural history and natural philosophy were regarded as ancillary subjects for all students of medicine. The final description of the Wormian collection from 1655 covered a total of 1,693 items, 27% of which belonged to the plant kingdom (*de vegetabilibus*) and 20% each to the animal kingdom (*de animalibus*) and the mineral kingdom (*de lapidibus*). The art objects (*de artificiosis*) accounted for 17%.

Among other natural wonders, the museum contained the skull of a narwhal that had led Worm to conclude in 1638 that this whale's long, spiralled tusk was identical with the "horn of a unicorn" included in several other collections, and highly valued by the royal houses. Worm had also received a bird of paradise from his nephew and fellow scholar Thomas Bartholin. Tradition had it that this bird was legless and would build its nest in the clouds, but the gift enabled Worm to establish firsthand that his specimen of the exotic creature actually did possess a pair of legs, which he also depicted in his *Museum Wormianum*. He used his own preface to this work to instruct the reader on the appropriate way of studying Nature, deriding the sterile method of book-learning with its "empty quibbling as to the first matter, ... of movement and the eternity of the World, the impenetrability of the celestial spheres and other such sophistries". He emphasized instead the necessity of personal observation, and of taking a critical-empirical approach to the information one found in the works of scholars. Worm's rhetoric was typical of his time, and yet it was equally typical that the principles expressed in Worm's own rhetoric were not fully reflected in his scientific practices.

Although Worm's Museum was a private collection that did not grant access to the public at large, it became renowned far beyond the borders of Denmark, and its wonders were displayed to the inquisitive eyes of foreign and Danish guests. Worm generally took a rational, analytical approach to his material, but like many of his contemporaries he believed that stone artefacts found in the earth had fallen from stormy skies as "thunderstones", and also believed that prehistoric objects were magical. He was prudently sceptical towards phenomena that others claimed to be "natural wonders" or "sports of nature" (*mirabilia*), yet he had no doubt that nature could indeed be creative and playful, a *natura ludens* that could reveal facets arbitrary and unpredictable.[30] His collection included many examples of strange minerals, plants and animals that he classified as *mirabilia*: stones looking deceptively like shells, or fossils displaying figures. Not only fossils, but also glassy ores and compound nodules of flint appealed to Worm's sense of wonder because of their occasional resemblance to animals or parts of the human body. "Nature has made uncommon jest in all the outward appearances of natural things," as he wrote in *Museum Wormianum*.

Worm's Museum was the foremost collection in Denmark during the 1600s, but other Danish scholars also created similar, albeit considerably smaller, collections. Thomas Bartholin was a passionate promoter of museums and collections for study purposes, and in his *Cista medica* he listed the contents of his own collection, featuring such rarities as the horn of a unicorn, the mouth of an octopus, and remains from a mermaid. The king's cabinet of art and curiosities – the Royal Kunstkammer – was of greater and more lasting importance. Founded by Frederik III in the early 1650s, the Royal Kunstkammer was, to a degree, inspired by the collection that the monarch's cousin, Duke Friedrich III of Schleswig-Holstein-Gottorp, had assembled at Gottorp Castle.[31] The king's collection was initially set up in 1653 in apartments at his castle in Copenhagen, and two years later it swelled appreciably when the king acquired the contents of Worm's Museum after the death of its founder. The Royal Kunstkammer grew, was relocated, and was once again expanded and augmented in the 1740s when the valuable Gottorp collection was transferred to Copenhagen.

The Royal Kunstkammer was a universal museum that contained natural-history specimens, antique and ethnographic objects, an extensive coin and medal collection, artware, and mathematical and mechanical apparatuses. The collection's first inventory list, prepared by Thomas Bartholin in 1674, states that among much else the collection contained a "Mathematical Chamber" dedicated to scientific instruments, watches and inventions such as telescopes, astrolabes and microscopes. In 1687 a professor of anatomy named Holger Jacobæus was requested to exhaustively catalogue the Royal Kunstkammer, and in 1696 the result was published in Latin as a splendidly decorated work entitled *Museum regium*.

Educational journeys

Although in scholarly terms Denmark was located on the fringe of Europe, the scholarly life in the country during the sixteenth and seventeenth centuries was by no means provincial. The main reason was that the vast majority of the country's scholars had made educational journeys to the European university cities south of the border, there coming into contact with the latest cultural trends and ideas. Such journeys were normally undertaken in one's youth, and were a once-in-a-lifetime experience – though often a protracted one. When a young man's travels were over and he returned to Denmark laden with learning and new impressions there was rarely funding or motivation for further travels. It was customary to keep in touch with one's colleagues abroad after returning, but only by written correspondence.

These prolonged journeys would often cover a huge geographical area, with the student travelling from one university city to another in search of learning and wisdom, and perhaps a suitable place to earn a master's degree or a doctorate. The amazing amount of travelling done by these young scholars is well exemplified by two of Denmark's most eminent intellectuals in the 1600s: Thomas Bartholin, and Niels Stensen, who was some years younger. Bartholin's journey began in 1637, initially taking him to Leiden, which he left in 1640 for Paris, moving on the following year to Orléans and Montpellier. From there he travelled to Genoa, ending up in Padua, where he was matriculated at the university. Late 1643 found him in Rome, and from there he continued south to Naples, Messina and Malta before returning to Padua in 1644. From Padua he slowly made his way north, remaining in Basle long enough to complete a doctoral degree, and lingering in Paris and Leiden before finally returning to Copenhagen in 1646. Thirty years later Thomas Bartholin, who had become a renowned anatomist, published an account of his travels as *De peregrinatione medica.* Even more impressive were the journeys of Niels Stensen – better known as Steno – who spent decades travelling far and wide. He initially went on an educational journey lasting from 1659 to 1664, with stops including Rostock, Leiden and Amsterdam. Other, later trips took him to well-known places such as Paris, Montpellier, Florence, Rome and Dresden, not to mention Venice, Prague, Vienna and Schemnitz. It is estimated that between 1659 and his death in 1686, Steno's combined travel itinerary conveyed him a total distance of no less than 27,200 km.[32]

During the reigns of Christian III and Frederik II it was considered essential that Danish students travel to universities abroad, as this would subsequently enable them to occupy the chairs at the university in Copenhagen. These kings and their administrators therefore sought to actively stimulate such foreign travel, for instance by sponsoring a number of scholarships. During the reign of Christian IV, however, it was far easier to recruit priests, civil servants and scholars domestically, and because of that development such royal scholarships were drastically reduced. What is more, the option of earn-

3.3 *The University of Leiden,* Academia Lugdunensis *in Latin, here in an engraving dated 1612, was the Danish students' university of choice for much of the seventeenth century. Not only did Leiden offer religious tolerance combined with a lively scientific environment. The university was also one of the first to boast a botanical garden, an anatomical theatre and an astronomical observatory.*

ing academic degrees in other countries was subject to strict limitations. One reason for the shifting attitude towards such formative journeys abroad was the religious orthodoxy that existed: There was a constant fear of the effect that religion, particularly Catholicism, could have on a young, tender mind left to its own devices in the wide world. It was nevertheless fairly common for Danish students to frequent the Jesuit schools and universities scattered across Catholic Europe, notably in Paris, Montpellier, Siena and Padua. In 1604, Jesuits were prohibited, by royal order, from preaching and teaching in Denmark, and by the same token Danes who had studied at Jesuit-run institutions could not be granted an educational, academic or clerical position in Denmark. This order later became even more severe, when Christian IV completely forbade students to travel to the Jesuit schools in southern Europe and barred Catholics from all public positions in the realm.[33] But in spite of these orders and various other constraints, contact between Danish scholars and colleagues in the Catholic world were maintained.

The universities most often frequented by Danish students between 1536 and 1660 were those in Rostock and Wittenberg, which not only were within a reasonable distance from their native land, but which were also bastions of the true Lutheran doctrine. Especially in the field of medical studies, however, students preferred the more southerly universities such as Basle, Montpellier, Bologna and Padua. What made Padua so very attractive to students of medicine was not just the university's proud traditions and high professional standards. It was also the fact that students did not necessarily have to be Catholics to take classes there. In contrast, no Protestant student could be matriculated or take a degree at the famous Sorbonne university in Paris – which by no means prevented Danish students from being attracted to the French capital as well. Even though they could not participate in the regular classes held at the faculty of philosophy, they were able to attend scientific lecture and defences of theses and dissertations, and they could make the most of the city's extensive and eminent library. But the leading destination for Danish students interested in medicine and the natural sciences was the university in Leiden – Academia Lugdundensis – which was probably the most important and the most influential centre of learning in seventeenth-century Protestant Europe. Although associated with the Calvinistic church, this university was relatively tolerant and did not have its teachers swear an oath of loyalty, for instance. From 1601 to 1690 a total of 740 Danish students were matriculated at the university in Leiden, while the corresponding figures for Rostock and Wittenberg were 511 and 513, respectively.

During the period 1536–1660 a total of 3,586 Danish-Norwegian students were matriculated at foreign universities, with about 13% coming from Norway proper.[34] These activities peaked with a total of 417 during the decade 1591–1600, dropping 60 years later to a mere 290. Interest continued to wane, and in 1711–1720 only 85 Danish university students were matriculated abroad. The significant numbers from the 1600s show that quite a large proportion of the students in the kingdom chose to study at foreign universities. In the mid-1600s this was true of about 13% of all students at the university in Copenhagen, whereas the incidence among those who had already obtained a master's degree was significantly higher, in the range of 65%.

The many journeys that Danish scholars undertook to visit southern Europe were by no means counterbalanced by visitors coming from those regions to the Danish-Norwegian kingdom.[35] A few hardy travellers did find their way to the frigid North, among them an Italian priest named Francesco Negri, who travelled around Denmark, Sweden and Norway for three years, in 1663–1665, and who recounted his travels in 1700 in his *Viaggio settentrionale di Francesco Negri*. Another example worth mentioning is the English diplomat Thomas Henshaw, who was a member of the Royal Society, and who lived in Copenhagen in 1672–1675. He wrote to the Society's secretary, Henry Oldenburg, telling of his life in the Danish capital. Henshaw himself was not impressed with "this unpleasant corner of ye world". There were exceptions to the general paucity of culture, and these included his acquaintance with professors such as Thomas and Rasmus Bartholin, with whom he would often dine. "It is ye best recreation I have, wee have a great deale of good discourse and talke often of ye Royall Society for whom they have great respect, and for y^{r} self [Oldenburg] in particular". While staying in Copenhagen, Henshaw also met the royal apothecary Johann Gottfried Becker, and the chemist and philologist Ole Borch, whom he described as "a very good philosopher as well as an excellent humanist".[36]

We get another impression of the international respect Danish science enjoyed during that period from a letter that Oldenburg sent to Robert Boyle in 1666. According to Oldenburg, a French correspondent had suggested that the conditions were propitious for the formation of a scientific society in Denmark "since there are able and very intelligent men to be found there". Oldenburg agreed, remarking, "Indeed they have some there, that are very learned. Men, as Erasmus Bartholin the Mathematician, Thomas Bartholin the Physician, and Steno the Anatomist."[37] Leibniz, too, was interested in scholarly life in Copenhagen, asking after news of it in a letter dated 1683. The German philosopher wished to know what had happened to Steno's geological and anatomical records, and was also curious as to what "the excellent mathematicians Rasmus Bartholin and Rømer" were working on. Leibniz, too, spoke with admiration of Borch, whom he referred to as "extremely competent in matters of chemistry".[38] Around that time the halcyon days of Danish natural philosophy were drawing to a close, but Copenhagen was still a city that could capture the attention of the international scientific community.

Scientific correspondence and literature

The beginnings of a Danish book culture were slow and late in coming, as already described: The years 1482 to 1550 saw the printing of a mere 226 book in the Danish-Norwegian kingdom. But the pace began to accelerate, and during the second half of the 1600s about 1,400 titles were published, roughly half in Latin and half in Danish.[39] One of the most important printing works in Denmark was the university's own printing shop, which was set up in the early 1540s. Its main activity was to issue dissertations, and in 1636 it was granted the exclusive right to print and sell almanacs. This particular type of publication was very lucrative, as a large number of each almanac was printed every year. Impressions of several thousands were not unusual, which presumably made almanacs the single most popular article on the Danish book market in the 1600s. The first almanac known to have been printed in Denmark covered the year 1549 and was prepared by Peter Capeteyn, a professor of medicine who was also physician-in-ordinary to the king. Capeteyn, a native of the Netherlands, wrote his almanacs in Low German, and the first almanac in Danish was printed in 1555 by Mads Vingaard. Those living in Norway had to wait until 1644 for the first local almanac to appear – which was, in fact, the first Norwegian book ever printed.

Tycho Brahe played a key role in early Danish and international book history, being one of the first to fully comprehend the possibilities for science that this new medium embodied. His own printing press on the island of Hven has been referred to as the world's first scientific printing shop.[40] Tycho's first scientific publication, *De nova stella* from 1573, was printed in Copenhagen in the establishment of Lorentz Benedicht, but after settling in at the newly constructed Uraniborg on Hven, he realized that his ambitious literary-scientific plans would require him to have his own press. In order to ensure precision in the contents, the quality of the illustrations, and the most attractive printing and binding, Tycho had to control the entire production process. He was exceedingly conscious of the value of his publications, which were more than merely scientific announcements. They were also symbols of his high class and his erudition, and as such they were essential to his self-esteem and relationship with his patrons and friends. The great significance Tycho attributed to his own ability to publish is reflected in his decision to bring along his printing press, types and paper upon leaving Hven in 1597. This equipment was just as important to him as his astronomical and chemical instruments, and it accompanied him from Copenhagen, through Wandsbeck and Wittenberg, all the way to Prague.

Besides setting up a printing shop on Hven, Tycho also built a paper mill – one of the first in Scandinavia. The new mill, ready for use in 1592, was later supplemented with a bindery. The first publications printed on Hven are from the mid-1580s, and in 1588 Tycho was able to publish 1,500 copies of his valuable work on the great comet observed in 1577, *De mundi aetherei recentioribus phaenomenis*. His many ambitious book projects included the idea of publishing his own extensive correspondence on astronomical topics. The title of this envisioned work was *Epistolarum astronomicarum*, but he only succeeded in printing a small part of the immense material, doing so in a single volume from 1596. His published correspondence was of considerable importance,

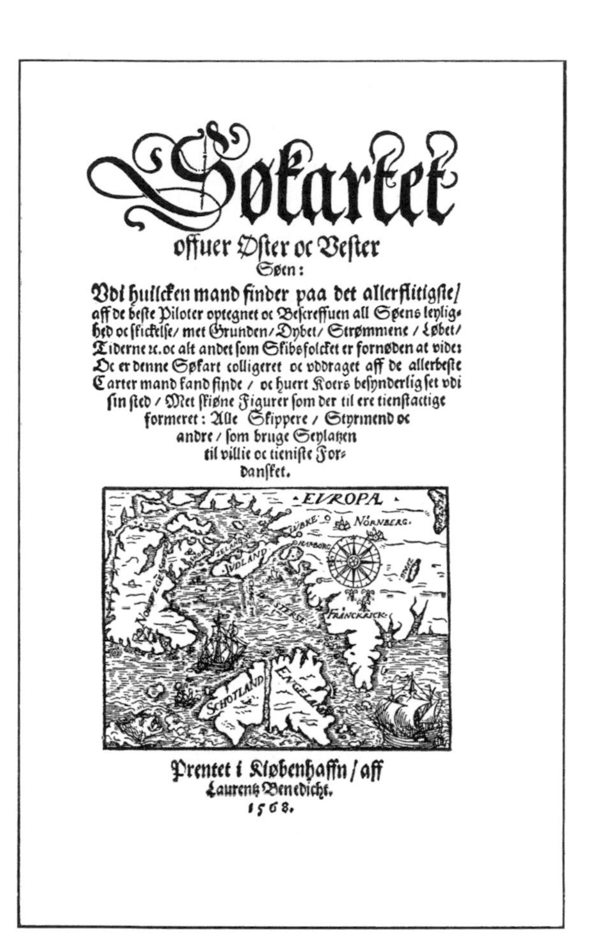

Søkartet
offuer Øſter oc Veſter
Søen:
Vdi huilcken mand finder paa det allerflitigſte/
aff de beſte Piloter optegnet oc Beſcreffuen all Søens leylig-
hed oc ſkickelſe/ met Grunden/ Dybet/ Strømmene / Løbet/
Tiderne ꝛc. oc alt andet ſom Skibsfolcket er fornøden at vide:
Oc er denne Søkart colligeret oc vddraget aff de allerbeſte
Carter mand kand finde / oc huert Koers beſynderlig ſet vdi
ſin ſted / Met ſkiøne Figurer ſom der til ere tienſtactige
formeret: Alle Skippere / Styrmend oc
andre / ſom bruge Seylatzen
til villie oc tieniſte For-
danſket.

Prentet i Kiøbenhaffn / aff
Laurentz Benedicht.
1568.

3.4 *Title page of one of the oldest technical publications written in Danish: the printer Lorenz Benedicht's textbook on navigation,* Søkartet offuer Øster oc Vester Søen, *from 1568. This instructional work, aimed at those who sailed Danish waters, was among the first of its kind. Benedicht also wrote other books, including his large* Astronomische Bescriffuelse *from 1594, which mainly dealt with astronomy.*

nonetheless. More than anything the text resembled a collection of scientific articles, making this work one of the earliest examples of a genre that would eventually develop into the scientific journal.[41]

The booksellers played an essential role in the 1600s, even when it came to scholarly and scientific literature. Noteworthy booksellers in Copenhagen include Jørgen Holst and Joachim Moltke, who were in charge of publishing and selling two of the most important scientific works of their time: Simon Paulli's *Flora Danica*, and the third edition of Caspar and Thomas Bartholin's *Institutiones anatomicae*, both appearing in 1648. Both Holst and Moltke had their bookstalls in a chapel at the Church of Our Lady cathedral, which, like several other churches in the city, served as a bookshop as well as a place of worship. Such arrangements were widely accepted, as the bulk of the literature in the 1500s was of a religious nature. Gradually, however, as the growing assortment of books and other printed matter became increasingly worldly, the established custom came to be seen as quite ungodly, and bookselling was finally banned from the Danish churches in 1658.

As for the university library in Copenhagen, it is difficult to overestimate its importance to the scholarly community in Denmark.[42] The library began with a donation made by the university's first rector, Peder Albertsen, in 1482, and was initially housed in the city's Monastery of the Holy Spirit. The book collection was moved several times before it found a more permanent home in 1652 in the brand-new Church of the Holy Trinity, as described elsewhere. The university library's importance is further evidenced in the list of librarians, which includes several of the most prominent scientists in the land. One was Thomas Bartholin, who was chief librarian from 1671 until his death in 1680. Another was Ole Borch, who held the position from 1681 to 1690, as did Ole Rømer. Caspar Bartholin the Younger served as assistant librarian between 1690 and 1721 – despite the fact that he had already held the post of university rector more than once. For many years the selection of books at the library was quite small, and completely dominated by theological works. The library's catalogue from 1603 shows a total stock of 675 works, 327 dealing with theological subject matter and 25 dealing with medical topics, and natural science was not featured as an independent category. In 1605, Christian IV transferred a large book collection to the library, and from that point on the library continued to expand. So it was that shortly before absolute monarchy was introduced in Denmark in 1660, the catalogue had swelled to about 11,000 tomes. By Danish standards this was indeed large, but there were also a few private collections of comparable size. By the early 1700s the university library had grown quite impressively, but sadly the great fire of Copenhagen in 1728 forced the university to begin collecting again: Virtually the entire library – about 35,000 volumes – had been lost in the conflagration.

The scientific literature published in Denmark up to 1730 counted a mere 238 titles, not including books and other texts dealing with topics of a medical nature. The vast majority of these publications were in Latin (86%) and published in Copenhagen (80%). Approximately half dealt with the field referred to at the time as "physics", and most of these were brief dissertations. Even though the scholarly and scientific literature was completely dominated by the Latin language, there was an ever-increasing interest during the 1600s in the local Danish language. Still, it was extremely rare to see Danish used in the scholarly community. The publications appearing in Danish were chiefly elementary textbooks and technical guidebooks, practical texts catering to local interests – not to forget the sensationalist and often anonymous descriptions of weather anomalies and natural phenomena that can be grouped in this

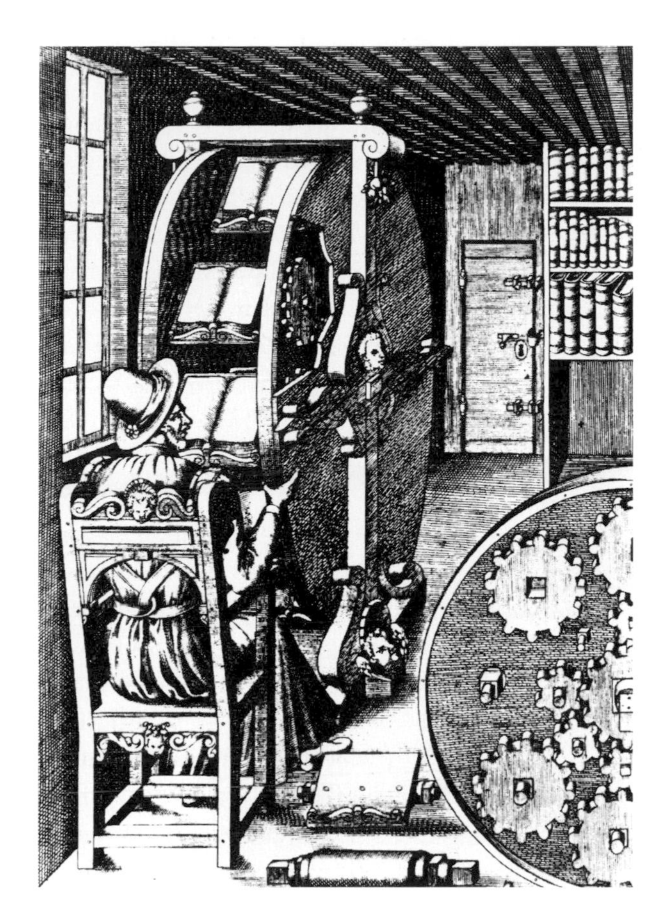

3.5 *In 1653 the university library in Copenhagen acquired a large book wheel, or "reading wheel", resembling the one shown here. Danes and foreigners alike marvelled at the ingenious contraption, and visitors often mentioned it when writing about their travels. The wondrous book wheel was destroyed in the great fire of 1728.*

genre. Generally speaking, the learned community found it difficult to imagine the Danish language serving as a means of scientific communication. It was therefore typical that when the eminent scholar Ole Borch published a book in 1688 that aimed to instruct the general public on the uses of domestic Danish plants for medicinal purposes, he wrote the text in Latin – the logical consequence being that his *De usu plantarum indigenarum* had to be translated into Danish.

There were nevertheless occasional suggestions of using Danish as a medium for disseminating technical and scientific information. This was, in fact, the message in a speech that Rasmus Bartholin delivered in 1657, and which was printed in 1674 as *De studio linguae Danicae*.[13] Bartholin eloquently argued that it was wrong for the Danes to ignore their native tongue in favour of Latin and other foreign languages, and that the extension of learning to the common Danish-speaking population would benefit the advancement of science. Not only would the number of potential scientific contributors be drastically increased. It would also result in fruitful input from those with practical life experience, "For those who each day are occupied in the tangible sense with a task will notice much that can shed light, and not insignificantly so, upon our knowledge of nature". Apparently, then, Bartholin was arguing in favour of providing more democratic access to the scholarly world by expanding the sphere of learning to also include the local Danish language. Still, the question remains of how seriously he took his own linguistic patriotism – expressing the very idea in Latin instead of Danish. Ironically, despite his rhetoric of linking science with the common folk, Bartholin published every one of his academic papers in Latin.

Two works do serve to illustrate the type of text, written in Danish, that could justifiably be termed "popular science", and both date from the reign of Christian IV. The first example was written by Hans Nansen, a wealthy merchant and seasoned traveller who was one of Copenhagen's most prominent citizens. Nansen became mayor of the city in 1644, and in this position he came to play a pivotal role in introducing absolute monarchy to Denmark in 1660. However, well before then, in 1633, he published a widely read *Compendium cosmographicum*, which (Latin title aside) was written in Danish and aimed at the country's non-scholarly citizenry.[44] The book covered not only the geography of the Earth and the astronomical world view – based on a mixture of Aristotle and Tycho Brahe – but also included a historical chronology and a manual for navigating with the aid of the stars. The text was distinctly factual and informative, directed towards the material world and demonstrating a conspicuous absence of such religious and moral passages as were customarily, and unfailingly, included in the technical literature of that era. God was mentioned only in passing, and the Devil not at all, and Nansen completely refrained from referring to supernatural causes when explaining natural phenomena such as earthquakes, comets and thunderstorms.

Whereas Nansen's work focused on the Earth, the same cannot be said of minister Anders Arrebo's poetic work in Danish, *Hexaëmeron rhytmico-Danicum*. Published in 1661, although written around 1635, this book was mainly theological, describing in poetic verse God's creation of the world as outlined in the Book of Genesis.[45] Yet Arrebo's *Hexaëmeron* still offers a good many descriptions of nature, and of natural philosophy, particularly in the fields of astronomy, geography, natural history and anatomy – and the author clearly had a solid grasp of them all. There are several obvious similarities between this work and Nansen's contemporary cosmography, notably in the generally Aristotelian world view mixed with Tychonic elements. Arrebo was acquainted with Copernicus' heliocentric alternative, but rejected as "a phantasm of the brain" the absurd notion that the Earth rotated on its own axis (though he made no comment on the Earth's annual journey round the Sun). He is also assumed to be the first person to refer, in Danish, to Galilei's telescopic observations:

> No longer must we here aspire on Milky Way to tread
> to gaze upon *Galacti's* seat with raised yet humble head;
> For that Way is but stars, so small in size but great in number,
> and so the Milky Path we need not venture to encumber.
> And if upon this truth you doubt, use Galilei's eye
> for thus you may yourself behold the twinkling stars on high.

The correspondence of Tycho Brahe's was extensive and of the highest professional (and artistic) quality, yet many of his contemporaries also excelled in the art of letter-writing. Exchanging information by means of letters was actually a key element in the scientific revolution taking place in Denmark and abroad. Ole Worm, for instance, wrote an astounding 1,800 letters between 1607 and 1654, but they were only published long after his death.[46] The copious correspondence that went on between scholars was semi-public, and often the letters were not aimed solely at the recipient in a personal sense. Nor was it unusual for such letters to be printed and made public, with or without the permission of the sender. Sometimes collections of published letters were almost scientific works in their own right, such as the four volumes of edited correspondence on medical issues that Thomas Bartholin published from 1663 to 1667 under the title of *Epistolarum medicinalium.* This work, which was widely read and greatly praised, even abroad, consisted of letters supplemented with Bartholin's own epistles. In one sense these volumes perpetuated a tradition already exemplified by Tycho, and in another sense they presaged a completely new phenomenon: the scientific journal.

Acta medica

Several of the earliest journals, such as the French *Journal des sçavans*, the Italian *Giornale dei letterati d'Italia*, and the German *Acta eruditorum*, were chiefly structured around published correspondence. These journals cannot be considered strictly scientific, however, since considerably more of their subject matter dealt with historical, literary, philosophical and theological issues than

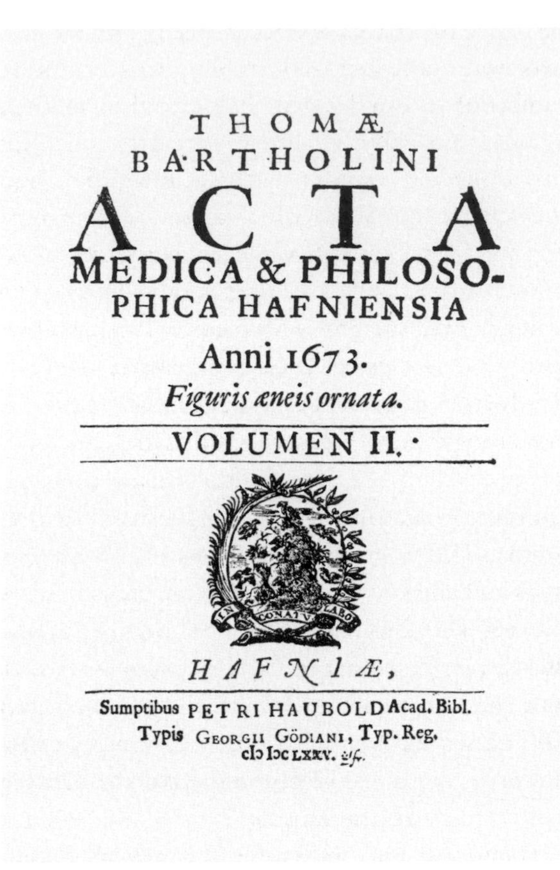
THOMÆ
BARTHOLINI
ACTA
MEDICA & PHILOSOPHICA HAFNIENSIA
Anni 1673.
Figuris æneis ornata.
VOLUMEN II.
HAFNIÆ,
Sumptibus PETRI HAUBOLD Acad. Bibl.
Typis GEORGII GÖDIANI, Typ. Reg.
cIↄ Iↄc LXXV.

3.6 Acta medica, *published by Thomas Bartholin in 1673–1680, was Scandinavia's first scientific journal, and one of the first of its kind. Unlike most of the foreign journals, it focused on medical science and related topics, and so* Acta medica *has justifiably been hailed as the first professional journal in the world. It was, of course, written in Latin.*

with medical and scientific issues.[47] They were all scholarly journals in the broadest terms, and to a great extent they consisted of reviews of new publications, whereas original articles of a scientific nature did not have a high priority. The first issue of *Journal des sçavans* became available on 5 January 1665. It was followed four months later by the first issue of the English journal *Philosophical Transactions*, founded by Henry Oldenburg and unofficially associated with the Royal Society. The profile of Oldenburg's journal was plainly more scientific, and most of the text was in English, which proved to be a problem for many foreigners, as English was not yet a recognized scientific language. In 1672, Rasmus Bartholin, in a letter to Oldenburg, expressed his "great longing for the Latin edition of your *Philosophical Transactions*, in order that everyone may read them."[48] A series of volumes was actually published in the course of the 1670s with Latin translations of the English journal. The plans for the *Acta philosophica* (as the Latin version was called) had close ties to Copenhagen, where the job was accepted in 1669 by Daniel Paulli, who was the Court Bookseller and son of the physician and botanist Simon Paulli. The texts were translated by Jean Sterpin, a French scholar who had lived for a number of years in England before settling in Copenhagen.

As far as scientific journals were concerned, Denmark became active very early on, when Thomas Bartholin established and edited the *Acta medica et philosophica Hafniensia*, which appeared five times in annual volumes between 1673 and 1680. Not only was this a very early journal – and the first of its kind in Scandinavia. It was also remarkable by virtue of its focus areas, in that it dealt first and foremost with medical issues, and only secondarily with natural history. Some even refer to *Acta medica* as the first scientific journal ever, as it mainly treated the medical questions with which Bartholin was so occupied.[49] The strong influence that Bartholin's ideas and his personality had on the journal are evidenced not only in its contents, but also by the fact that the journal had to be discontinued when its founder died. *Acta medica* was well-known internationally, and large portions of it were translated into French in the 1770s and appeared in the series *Collection académique*. The contents of these volumes were also regularly mentioned in *Philosophical Transactions*. During his years in Copenhagen (1672–1675, as noted), the diplomat Thomas Henshaw received copies of *Acta medica* for the library of the Royal Society. "Y^{u} will find some things will please y^{u}, for it is a collection of severall new observations," he wrote to Oldenburg, "and y^{u} will meet wth other things will move y^{r} pitty for theyr insipidnesse."[50]

Danish scholars who were interested in the medical and natural sciences found this journal – Danish, though written in Latin – a natural place to publish their observations and ideas. During its short lifespan, no fewer than 48 physicians, natural scientists and scholars from Denmark and Norway contributed to this journal. The list includes the foremost scholars from the latter half of the seventeenth century, among whom Steno, Borch and several Bartholins shine. Although *Acta medica* was a Danish journal (or perhaps more rightly a Copenhagen journal), it also received a few contributions from scholars outside the kingdom, such as the English physician and anatomist Edward Tyson. A large percentage of the articles, often written as brief "observations", were written by Thomas Bartholin himself, or by Ole Borch. Among other things, Borch wrote pieces on opium, on the harmful effects of mercury, and on the existence of a sulphur spring on Iceland. In the last volume of *Acta medica* from 1680, Borch in his article *Nitrum non inflammari* reported on his experiments with the heating of potassium nitrate.

After the death of Thomas Bartholin, Denmark went many years without any scholarly or medical journals, and only very few Danes had contributions, journals abroad included. It was not until 1743 that this situation changed, when a journal was established under the auspices of the new Royal Danish Academy of Sciences and Letters, which is treated elsewhere. A decade later, in 1753, the Danish Collegium Medicum did try to revive *Acta medica*, but given that only a single volume was published, the attempt can hardly be termed a success.

The religious dimension 4

Throughout the entire period from 1536 until the early 1700s, religion was the absolute determining factor when it came to developing the nascent field of science.[51] The contours of a secularized science emerged in Denmark in the late 1600s, but generally the development that took place was guided and governed by the country's Lutheran state church. The entire *raison d'être* of science was linked so closely to the true faith that it was virtually impossible to imagine any actual separation between the two. Yet even the immense influence of the faith did not mean that theologians could dictate the activities or opinions of scientists operating in this new era – provided, of course, that these did not disagree with the true faith. The natural and medical sciences were, by and large, seen as belonging to the secular sphere, which quite logically held little interest for the theologians, and where they wielded little authority.

Natural philosophy, like all other branches of learning, was a gift of God and an integrated part of the true Christian faith, and hence deserving of protection. Indeed, the study of nature in all its richness and diversity was widely regarded as a Christian duty and a pursuit that could by no means oppose the holy scripture. The interdependence between faith and knowledge was evident, for instance in the "Mosaic physics" – the attempt to create a system of physics based on the Old Testament. It was also manifest, though in a different manner, in the "natural theology" that enabled the scientists to praise God and his Creation through their studies of nature. In both cases science served the Lord, which gave it a considerable degree of religious and therefore social legitimacy. The orthodoxy that typified theology from the closing years of the 1500s onwards also led to a kind of separation between theology and science, though only in the sense that at this point the theologians began to concentrate on issues of faith and increasingly allowed the scientists to mind their own affairs. Although Denmark did see the occasional disagreement between science and theology in the 1600s, such disputes were not serious, and the development of the former largely remained free of interference from the latter.

The Reformation and the University*

The foundation of the resurrected, and now Lutheran, university in Copenhagen was based on the charter dated 10 June 1539 and entitled *Fundatio et ordinatio vniuersalis schole Haffniensis*. From the outset it was very clear that all subjects were to be perceived as ultimately belonging within the ecclesiastical sphere, but it lay in the nature of things that the Church chiefly focused on the study of theology, whereas its vigilance was less keen in the academic fields of natural philosophy and medicine. Luther had wished to see the new universities completely purged of Aristotle – that minion of the Devil, as Luther said – but the reality was that when it came to the natural sciences and medicine, there was no alternative, and so the Aristotelian tradition was allowed to live on, and was even strengthened. Paradoxically the new university was a bastion not only of Lutheran thinking but also of Aristotelian thinking. The university was still by no means an institution that promoted scientific study, and even less of a centre where research efforts were pooled, as in today's scientific communities. The university had been founded for the sake of the State and its religion. Not only did the scholarly world acquiesce; it found this state of affairs quite natural.[52]

The university was resurrected as a fully valid academy offering instruction in all of the traditional faculties taken over from the Catholic era: philosophy (*artes liberales*), law, medicine and theology. The difference was that the new university was organized as a Lutheran-Evangelical establishment with the explicit goal of supplying educated men who could reinforce the State and the Church. The charter proclaimed that the knowledge provided by the university was to be perceived as a gift from God, and hence it would be un-Christian for those in power to meddle in the affairs of the university and its instruction. This did not mean, however, that free research had been considered as part of the plan, for any scientific freedom that might exist or arise would necessarily have to be in accordance with the true religion. There was an awareness of the risk that science might venture into areas that were problematic from a theological point of view, and that risk had to be addressed. To simply prohibit all scientific activity was not a viable solution. On the contrary, the view was that science must be purged and purified, and thereby rendered theologically ac-

---

* This section, and the next two, are based on Morten Fink Jensen's contribution to Helge Kragh (ed.), *Fra Middelalderlærdom til Den Nye Videnskab*, which is volume 1 of *Dansk Naturvidenskabs Historie* – this work's Danish predecessor.

4.1 *Luther and his faithful associate Melanchthon were the two most important sources of inspiration for the new Lutheran university in Copenhagen. Its charter, written in 1539, was steeped in their thoughts and writings, with Melanchthon carrying great weight. This double portrait, painted by Lucas Cranach the Elder in 1543, hangs in the Uffizi Gallery in Florence. © Photo: Scala Firenze. 1990.*

ceptable. The original source of all science was divine, and so its main aim was to help spread the word of the Gospel.

Subjects relating to natural philosophy were placed under the faculties of medicine and philosophy, the latter having two chairs. The university's charter stipulated whose writings the professors of medicine were to teach, and here one can plainly see the continuity from the pre-Reformation university, as the authors carried over from the old curriculum included Hippocrates, Galen and Avicenna (as Ibn Sina was known in Europe). These few provisions alone make it clear that the Reformation had not changed what was looked upon as the fundamentals of medicine. Despite the preservation of an age-old tradition of medicine that was typical of the mid-1500s, there is no indication that the physicians of that time were reactionary or afraid of new ideas, since the Antique tradition did not necessarily involve stagnation. In fact, the years around the Reformation gave rise to new ways of reading and understanding Hippocrates and, most notably, Galen. It is even fair to say that a scientific progression took place in the study of Galen, who was greatly admired.[53]

Physics and mathematics were placed under the faculty of medicine, where the conservative trend of relying on the authors of Antiquity was also evident. Physics was to be explained solely by means of Aristotle's *Physica*, and the subject of physics was also to include basic studies of medicine and anatomy. Subsuming anatomy under the field of physics was meant to maximize the number of students possessing at least a rudimentary knowledge of the subject, which was also deemed useful for the many who did not pursue a career as physicians. Melanchthon had propounded that the study of the human body proved the existence of Divine Providence, and of the Lord's omnipotence and wisdom, and such things were not to be taken lightly. The human body, created in God's image, was regarded as an excellent means to inculcate upon the students an admiration and fear of the almighty Creator.[54]

Instruction in mathematics and physics was given at the faculties of medicine as well as philosophy. Basic physics per se was taught at the faculty of philosophy, and both faculties built their instruction around Aristotle, although they did use different books. The mathematics taught at the faculty of medicine was mainly directed towards the study of astrology, astronomy and geometry, whereas mathematics at the faculty of philosophy was directed towards elementary arithmetic. Knowledge of the movements of the planets was seen as very important, however, and the university used Johannes de Sacrobosco's hard-wearing *De sphaera*, which dated back to the 1200s. Although the work was outdated, a new addition appeared in 1531 with a foreword by Melanchthon, who took the opportunity to extol the study of the celestial bodies' movements as a sacred duty – declaring that those who spoke ill of astronomy deserved to have their eyes torn out. All of the recommended books on astronomy were based on Ptolemy and his Arabic commentators, which meant that the textbooks were in accordance with the medieval view of the world. And how could they have been anything else, since no serious alternative had yet been presented? The Copernican world view was not put forward until the 1540s, and it was some time before it became widely known.

Astrology, which was part of the faculty of medicine, was an important yet problematic field that had to be treated with due consideration for the theological dogmas. The university's charter expressly stated that a good Christian was to eschew those areas of astrology that concerned superstition and predictions, since predicting a concrete event was the same as encroaching upon God's domain. The charter nevertheless defended astrology, arguing that it was, after all, one of God's gifts.[55] One had to strike a balance between the good and evil uses of astrology, but apart from such vague wordings the charter gave no guidelines as to how the subject was to be taught. Astrology was too well-established a field to simply be eliminated, and probably the positive evaluation in the charter reflected Melanchthon's attitude towards the field. In his view astrology and astronomy were inseparable, and the two sciences were valuable because they proved and interpreted Divine Providence and the God-created world. One reason why the charter mentioned the great popularity of astrology was its required use in the preparation of almanacs, a duty that fell to the professors of medicine. It was also hoped that there would be an addition al benefit, with a popular subject like astrology egging the students on to further studies in mathematics and astronomy.

Gradually, however, the relationship between astrology, theology and politics became strained, and in 1633 a ban was issued on almanacs that mentioned specific prophesies portending occurrences like war or plague. This was not a general ban on astrology, however, and certainly not a rejection of the influence that the stars and planets exerted upon earthly events. What had to be rejected was Man's possibility of predicting the consequences or outcome of these events. Generally speaking, this ban against prophesying was not a conscious move by the State to promote rational reasoning. It was, rather, part of the battle *against* reason, and an attempt to impose blind faith in God's mercy, His mysterious ways and His divine will.

The science and learning that was cultivated at the new university was linked not only to religion but also very much to the power of the State, which stood as guarantor for the position of the Church. Whatever went on at the university had to be consistent with the teachings of the Church as well as the attitudes prevalent among the social elite and the political establishment, not least the king and his advisers. This interdependency manifested itself in various ways, for instance in the frequent appointing of persons to certain positions based, more than anything, on the king's wishes. When a professor's chair fell vacant, the governing body of the university would recommend a candidate for the position, but usually only after obtaining prior approval from the king's chancellor. Sometimes the king even demonstrated his power by demanding that a professor be fired to make room for his own preferred candidate, as when, in 1577, Frederik II wished to have Jørgen Dybvad employed as professor of mathematics even though the position was already occupied.[56] The king demanded the dismissal of the incumbent professor, and despite opposition from the university the monarch had his way, and Dybvad had his chair. For several centuries the university had no formal influence on who was employed there to teach, and appointments were by no means the sole prerogative of the university's governing body.

The censorship with which the university was empowered played an important role in the relationship between the university, the Church and the State. Their main fear was the appearance of writings that could call into question the true faith or destabilize the State, but censoring applied to all books, including scientific works. During the 1500s and 1600s, the rules of censorship were reinforced and reiterated at regular intervals, which shows that the various bans and prohibitions did not achieve the desired effect. Monitoring the books produced in Denmark was a doable task: Only a modest number of printing houses were given a licence, and the printers were ordered to only print publications approved by the censors. Monitoring imported books was rather more demanding, however, not to mention monitoring books in Danish that were published outside Denmark's borders. The fact that this posed a problem is plain, considering the unconditional ban on importing Danish-language literature that was introduced in 1562.

Most of the scientific literature in Denmark in the 1500s consisted of dissertations, which according to provisions in the charter were subject to internal censorship within the university itself. What is more, self-censorship was widespread, since every author knew that certain opinions and turns of phrase might be considered offensive and could cause the censors to ban his work. Tycho Brahe would not submit to any censorship from the university's professors, and to make completely sure that no one would interfere with his work Tycho had his books done in his own printing shop on Hven, as described earlier. Admittedly, he could not claim to have any formal right to do so, but as long as the books he printed gave no offence there was no reason to make an issue out of his insubordination. Generally the authorities were quite pragmatic when such cases involved scientific literature.

In practice, sometimes scientific products that were not dissertations were also censored, the manuscripts being subjected to a review by a professor knowledgeable in the relevant field. This also applied to scientific works, which had to comply with the frameworks defined by the religious and political norms. At times this censorship was quite effective, but at other times less so, and as far as the foreign scientific literature was concerned there are no indications that they faced any special obstacles. Theoretically, before any such book could be procured the expert professors, theologians and secular authorities had to ensure that it did not contradict the true religion. In practice, however, it was impossible for the censors to possess knowledge that was sufficiently profound and comprehensive to discharge this duty intended, so scientific literature from abroad frequently came into Denmark without first receiving censorial approval.

Unlike the Catholic regions in Europe, Denmark and other Protestant countries made no attempts to forbid entire bodies of work. The Catholic index of forbidden books, *Index librorum prohibitorum*, included several works by Protestant scientists, even though on the face of it these works did not present any religious problems. Scientific works could end up in the *Index* simply because the author was a Protestant or a Calvinist, as several examples show. This sort of thing would be unthinkable in Copenhagen, and at no time was there ever any question of preparing a Danish-Protestant version of the *Index*.

The university and religion under Christian IV

Near the end of the 1500s the contours of a Lutheran orthodoxy were swiftly growing on the religious horizon. The trend came from Germany, where the so-called Gnesio-Lutherans (or "true Lutherans") were doing away with the Calvinistic inclinations that had been allowed to develop against the backdrop of Melanchthon's brand of Lutheranism. This essentially led to an intense confrontation with reason and its potential within the realm of faith. It also rubbed off on the attitude towards science, which became increasingly segregated from theology.

Christian IV's ascension to the throne in 1596 did not immediately mark a new era for religion and science, but compared to previous times his reign would involve far more rigorous legislation on discipline and monitoring. Regulatory measures were introduced for large issues and small, and the relationship between State and Church grew ever more intimate. Uniformity and centralization were catchwords in all areas of society, including the Church, education and science. Only a single, unchallenged faith was allowed to exist, and this intense focus on the purity of the religious teachings evolved into the definition of Lutheranism that is referred to as "Lutheran orthodoxy". The university in Copenhagen was a vital tool in this process, which had a profound and far-reaching impact on the attitude towards science and scholarship in general. The university was perceived as being a single organic entity, so not only the theologians but the university as such was expected, indeed required, to act as a spearhead in the battle against all falsehood in the faith.

This approach had an effect on the instruction given at the faculties of philosophy and medicine alike. The basic view was that the Almighty God was the maker of all branches of learning, and that ultimately all learning was justified by its capacity to serve the Lord. Of this there could be no doubt. On the other hand, as long as this overarching goal was fulfilled, the theological restrictions imposed on the other academic and scientific disciplines were not inordinately harsh. The reason was that even as the theologians became more and more orthodox, they increasingly came to concentrate on the true teaching, almost to the exclusion of all else, permitting *neither* deviating theological interpretations *nor* arguments based on reasoning or empirical findings to pass through the wall of religious doctrine. Clearly the deviating interpretations were considered the more dangerous of the two, and were therefore mercilessly stamped out. Science, however, was permitted to simply remain outside the wall, and as long as it stayed there the State and the Church could even benefit from it. The validation for this view was that although the premises underlying the science of nature differed from those underlying the science of God, the two were still interrelated. The professor of medicine Ole Worm expressed this mutual link in his inaugural lecture in 1613 by describing nature as "the divinely instituted power and order, or all-governing divine spirit, which lies within the All and all of its parts. He who follows Nature follows God, as truly Nature obeys God, or is the very spirit of God."[57]

On the whole, the teaching guidelines set out in the university's charter of 1539 were respected. The professors adhered to the stipulated textbooks and topics while at the same time avoiding the total petrifaction of their teaching. It was possible to move beyond the mandated curriculum and include new authors, which was also done to a certain extent. As time passed it became fairly common to offer lectures on topics that the charter had not envisioned. In 1598, for instance, Johannes Stephanius, a profesor of philosophy, announced lectures on Ramist philosophy, and Jon Jakobsen Venusin's physics lectures in 1600 included studies of magnetism. Not quite so at the faculty of medicine, where the charter-bound authority of Hippocrates and Galen was upheld, and where even in the early 1600s it was difficult to detect any novel thinking relative to the framework of the 1539 charter. Even in 1603 the mathematics professor Christian Hansen Riber was still lecturing on Sacrobosco, now finally outdated, and as late as 1632–1633 Worm's medical lectures were still reviewing the aphorisms of Hippocrates.

The scientific method propagated at the university was Aristotelian. The

4.2 *In the 1600s it was crucial for scientists to be well-connected to a prominent patron – preferably the king or someone in his inner circle. This engraving from* Quadripartitum botanicum *from 1667 shows Simon Paulli handing over his work to the absolute monarch Frederik III. The book is dedicated to God and King. The other men on the picture are the king's most trusted advisers.*

lecture catalogue for 1623 shows that the basic textbook on logics was Aristotle's *Organon*, and that Worm also used the ancient scholar in his physics lessons, while others used him as the basis for the university's instruction in rhetoric and ethics. Although Aristotle's standing at the university remained undiminished in the 1620s, it did not mean that newer authors were excluded. Caspar Bartholin's numerous textbooks are packed with references to the scientific literature from the sixteenth and seventeenth centuries, and according to the lecture catalogue for 1660 the professor of mathematics Villum Lange used works by the Dutchman Simon Stevin as well as the Englishman Henry Briggs and the Scotsman John Napier. These authors were not the very newest, but their presence in the catalogue reflects the gradual seeping of international scientific findings through cracks in the wall and into the University of Copenhagen.

Occasionally, though rarely, the works of the Danish professors themselves would be part of the international scientific literature. One example was Thomas Fincke's textbook entitled *Geometriae rotundi libri XIIII*, which was published in 1583 and for years remained one of the better known works in the European mathematical literature.[58] Fincke wrote the book when he was still a young man, before his appointment as a professor of mathematics in Copenhagen in 1591. His subsequent promotion in 1603 to the more prestigious chair of medicine was quite in keeping with university tradition, as was the low professional profile he kept before his promotion, writing just a few dissertations during his years as a mathematics professor. The concept of logarithms, introduced by the above-mentioned Briggs and Napier in the early 1600s, came to Denmark fairly quickly, appearing in a textbook from 1628 written by Hans Lauremberg, who was not a professor in Copenhagen but at the new academy in Sorø, where he taught mathematics, cartography and fortification. In his book Lauremberg presented logarithms as a practical tool, emphasizing their usefulness in trigonometry as applied in cartography and the study of fortification.

Caspar Bartholin – a Christian physician

Many of the trends affecting Danish natural philosophy and medicine during the first decade of the 1600s are illustrated in the life and deeds of one of the most important scholars of that period: Caspar Bartholin (or Jesper Berthelsen, as he was actually named in Danish).[59] Having studied briefly at the University of Copenhagen he set out, as was customary, on an educational journey around Europe. His stops included Wittenberg, Basle and Padua, and while in that Italian city he wrote *Institutiones anatomicae*, his successful textbook on elementary anatomy first published in 1611, the very year Casper Bartholin was appointed professor of pedagogy at the University of Copenhagen.

4.3 *Caspar Bartholin the Elder was one of the most important professors of the early 1600s, particularly in the field of medical science. Not only was he a prolific and popular writer of textbooks; he was also a pivotal figure in the Bartholin dynasty, which came to dominate the natural and medical sciences at the University of Copenhagen for the better part of the seventeenth century. Engraving from 1625.*

The following year he married one of Thomas Fincke's daughters, which brought him into the academic family that came to dominate university life throughout the 1600s. Bartholin was later appointed professor of medicine, and in 1624 he became professor of theology and came to hold the university's finest chair.

Bartholin was first and foremost a writer of textbooks, and a very productive one at that. He had an excellent reputation in Denmark and abroad, thanks not least to his very popular *Institutiones anatomicae*, which appeared in about 30 editions all told (most of them posthumously). There is no doubt that during his active years as a professor in Copenhagen he was read more widely than any other Danish author of natural philosophy. Some of his books were aimed at the grammar schools, *gymnasier*, that were established across Denmark in 1619, as he was given the task of writing new textbooks in logic, rhetoric, ethics, physics and metaphysics. One can gain a good impression of his sources and his preferences among the learned scholars by looking at the works and the total of 100 authors that he recommended in his *De studio medico* from 1626, a work offering guidance to students of the medical and physical sciences. The vast majority of these 100 authors – 72 to be exact – were from Germany, Italy or France, whereas England and Holland – the new but rapidly rising powers in the world of natural philosophy – were only represented with one author each. It is also remarkable that Bartholin included roughly the same number of Catholic and Protestant authors, indicating that when it came to science, an author's faith was of no import to the pious Danish scholar.

Caspar Bartholin was generally conservative, but he did not categorically reject new impulses. In astronomy he was against the Copernican ideas and staunchly refused the suggestion that the Sun was located at the centre of the universe. This view was in conflict not only with Aristotelian physics but more importantly with the word of the Bible. Still, as early as 1610, while he was living in Padua, Bartholin had observed the night sky through Galilei's brand-new telescope, an experience he mentions in his *De mundo* from 1617. Despite his somewhat conservative views, Bartholin did seek, at least to some extent, to include new ideas in his natural philosophy, and his *De studio medica* calls attention to three astronomical authorities: Tycho Brahe, Kepler and Longomontanus.

Bartholin's viewpoints were significantly influenced by his Christian, orthodox Lutheran world view, which affected his outlook on physics, astronomy and even medicine. He explained in a handbook entitled *Astrologia* how the positions of the stars and planets could influence the outbreak and later course of a disease. There was nothing odd in this, as it was common knowledge that astrology was an indispensable auxiliary science to the field of medicine. To Bartholin the art of medicine without faith and piety would be inconceivable. Only such medicine as was based on the true faith, and practised by a God-fearing physician, could be effective, and the empirical study of nature and the human body therefore had to be undergirded by continuous Bible studies. As he admonished in *De studio medica*: "Everyone who studies this art will see it as an advantage to read, each morning after prayer, one or two chapters of the Holy Scripture, not superficially but carefully and thoughtfully; and in such a way that each year he completely reads through the entire Bible."

The physician also had to study the basics of minerals, metals and stones, which fell into the domain of physics but were often important ingredients in medical preparations. Gems and precious stones were regarded as particularly effective in treating diseases. The reason they were so beautiful and so rare was that God had endowed them with certain extraordinary properties, the physician's task being to release and harness the powers that lay within them. Bartholin's *De lapide nephritico opusculum* from 1628 describes how the mineral nephrite, a type of jade, could be used for medicinal purposes, and also discusses various mineralogical aspects of metals, which the author – according to tradition – believed grew from the earth.

Natural theology and Mosaic physics

Natural theology, or "physico-theology", is the term used to express the idea that God's existence or qualities, at least to some degree, can be discerned through the study of the nature he created to fulfil a specific purpose. During the 1600s this ancient idea, which in fact goes back to Antiquity, was of vital importance to the scientific revolution spearheaded by pioneers like Robert Boyle, John Ray and Isaac Newton. The idea was formulated as a theological theory by the German Lutheran-orthodox theologian Johann Arndt, whose grand oeuvre *Vier Bücher vom wahren Christentum* appeared in 1605–1609. The fourth book was translated into Danish in 1618. Arndt's thoughts on natural theology became quite influential among theologians and other learned scholars in Denmark, inspiring the likes of Holger Rosenkrantz and Caspar Bartholin. Nature was a reflection of God's thoughts and a source of information on things divine, and consequently natural theology came to function in perfect harmony with, and as an ideal legitimization of, the new field of natural science.

An early example of Danish natural-theological thinking is found in a speech delivered in 1545 by the university's rector Jens Sinning, a professor of theology, in which he emphasized the theologically valuable aspects of philosophy, including the natural sciences. He was, he said, quite aware of the problematic relationship between Athens and Jerusalem, but pointed out that good philosophy was the servant and the indispensable ally of theology. Indeed, the Holy Scripture contained numerous references to natural phenomena that only a reader schooled in natural philosophy would be capable of understanding. The apostle Paul had compared the Church with a human body and the individual Christians with its limbs, and Sinning employed this as an important argument for familiarizing oneself with the human anatomy. He also stressed that the study of nature inspired awe of God, thereby making people better Christians. Physiology, mathematics and astronomy were not the enemies of faith but its assistants. God's abundant Creation could only be fully understood and appreciated by "those people who through mathematical studies have acquired knowledge of the size, effects and movements of the heavenly bodies ... and know that the Sun, which delights us daily, is larger than the entire Earth."[60] In a day and age when the true Christian faith was decisive, the line that Sinning drew to link science and religion was an important argument for justifying the natural and scientific branches of study and upgrading their priority.

It was commonly recognized that there could be no real contradiction between the Bible and God's other great book, nature, and so the true understanding of nature could obviously not be in opposition to faith. What is more, natural philosophy could be useful to theology, for example in determining which of the narratives in the Bible were genuine miracles and which had a natural explanation. This was the line of reasoning used by Peder Winstrup, a professor of physics, in a dissertation he published in 1633. Here he not only emphasized the close ties that philosophy and philology had with theology, but also stated that physics could assist the theologians in their battle against heretic ideas, and in other ways strengthen knowledge of the Bible. To demonstrate the usefulness of physics and mathematics he cited such examples as gauging the size of Noah's Ark and King Solomon's Temple. A different and slightly later variation on such natural theological thinking was expressed by the young Niels Stensen (Steno), who wrote the following in his private notes from 1659:

> One sins against the majesty of God by being unwilling to look at nature's own works and contenting oneself with reading others; in this way one forms and creates for oneself various fanciful notions and thus not only does one not enjoy the pleasure of looking into God's wonders but also wastes time that should be spent on necessities and to the benefit of one's neighbour and states many things which are unworthy of God.[61]

A particular brand of natural theology was referred to as *Mosaic physics*, which attempted to establish a physics and cosmology founded on the account of Creation given in the Old Testament's book of Genesis. In Denmark this tradition had a prominent representative in the Norwegian-born sage Cort Aslaksen. For a time Aslaksen had been Tycho Brahe's assistant on Hven, and he was appointed a professor in Copenhagen in 1599, gaining a chair in theology in 1607.[62] In 1597 he wrote a work entitled *De natura caeli triplicis*, in which he expressed many of the ideas behind Mosaic physics. Among other things he discussed in some detail the difficult issue of the stars and the firmament they occupied, concluding that the eternal Heaven did not belong to the realm of natural philosophy but to that of theology. Incidentally, *De natura caeli triplicis* is interesting in an astronomical context by being the first publication in Denmark-Norway to recognize the Earth's daily rotation around its own axis. Aslaksen believed, based on his perception of the subtle material of which the starry firmament consisted, that it was reasonable to follow Copernicus, "who gave up the heavens and sought his refuge on Earth, as he believed that it was less foolish to assume the Earth to be movable than the firmament."

Another notable Danish-Norwegian representative of Mosaic natural philosophy was Jens Nilssøn, a bishop in Oslo who belonged to the circle around Tycho Brahe, with whom he would correspond on astrological subjects. Nilssøn was one of those who observed the passing of the great comet in 1577, and he later penned a poem about the fearsome phenomenon entitled *De portentoso*, which was published in Rostock. In a later publication from 1597, likewise printed in Rostock, Nilssøn analysed the Book of Genesis from the perspective of the Mosaic-physics tradition, perceiving the word of the Bible as a "divine sermon" that embodied the most profound knowledge of mankind and of nature.

Of Tycho and his time 5

The authoritative 18-volume work *Dictionary of Scientific Biography* contains articles on about 5,800 prominent scientists, 66 (or 1.1%) of whom can be characterized as Danish. Given that 630 of those described were active during the scientific revolution (being born between 1470 and 1680), and 14 (or 2.2%) of these 630 were Danish, the Danes were significantly overrepresented relative to the size of the country's population.[63] Although these numbers should not be taken too seriously, they do hint at the vitality and vigour of Danish science during that period. There is no question that the Danish scientist figuring most prominently in the *Dictionary* is Tycho Brahe, who has been allotted three times as much space as Steno and ten times as much as Thomas Bartholin (and even a bit more than Niels Bohr). At any rate, Tycho Brahe holds a unique position in the Danish and international history of science, not only because of his important contributions to astronomical science, but also because of the exceptional institution he created at Uraniborg, his castle on the island of Hven.[64] Although Tycho's dealings with his native Denmark became strained and he was obliged to spend his later years in voluntary exile, his efforts continued to affect Danish astronomy throughout the 1600s.

Today chemistry – that most typical of laboratory sciences – seems to be a world away from astronomy – the classic observational science. This was also the case in the 1500s, but there were nonetheless links between the two that arose in connection with the so-called Hermetic world view, which was characteristic of the Renaissance, and which sought to apply convergent thinking to microcosmos and macrocosmos; to things worldly and divine. Tycho Brahe, who was an eminent representative of this world view, was an enthusiastic practitioner of chemistry and alchemy, even though he kept his results to himself. Another Danish representative of such Hermetic chemistry was the internationally respected physician Petrus Severinus, and traces of it are also discernible in the work of Ole Borch from the 1670s. It is therefore meaningful to use Tycho as the fulcrum for an inclusive review of key elements in Danish astronomy and chemistry during the period in which the scientific revolution took place.

Tycho Brahe's world view

The celebrated astronomer and Renaissance man Tycho Brahe – who was actually christened Tyge Ottesen Brahe – hardly requires an introduction. Still, it is apposite to mention that he was the first scientist to become the subject of a full biography. In 1654, following several years of research, the renowned French natural philosopher Pierre Gassendi published his *Tychonis Brahe, equitis dani, astronomorum coryphaei, vita*, and this work would set the standard for much of the later literature on Tycho.[65] Because treating the many facets of Tycho's life and work goes far beyond the scope of this publication, and because a multitude of other sources have already done so, this chapter concentrates on Tycho's perception of the astronomical world view.

Tycho had become familiar with the new Copernican theory of the world system very early on, admiring its greater economy and its geometrical superiority. One source that clearly reflects this admiration is the lectures he gave, upon request, at the University of Copenhagen in 1574–1575, two years after he had analysed the new star identified in Cassiopeia and created a name for himself in the astronomical community with his *De nova stella*. In his introductory lecture Tycho referred to Copernicus as "the second Ptolemy", and he subsequently treated the movements of the planets using Copernicus's mathematical theory, though taking a geocentric approach. Tycho was not the first learned Dane to comment on Copernican theory, however. The astrologer and mathematician Jørgen Dybvad (or Georgius Dibuadius) had already done so in 1569 in a small book published in Wittenberg, although like Tycho he was no Copernican by conviction. He found it difficult to accept the movement of the Earth as a physical reality and wrote that it was easier and more natural to explain the celestial phenomena based on the traditional presumption of a steady and central Earth.

Tycho remained fascinated by "the second Ptolemy", yet despite his appreciation of the man who devised it Tycho could not accept the Copernican system, and finding the Ptolemaic system equally unacceptable he sought an alternative to both models. This alternative would still have to have the Earth resting at the centre of the universe, even while reflecting the immense progress that he believed the Copernican model represented. Tycho's ideas of a geocentric but non-Ptolemaic world view were probably rooted in his analysis of the comet observed in 1577, which had conclusively convinced him that the Aristotelian-Ptolemaic view of the universe was untenable.[66] Around 1583, Tycho had worked out a system that he believed combined the best of the two existing models, but he did not announce it before publishing his treatise from

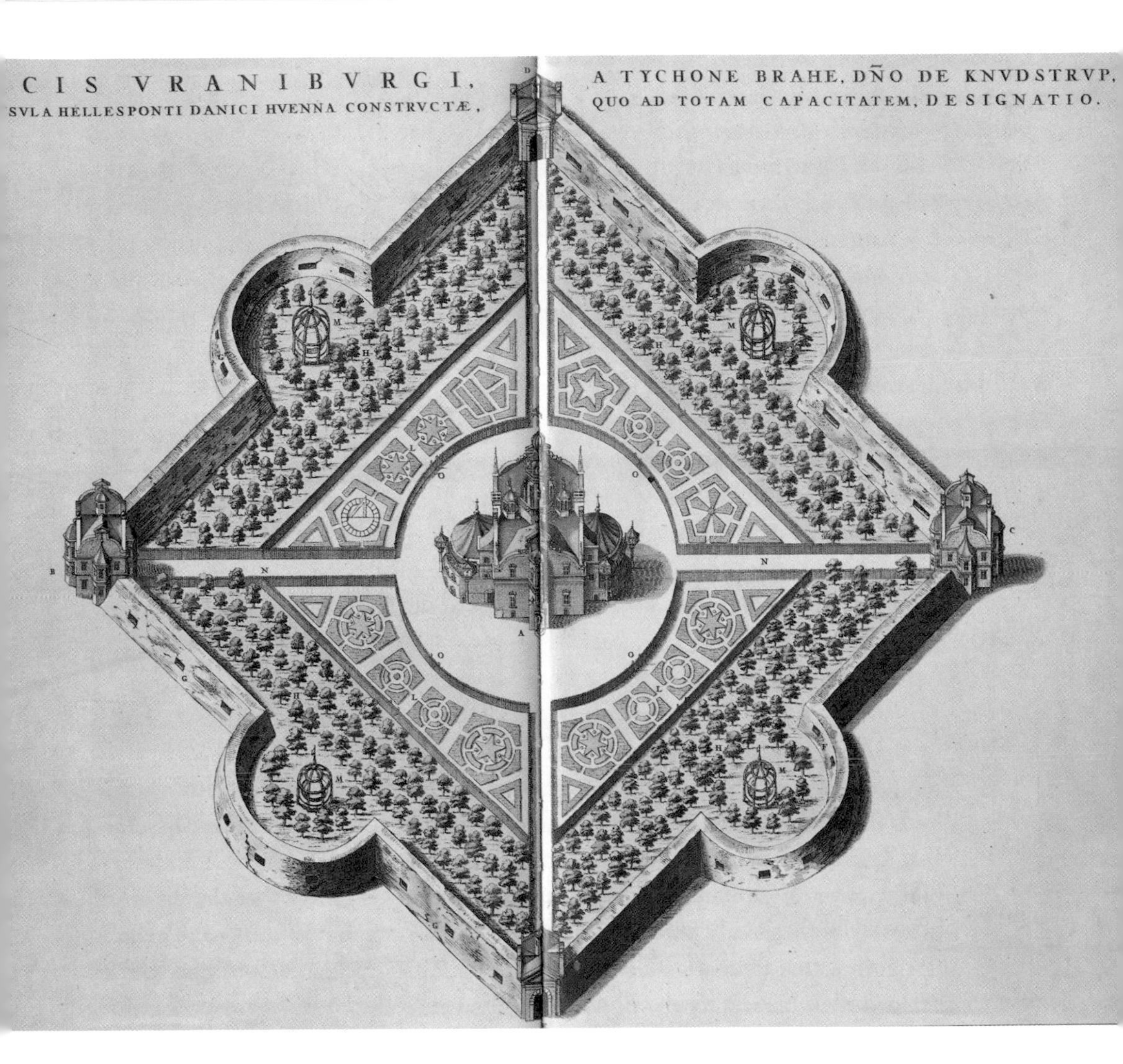

CIS VRANIBVRGI,
SVLA HELLESPONTI DANICI HVENNA CONSTRVCTÆ,
A TYCHONE BRAHE, DÑO DE KNVDSTRVP,
QVO AD TOTAM CAPACITATEM, DESIGNATIO.

5.1 *Uraniborg with surrounding park and gardens. Tycho Brahe's castle-cum-observatory was only a small part of the complex that was constructed in the late 1570s. The quadrangular grounds also contained a medicinal herb garden. Besides being an astronomer Tycho also had a reputation as a physician, and he prepared medicines using his own herbs.*

1588 on the great comet entitled *De mundi aetherei recentioribus phaenomenis.* According to Tycho's system the Moon and the Sun travelled around the Earth, which held a fixed position, while all of the other planets travelled around the Sun. In order to maintain the planetary distances found by Copernicus, Tycho was obliged to let the orbit of Mars intersect that of the

5.3 *The cosmological dispute as illustrated in Riccioli's* Almagestum novum *from 1651. On the right Urania, goddess of the heavens, weighing the Copernican system against Riccioli's version of the Tychonic system. The man at left, bespeckled with eyes, is Argus Panoptes from the Greek mythology, holding a telescope – a "Galilean tube" – in his left hand. He symbolises the starry sky and empirical science. Near the bottom of the picture sits Ptolemy, whose system lies discarded at*

acceptance of the Copernican system – in other words, between roughly 1590 and 1680.[67] One major reason for this was that Tycho's system seemed more acceptable from a theological point of view, whether founded on Catholicism or Lutheranism. Following the persecution of Galilei in 1633 and the Catholic Church's condemnation of the Copernican system, Tycho's alternative gained a large following among Catholic astronomers. The above-mentioned Gassendi, for instance, who was also a Catholic priest, could not openly back the Copernican world view that he actually supported, which is why he publicly spoke in favour of Tycho's alternative. The prominent Jesuit astronomer Giovanni Riccioli, a contemporary of Gassendi, preferred Tycho's world view to both the Ptolemaic and the Copernican model. In his widely read *Almagestum novum* from 1651, Riccioli argued for an astronomical system based on Tycho's model. However, it was not only across Catholic Europe that Tycho's model enjoyed a widespread if fleeting popularity. In Sweden both the Copernican and the Tychonic system were first described in the late 1640s by Martinus Nycopensis, a professor of optics and mechanics at the university in Uppsala.[68] Of the two models, Nycopensis regarded Tycho's as the more acceptable.

Tycho Brahe's astronomical and cosmological ideas also represented a sort of "physical cosmology" that went beyond traditional positional astronomy. One place these ideas appear is in his first report on the comet of 1577, which was published the following year and which, for once, was not written in Latin. Not only is his *Von der Cometten Uhrsprung* important because it is Tycho's only contemporary description of the comet; it is also significant because it shows the influence from the Paracelsian world view. Tycho believed that the advent of new heavenly bodies could be understood based on Paracelsian natural philosophy:

> The Paracelsians hold and recognize the heavens to be the fourth element of fire, in which generation and corruption may also occur, and thus it is not impossible, according to their philosophy, for comets to be born in the heavens, just as occasional fabulous excrescences are sometimes found in the earth and in metals, and monsters among animals. For Paracelsus is of the opinion that the Superior Penates, which have their abode in the heavens and stars, at certain times ordained by God, fabricate such new stars and comets out of the plentiful celestial matter ...[69]

In a Danish-language almanac offering weather predictions and advice on practical meteorology and prepared in 1591 by Tycho's assistant Peder Jacobsen Flemløse, Tycho himself – anonymously – expresses his understanding of the physics of the celestial space. He mentions the three elements (earth, water

Longomontanus's work through the Jesuits, and they used it as the basis for the first Chinese representation of Western astronomy, which appeared in 1635. The rebuilding around 1670 of the Imperial Observatory, the result of a Jesuit initiative, was done by building on the instruments Tycho had described in his *Astronomiae instauratae mechanica*, published in 1598. The technology the Chinese used meticulously followed Tycho's models, except that the instruments were richly decorated with dragons, flaming pearls and other icons, signalling a fascinating blend of European and Chinese culture.[71] Had Tycho been able to remove himself in time and place from Uraniborg to the observatory in Beijing, he would have experienced more than a hint of déjà vu.

Not only was Longomontanus a well-known astronomer, he was also an active mathematical writer and educator. One of his more unfortunate exploits was claiming to have solved the classic (but unsolvable) problem of the quadrature of the circle, for which he earned notoriety among the scholars of Europe. In 1631, he discussed the problem with Descartes in Copenhagen, but the French mathematician and philosopher was unable to persuade him that his proof of the quadrature was invalid.[72] Longomontanus also carried out calculations to determine when the Earth had been created, ultimately arriving at the year 3967 BC, which was generally in accordance with the results deducible through studies of the chronology in the Bible.[73] Attempting to calculate the age of the world was a valuable and reputable activity among astronomers and mathematicians, and Longomontanus was neither the first nor the last Danish scholar to do so. The oldest almanacs, dating from the 1500s, stated that God created the world in 3962 BC, whereas Villum Lange, who was appointed to the astronomy chair at the University of Copenhagen in 1652, concluded that the world had, in fact, been created on Monday the 30th of April 4042 BC. Even into the early twentieth century, the university's almanacs supplied information on the age of the world, still more or less in keeping with the findings of Longomontanus and Lange.

The Copernican system gradually found acceptance in Denmark from about 1660 onwards, and as in several other countries this came about by means of Descartes's physics and cosmology. For much of his academic journey from 1646 to 1651, Rasmus Bartholin (one of Caspar Bartholin's sons) had stayed in Leiden. Here he had met Descartes, whose new geometrical theories Bartholin found extremely interesting; so much so that he ultimately contributed to their further development. He also gradually came to adhere to Descartes's physical-cosmological ideas, which were based on the hypothesis of cosmic vortices within a sort of world ether, and which were linked to a Copernican perception of the solar system. Works written by Rasmus

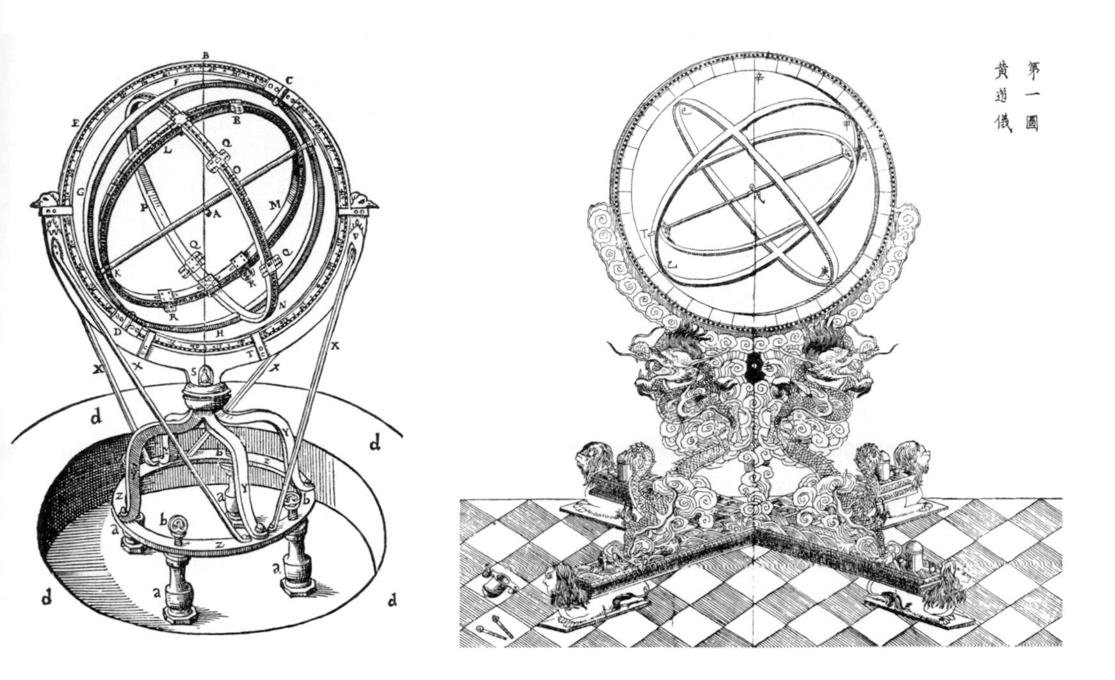

5.5 *The instrument at right is Tycho's armillary sphere as presented in his* Astronomiae instauratae mechanica *from 1598. The corresponding instrument at left, drawn around 1670, was the armillary sphere from the newly equipped imperial observatory in Beijing. Unlike Tycho's functional instrument, its later Chinese counterpart was richly decorated.*

Bartholin in the 1660s, while he was a professor in Copenhagen, reveal him as a spokesman for the Cartesian vortex theory and hence also for the Copernican world view. His collection of articles from 1674, *De naturae mirabilibus*, was decidedly Cartesian in nature and the collection shows that he was a supporter of the cosmological model attributable to "Aristarch, Copernicus or René Descartes".

From that time onwards the heliocentric system gained ever wider use and recognition, even though the process was very slow and it remained customary for many years to present the Tychonic and Copernican systems as if they had equal validity. The heliocentric system gave rise to some debate outside the circle of professional astronomers, and a few theologians were vocal in their criticism of it, but generally the issue was of little interest to the Church. In terms of Danish astronomy, the most eminent Cartesian of the period was Ole Rømer, who served as an assistant to Rasmus Bartholin in the 1660s, thereby gaining a thorough knowledge of modern astronomy and mathematics.[74]

After returning from Paris in 1681, Rømer was appointed professor of mathematics in Copenhagen, and from 1686 he was professor of astronomy. Rømer's numerous interests included studies of the world system: Though remaining a confident supporter of the Copernican model, Rømer realized that it still lacked unquestionable verification in the form of observations proving that the Earth moved in a circle around the Sun. An observation of the annual parallax would provide such verification, which is why Rømer, like so many of his contemporaries, strove to demonstrate this effect, which Copernicus had predicted more than a century before.

Around 1700, Rømer believed that he had succeeded in demonstrating the stellar parallax, as described in a letter he wrote to Leibniz in 1703: "As concerns the *parallaxis* of the fixed stars, I became sure ten years ago, and have continually gained confirmation of this throughout the past four years, thus being able to satisfy the general public."[75] Rømer did not publish his work, however, possibly because he intuitively felt that the findings based on his measurements were not sufficiently reliable after all. Administrative work and a number of practical tasks made heavy demands on his time, and moreover he was a perfectionist when it came to astronomical measurements, and therefore inherently cautious when it came to publishing his findings. The most probable reason for his reluctance to publish on the stellar parallax is that he had become aware of yet another significant source of error. Rømer's method called for the use of pendulum clocks, and he had realized that these clocks ticked with a slightly different rhythm during the day and at night due to variations in ambient temperature. His scientific diary, the *Adversaria*, shows that he sought to compensate for this new source of error, as he experimented with the effect temperature variation had on the lengthwise dilation of various metals.[76] This in turn led him to experiment with thermometers, and to suggest a thermometric scale with two fixed points – which, once again, he refrained from publishing. The scale Rømer proposed later became known as the Fahrenheit scale, named after the German-Dutch natural scientist Daniel Gabriel Fahrenheit, who had visited Rømer in Copenhagen in 1708 and there learned how to calibrate thermometers using Rømer's method.[77]

Rømer is above all famous for having demonstrated that light travels at a finite speed, which is one of the most important discoveries ever made in the field of physical science.[78] Having arrived in Paris in 1672, Rømer later became a member of France's prestigious Académie royale des sciences, chiefly concentrating on observing lunar eclipses involving the moons of Jupiter. He presented his observations in a brief treatise published in the academy's own *Journal des sçavans*, in the issue dated 7 December 1676, using them to argue that light

5.6 *This portrait of Ole Rømer, the work of an unknown painter, now hangs in the Round Tower. Around 1700 Rømer was the only Danish scientist of international calibre. He was in contact with such prominent philosophers and scientists as John Locke, Christiaan Huygens and G. W. Leibniz.*

takes 22 minutes to traverse the distance of an Earth orbit diameter (implying an approximate speed of 227,000 km/s). His claim aroused considerable attention, not least because the idea of a finite speed of light contradicted Cartesian physics. Not surprisingly, French astronomers eagerly sought to disprove the

king, retaining that position until his death in 1602. Severinus, little known as a chemical-medical practitioner, was far better known as a theoretician and systematician in the Paracelsian tradition. He was full of admiration for the legendary Paracelsus, as is apparent in the eulogy entitled *Epistola scripta Theophrasto Paracelso*, which he wrote as a young man around 1570.[83]

Severinus's most outstanding work, *Idea medicinae philosophicae* published in Basle in 1571 and subsequently in 1616 and 1660 as well, is also a product of his studious years abroad. The main reason why this work earned renown among the scholars of Europe was that it offered a systematic, well-balanced and fairly clear presentation of Paracelsian medicine while leaving out the more bizarre claims often encountered in the literature. Severinus admittedly had few kind words for Galen, the great Greek physician who was still the authority when it came to traditional medicine. On the other hand he successfully pulled off the feat of making the teachings of Paracelsus (more or less) consistent with the most important Hippocratic and Aristotelian doctrines.

The world view described in *Idea medicinae* was completely in accordance with Hermetic philosophy, in that it was lodged within a neo-Platonic and vitalistic framework in which matter is endowed with soul. Severinus's medical theory was based on the existence of a universal "balsam", an *ultima materia* that was spiritual in nature and was, purportedly, the origin and source of all natural and living phenomena. He believed that all organs in the human body contained "spirit-like substances", and that these life-spirits corresponded to the bodily organs, which he described as storerooms for the spirits. The popular Paracelsian idea of a profound correlation or correspondence between microcosm and macrocosm can also be found in the work of Severinus, who wrote that "even the tiniest seed grains are like to the universe in their construction."[84] His version of the Paracelsian ideas was less radical than the original teachings, being more eclectic and open towards the classical authorities, and it was precisely because of this that Severinus's *Idea medicinae* helped to make Paracelsian views more palatable to his peers.

Even Francis Bacon, the famous English natural philosopher and statesman, praised the book, despite the fact that he generally rebuffed the Paracelsian tradition with its "empty delusions". Providing further evidence that the ideas of Severinus enjoyed recognition in the highest places, both Christian IV and his chancellor, Christian Friis til Kragerup, were very keen to promote the publication of Severinus's posthumous works. The chancellor tried to get Ole Worm and Severinus's son, Frederik Sørensen, to take on the task, but unfortunately his efforts came to nothing. The extensive influence of Severinus's work far into the 1600s is exemplified by a Scotsman named

William Davidson, who was the first to hold the chair in chemistry at the famous Jardin du Roi in Paris. *Idea medicinae* made a great impression on Davidson, who took it upon himself to have the work republished in 1660 and supplemented it with comprehensive annotations entitled *Commentaria in ideam medicinae philosophicae Petri Dani.*

During the theologico-political struggle that unfolded in Denmark in the 1590s between the orthodox or "genuine" Lutherans and the more liberal "Philippistic" theologians, the Hermetic and Paracelsian ideas played a role, albeit an indirect one: The orthodox group would routinely criticize Paracelcianism in terms of religion for being unorthodox, perhaps even heretic. There was general concern among theologians and scientists alike as to the potential theological and moral consequences of Paracelsianism if it was not kept in check.

In this context Ole Worm can be used to represent the new, wary attitude towards Hermetic chemistry that resulted from the orthodoxy in Denmark. This attitude was not a product of ignorance, as Worm was well acquainted with Paracelsianism from his studies abroad, having encountered it first-hand in Johannes Hartmann's laboratory in Marburg and elsewhere. As a Danish professor Worm's position on Paracelsian medicine, which he either rejected or ignored, was one of scepticism if not hostility. Doubtless he had several motives, one of which was his reasoned preference for the neo-Aristotelian approach to medicine and chemistry that was gaining ground in the first half of the 1600s. There were also more "ideological" reasons, however: Worm could not help inferring connections between Paracelsian natural philosophy of the more radical sort and the dubious religious views associated with the Rosicrucian brotherhood. A letter that Worm wrote in 1620 scoffed at "this whole Rosicrucian society, if indeed it be anything at all", stating that several of its writings "smack of fanaticism or anabaptism watered down with Paracelsus."[85]

But who were these Rosicrucians? As Worm intimates, no one really knew who they were and what the gist of their thinking was, and it was even unclear whether they really existed as a kind of group or organization. Nevertheless, everyone agreed that the Rosicrucians were dangerous and had to be opposed. They were seen as being not only fanatical but radical as well; as being a destabilizing movement whose religious views in certain important areas contradicted Lutheranism. In the early 1600s that was more than enough to blacklist any movement, and if Paracelsianism could be linked to the Rosicrucian order, then the condemnation would also apply to Paracelsian ideas and practices. Worm was even obliged to criticize such a moderate Paracelsian as Severinus, though not unconditionally in his case. Worm has one passage where he in fact refers to Severinus as the leading figure in the field of chemical science.

Severinus himself was seemingly aware that certain parts of his theory of chemistry were dangerously close to theology. His nightly studies had brought him closer to understanding "the true hidden course of the creation process, as I came to realize the original and eternal principles of the bodies and of nature", as he wrote in 1583 to the Swiss physician Theodor Zwinger.[86] Creation was the domain of the Bible, not the laboratory, a fact that was asserted by Hans Poulsen Resen, an orthodox theologian and powerful bishop who criticized the Paracelsian-Severinian teaching. In a book published in 1614, Resen attacked the "fantasies about the secrets of Creation and things hidden" that typified the Paracelsian iatrochemical system. Copenhagen's other professor of medicine, Caspar Bartholin, largely agreed with Worm. While rejecting the Paracelsian-Severinian notion of illness, Bartholin agreed with the Paracelsians on other things, such as their perception of nature as reflecting divine forces and intentions. And in his guidelines on medical studies in Copenhagen from 1628, *De studio medico*, Bartholin had no compunction about recommending Paracelsian and Severinian writers as further reading.

Ole Borch – chemist and alchemist

Ole Borch, latinized into Olaus Borrichius, was appointed to a special chair in Copenhagen in 1660. As a professor he was chiefly to concern himself with philology and poetry, but he also had extraordinary duties involving botany and chemistry. Just after his appointment he left on an academic odyssey that would last six years and bring him into contact with many of the leading physicians and chemists in England, France, Italy and the Netherlands. In the course of his travels Borch became acquainted with the new chemical theories and methods. Although these emphasized the experimental element in chemistry, Borch took them to heart without rejecting the older Hermetic-Paracelsian tradition that he held in such high esteem.[87]

Borch did not practise the art of alchemy himself, but he did not reject the idea that transmuting base metals to precious metals was possible. Not only would such a rejection have lacked a theoretical foundation; a number of credible persons had also verified the transformation of metals in a laboratory. While in Amsterdam, Borch himself had witnessed the transformation of a silver coin into gold. The famous natural philosopher Robert Boyle refers to this episode in his alchemical manuscripts, finding no reason to doubt Borch's observation.[88] Borch had the privilege of meeting Boyle in London in 1663,

5.9 *Ole Borch in an engraving from 1754, done many years after his death. Borch was a professor of classical philology, but this picture emphasizes his scientific interests. Leaning on the oven in a chemical laboratory, the scholar stands above an herbal or botanical volume that lies open at his feet. Borch did research in chemistry and botany, and he taught both subjects.*

where the two conversed on topics of chemistry and natural philosophy, and Borch later wrote Boyle a long letter discussing, among other things, his chemical experiments. In his posthumous work *Conspectus scriptorum chemicorum illustriorum* from 1696 Borch praised the alchemical author Eirenaeus Philalethes – a pseudonym for George Starkey, an American alchemist who practised the transmutation of metals.

Borch's reputation as a learned chemical scientist was largely based on two works he wrote about the history of chemistry and its roots in Antiquity. *De ortu et progressu chemiae dissertatio*, which appeared in 1668, may well be the first book to give a comprehensive presentation of the history of chemistry. It was followed in 1674 by *Hermetis, Ægyptiorum, et chemichorum sapientia*, in which Borch examined the ancient Egyptian wisdom and its links to the Hermetic and Paracelsian understanding of chemistry. Both books contained a series of polemical attacks on the German physician Hermann Conring, who had himself criticized Paracelsianism as being devoid of scientific value. Borch's *Hermetis* received thorough treatment in the leading scientific journals of that period – the French *Journal des Sçavans* and England's *Philosophical Transactions.* The anonymous reviewer in the latter referred to various experiments that Borch had described in his book, such as the following:

> And *Conringius* being positive in asserting, that no force of fire is able to dissolve Gold, our Author mentions a way to perform it with a heat, at first scarce sensible, which he affirms to have been experimented by his late Majesty of Denmark, *Frederick* III; who commanded a thin plate of very pure Gold to be ground in a mortar, until it was reduced into a darkish powder, which being afterwards put into a Glass-retort, and driven by a strong heat, yielded a very red liquor, which tinged spirit of Wine, and became a good potable Gold.[89]

Borch was a Hermetic of sorts, though without adhering to all aspects of the Hermetic-Paracelsian world view – which, at all events, was very loosely defined. His early works at least seem to indicate that he had accepted the idea of occult powers that connect humans to their surroundings, earthly and heavenly, at some fundamental level. Even if these speculations on the harmony existing between microcosm and macrocosm played no role in his later works, there is still no reason to assume that this was because he had abandoned the Paracelsian philosophy of chemistry.

Borch, in his polemical discussions with Conring, took great pains to distinguish clearly between what he perceived as the genuine Hermetic-Paracelsian literature and its more magical, and therefore often heretical and supernatural, contributions. Rejecting the latter, Borch characterized such works as aberrations and fallacies. And yet he stood by his claim that the Paracelsians and their alchemist predecessors represented an important empirical approach to natural philosophy, and that this approach was firmly anchored in observable phenomena rather than in Aristotelian speculations. There was a governing idea than ran through Borch's works on chemistry: the rejection of black magic and the desire to reshape Hermetic chemistry into a non-magical science, which would include purging the Paracelsian teachings of all magical elements. Borch looked upon chemistry as both a theory of matter that was Paracelsian in design, and as an empirical laboratory science – and chemistry in its scientific incarnation was not greatly dependent on the Paracelsian ideas. By avoiding all aspects of the Paracelsian philosophy that orthodox Lutherans might find problematic, or even heretical, Borch was able to adapt it to a

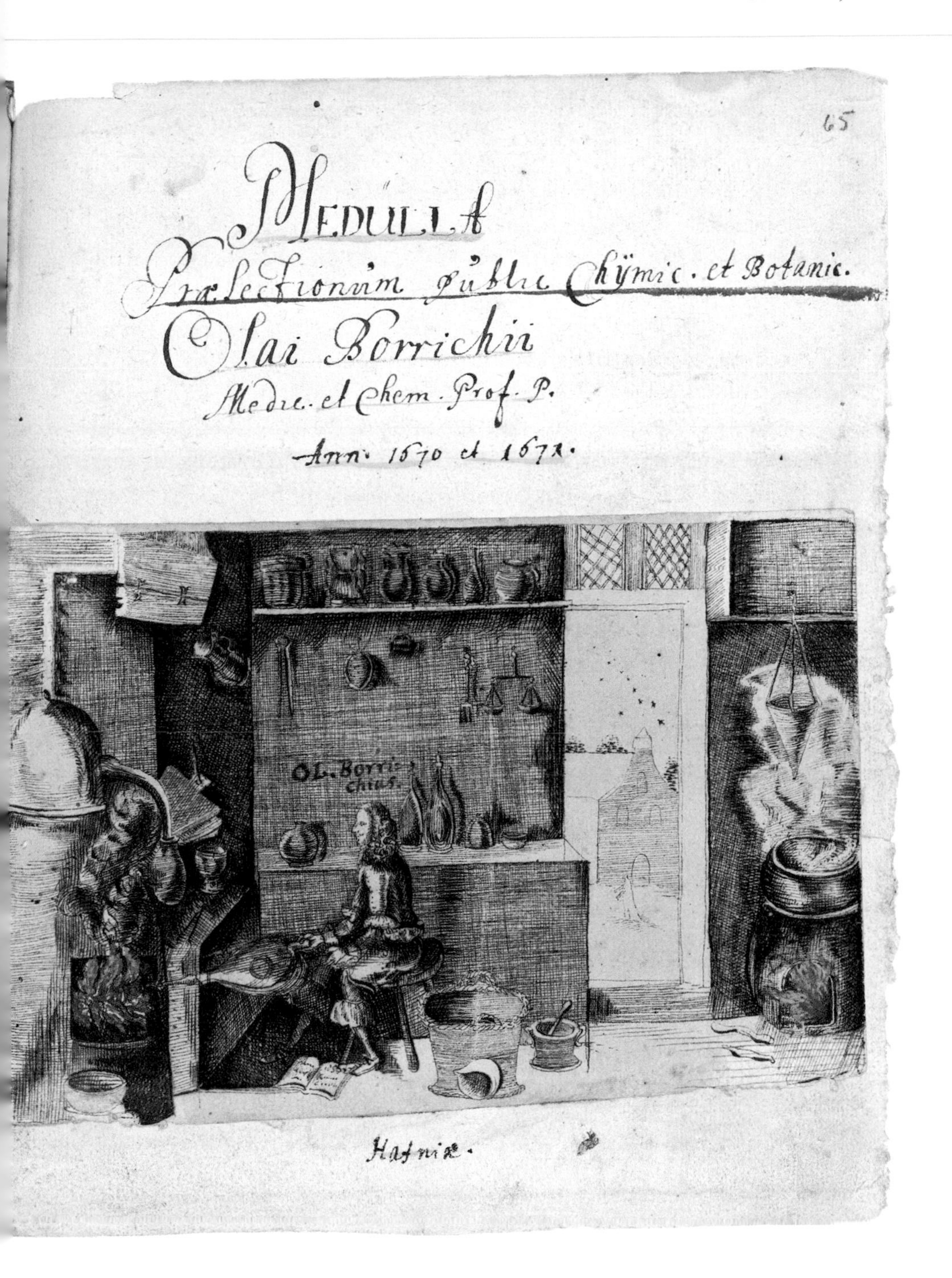

5.10 *Holger Jacobæus's drawing of Ole Borch at work. This charming portrayal by Borch's junior colleague shows the venerable professor in his laboratory, busily tending the oven to maintain the proper temperature. The exact location of Borch's Copenhagen laboratory remains unknown.*

broader and more eclectic iatrochemistry, just as others were doing elsewhere in Europe.

The new perception of matter that (unlike Aristotelian and Cartesian perceptions) was breaking through in the mid-1600s, was based on a corpuscular theory, which implied that matter was imagined as consisting of its smallest parts. The atomists, like Gassendi and Boyle, held that these small parts were elemental and indivisible "atoms", which moved within an empty space or vacuum. At no point did Borch become explicitly involved in discussions as to the structure of matter, but there is hardly any doubt that he accepted a corpuscular hypothesis. It is less clear whether he supported the theory of atoms or imagined the particles taking some other form, and there is no hint or indication that he accepted a vacuum in the sense that Boyle and others did. In his important treatise *Nitrum non inflammari* from 1678, Borch demonstrated that saltpetre when heated will release certain particles (later recognized to consist of oxygen). His very description of the experiments reveals traces of a mechanical, corpuscular perception. The hot saltpetre "acts as a bellows", he writes, describing the salt as "wholly full of innumerable blowing particles".[90] Borch's botanical works also contain the same sort of theme, found for instance in his attempts to explain how plants absorb nutrients in *De natura dulcidinis* from 1684. Here he saw the sense of taste as resulting from a mechanical process in which the particles released by the tasting agent affected the "warts" on the tongue, depending on the shape of the particle.

Borch was the first Nordic chemist to produce phosphorus, the new, astounding substance that a German alchemist had produced in 1669 by evaporating urine. The phosphorescent material was all the rage, and Leibniz, like so many others, found it absolutely fascinating. He carried out experiments on the substance, and at the request of the Danish king, Christian V, he sent his assistant Jobst Dietrich Brandshagen to Copenhagen, where he demonstrated the properties of phosphorus to the King and his court in January 1682. It is not hard to imagine that Borch attended Brandshagen's demonstration: In a university speech he gave in May 1683, Borch spoke of phosphorus and described in detail how it was produced. He explained how one could prepare a phosphoric oil, which could then be diluted and "rubbed into the skin and will then penetrate it with a marvellous light in the darkness". And as he evocatively added: "If, in a dark night, you meet someone with his face rubbed in this manner, you may become scared and believe to have met a ghost which, like an apparition, threatens with all kinds of misfortunes."[91]

The age of the Bartholins 6

Medical science had an exceptional status in Denmark in the 1600s in that it held multiple roles: It was an educational discipline, a branch of practical medicine, and the only scientific field cultivated at a higher faculty of the university (and therefore the only one that could result in a tenured position). What is more, medical science was closely linked to physics, botany and chemistry, and even to astronomy – which seems most odd today, but comes about through its association with astrology. Because of these various circumstances, medically educated scientists largely made up the backbone of Danish science in the 1600s. That is why a closer look at their work can give us a good impression of the unrest and upheavals that typified the entire period.

The learned physicians and natural philosophers behaved like heralds of a new age, criticizing the classical authorities like Aristotle and Galen. Nevertheless, in the early 1600s Aristotelianism and Galenism were thriving as strongly as ever, and they continued to serve as the foundation for most scholars in Denmark and abroad. The challenge to tradition was ambivalent. Caspar Bartholin the Elder still had both feet firmly planted in the soil of Aristotelian-Galenist science, whereas in the 1660s there were signs of upheaval even in the scientific dialogue in Denmark. People like Ole Worm and Thomas Bartholin were transitional figures who had thrown one leg, if not both, over the fence and touched down on the cultural loam of Renaissance humanism, but who definitely continued to find the scholarly literature of Antiquity both engaging and relevant.

It can be quite difficult to navigate around this rather ambivalent challenging of the old world view, since so much of the contemporary rhetoric did not reflect realities. We have sources dating back to Severinus's day that recommend practical knowledge rather than book-learning; words of encouragement that wax poetic and extol the virtues of learning by experience rather than relying entirely on the ancient authorities – but much of it was just that: pure rhetoric. Throughout the 1600s we encounter, time and again, words of praise directed at the empirically based natural science and its progress, compared to the Antique wisdom that the humanists so greatly admired. However sincere their intentions, though, such lyrical passages elegantly delivered by

6.1 *Erasmus Bartholin. Although best known for his important discovery of the double refraction of light in Iceland spar, in his youth he also did scientific work in the field of mathematics. Having met Descartes and his peers as a young man, Bartholin's scientific work was greatly influenced by the ideals on scientific method to which the French philosopher's thinking gave rise.*

men like Ole Worm and Thomas Bartholin still did not reflect any real revolution in the perception of nature and science. If there was a scientific revolution in Denmark of any perceptible magnitude, it would have to be dated to the period after 1670 and associated above all with scientists like Steno, Rasmus Bartholin and Ole Rømer.

The Bartholin dynasty

One feature that is particularly remarkable in Denmark's history of seventeenth-century science and scholarship, as opposed to that of other countries, is that a very large percentage of the development sprang from, or revolved around, a few intermarried professorial families. The original ancestor of the most powerful family was the mathematician and physician Thomas Fincke, but this academic dynasty was actually dominated by the Bartholins. To quote the historian of science Richard Westfall on the Bartholins and their clan: "I do not know of any other national scientific community so tightly organized around one family."[92] In addition to the network built around the Fincke, Bartholin and Worm families there was also another highly influential dynasty at the University of Copenhagen. Its progenitor was the bishop Peder Winstrup and it encompassed the Resen and Wandal families, but it played only a minor role in the fields of medicine and science. The group of professors active at the university between 1621 and 1732 numbered about 100, and only about one third of them were unrelated to these two major clans.

It would be far too confusing to enumerate all of the connections in the Fincke-Bartholin-Worm dynasty, but the most important links illustrate their close kinship. Thomas Fincke's children included Jacob Fincke, who became a professor of physics, and two daughters who were married to Ole Worm and Caspar Bartholin, respectively. The offspring of Worm's second marriage included his son Willum, who became a professor of medicine.

Caspar Bartholin was the brother-in-law of Longomontanus the astronomer, as well as the uncle of the physician and botanist Jørgen Fuiren. Bartholin's most famous children are undoubtedly Thomas Bartholin and Rasmus, his nine-year younger brother. The former sired a son named Caspar Thomesen Bartholin (referred to as "the Younger"), later a professor of medicine, while the latter sired a daughter who married Ole Rømer – and when she died in 1694, Rømer took as his second wife a daughter of Caspar Bartholin the Younger. They definitely kept things in the family. Interestingly, the celebrated Danish discoverer of electromagnetism, H. C. Ørsted, was also associated, if only peripherally, with this dynasty: His great-great-great-grandfather was Ole Worm's uncle.

The professorial families in Copenhagen became established in the 1600s, not only as a powerful academic aristocracy but also as an extremely wealthy group, thanks to financial transactions and links to the capital's rich and powerful merchants. Minerva and Pluto walked blissfully hand in hand. In 1638 the aggregate wealth at the disposal of the professorial college in Copenhagen was recorded at 207,600 rixdollars, 35,000 of which were attributable to Thomas Fincke. Thomas Bartholin was more than a scientific heavyweight; he was also quite successful financially. So much so that in 1657 his personal fortune accounted for 295,000 rixdollars – more than half the aggregate wealth of the professors in Copenhagen.[93]

The social success enjoyed by the Bartholins and their associates hinged on the political manoeuvring of its members and their appreciation of how important it was to avoid disputes with the orthodox Lutheranism that held sway, and with the Aristotelianism with which it was associated. As a result the Danish scientists were acutely aware how essential it was to maintain a line of demarcation between natural science and theology. That is partly why major questions of a religious or metaphysical nature were so rarely discussed in Denmark, even while being passionately debated in many other European countries. The Bartholins would typically focus on observations and stay away from philosophical speculation, enabling them to create a certain freedom of movement for the natural sciences. From the mid-1600s or so the scientific life unfolding in Copenhagen was vibrant and diverse. One might have thought that the basic incompatibility of the metaphysical basis of Cartesianism on the one hand and Hermeticism on the other – the former with its rigid distinction between spirit and matter, and the latter with its conspicuous tendency towards animism – would ignite heated discussions, but this was not the case. Not only was there a general lack of interest, as mentioned earlier, in determining the broader philosophical consequences of sciences; the two traditions were also associated with different subjects at the university. It was quite all right for Rasmus Bartholin, for instance, to be Cartesian in his mathematics even while Ole Borch in his chemistry was leaning towards the Hermetic tradition.

There are distinct traces of a development towards a new view of nature and science in the years around and just after the introduction of Absolute Monarchy in Denmark, particularly in the work of scientists such as Rasmus Bartholin, Steno and Ole Rømer. Their scientific style and interests differed in several respects from the way in which science was cultivated by the previous generation, which included Ole Worm, Thomas Bartholin and Ole Borch – all of whom were steeped in the spirit of Renaissance humanism with its boundless admiration for classical knowledge and learning. That is why they found it necessary and natural that a scholar-scientist should demonstrate his intimate knowledge of the Bible and of the classical Graeco-Roman literature. In contrast, the works of the new generations were much more factual and direct with far fewer references to that sort of literature. Consider the hundred or so references that Steno gives in his famous *Chaos* manuscript from 1659, of which only one fifth predate 1600.

A beautiful example of the new scientific style is Rasmus Bartholin's famous treatise on the double refraction of light in crystals of Iceland spar (calcite), published in 1669 in a slender volume entitled *Experimenta crystalli Islandici disdiaclastici.* Bartholin emphasized in his introduction that his interest was unrelated to the use of the crystal as an amulet or gemstone and therefore purely scientific in nature:

> Since ... so much pleasure is taken in the display of wealth's various ornaments, gems and pearls, that they may be worn around finger or neck, I might hope that those whose faculty of curiosity is more powerful than their sensuality will be no less affected by the delightful novelty of a certain body in the form of a translucent crystal, which has recently been conveyed to us from Iceland. ... I have long exerted myself in the contemplation of this body, and various experiments suggested themselves. These I wanted to communicate freely to the public because, by promoting human understanding, they seem suited to provide some use or pleasure to investigators of nature and to cultivated minds.[94]

Bartholin proceeded with his investigations systematically and pedagogically, conducting a series of experiments according to a predefined strategy that broke the scientific task down into simpler, more straightforward examinations, and finally using inductive reasoning to propose a mechanism that could explain the "birefringence" he had observed. Clearly Bartholin's strategy was greatly inspired by Descartes, who was the author's methodological and scientific ideal. The little book was a consummate model of excellent scientific exposition in its day. A significant part of Bartholin's analytical work consisted of chemical analyses in which, for instance, he heated the substance and proved that it could be electrified. Attempting to explain his observations, Bartholin split the refraction into two parts: one ordinary (*refractionis solita*) and the other extraordinary (*refractionis insolita*). That in turn posed the problem of defining a mathematical law that could describe the latter phenomenon. Although Bartholin himself did not succeed in doing so (whereas Huygens did, in 1690), his investigations were of great scientific value and were recognized as such by his contemporaries. Thanks to Henry Oldenburg in London, Bartholin's *Experimenta* became even more widely known when Oldenburg caused a lengthy English summary of the book to appear in *Philosophical Transactions.*

The sort of science represented by Bartholin's investigations of double refraction was not typical of the Danish scientific community, however. Despite its auspicious beginnings, the scientific revolution never really flourished in seventeenth-century Denmark, or rather, it did so only briefly and to a limited extent. In Copenhagen the first promising shoots quickly withered, casualties of the general deterioration in the scientific climate that had already begun around 1680.[95] This decline cannot be attributed solely to the nepotism of the Bartholin family, as has previously been done. On the contrary, it is only fair to point out that the large clan's careful nurturing of the buds on its own family tree, obviously combined with its self-awareness and shrewd political manoeuvring, played a vital part in priming Denmark for the scientific blossoming in the 1600s – however short-lived. That is not to say this nepotism had no downside, particularly in the late seventeenth century when no one in the Bartholin dynasty exhibited any notable scientific talent or interest. The author of an anonymous complaint from 1707 bemoans the academic state of the university, which he ascribes to the nepotism that had long been an accepted part of the academic environment: "By intrigues such as these, and countless others, the Professores, and most especially the Bartholins, who are among the strongest, have directed towards their families, as if by inheritance, some of the best stipends at the Academy … for which reason it would be highly desirable to see their powers weakened somewhat."[96]

Much of the decline in Danish science is doubtless attributable to factors lying outside scientific circles, externally affecting the status of science and hence its opportunities. Following the introduction of the Absolute Monarchy in Denmark, the country's class system was radically changed. Family ties came to play a much smaller role than before, whereas military and administrative positions gained a new weight. The monarchs Christian IV and Frederik III had felt for, and been interested in, scholarship and science, whereas Christian V and his administrators did not share these sympathies. The new system of Absolute Monarchy had no room for scientists and scholars. In the old view, which was consistent with the self-perception of the Bartholin clan, the "republic of scholarship" was theoretically autonomous relative to society at large. In the new system, on the other hand, scholarship was not considered valuable for its own sake. Any man aiming for a career within the system had to secure himself a position as an official, either in the military or the administration.

Absolute Monarchy robbed science, and scholarship in general, of its opportunities to operate as an independent culture with its own values and norms, and scientists experienced a sudden loss of prestige, identity and power.[97] This state of affairs led to a virtual mass-metamorphosis among Danish scientists, and almost all of them became administrators and practical problem-solvers for the State. Borch and Rasmus Bartholin both held numerous administrative positions, as did Caspar Bartholin the Younger – who even became a privy counsellor and was raised to the nobility in 1731. We also know that Ole Rømer was swamped with practical and administrative tasks after he returned from Paris, although in his case even that did not prevent him from continuing his excellent scientific work. While the transformation of Denmark's natural scientists into civil servants and practical advisers undoubtedly benefited the Absolutist nation, it was also a devastating blow to the country's scientific culture. Most of the prominent scientists were still professors at the university, and most of them found time for at least some scientific pursuits, but during Absolutism these elements were reduced to a secondary role. There is no doubt that their many other obligations impaired the quality of their teaching and their research.

How ironic that in the mid-1600s Copenhagen was bursting with talent and even attracting foreign students to Denmark, and just decades later, near the turn of the century, the university was hardly capable of recruiting students, let alone teachers. After the scientists' loss of identity in the 1670s, the general interest in investigating the secrets of nature quickly waned. Talents that might previously have been attracted to a scientific career, or at least taken an interest in natural philosophy, now sought greener and more lucrative pastures, primarily within the country's central administration.

Medicine and anatomy

In Denmark, medical science and education in the 1600s was by and large identical with the University of Copenhagen and the professors employed there. After the post-Reformation restructuring the faculty of medicine had two professorships: medicus primus and medicus secundus. During the 1600s these were frequently supplemented with temporary chairs referred to as medicus tertius and, if need be, medicus quartus. As described earlier, throughout the entire period the field of medicine was massively dominated by the intri-

The dissections conducted at the *Domus anatomica* were public, and after 1652 they were even free. In addition to their educational function for the anatomy students at the university, the dissections were also meant to show God's supreme creative power by demonstrating the marvellous manner in which he had arranged man and beast. It was vital for the reputation of anatomy as an academic field that the dissections came across as being not only practical acts but also pious and moral acts, so it was crucial that they be attended by the most prominent people in the realm. The presence of noblemen, clergymen and other persons of high repute was also meant to prevent any improprieties during the dissections.

The *Domus anatomica* had its golden days under the leadership of Thomas Bartholin, who debuted as a public anatomist in 1649, on which occasion Frederik III, the new king, was also in attendance. Soon after Bartholin's retirement the anatomical theatre also began to wane, though its popularity was briefly revived in 1673 during Steno's public dissection of an executed female convict. In that context Steno stated that anatomy, although in itself a useful and respectable science, also served a higher and nobler purpose, namely to praise the glory of God: "True anatomy, ... that which can be adapted to all onlookers, is the way along which God, by the hand of the anatomist, leads us into knowledge, first about the animal body, then about His own nature."[100] After Steno left Copenhagen, the institution quickly went downhill and was used only rarely for public dissections. Finally in 1728 the *Domus anatomica*, now itself a decaying shell, fell prey to the great fire that destroyed much of Copenhagen.

The State was interested in having the professional medical community serve other interests than those that were purely academic. Medical knowledge was to be put to practical use in fighting diseases and raising the standard of public health. Around 1645 Denmark was plagued by epidemics, in addition to its financial problems and shrinking population, all of which understandably caused the government some concern. Paulli's appointment as a royal professor was due not least to his expertise in the appropriate use of medicinal herbs as demonstrated in his botanical work from 1639, commonly referred to as *Quadripartitum*. One reason the enormous job of preparing a Danish flora was commissioned – a task placed in Paulli's competent hands – was the desire to create a handbook of medicinal herbs and their uses. The publication in 1648 of the voluminous and important work *Flora Danica* did indeed give the Danes a book that dealt, at least in part, with practical medicine. The article for each plant had a segment on "Powers and Application" offering advice on the properties and uses of the plant for either promoting health or fighting illness.

The same interest in public health is observable in the field of pharmacy, where the medical professors prepared various lists of medicinal plants. Thomas Bartholin wrote a book in Danish instructing readers on the preparation and mixing of medicines – a "dispensatorium" – that was published in 1658 as *Dispensatorium Hafniense*. This book, intended for Denmark's apothecaries, contained recipes for more than 600 medical preparations, many of which were taken from the Augsburgian *Pharmacopoeia Augustana*. Bartholin also sought to promote public health by arguing that the people of Denmark ought to use local medicines and take nourishment through traditional Danish food and drink – as he advocated in his book *De medicina Danorum domestica* from 1666. This patriotic publication criticized the use of expensive imported goods and praised the virtues of domestic food products, for had not the Lord in His wisdom made Denmark a veritable Garden of Eden, rich and self-sustaining? The Danes, he wrote, had "no cause to search for a temple of Asclepius outside their own kingdom." The sort of nationally oriented health science that Paulli and Bartholin were promoting must be seen partly as an attempt to assert the privileges of the medical profession, and partly as a patriotic mobilization for the benefit of King and Crown at a time when the realm was in dire economic straits.[101]

William Harvey's presentation of the cardiovascular system in his famous treatise *De motu cordis* from 1628 may have been the most important anatomico-physiological discovery of the seventeenth century. Tracing the Danish reception and subsequent discussion of his controversial theory also allows us to follow developments in significant parts of the country's medical-science community.[102] Ole Worm had learned of Harvey's theory in 1631 but was not convinced. In a treatise he wrote the following year he concluded that not only was the theory wrong, contradicting Galen's perception and leading to "absurdities"; it also went against long years of experience.[103] Thomas Bartholin, who in 1640 was in the process of publishing a modernized edition of his father's textbook *Institutiones anatomicae*, decided on a compromise according to which the text itself followed Caspar Bartholin's Galenic representation, while Harvey's theory was incorporated as a sort of appendix written by Professor Johannes Walæus of Leiden, one of the new theory's leading proponents. The 1641 edition of *Institutiones anatomicae*, which was popular and widely used, was the first medical textbook in Europe to explain the theory of the circulatory system.

Personally, Thomas Bartholin preferred during the 1640s to remain neutral, despite feeling a certain sympathy towards Harvey's teachings. Later, as is obvious from the much-revised editions of the anatomical handbook dating from 1651 and 1675, Bartholin essentially came to agree with the theory, although even then it remained controversial. He had not quite given up on the teachings of Galen, however, as evidenced by his scepticism regarding the nature of the heart. When, in 1663, the young Steno wrote to Bartholin that the heart was nothing more than a strong muscle, the Copenhagen professor found himself unable to agree, protesting that according to Galen the movement of the heart must be different from that of the body's other muscles. He still recognized the traditional view of muscles as organs meant for voluntary movement – a view that Steno had renounced. While studying in Leiden, Steno had become aware of the mechanical-physical view of the muscular structure, which he initially published in 1664 as *De musculis et glandulis*, later publishing a more detailed and ambitious version of his findings in 1667 as *Elementorum myologiae specimen*. According to Steno, the heart was merely an ordinary muscle. As he wrote in his 1663 letter to Bartholin, "As concerns the substance of the heart, I shall, so I believe, supply compelling evidence that nothing is found in the heart that is not found in the muscle; nor that anything is missed in the heart that is found in the muscle, if one speaks only of the muscle's nature."[104]

Thomas Bartholin's reputation as one of Europe's finest physicians and scientists was principally based on his anatomical discoveries from the early 1650s, which were associated with the controversial issue of the liver's role in producing blood.[105] The French physician Jean Pecquet had claimed in 1651, going directly against Galen, that all of the body's nutritional fluids were led to the heart through the *ductus thoriacus* – which Pecquet himself had discovered. He had only demonstrated the existence of the thoracic duct in dogs, and so the question of whether this organ was also found in the human body remained open until 1652, when Bartholin was able to verify that it was. As an offshoot of this work Bartholin, by means of dissection and vivisection, found a new type of serous vessel that led lymph from the liver to the circulatory system. He presented his findings in the treatise *Vasa lymphatica nuper Hafniae in animalibus inventa* from 1653, and the following year he was also able to prove the existence of lymph vessels in humans. Bartholin's discoveries in 1652 and 1654 were based on the dissection of human corpses, and the results of this work was what finally enabled Bartholin to free himself from the Galenic theory of the liver and embrace Pecquet's new ideas.

The rather younger Steno's earliest works as a medical scientist dealt with the same sort of areas as Bartholin had examined: the more delicate glands and vessels in the human body.[106] While living in Leiden in 1661, Steno was already set to publish his first scientific discovery: *ductus Stenonianus*, the salivary duct

THOMAS BARTHOLINUS, Casp. Fil. D.
MED. et ANATOM. IN ACADEM. HAFNIENSI
PROFESS. REGIUS. Ætatis 35. A°. 1651.
Carl van Mander pinxit.
Ionas Suiderhoef sculpsit.

6.3 *Thomas Bartholin, here portrayed at 35, may have been Denmark's most important physician and scientist in the 1600s. He gained international renown as an anatomist and was a prolific writer of textbooks and scientific works.*

leading from the parotid gland and known even today as Stensen's duct. The resulting publication was called *De glandulis oris*, and the following year Steno investigated the structure and functioning of the lacrimal apparatus. It had previously been assumed that lacrimal fluid – tears – originated in the brain, which would contract when experiencing strong emotions, thereby pressing the fluid out by way of the nerves. Steno, however, found that tears were created in the lacrimal glands, passing over the surface of the eye and draining off through tiny ducts. He regarded tears as a lubricant for the mechanical parts that constituted the human organism, fulfilling a divinely preordained purpose, as expressed in his *De glandulis oculorum* from 1662.

Returning now to Thomas Bartholin, he was a dedicated scientist whose diverse interests were by no means limited to anatomy. His first book, printed in Padua in 1645, dealt with the legendary unicorn, the creature whose horn was known to be a universal remedy against disease. It is true that in 1638 his

TH: BARTHOLINI
DE
UNICORNU
OBSERVATIONES.

6.4 *Title page from Thomas Bartholin's* De unicornu *from 1645. The horn of the unicorn was generally regarded as a panacea, and Bartholin saw no reason to doubt the assumption. This publication from his youth on the legendary creatures was just one of many contributions to the extensive unicorn literature.*

uncle, Ole Worm, had published a treatise establishing that the horns possessed by the noble and wealthy were nothing but the teeth of the narwhal, but even so the myth of the unicorn lived on. Neither Bartholin nor Worm was prepared to rule out the existence of other types of unicorn – resembling, for instance, a horse or a stag – and in all events they continued to believe in the healing of power of such horns, or teeth. A letter Bartholin wrote in 1655 told of his own success in using the celebrated panacea to cure patients suffering from dangerous fevers. Bartholin's brand of knowledge, scholarship and science contained a significant element of Renaissance ideals, which in his works coincided with the experimental standards of the new age. His first major work was *De luce animalium*, a comprehensive investigation of organic light phenomena. It appeared in 1647 and eruditely described numerous observations of what would later come to be known as phosphorescence, fluorescence and bioluminescence. Judged by later standards the book is a bizarre blend of raving superstition and reasoned science, but it was well received and long remained the standard work on that topic (additionally being published again in 1667 under the new title *De luce hominum et brutorum*).[107]

Another topic that Bartholin found interesting was low-temperature phenomena and their medical uses, on which he wrote a book entitled *De nivis usu medico* in 1661. At that time, cold was still generally regarded as an independent property on a par with warmth, rather than simply understood as an absence of warmth. Around that time Ole Borch was also experimenting with cold and the freezing of water, and he believed that cold was an active principle. The work that both Borch and Bartholin did on cold phenomena was well known outside Denmark. Another scientist interested in the topic was Boyle in England, who unlike the two Danes perceived heat as expressing a movement of the particles in a given substance. A book Boyle wrote on the topic, dating from 1665, referred to "the industrious Bartholinus", but the author did not support all of the Danish physician's claims – such as his description of how if one boiled, say, a cabbage and froze the resulting decoction, the ice would take on the cabbage's shape. The explanation, Bartholin suggested, was that the cold would concentrate the vegetative principles of the boiled food item. Boyle duplicated the experiment but was unable to observe the phenomenon Bartholin had described.[108]

Nicolaus Stenonis and the birth of geology

Niels Stensen, who called himself Nicolaus Stenonis but is commonly known simply as Steno, began studying at the University of Copenhagen in the late 1650s. Steno's thoughts on his studies and the many books he read are portrayed with fascinating clarity in his so-called *Chaos* manuscript, which he compiled in the spring of 1659.[109] Although a mere 21 years old at the time, Steno was extremely well versed in both classical and contemporary literature, and his interests were broad and diverse. The works he annotated and commented on were written by such notable authors as Paracelsus, Tycho, Francis Bacon, Kepler, Descartes, Gassendi, Athanasius Kircher and Libavius. In early 1660 Steno set out on his great educational journey, which would introduce him to prominent circles within Europe's academic elite. In Holland he became close friends with the microscopist Jan Swammerdam and the anatomist Franciscus Sylvius and conversed with the Jewish philosopher Baruch Spinoza. Later, in Germany, he met Leibniz, the famous philosopher. On his travels in Italy he became acquainted with several leading contemporary anatomists and naturalists, including Malpighi, who was a native Italian, and the Englishmen William Croone, John Ray and Martin Lister.

Initially, Steno's most important scientific work during his years abroad was in the field of anatomy, focusing mainly on the functioning of glands and muscles and on the arrangement of the human brain. On the latter topic he gave a famous lecture in Paris that was published in 1669 as *Discours sur l'anatomie du cerveau*, a classic work in the brain research literature.[110] Although *Discours* was based on independent dissections, it was more than anything a methodological and programmatic exposition that criticized the existing claims and methods. Descartes's book *De homine* had been posthumously published in 1662, and Steno took the opportunity to point out that although the work was brilliant, the way it portrayed the structure and functioning of the brain had very little to do with the real brain. Criticism aside, however, Steno was indebted to the great French thinker for his methodology, for Steno's very formulation of his research programme concerning a "science of the brain" was decisively influenced by Descartes's *Discours de la méthode*.

Steno's most seminal contribution to the development of the geological

6.5 *Portrait of Steno, probably by the Flemish painter Justus Sustermans, who was court painter to Ferdinand II, grand duke of Tuscany. Niels Stensen, as he was christened, never used the name "Steno" himself, but consistently preferred the Latinized form Nicolaus Stenonis.*

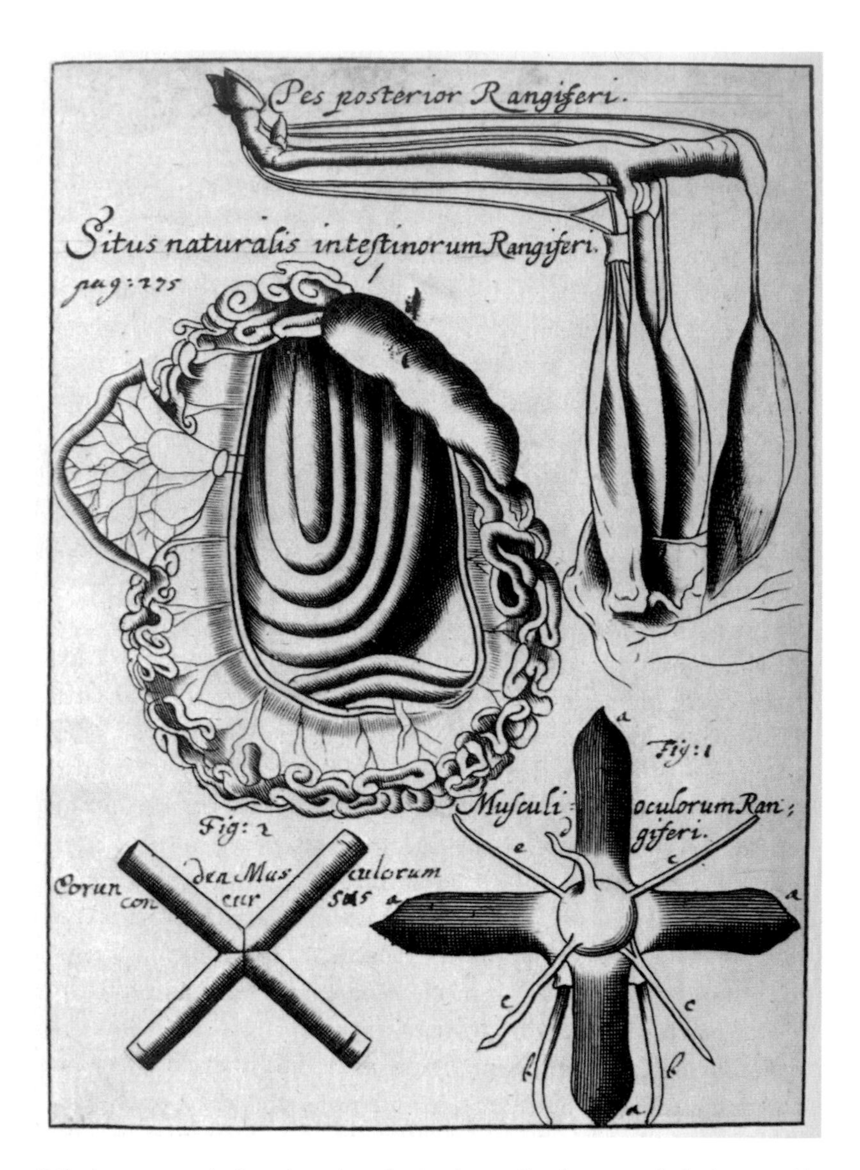

6.6 *As an anatomist Steno investigated animals as well as humans, which also earned him a place in the Danish history of zoology. Here, from* Acta medica *1673, reproductions of his drawings from dissections of a reindeer and a squirrel. Like many physicians and scientists of that era, Steno was a competent artist.*

and mineralogical sciences was set in motion in 1666, when he was ordered by Ferdinand II, Grand Duke of Tuscany, to dissect the head of an enormous shark caught off the coast of Livorno. While doing so Steno compared the shark's teeth with the "tongue stones" or *glossopetrae* that were found most notably on Malta and caused quite a sensation at the time. Steno concluded that the Maltese tongue stones were fossilized shark's teeth deposited long ago when Malta was covered by the sea. This conclusion was not entirely new, but it was very controversial and remained unaccepted by many, including Thomas Bartholin, who had himself been on Malta and was quite familiar with tongue stones and other stones with embedded objects or markings on their surface. Bartholin did not believe these resulted from the petrifaction of originally organic matter, but saw them as examples of nature's jests or whims. In 1671 Bartholin wrote a piece in which he mentioned a stone containing something that looked exactly like a fish skeleton, although, as he wrote "I do believe it is due to chance, when something resembling animals is found in stones."[III]

As mentioned, claiming in the 1660s that *glossopetrae* were fossilized shark's teeth was by no means revolutionary, for others had already done so. Steno, however, used the phenomenon to develop a completely new perspective on geology and palaeontology. Applying a series of arguments based on observed facts (*historiae*) and assumptions (*conjecturae*), Steno proposed that the Earth must have changed significantly over time, and that the geological formations themselves must be some sort of historical document from a distant and very different past. The Earth's crust, he claimed, must be made up of layers that were sedimentary in nature and consequently must have originated as deposits in the sea or in fresh-water lakes and rivers.

Writing about the shark's head motivated Steno to direct his interests away from anatomy and apply himself to geological matters, which gave rise to his important work *De solido intra solidum naturaliter contento dissertationis prodromus* (The Prodromus of Nicolaus Steno's Dissertation Concerning a Solid Body Enclosed by Process of Nature Within a Solid). Here Steno expanded on his proposed idea of a historical or evolutionary geology, arguing that the differences in the various layers of earth must be attributable to differences in their age; that they must be the result of successive periods of sedimentation in a distant past. On a more general level, Steno's thinking on this issue offered an answer to the question of how the current state of a thing can reveal its previous state, in other words how one can infer or deduce from present to past. Steno is usually recognized as the founder of stratigraphy, the branch of geology concerned with the age, order and relative position of the layers in the Earth's

PART II

Naturalism, Knowledge and the Public Good

1730–1850

Helge Kragh

An era of progress and upheaval 7

Moving forward in time, let us look at science and the conditions under which it existed in Denmark – and to some extent in Norway – during the period from 1730 to about 1850. For most of these 120 years the kingdom of Denmark was still ruled by monarchs wielding absolute power (as they had done since about 1660). Also, until 1814 Denmark and Norway were bound together in what was known as the Dual Monarchy. Whereas the latter half of the 1700s had generally brought economic prosperity, the country's problems began piling up soon after the turn of the century. The most obvious reasons were the war against England and the Danish government's decision to join forces with France, which was ruled by Napoleon. The period reviewed in these next six chapters runs up to the end of the Three Years War. Although Denmark had successfully quelled the Prussian-backed rebellion in Schleswig-Holstein, thereby keeping the nation's borders intact, it was only postponing the inevitable conflict over the contested region. No less significant was the development that after the new constitution of 1849, Denmark, for the first time, was to be ruled by a democratically elected government.

In terms of the country's intellectual and scientific history it is natural to see this period as falling into two parts: the Enlightenment, which dominated the latter half of the 1700s, and Romanticism, which peaked in Denmark during the first decades of the 1800s. Although Romanticism set its distinctive mark on the country's science, particularly in the work of Hans Christian Ørsted, it was by no means true that Danish scientists generally subscribed to the novel German ideas of a *Naturphilosophie,* a "philosophy of nature", or turned their backs on the previous ideas of the Enlightenment. Despite all the Romantic rhetoric there was still a high degree of continuity across the turn of the century. Ørsted himself is an excellent example of this, since in many ways the ideas originating with the Enlightenment were synthesized into the new trend of natural Romanticism.

Outlining the period 1730–1850

During the 1700s the population of Denmark began to grow at a steady pace, for the first time in history. Rising from a total of some 600,000 in the year 1700 up to about 718,000 in 1735, the first Danish census carried out in 1769 set the population at 797,584. This growth rate continued to increase over subsequent decades, and in 1850 Denmark proper had a population of 1.42 million, to which could be added the citizens in Norway and the duchies of Schleswig-Holstein. Based on this near-doubling of the population from 1730 to 1850 one might well have expected an analogous rise in the number of university students, whose swelling ranks in turn would provide a larger recruiting base for the new generation of scientists. Yet this was not the case. Throughout the entire period the enrolment of students at the University of Copenhagen was small, corresponding to a mere 1–2% of each birth-year cohort. The vast majority, including many sons of the clergy, came from the highest echelons of society, whereas virtually no students came from the rural population, which constituted 80% of all Danes.

The growth in the number of academic students, which we perceive today as a natural development, is really quite a recent one. During the 110 years from 1740 to 1850, the number of students matriculated at the University of Copenhagen varied between 209 (for the decade 1741–1750) and 111 (for 1811–1820). What is more the university, the country's largest and most important scholarly institution, had only a small number of full-fledged graduates, the vast majority of whom became priests or bureaucrats in the State administration.[117] Between the two university reforms of 1732 and 1788, an astounding 97.3% of those earning a master's degree were graduates in theology or law. Given that theology alone accounted for 66.7%, the frequent use today of the expression a "pastoral school" to describe the university is more than just a turn of phrase. Theology long remained the most important subject, but over the years it became less dominant. The decade of graduate statistics for 1741–1750 shows 144 theologians and 4 lawyers, while the corresponding figures for 1841–1850 are 63 and 20, plus 15 physicians. The decade 1861–1870 was the first that did not show theology as the largest subject, as the 201 recorded graduates included 44 lawyers, 35 theologians and 26 physicians.

As part I of this book has already explained, science and scholarship in the 1600s were closely linked to religion, and they were often justified as being part of a religious project. These intimate ties existed for most of the 1700s, though they became less rigid and were upheld by other forces, especially the ideas relating to natural theology. There was no contradiction between religion and science, and (at least in public) the vast majority of scientists declared themselves not only Christians, but also believers in some sort of natural theology. It is quite typical that both the brilliant writer Ludvig Holberg and the physics professor Jens Kraft immediately and vehemently opposed the materialistic trends that arrived from France with the novel ideas of the physician and philosopher Julien La Mettrie. As late as the 1850s the polytechnician Ludvig Colding used a spiritual and religious context to justify his theory about the conservation of energy. All the same, during the period under review the power of the Church did diminish, as did the influence that religion exerted upon science. By 1850 this factor was making itself felt with no more than a fraction of the considerable weight it had carried around 1730.

Just as the role of religion waned with respect to science, so too did the significance of Latin, the traditional academic language. Had anyone told Thomas Bartholin that 150 years after his time most of Denmark's scientific literature would be written in Danish, he would have thought it an offensive joke. The first signs of this development appeared as early as 1745, when the newly established Royal Danish Academy of Sciences and Letters decided to use the country's own language in its series of publications entitled *Skrifter* – despite the obvious fact that its members were fluent in Latin and were used to writing in that language. One unfortunate result of this policy was that many of the contributions in the *Skrifter* series remained unknown outside Scandinavia, as several examples in the following will show. Latin long prevailed as the scientific language at the university, especially for those writing dissertations, but here, too, there was a move towards using Danish. Even into the 1840s some people were still writing scientific and medical treatises in Latin, although by that time academic articles were usually written in one of the living languages. In Denmark this generally meant Danish, with German as a strong alternative and a natural choice for several scientists of German or Schleswigian extraction, who also hoped it might help their contributions to become known internationally.

The 120 years of scientific development in Denmark covered in the following chapters fall naturally within two intervals: 1730–1800 and 1800–1850. The boundary is defined by internal and external factors, with the years around 1800 bringing momentous changes. Internally the country was rocked by the political and economic impacts of Denmark's wars with England: the end of Denmark's heady days as a major overseas trading nation, violently and abruptly culminating in the bombardment of Copenhagen (in 1807), then the national bankruptcy (in 1813) and the loss of Norway (in 1814). Externally it was influence by the Romantic wave that heralded a new cultural atmosphere and the beginning of what is often referred to as Denmark's "Golden Age".

Much of Copenhagen was reduced to ashes in the great fire of 1728, including the university's library and several other university institutions, and this severely crippled the city's academic community. But truth be told, at the time Danish science was at such a low point that the fire itself hardly made much difference: Twenty years after the death of Ole Rømer, science in Denmark was all but non-existent. Physics and medicine were certainly represented at the university, but in a form that can best be described as petrified, and to which research was virtually unknown. The general attitude of the Crown and the civil administration toward science was one of indifference. After this dismal situation had lasted for 10 to 20 years, a new attitude began to emerge, undoubtedly in response to the country's improved economic situation. One of the earliest initiatives that show a renewed interest in natural science in Denmark was the establishment in 1742 of the Royal Danish Academy of Sciences and Letters. This institution, which was private despite public funding, became very influential and left its mark on the scientific development throughout the period. The 1750s led up to a period known in Danish history as "the flourishing years", spanning about five decades of unbroken economic growth fuelled mainly by shipping, as Denmark's foreign trade blossomed by virtue of the country's status as a neutral haven in a sea of warring nations. The influx of money to the kingdom grew, and progressive landowners and leading members of government increasingly recognized that the way to advance and augment agricultural production was to make it more scientific.

At all events, the decades following 1750 put an end to the stagnation in Danish science. The country's establishment and visionary landowners took the lead to create a new atmosphere of progress, whereas the university in Copenhagen remained a conservative bastion that was opposed to change and regarded science with a curious blend of suspicion and indifference. It was characteristic of the development well into the 1800s that the slowly improving conditions for science came about not because of, but *in spite of* a university that was intransigent and reactionary. The initiatives that took place resulted in the founding of a series of new institutions that in 1760, for the first time ever, created a sort of scientific infrastructure in Denmark. The new scientific system did include the University of Copenhagen, but most activities went on between other players and institutions that were either separate from or only loosely associated with the once-revered seat of learning.

In the late 1700s the University of Copenhagen began the first small, slow steps towards reform, and gradually these efforts helped the situation of the natural sciences to improve. A long, drawn-out process, which took place in the difficult political and economic conditions of early nineteenth-century Denmark, led to the establishment of several new professorial chairs, though it was not the University but rather the State that took the lead. Initiatives would normally originate with the university's patron or its directors, with the State usually being called upon to remunerate the new teachers in the sciences. But between 1821 and 1831, for instance, positions were created in botany, chemistry and mineralogy, typically at the behest of the State and against the university's wishes. At that time all mathematical and scientific subjects were gathered under the faculty of philosophy, which was, however, a highly inappropriate and dissatisfactory arrangement. This led to the establishment in 1850 of an independent faculty of mathematics and science – the first of its kind in the northern countries, and unparalleled at the universities in Lund and Uppsala until 1956. The new faculty enhanced the status of science in Copenhagen, simultaneously creating a basis for continued growth, which did in fact materialize during the second half of the nineteenth century.

From Enlightenment to Romanticism

As far as the natural sciences and natural philosophy were concerned, the Enlightenment came to Denmark by means of the comprehensive, Leibniz-inspired thought system constructed by the German philosopher and polyhistor Christian Wolff in the 1720s, which became extremely popular in Denmark and in Sweden. Incidentally, at one point Wolff could have secured a professorship in Copenhagen, which might in turn have positioned the Danish capital at the very heart of European scholarship: In 1710, when Wolff was a professor of mathematics in Halle, Leibniz wrote to Rømer that the young German mathematician was interested in taking over the vacant mathematics chair in Copenhagen.[118] In short, Wolff sought Rømer's support, but nothing ever came of Leibniz's inquiry because Rømer died shortly after the letter was sent, possibly without even reading it. And so Wolff remained in Halle and history took its course.

Wolff's system was decidedly eclectic, and although he was intensely preoccupied with mathematical-deductive methods he was not a rationalist in the literal sense of the word. He also emphasized the significance of experience as a means to cognition, and this particular combination of rationalism and empiricism is what made Wolff's ideas appealing to men of science. One major reason why Wolffianism became influential was its claim that the Christian teachings of the Revelation were consistent with the rational or scientific cognition of God. Human reason was capable of comprehending both, and God would only reveal what the human senses were able to perceive. This problem - the relationship between revelation, reason and sensory perception - was also scientifically relevant, namely in connection with the so-called natural theology, or physico-theology. Whereas this view had been routinely rejected by theologians as deism or potential atheism in the early 1700s, the Wolffian variety of natural theology was not merely acceptable; it was rapidly incorporated into the Lutheran Evangelical doctrines of the Danish state church.

The immense influence of Wolff's ideas disappeared by the end of the eighteenth century. One important reason was that in a 1762 dissertation Immanuel Kant had pointed to decisive errors in the very basis of Wolff's theoretical construction. In his *Kritik der reinen Vernunft* from 1781, Kant had further directed harsh criticism against natural theology, causing its influence to slowly wane. Kant's alternative ideas opened a new chapter in the history of philosophy and were also very important to the natural sciences. The Danes were introduced to his thinking in the 1790s, and in various way this served as a transition to the ideas of natural Romaniticism.

As the remaining chapters in part II will explain, Romaniticism was a powerful force, in Danish science as elsewhere, not least because of the great and lasting influence of physicist H. C. Ørsted. To a great extent Romanticism was a reaction to the way the Enlightenment had understood culture, nature and humankind. It would, however, be misleading to portray the years around 1800 as an actual break where a whole new view of science and culture replaced an old one. In spite of all the romantic rhetoric, there was a high degree of continuity throughout the entire period, and actually Romanticism did not repudiate all of the Enlightenment's ideals. As eminently exemplified by Ørsted, a person could be a natural Romanticist *and* a man of the Enlightenment as well. The enormous interest during the late 1700s in disseminating science to much larger population groups by no means waned in the early 1800s, nor was it inhibited by the elitist arrogance and cultivation of genius that characterized certain parts of the Romantic movement. On the contrary, enlightenment and the transference of knowledge was taken even more seriously, quite in keeping with Enlightenment ideals. The Danish efforts climaxed with the creation of bodies such as the Danish Society for the Dissemination of Natural Science (founded in 1824) and the Danish Natural History Society (established in 1833). Both were crucial in creating a Danish science community and remain active to this day. Another point to remember when assessing the significance of Romanticism is that unadulterated natural-Romantic views were limited to

7.1 Christian Wolff, influential mathematician and philosopher of the Enlightenment, from the University of Halle. Wolff's eclectic philosophy and systematic thinking, which combined rationalism and natural theology, made a great impression on Danish scholars in the sciences and the humanities.

a fairly small circle of people, and that they had no influence at all on either the national administration or the university's directors.

Moving on to the end of the period, by the mid-1800s Romanticism was a relic of the past, and a sort of Positivism was on its way to becoming the dominant scientific ideology. In Paris during the 1830s, Auguste Comte had given a series of famous lectures in which he introduced his system of Positivism, and which led him to write his great work *Cours de philosophie positive* between 1830 and 1842. Even though Comte's thoughts and the similarly empiricist ideas of others did not play a role until the latter half of the nineteenth century, at the time of Ørsted's death in 1851 another major shift was already under way. Reading a lecture given by the 64-year-old geologist Johan Forchhammer in 1858, one can almost feel the winds of change stirring as he expresses his concern at the materialistic and atheistic tendencies stimulated by certain parts of modern science. In his view, scientific progress gave no cause to abandon the

Christian standpoint, and he therefore stated his opposition to the "scientists [who] in the development of science see an attack on our highest religious truths". This, he believed, was unjustified, for the borderline between religion and science had merely been moved, and that in itself did not validate an attack on the religious truths. "No chemist, physiologist or geognostic has, as yet, been able to explain to us the manner in which inorganic matter could be transmuted into an organic being with its own particular, limited task of life, and no one has been able to fully enlighten us as to how, each day, stone is transformed into bread."[119]

Three publications from the mid-1800s are particularly fitting as examples of the new age and of the trends that were making Forchhammer feel apprehensive. In 1847, Hermann von Helmholtz published his definitive formulation of the principle of energy conservation in his treatise *Ueber die Erhaltung der Kraft*. Then in 1855 came the philosopher Ludwig Büchner's materialistic manifesto *Kraft und Stoff*, followed four years later by Charles Darwin's groundbreaking work *The Origin of Species*. During the half-century that followed those very topics – energy, materialism and evolution – would become leading themes in science, and in the philosophical discussions to which scientific progress gave rise.

Institutions, patronage and funding

8

Back in the 1600s, certain parts of Danish science had lived up to very high standards and in some cases even earned international recognition. Generally speaking, however, any sort of institutional foundation outside the University of Copenhagen had been sorely lacking. The ability of science to regain its footing in Denmark from about 1740 onwards was linked to a surge in the creation of new social and institutional entities, and for the first time the country found itself equipped with an actual scientific infrastructure; a network of societies, educational institutions and other organizations that would serve as a framework for further scientific and technological development. The Royal Danish Academy of Sciences and Letters, located in Copenhagen, became an important player and soon came to host and handle several large state-run projects of a technical or scientific nature.

The University of Copenhagen was still the pre-eminent central institution for education and research in Denmark, but from the 1780s onwards various other institutions, some short-lived, were more innovative and unfettered by the conservatism so characteristic of the university at the time. Around 1830 there was another spurt of infrastructural growth involving the establishment of the Royal Danish Military Academy and, no less significantly, the Polytechnical College. At one point the latter was closely associated with the university, serving as a sort of unofficial faculty of science for several years. Throughout the period under review here, the king's decisions and fields of personal interest were of paramount importance, as reflected in his partiality towards the German astronomer H. C. Schumacher. This remarkable example of blatant favouritism, occurring during the 1820s and 1830s, meant that for quite some time Danish astronomy was effectively outsourced to the observatory in Altona.

Universities and research

Pursuant to the new university charter of 1732 the number of professors was reduced to 15, with the faculty of philosophy losing 2 chairs and the faculty of

8.1 *Partial view of the University of Copenhagen, c. 1750. This picture shows the building that housed the university's new natural theatre from 1770 to 1807. Behind it stands the main building, constructed after the great fire of 1728 and used to house Georg Detharding's extensive collection of natural specimens from 1740 until around 1770.*

law gaining 1 new chair in Danish law, raising its total to 2. In addition to the 15 professorships defined in the charter, the university had a number of unremunerated, titular or extraordinary professorships, some of which were *designatus* – a term that implied automatic access to an available chair, usually after one's predecessor had died.[120] Such designated professorships hampered the dynamics of the university and the quality of the scientific work done there, since professional qualifications were not necessarily a key criterion for appointing tenured professors. The unfortunate practice was not halted until the charter was amended in 1788, the new rule being that available professorships had to be publicly announced and awarded to the applicant deemed best qualified. Bear in mind that throughout the 1700s the University of Copenhagen was still primarily a school for educating clerics, and secondarily a school for hatching judges and other bureaucrats who would enter into the service of the Danish state. Of the 8,176 graduates leaving the University of Copenhagen between 1732 and 1788, no fewer than 5,455 (or 66.7%) graduated in theology/divinity and 2,504 (30.6%) in law, whereas only 121 (1.5%) graduated in philology and 96 (1.2%) in medicine.

The reformed university charter of 1732 abolished the chair in natural philosophy and experimental physics, which dated back to 1539, and it completely ignored geology, zoology, botany and chemistry. To the extent that these subjects were taught at all, they were handled by professors in other fields (medicine and mathematics) or explained piecemeal and at random as part of the instruction students received for their *examen philosophicum*. Regarding the professors of medicine and their areas of work, the charter stated that these included some instruction in chemistry, botany and pharmacy, and that one of the professors was to "conduct a number of *Experimenta Physica*". On physics or natural philosophy the charter also summarily mentioned that "*Philosophia Naturalis* must be taught either by one of *Proffesoribus Medicinæ*, or by one of *Proffesoribus Mathematum*, so that weekly he shall instruct one day on *Physicam*, but the remaining days on *Medicinam* or *Mathesin*, whichever of these is his profession."[121] Such was the wording of the charter, but in practice it was rarely followed, and so for years the natural sciences were simply not represented at the University of Copenhagen.

Despite this adversity, the sources show a growing interest in certain fields of science from around 1740, and that trend had two root causes. Firstly, the cultivation of these sciences was seen as a means of increasing the international prestige of the Danish-Norwegian kingdom. Secondly, and no doubt more significantly, leading circles in the Danish establishment saw the natural sciences as useful, indeed indispensable instruments in pursuing the mercantilist policy that was the prevalent economic ideology of that era. The state bureaucracy was a firm believer in science as the driving force behind progress, and accordingly several plans appeared that were intended to reform the country's system of higher education – none of which were ever realized. During the latter half of the 1700s it was mainly natural-history subjects that were seen as interesting, the underlying idea being that they should be cultivated as "oeconomic sciences"; concrete sciences that could elevate manufacturing, agriculture and forestry to a higher level. The Danish state was extremely keen on conducting natural exploration throughout the kingdom, not least in the hopes of finding useful raw materials. However, no one at the university was interested in – let alone capable of – taking on that sort of task.

The country's lack of the necessary scientific expertise could be remedied either by sending talented students abroad to study or by calling in foreign experts. An attempt was made as early as 1732, in connection with the university reform, to bring the renowned Dutch natural philosopher Pieter van

Musschenbroek to a chair in Copenhagen, but in vain.[122] Similar offers from Denmark in the early 1750s were more successful: The influential Count Bernstorff called upon Georg Christian Oeder of Göttingen to become professor of economics, and Christian Gottlieb Kratzenstein, another German scientist, came from St Petersburg to be appointed extraordinary professor of experimental physics. The policy that was followed in the cases of Oeder, Kratzenstein and others actually went against the reformed charter from 1732, which had set the stage for a "de-Germanification" of the university. The king had promised to give preference to Danes rather than foreigners when appointing professors, but this sort of preferential treatment was not practically possible. There was, quite simply, a paucity of qualified Danes, and an overabundance of Germans who fit the bill. Between 1732 and 1788 a total of 70 professors were appointed in Copenhagen, 13 from Germany and 5 from Holstein, corresponding to 26%. Throughout the eighteenth century, the substantial German presence at the university continued to be a source of controversy and dissatisfaction.

Oeder, who went on to become a "professor by royal appointment", was also created director of the newly established royal botanical gardens at Amalienborg castle. His greatest achievement was beginning the actual compilation of an authoritative national flora for the Danish-Norwegian realm, the exquisite *Flora Danica.* Finally in 1761 after several study trips to botanical gardens abroad, Oeder succeeded in publishing the first fascicle of the *Flora Danica*, handing the laborious undertaking over to fellow botanists in 1772.[123] His compatriot Kratzenstein made important contributions to Danish scholarship and research covering a variety of fields, most notably medicine and physics. He chiefly earned renown for his work in practical and theoretical electricity, and as an acclaimed expert in the new field of electrotherapy he came to be seen as a real benefit to Denmark. An incredibly versatile teacher and a tireless researcher, Kratzenstein produced a body of scientific work that was enormous and wide-ranging, although it was neither very original nor very profound.[124] In addition to his obligatory lectures he was also a private tutor for the nobility and engaged in scientific colloquies with the gentry – at a price. This was such a lucrative business that several years before his death he was able to make an endowment of 12,000 rixdollars to the University of Copenhagen, stating as his wish that the amount be used to found a new chair in physics.

Kratzenstein's greatest influence on the scientific climate in Denmark was exerted through a series of popular physics lectures he gave between 1754 and 1786. These lectures, delivered in German or Latin, included experimental demonstrations that the most attentive spectators could also follow in his text-

8.2 *Plate from the first fascicle of the new, ambitious* Flora Danica, *showing a hitherto undescribed plant: Oeder's lousewort (*Pedicularis oederi*), named after Georg Christian Oeder, a German professor who moved to Copenhagen. The* Flora Danica *is considered Oeder's most important contribution to Danish botany, though he also worked with other, more theoretical aspects of his profession. Meticulous reproductions adorn the exquisite Flora Danica porcelain used at royal banquets and still in production today.*

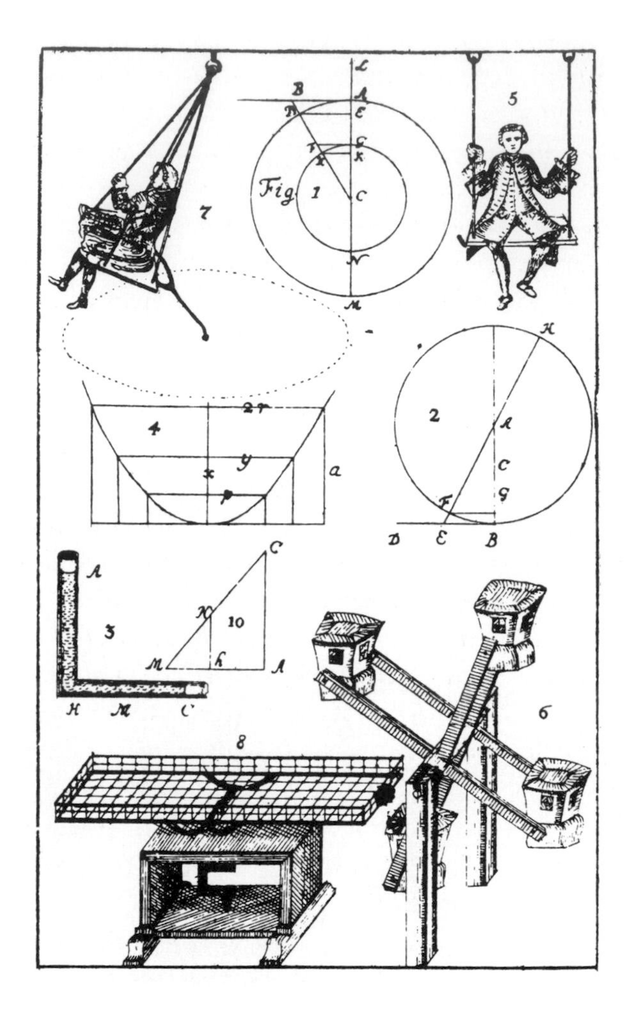

8.3 *Christian Gottlieb Kratzenstein's illustration of his proposed cure, which involved subjecting the patient to massive centrifugal force. He presented his ideas in a medical dissertation from 1765 entitled* Vis centrifuga ad morbos sanandos applicata. *Fortunately for prospective patients, Kratzenstein's novel idea was never put into practice.*

book. He had an impressive collection of physics apparatuses, most of which were sadly lost in the great fire that ravaged Copenhagen in 1795, shortly before his death. Kratzenstein published a textbook in Latin for university students, as well as a book that targeted a wider audience, entitled *Vorlesungen über die Experimental-Physik* (1758 and later editions). An abridged Danish version of his Latin-language textbook also appeared in 1791, and unlike the Sorø professor Jens Kraft's mechanical works it contained only a little elementary mathematics. It did, however, deal with electricity, magnetism, sound, gases and meteorology in quite some depth. What is more, like Kratzenstein's other textbooks, his *Vorlesungen* also contained a good bit of chemistry, chiefly presented within the framework of the phlogiston theory he had learned as a young man in Halle.[125]

Despite the efforts of Kratzenstein and a few other professors, Danish science was definitely not thriving in the latter half of the 1700s. To the modest extent that it was cultivated in Denmark, science rarely moved beyond serving as an ancillary subject at the university or appearing in descriptions of economically useful initiatives set in motion by the State. Although not unknown, systematic basic research founded on quantitative experiments and mathematically formulated theories had not yet caught on in the kingdom. Scholarship still counted for more than research, and utility for more than truth. There was often dissatisfaction with the quality of Danish scholarship and science, not least in comparison with foreign standards, which at times made Danish efforts seem painfully inadequate. As the powerful university patron Count Johan Ludvig Holstein pointed out in 1754, the professors in Copenhagen only rarely presented scientific treatises that earned international attention. The professors immediately objected to this criticism, asserting that they saw no significant reason for becoming involved in research. They were primarily employed to teach – and quite apart from that, the concept of research-based teaching had yet to be conceived. The instruction given at the university mainly consisted in passing on existing knowledge, as it had always done, and personal research contributions were not prerequisite for doing this. Naturally, that does not mean none of the professors ever did any research. Some did, after a fashion, typically publishing their results as dissertations or literary works. But they did not consider such activities to be an integral part of their tenure obligations – nor were they, according to the reformed university charter from 1732.

Between 1773 and 1864 the university in Kiel was under Danish administration, although in practical terms it remained German in many ways and was regarded by Copenhagen as something of a foreign body embedded in the realm. The institution was mainly important to the still-Danish region of Schleswig-Holstein, and its significance to science in Denmark was limited. There was one noteworthy scientist there, however: the professor Christoph Heinrich Pfaff, who took up his chair in 1798 and remained a leading figure in German and Danish science until his death in 1852. It was Pfaff who made sure, in 1802, that the university in Kiel was equipped with a modern physics and chemistry laboratory, which later became the primary location of Pfaff's own seminal investigations into galvanic electricity. He later summarized his findings on this topic in the great work *Revision der Lehre vom Galvano-Voltaismus* from 1837.[126] Pfaff was also an important figure in the new analytical chemistry, and was especially well known for his *Handbuch der analytischen Chemie* from 1821–1822. Although Pfaff was recognized as one of Europe's

8.4 Christoph Heinrich Pfaff, the most prominent scientist at the University of Kiel in the early 1800s. Pfaff was a ground-breaking researcher in the budding field of electricity as it developed after Volta's invention of the battery in 1800. Through assistants like J. G. Forchhammer, Pfaff became associated with the scientific community in the Danish-speaking part of Germany.

leading scientists and, thanks to his many travels, had excellent relations with some of the most renowned chemists and physicists of his day, his contributions had little influence on science in Copenhagen.

A local movement known as "Schleswig-Holsteinism", springing from the German movement for unity and freedom, arose around 1815. Its goal was independence. This new movement quickly became entrenched at the university in Kiel, which would be at the centre of the region's pro-German rebellion in 1848. The connection between the university and the German nationalist sentiments had dire consequences: The administration in Copenhagen lost interest in the university, which therefore ceased to develop. The University of Copenhagen was favoured financially, while the north-German city of Kiel faced a bleak academic future. In 1852 the Danish government fired one-third of the professors there, additionally causing the number of students to plummet.

Finally it should be mentioned that for a few short years (1811–1814) the Danish kingdom could pride itself on having no less than three universities, after the founding of the Royal Frederiks University in Christiania (today's University of Oslo) in 1811 and until Denmark lost Norway to the Swedish crown three years later. The Norwegian-Danish university's early years were

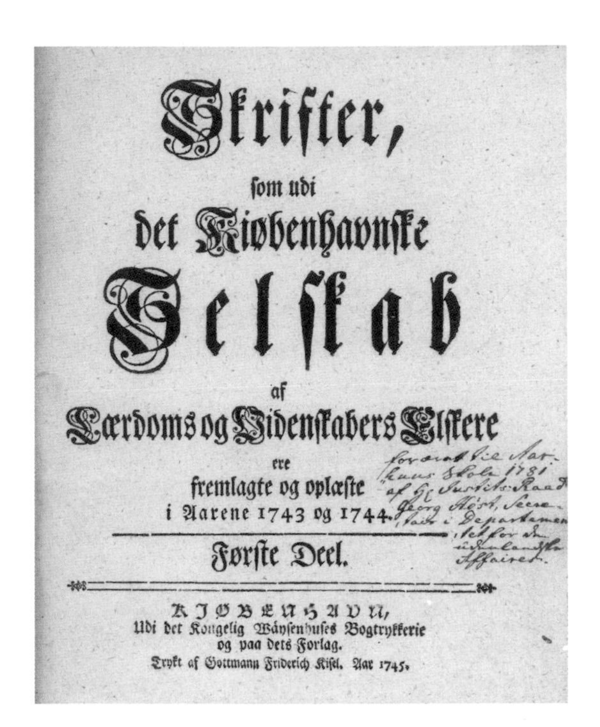

Skrifter,
som udi
det Kiøbenhavnske
Selskab
af
Lærdoms og Videnskabers Elskere
ere
fremlagte og oplæste
i Aarene 1743 og 1744.
Første Deel.

KIØBENHAVN,
Udi det Kongelig Wäysenhuses Bogtrykkerie
og paa Dets Forlag.
Trykt af Gottmann Friderich Kisel, Aar 1745.

8.5 *Title pages from the first volumes of writings published by Copenhagen's and Trondheim's academies of sciences and letters, appearing in 1745 and 1761, respectively. While the Danish academy was exclusive, its Norwegian counterpart was less particular about its membership. This difference was reflected in the contents of the two journals.*

marked by hardship, the country being at war with England and therefore desperately short of funds. It was not until 1813 that the first professors were appointed and construction work begun. From a scientific point of view, the most important professor was the astronomer and physicist Christopher Hansteen, who specialized in geomagnetism and was a good friend of Ørsted. The close contact between these two men was one factor that sustained the bond between the universities in Copenhagen and Oslo through the split in 1814 and beyond.

Societies and academies

In the early 1700s there had been a few attempts to establish a Danish learned society, but all were unsuccessful until 1742. In fact, the founding of the Royal Danish Academy of Sciences and Letters that year was not originally rooted in a wish to promote scientific activity. The new Royal Academy was an offshoot of the coin and medal commission, set up in 1739 mainly to describe and catalogue the king's collection of coins and medals. This commission envisioned a learned society, a *Collegium Antiquitatum*, whose main concern would be the nation's history and ancient artefacts. The actual proposal came from the historian and philologist Hans Gram, but at the founding meeting held on 13 November 1742 it was decided that the scope of the society would be extended to include science and mathematics as well. The body, which for the first few years was referred to simply as "the Academy" or "the Commission", was immediately given an official status when Christian VI not only approved it, but also provided financial support and offered his personal involvement as royal patron. The new academy's status as an unofficial part of the establishment was underscored by the choice of the influential academy co-founder Count Holstein, a member of the king's inner circle, as its first president. This also ensured the Royal Academy's informal association with the University of Copenhagen, since both Holstein and his successor as academy president, Count Otto Thott, were both patrons of the university.

The Royal Danish Academy of Sciences and Letters had initially been conceived of as a humanist academy, as is also evident from its original eight members. Only one of these eight belonged to the world of science: Joachim F. Ramus, who was a professor of mathematics. The initial preponderance of members from the humanities was brief, however, for the Royal Academy was quickly sought out by scholars in the fields of mathematics, medicine and the natural sciences. A central element in virtually all learned societies of that era was the creation of a journal to which members could contribute scientific articles, normally written in the country's own language. It was certainly clear that the new Royal Academy in Copenhagen needed a journal of its own, and the first volume appeared in 1745 under the abbreviated title of *Skrifter* – "writings". The language was Danish, even though Gram had wanted the articles published in Latin so they could be read by foreign scholars as well as Scandinavians. The majority of the academy's members did not believe this was a decisive point, however, given that the journal was perceived above all as a means of promoting science in Denmark and reaching out to the Danish population. Nevertheless, the first volumes did appear with a parallel edition in Latin entitled *Scriptorum*, though the practice was discontinued soon after Gram's death in 1748. Of the 160 articles appearing in the annual volumes for 1745–1779, a full 89 belonged within the natural sciences at large. The articles were extremely diverse in topic and in content, and most were intended more as information for Danish readers than as original scientific contributions aimed at European scholars and scientists. Even so, the quality and the originality of some of these articles was such that had they not been written in Danish, they would undoubtedly have attracted international attention.

Besides publishing its journal, the Royal Academy's most important activity was to initiate and judge high-level academic competitions. These competitions, and the gold medal awarded to each winner, were based on a royal endowment from 1767 of 8,000 rixdollars. The interest accrued on the endowment would be used for prizes awarded "within History, Physics and Mathematics, and Sciences subsumed into these fields". The Royal Academy arranged a large number of competitions, which were also publicized in foreign societies and journals, but in many cases, despite repeated calls for entries, no papers were submitted in response. The intention of these academic competitions was not only to make the academy in Denmark known abroad, but also, and perhaps just as importantly, to encourage Danes to study and write on the relevant topics – and so it was quite embarrassing that not a single prize was awarded to a Dane or a Norwegian until 1775. The Royal Academy also received two sizeable private endowments, one in 1785 and the other in 1792. Both were to be used for awarding prizes in academic competitions, and both were directed towards practical, technological and economic issues, especially those relating to agriculture. Thus, in the late eighteenth century the academy's competitions increasingly came to focus on issues pertaining to agricultural economics.

The scholars living in Norway felt the need for a Norwegian society similar to the new academy in Denmark. This led to the formation of the Trondheim Learned Society in 1760, which was officially recognized seven years later as the Royal Norwegian Society of Sciences and Letters.[127] The real prime mover in the new Norwegian society was Johan Ernst Gunnerus, who was bishop of Trondheim. In addition to his theological and philosophical qualifications, Gunnerus was immensely interested in natural history, and he played a crucial role in conveying scientific knowledge to a wider audience in Norway, and generally promoting public interest in science. He first proposed the establishment of a learned Norwegian society in a pastoral letter dated 1758, which listed the fields he imagined such a society would encompass: "oratory, poetry, natural and other history, physics, the object and Divine Intention of natural things, oeconomics, empirical psychology and other matters which, as far as this is possible, can be discussed in an attractive and pleasing Manner and do not surpass the Horizons of the sagacious non-academic Reader."[128] Much like its Danish sister society, the Trondheim Society published its own journal, putting out the first volume in 1761. Many of the articles published or described in the Norwegian *Skrifter* were about natural observations done by Norwegian ministers, a group that remained foremost in Gunnerus's mind and was prominently represented in the Trondheim Society. Actually, it was more a regional society than a national one, and its status relative to the Royal

Academy in Copenhagen was that of a "junior partner" in more ways than one. Typifying this relationship, its series of journals was printed and published not in Norway, but in Denmark. A fully independent Norwegian society did not become a reality until 1857, when the university in Christiania (now Oslo) founded the Science Society in Christiania (today the Norwegian Academy of Science and Letters).[129]

The Royal Academy in Copenhagen continued to bring together the country's academic and scientific elite throughout the nineteenth century. During the years of hardship from 1801 to 1815 with their succession of national calamities – including the bombardment of Copenhagen, Britain's seizure and removal of the entire Danish fleet, and the loss of Norway – the astronomer Thomas Bugge served as the academy's secretary. He was succeeded by Ørsted, who held the position until his death in 1851. The academy still published its *Skrifter*, but they appeared rather sporadically, often with considerable delay, and sometimes with articles of substandard quality. Many scientists in Denmark preferred to publish their work in other journals, motivated among other things by their desire to reach an international audience. The language in *Skrifter* was still Danish, meaning that not only would its contents remain unknown to scientists abroad; foreign scientists would also rarely, if ever, choose to publish their work through the Danish academy. Not until 1902 did the academy allow other languages than Danish to appear in its publications.

While the Royal Academy was the most important new scientific and academic institution born in the 1700s, it was not the only one. A decision was made in 1747 to reopen the old academy at Sorø, but using a new model that was fundamentally different from the University of Copenhagen and offered an education that was novel in structure and in content. Rather than being an erudite institution focusing on theology, law and medicine, the school at Sorø was intended to provide higher education for civil servants aspiring to positions at the royal court or in the state administration. The courses of education at Sorø were therefore limited to three years and did not result in an actual academic degree, in addition to which the new Sorø Academy had no right to promote its graduates to pursue further academic studies towards a more advanced degree. On the other hand, the professors at Sorø were equal in status to their peers in Copenhagen. The curriculum was largely made up of philosophy, law and political science, physics, mathematics and history, but remarkably there was no room for classes in agronomics or natural history.

The most important science teacher at Sorø was plainly Jens Kraft, who took up a position there in early 1747 as a professor of philosophy and mathematics. That was also the year Kraft had a treatise published in the Royal

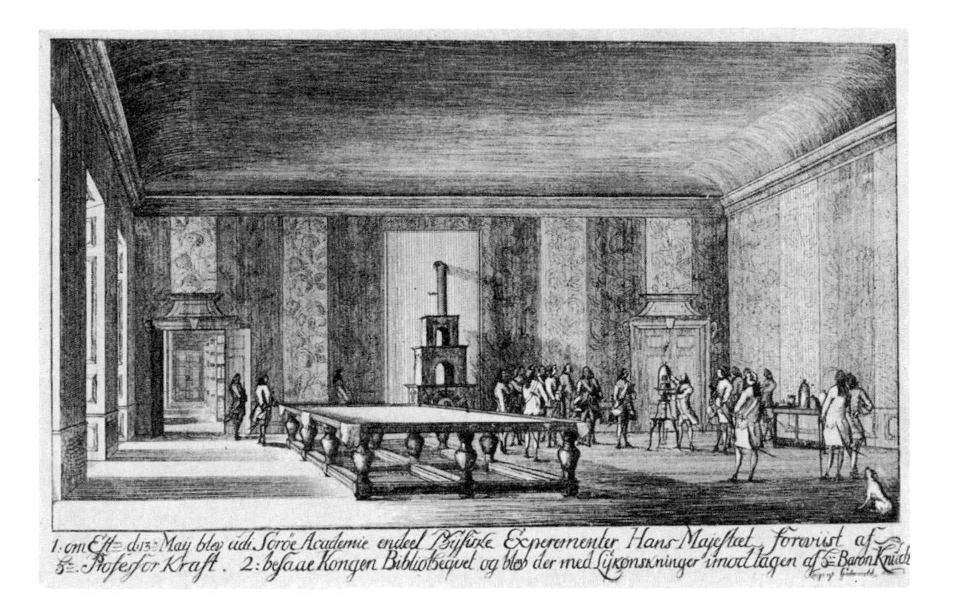

1: om Eft= d. 13= May blev udi Soröe Academie endeel Physiske Experimenter Hans Majestæt forviist af
S=. Professor Kraft. 2: besaae Kongen Bibliotheqvet og blev der med Lykonskninger imod tagen af S= Baron Knuth

8.6 *King Frederik V attending Jens Kraft's experiments with the air pump at Sorø Academy in 1749. Although the technology itself was almost 90 years old, it had never before been used for serious research in Denmark, and the air pump was still regarded as something of a magical instrument.*

Academy's *Skrifter* in which he systematically criticized the Cartesian physics and, conversely, argued in favour of the Newtonian world view. The Danish-language treatise entitled "Betænkninger over Neutons og Cartesii Systemata" ("Reflections upon the Newtonian and Cartesian Systemata") also appeared that year in a Latin translation – in the Latin version of the Royal Academy's journal. This treatise marks the real arrival of Newtonian physics to Denmark, where the various brands of Cartesianism had so far ruled supreme. From an international point of view Kraft's treatise cannot have been especially interesting, but to Denmark's scientific community it was nothing short of extraordinary. What Kraft delivered was actually a very clear and well-argued defence of Newton's physics in an environment that was neither familiar with nor approving of Newtonian thinking. This notwithstanding, Kraft's greatest scientific work was an impressive textbook on mechanics and its uses, which appeared in 1763–1764 and was based on his lectures at Sorø. The publication of Kraft's voluminous and richly illustrated book was supported by the Crown, as Frederik V deemed that "The work is not only useful to those studying at Our noble Academie, but useful, too, for all those who would uphold and promote the mathematical sciences."[130]

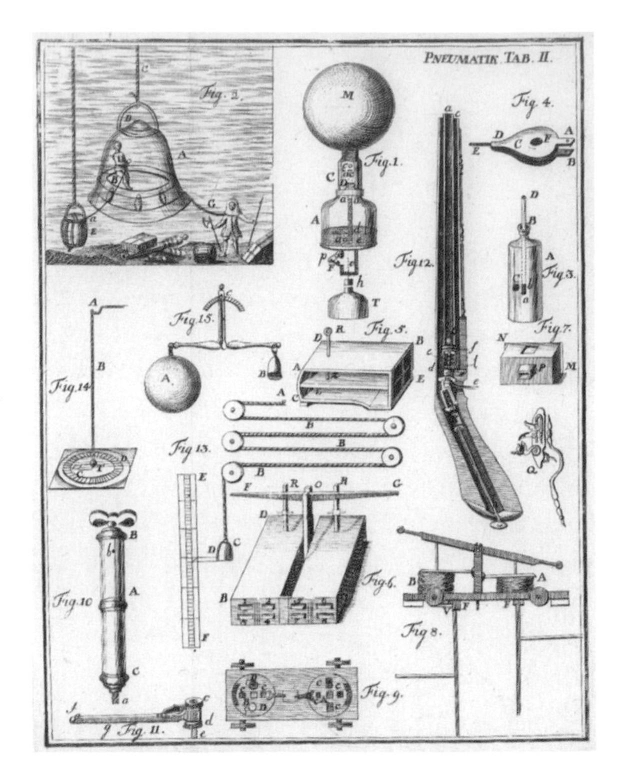

8.7 *Plate from the pneumatics section of Kraft's* Forelæsninger over Mekanik *(Lectures on Mechanics), showing such contraptions as a diving bell (fig. 3), bellows (fig. 4), an air rifle (fig. 12) and a compression pump (fig. 10). Kraft's textbook featured illustrations of all sorts of mechanical devices, the workings of which he explained using scientific principles.*

In Kraft's day, the instruction given in physics and mathematics at the Sorø Academy was highly scientific, but it must have seemed strangely irrelevant to the young noblemen who were mainly preparing for a career in the bureaucracy administrating the affairs of State and Crown. Whatever the case, not one of Kraft's students felt called to pursue further scientific studies. Kraft did not create a school of thinking, nor does he seem to have had much contact with his peers in Denmark or elsewhere. The oddest thing, given that Kratzenstein was located in Copenhagen, is that he and Kraft had no ongoing dialogue either, despite their shared interests in the new discipline of experimental physics. As far as Danish science was concerned, the great significance of Sorø Academy lay in Kraft's efforts – Kraft being the one notable exception in that respect. All teaching at Sorø stopped in the early 1790s, for the obvious reason that the academy was no longer able to attract new students.

Around the time the Sorø Academy began its long hibernation, the Danish veterinary scientist Peter Christian Abildgaard and others set up the Natural-History Society, which was to act as a sort of private university in the fields not adequately covered by the old university. This society, founded in 1789, had high ambitions, offering courses and examinations, giving lectures, and establishing a library and a museum-style collection that eventually grew to considerable size. The society's scientific and instructional activities were publicized in its new journal entitled *Skrivter af Naturhistorie-Selskabet*, which appeared between 1790 and 1810 in a series of 11 half-volumes each containing about 200 pages. *Skrivter* holds an important place in the Danish scientific literature, being the country's first natural-history journal, and one of the first to feature the natural sciences as more than merely an appendage to medicine. What is more, it also appeared in a German-language version as *Schriften der naturforschenden Gesellschaft zur Kopenhagen.* The journal was dominated by botany and zoology, but it also ran a fair number of articles on chemistry, geology and topography. Even though the Natural-History Society was a highly active and highly significant institution, it was also short-lived. Its means were not proportionate to its ambitions, and its entire existence was linked to the lack of natural-history instruction at the University of Copenhagen. When the university began to offer such courses, the society lost its very *raison d'être.*

The utility of science

During the period of enlightenment Denmark, like other European countries, was greatly interested in research that was potentially useful. Many of the projects that the State or private enterprise became involved in were utilitarian in nature, as exemplified by the two examples below, both from the last half of the 1700s.

One of the most comprehensive and lengthy projects in the history of Danish science was the mapping of Denmark, which began around 1760 and was not completed until well into the 1800s, roughly 80 years later. The original idea for the combined geodetic-cartographic project came from a young student named Peder Koefoed, whose original proposal was transferred in 1761 to the Royal Academy, which set up a special surveying commission. One of the first surveyors employed for the project was the young Thomas Bugge, a twenty-year-old theology student who would go on to become one of the most famous scientists of his time. Bugge was soon given responsibility for the project's trigonometric measuring and quickly became its driving force. Bugge was appointed professor of astronomy in 1777, and two years later the Danish publishing house Gyldendals Forlag released his great work on the theory and methodology behind the surveying project, *Beskrivelse over den Opmaalings Maade, som er Brugt ved de Danske geographiske Karter*, which appeared in a German translation ten years later. Soon after, in 1780, Bugge officially assumed executive responsibility for the project that he had, in effect, been running for several years already. He continued as project director until his death in 1815, by which time the triangulation of the country had long since been completed, the only unfinished maps being those of Schleswig and Holstein. The publication in 1841 of a comprehensive map covering the whole of Denmark marked the completion of the project. Even before that time, however, the Royal Academy had ceased its involvement, since by then the responsibility for making new topographic maps with contours had been transferred to the General Staff of the Danish armed forces.

Compared with other projects on the international scene, the Danish surveying project was not scientifically innovative. Nevertheless, it was quite important to Danish science, not only because it resulted in an immensely improved topographic knowledge of the country, but also because it generated an expertise in the field of surveying and cartography that had not previously existed – besides which it continued, during the 1770s, to sustain and give meaning to an otherwise ailing Royal Academy. The issue of its scientific quality aside, the project was ambitious, grandiose, and well executed. Initially the annual subsidy from the State, which was to cover all project costs, was 1,600 rixdollars, which soon proved to be absurdly inadequate. Wages and salaries alone during 1781–1820 amounted to about 180,000 rixdollars, or on average more than 3,000 rixdollars a year.

Another person centrally involved in the project was a young Norwegian named Caspar Wessel. Though already a much-admired surveyor in his own time, what Wessel is best remembered for today is his work as a mathematician, thanks to his introduction in 1799 of a geometrical representation of complex numbers.[131] The main aim of the treatise he published in the Royal Academy's journal was to develop arithmetic operations for oriented lines of the type found in algebra. His treatise went virtually unnoticed in Denmark, and not even Wessel himself seems to have realized the significance of his geometrical interpretation of complex numbers. And because, in compliance with the academy's rules, the article was written in Danish, no foreign mathematicians noticed it either. Years would go by before a similar geometrical interpretation appeared in the international mathematical literature, and nearly a century would pass before Wessel's earlier treatise earned the recognition it rightfully deserved, after being translated into French in 1897.

Thomas Bugge did not produce any scientific work of similar originality, but his overall contribution to the natural sciences in Denmark was far greater than Wessel's was.[132] After his appointment as a professor of astronomy, Bugge toured Germany, the Netherlands and Britain studying ways to modernize the

Herrn Thomas Bugge,
Königl. Dänischen Justizraths, Prof. der Mathematik und Astronomie bey der Universität zu Kopenhagen und der Königl. Marine; Mitglieds der Akademien der Wissenschaften zu Stockholm, Kopenhagen, Mannheim und Drontheim,
Beschreibung
der
Ausmessungs-Methode,
welche bey den Dänischen geographischen Karten angewendet worden.
Mit Kupfern.

Dresden, 1787.
In der Waltherischen Hofbuchhandlung.

8.8 *Title page of Thomas Bugge's description from 1787 of the great Danish surveying project, here from the German edition. Not only was the project interesting from a scientific and economic angle. It was also extremely prestigious, and meant to demonstrate the willingness and the ability of the Danish Crown to match projects undertaken by the foremost nations of Europe.*

observatory in the Round Tower. Bugge's travel journals give an intriguing portrayal of the scientific and technological state of Europe in the 1770s, particularly in the field of instrument technology.[133] Supported by a royal grant of 7,000 rixdollars, Bugge was able to obtain new and better instruments for the Round Tower, thereby bringing Copenhagen's observatory up to a reasonable standard. Still, it would hardly have been "able to hold its own among the

8.9 *Engraving of Thomas Bugge, looking rather worn and weary. Not surprising, since for many years Bugge was one of Denmark's most active and most famous scientists. His contributions to geodetics and astronomy were also well known outside Denmark.*

most eminent Observatories in Europe", as the royal ordinance of 1761 had stipulated.

In 1798–1799, Bugge was in Paris as a Danish member of the international but French-dominated commission that would proclaim the introduction of the metric system, and in 1800 he published an account of his stay in France during the revolution.[134] By virtue of his many prominent positions – for instance as a leading figure in the Royal Danish Agricultural Society, and secretary of the Royal Academy of Sciences and Letters between 1801 and 1815 – Bugge was, for a time, at the very nucleus of Danish science. His enormous contribution to the surveying project aside, most of his scientific work lay in the field of astronomy, studying Saturn's shape, for instance, and determining its period of rotation. Bugge was one of just a handful of Danish eighteenth-century scientists to publish treatises in *Philosophical Transactions*, the journal of the Royal Society in London, and generally he was more internationally oriented than was customary in Denmark during the last decades of the 1700s. His astronomical works also included studies of the Northern lights, comets, and various types of instruments.

Generally there was a great interest in seeking out and exploiting the mineral and energy resources believed to exist throughout the kingdom, notably on the Baltic island of Bornholm, and from the mid-1700s a number of projects were initiated in pursuit of this practical and patriotic goal. An expedition from 1798 concluded that Bornholm's coal had a particularly high capacity for generating heat, and specimens were sent to experts in Copenhagen to confirm this claim. Although the results were not conclusive, optimism in Denmark remained undimmed. Following the wars with England and the loss of Norway, questions concerning the country's energy supply were even more pressing, and in 1818 the government set up a commission entrusted with the task of examining Bornholm – yet again – for any undiscovered coal and metal deposits. As a result, Hans Christian Ørsted, Johan Georg Forchhammer and the seasoned surveyor Lauritz Esmarch travelled to the island to investigate the matter. The report from Ørsted and Esmarch was wildly optimistic, envisaging the rocky isle as a veritable industrial eldorado where the Danes would be able to extract not only coal, but iron, soda, alum and other minerals as well; indeed, an eldorado made all the more golden by that marvel of new technology, the steam engine. While their conclusions and aspirations did not reflect a desire to please those in power, these scientists were no doubt affected by their own wishful thinking. As the two men wrote in *Statstidende*, the Danish official gazette, in 1821: "But that every friend of the fatherland would willingly exert himself to obtain, for our country, iron and pitcoal, two auxiliary materials for so many kinds of industrial business and useful production, on our own land, of this there can hardly be any doubt." Undimmed optimism or not, the bedrock of Bornholm failed to reveal any sort of industrial eldorado to the Ørsted-Esmarch expedition – and similarly thwarted all later efforts as well.

Technological learning and the science of war

Even though a faculty of science at the University of Copenhagen did not become a reality until 1850, Denmark did have a formalized technical-scientific education at university level before that time – at the Polytechnical College founded in 1829 and known today as the Technical University of Denmark (DTU).[135] The first comprehensive plan for an actual technical education was presented in 1827 by the mathematician and astronomer Georg Frederik Ursin, who proposed founding an academy for craftsmen; a place of higher education where science and practical experience would blend into one programme, modelled after the German *Gewerbeschulen*, technical schools where science was taught at a fairly modest level. The school was to be fundamentally practical, and the instruction aimed at skilled craftsmen and similar groups who would benefit from both scientific and functional knowledge, and thus in turn benefit Denmark's trade and industry.

The idea of a technical college was brought to fruition, but not according to Ursin's plan. What happened instead was that Ørsted successfully made use of the original plan to elaborate a proposal for a very different sort of institution that would teach advanced technology. Ursin's plan was based on the technical colleges in Germany, whereas Ørsted's was inspired more by the École Polytechnique in France, and his proposal was much more scientifically oriented, offering little or no training in practical craftsmanship skills. Ørsted also envisioned his new construction as being closely linked with the university, and as a way of creating a technical-scientific graduate-level degree outside the university in Copenhagen, but still associated with it. Ørsted's proposal met with royal approval, and in November 1829 the Polytechnical College was officially inaugurated in the presence of King Frederik VI.

The stated goal of the new educational institution, as expressed in Ørsted's proposal, was "to instil, in young people possessing the necessary previous knowledge, such insight into mathematics and experimental natural science, and such skill at making use of these insights, that they may thereby become formidably useful in certain branches of State service, and in overseeing industrial enterprises."[136] The courses were to run for two years, with the students divided into two classes of candidates, one specializing in "mechanics" and the other in "applied natural science". The latter mainly referred to technical chemistry, whereas the former focused on mathematics, mechanical physics, machine construction, and similar subjects. The close links to the university were ensured not only by teachers serving on both staffs, but also by the new institution taking over two of the university's buildings in Copenhagen. The initial staff consisted of seven people, four of whom were associated with the university as either professors or lecturers. Ørsted himself taught physics and also served as the director of the new technological college, retaining that position until his death in 1851.

During its first decades the new college had quite a high drop-out rate, and consequently during that time very few students actually earned a degree in advanced technology. From 1832 to 1849 the total number of graduates was a mere 67, and one year there was not a single graduate in applied science. Producing graduates was not the Polytechnical College's only purpose, however. Many people attended the lectures of Ørsted and his colleagues without aspiring to take a degree, and in this way the new institution in its early years had quite a significant impact, culturally and economically. The audience would often include a young man named Jacob Christian – "J. C." – Jacobsen, later founder of the Carlsberg breweries, who was inspired by what he heard to take a more scientific approach to brewing beer.

Although Ørsted had been greatly inspired by the École Polytechnique, the new college of advanced technology in Copenhagen was by no means a school

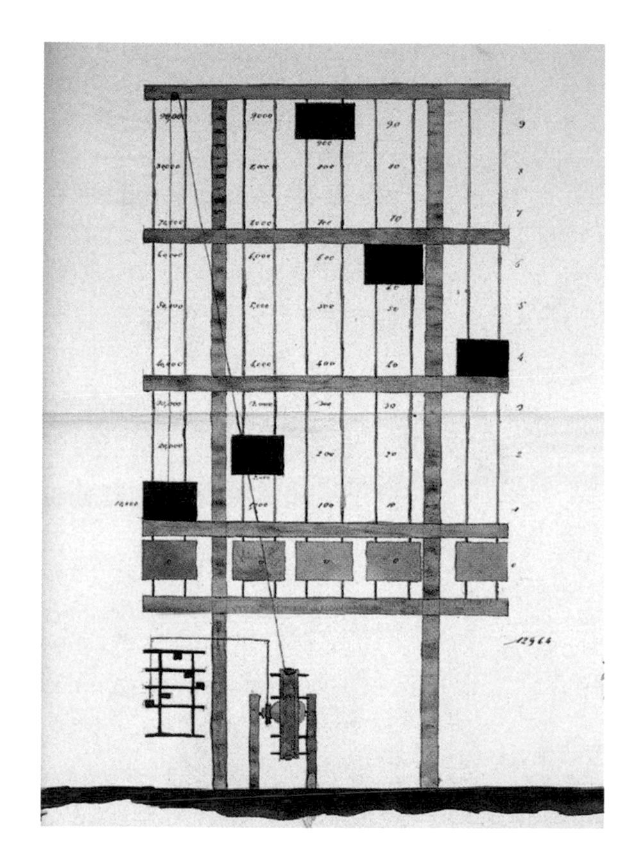

8.10 *Andreas Schumacher's frame telegraph from 1808 was mainly used to communicate across the Great Belt. The apparatus had a small repeat mechanism – also a Schumacher invention (seen at lower left) – enabling the telegraphist to read the boards on the large frame.*

of engineering during its early years. Later, from 1861 onwards, the college did develop to educate civil engineers, but under Ørsted's leadership it had nothing whatsoever to do with engineering subjects, being an academic institution whose purpose was to school technological graduates. While Ørsted's brainchild soon had a positive impact on Danish science, it did not stimulate the country's industrial development as its creators had intended.

Although the Polytechnical College did not produce engineering graduates, Denmark did have a high-level education in engineering – at the Danish school of military engineering. One of the few engineering officers who was favourably noticed in the early 1800s was Andreas A. F. Schumacher, the technically gifted brother of astronomer Heinrich Christian Schumacher. Around 1810, Andreas Schumacher devised a new system of optic telegraphy, which was used as a link across the Great Belt. Another of his projects sought to construct missiles similar to the Congreve rockets that had spread death and destruction in Copenhagen during the 1807 bombardment. The rocket-launching company he was appointed to lead in 1813 was the first military unit of its

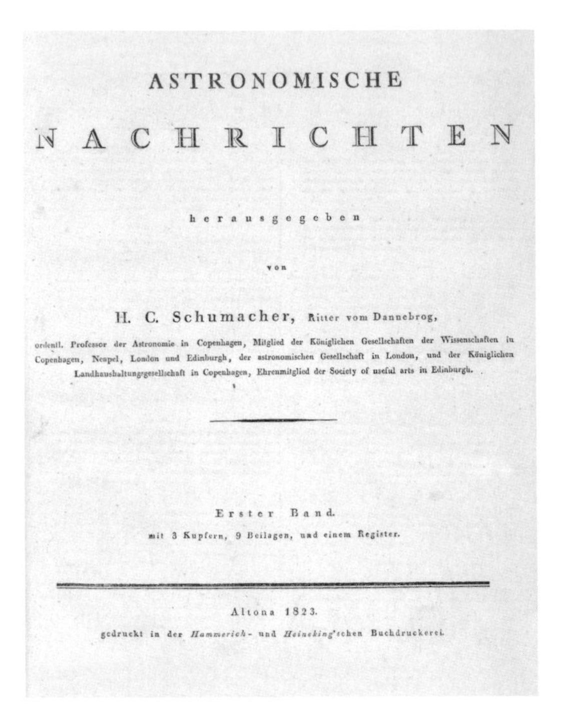

ASTRONOMISCHE

NACHRICHTEN

herausgegeben

von

H. C. Schumacher, Ritter vom Dannebrog,

ordentl. Professor der Astronomie in Copenhagen, Mitglied der Königlichen Gesellschaften der Wissenschaften in Copenhagen, Neapel, London und Edinburgh, der astronomischen Gesellschaft in London, und der Königlichen Landhaushaltungsgesellschaft in Copenhagen, Ehrenmitglied der Society of useful arts in Edinburgh.

Erster Band.

mit 3 Kupfern, 9 Beilagen, und einem Register.

Altona 1823.

gedruckt in der *Hammerich*- und *Heineking*'schen Buchdruckerei.

8.11 *Title page from the first issue of* Astronomische Nachrichten *from 1823. This important journal, founded and edited by H. C. Schumacher, served as an efficient means of communication for astronomical news, while at the same time helping to professionalize the science of astronomy.*

kind on the Continent.[137] Besides developing military missiles, Andreas Schumacher worked with his brother Heinrich on experiments on missiles intended to serve astronomical purposes, as described in the journal *Astronomische Nachrichten* in 1823.

From 1815 onwards there was a growing recognition of the shortcomings in Denmark's military engineering education compared with other countries, but no action was taken until 1829. That year a proposal was made that outlined an actual institution for military education, and just one year later this vision was fulfilled with the founding of the Royal Danish Military Academy. Here officers who would serve the military, in the corps of engineers and elsewhere, would receive a four-year education in two phases: the first broad and general, and the second more specialized. Subjects taught in the first phase included mathematical analysis, rational mechanics, descriptive geometry, chemistry, physics, and geodesy, while the specialized phase included courses in technical mechanics, machine construction and military technology. Although the military academy's instruction in mathematics and the sciences was less scientific than the college's instruction in advanced technology, it was still at quite a high level, and some of the academy's teachers also taught at the Polytechnical College.

Most of the graduate-level students at the military academy chose other career paths than the corps of engineers, and from 1830 to 1848 the academy produced no more than 30 engineering graduates. On the whole, the military academy was of very little significance to Danish science, though its staff did include a number of young scientists who were initially unable to find other employment. Two later professors who made ends meet during the 1850s by doing a stint at the military academy were the mathematician Adolph Steen and the chemist Julius Thomsen. Later in the century the prominent Danish physicist Ludvig Lorenz taught for many years at both the Royal Danish Military Academy and the Army Academy, its successor institution.

Danish astronomy – in Germany

After Thomas Bugge's death in 1815, his astronomy chair in Copenhagen went to Heinrich Christian Schumacher, a native of the Holstein region. Schumacher had studied for three years under the brilliant mathematician Karl Friedrich Gauss in Göttingen, supported financially by an important Danish fund. During its lifetime, this fund – Fonden ad usus publicos – set up by the Crown in 1765 and discontinued in 1842, was a key factor in Denmark's scientific research policy: Virtually all Danish scientists benefited from its support, some receiving travel grants and others publication support, contributions to scientific instruments, or extraordinary wages.

In 1821–1823 Schumacher – finding the astronomy in the capital of Copenhagen utterly uninteresting – set up a new observatory, funded by the Crown, in Altona, which at that time still lay on the outskirts of the Danish realm. The work done at the Round Tower observatory in the heart of the old city was mainly overseen by Ursin and his associate Erasmus G. F. Thune, a mathematics professor at the university. Not until 1832 was an extraordinary professor, Christian F. R. Olufsen, appointed to manage the work there. The contributions of Olufsen, who had studied under the eminent German astronomer Friedrich Wilhelm Bessel, lay mainly in the field of theoretical astronomy. Collaborating with the South Jutlander and fellow astronomer Peter Andreas Hansen, Olufsen published an important work in 1853 entitled *Tables du soleil,* investigating the Earth's movement around the Sun under the perturbing influence of the gravitational pull of other planets.

Around 1840 the world of science was in a transitional phase, slowly evolving into a kind of modern, professionalized science. It had previously been customary for scientists to have a theological background, as Bugge had, but this was becoming increasingly rare. Thune was a late example, in fact almost the last example, of a theology graduate making a career in Danish science. Another significant development in the first half of the 1800s was the striking change in the language of science, as Latin all but disappeared from the scientific literature. The first dissertation submitted in Danish to obtain the *magister* degree appeared in 1836, heralding a rapid and widespread shift from Latin to Danish that reached even the highest levels of academic writing.[138] Even so, well into the 1840s it was still possible, and even acceptable, to write in Latin when treating topics in science and other fields. Not surprisingly the last area to change over to Danish was that of the tradition-bound doctoral dissertations, as exemplified in Olufsen's dissertation from 1840, *Disqvisitio de parallexi lunae*. As for handing in dissertations in other major European languages, that practice was not permitted until 1921.

What enabled Schumacher, a professor at a Danish institution, to base his activities in Altona and build an observatory there with Danish funding was his exceptionally favourable standing with the king of Denmark, combined with his close relations with many wealthy patrons in Denmark. Just as Tycho Brahe had done 250 years before, Schumacher knew how to manoeuvre in the patronage system that still existed, albeit in a different form than it had in Tycho's day. His professorial appointment was partly a result of recommendations made by two powerful counts and ministers of state, Christian D. F. Rewentlow and Heinrich Ernst Schimmelmann. However, his most influential ally and advocate was Johan Sigismund von Møsting, who served as finance minister and prime minister and backed Schumacher's plans and financial requests. Møsting was interested in astronomy and a confidant of Frederik VI, and the king, who was himself quite enthusiastic about Schumacher, made sure the professor had ample access to the country's coffers.

The observatory in Altona – paid for by the Danish state – was completed in 1823, and although modest in size it was lavishly equipped.[139] Quite a costly business, and that at a time when Denmark's economy was far from healthy. And yet the powers in Copenhagen committed large sums of money to Schumacher's project in Germany rather than spending it on modernizing the observatory atop the Round Tower. Whatever his motives, Schumacher succeeded in transforming the observatory in Altona into one of Europe's leading astronomical institutions. Based in Altona, he became a central figure in the international astronomy community, creating a professional network the likes of which had not been seen since Tycho Brahe's day. He was a tireless letter-writer and corresponded frequently and at length with scientists such as Gauss, Bessel, Humboldt, Hansteen, George Airy, Heinrich Olbers and John

8.12 *Heinrich Christian Schumacher. Painting from 1839. Schumacher was a central figure in the network that organized European astronomy during the first half of the nineteenth century. The extensive privileges he was granted by the Danish state were very much a result of his special relationship with Fredrik VI.*

Herschel. He exchanged more than 1,319 letters with Gauss alone, and also helped several young Danes to study with the master in Göttingen.[140]

In the international history of astronomy Schumacher is particularly well known for establishing *Astronomische Nachrichten* – the most prominent astronomy journal of the 1800s. The first volume of the "astronomical news" appeared in 1823, and like Schumacher' other enterprises, this too was supported by Møsting and the Danish court. The publication was significant not only because it was one of the first astronomical journals in existence, but also, and perhaps even more so, because it was internationally oriented and served as an efficient communication channel for astronomers around the world. The journal was principally in German, but its international approach was evident from occasional articles in French and reproductions, in English, of several excerpts from English sister journals. *Astronomische Nachrichten*, which was largely created and produced by Schumacher himself, reflected and reinforced the increased specialization of astronomy as an independent science. Astronomy had previously included such subjects as geodesy, meteorology and certain aspects of geography, but gradually these and other non-astronomical areas were toned down in *Astronomische Nachrichten.*

Shortly after the accession of Frederik VII to the Danish throne, the growing German nationalism in Schleswig-Holstein led to demands for a free constitution for the region, and the long-smouldering rebellion, supported by Prussia, broke out on 23 March 1848. Schumacher may have been a German, but he was also a civil servant of Denmark and deeply loyal to the Danish king (though more dedicated to the monarch than to the nation). He resolutely declined invitations to join the rebel cause, refraining from any involvement in the political dispute. His situation was precarious, however, and financial difficulties forced Schumacher to take up with the rebellious pro-German government of Holstein, which promptly returned the favour by supplying the impoverished observatory with fresh funds. This turn of events took a heavy toll on the aged and ailing astronomer, who died less than six months after the rebellion failed.

One could discuss whether Schumacher was indeed Danish or German, and to which nation's history of science he more rightly belongs, but such a discussion is hardly meaningful today. The more significant issue is that contemporary Europe perceived him as a German in the employ of the Danes, but one who belonged more to the Danish scientific tradition. Interestingly, Gauss called him "a German astronomer", which is how Schumacher also described himself, and when in 1829 he was honoured by the Royal Astronomical Society in London, the letter of commendation paid tribute to the great German nation while not mentioning Denmark at all.

Science, culture and education 9

From the mid-1700s, in keeping with the ideals of the Enlightenment, there spread across Europe an effort to popularize knowledge and convey it broadly to landowners and well-informed citizens. Men and women, young and old; all must now partake of the fruits of reason, the holy gospel of the new age as read by those who studied nature. This quickly created a market for a wide variety of popular-science literature, ranging from the purely entertaining (in the vein of "natural magic") through the morally edifying, to the practically applicable in the form of handicraft and manufacturing manuals. Many of the readers were ministers who found that such books served a dual purpose: By keeping abreast of scientific progress they could pass on useful information to their parishioners, even while continuing to find new examples of how ingeniously the Lord had fashioned his Creation to accommodate Man's needs.

In Denmark, as in other countries, a plethora of publications appeared that targeted the "well-informed citizenry". These were either original texts or translations, and many of the leading scientists became involved in public writing and lecturing, or in the countless enlightenment societies that were popping up all over Denmark. It may seem strange that people like P. C. Abildgaard, H. C. Ørsted and J. F. Schouw spent so much of their time supporting enlightenment activities, and it may seem valid to argue that their talents and resources would have been better spent on teaching and research. However, what happened was that by generating an interest in science outside university circles, they created a basis for recruiting a new generation of scientists. In the long term, the massive enlightenment effort was an excellent investment in the future of Danish science.

As the nation's scientifically interested audience gradually grew in size and diversity, so the cultivators of science grew into a larger and much more varied group than their predecessors in the 1600s – and indeed than their successors in the late 1800s. The scientific community certainly had a core consisting of academically educated scientists, but they were not necessarily its most significant element, or its most productive. In fact, their academic background was often in subjects (such as theology and law) that were not even remotely related to science. University positions and academic degrees were no precondition for entering the scientific elite. Neither Martin Vahl nor Otto Müller had graduated from university – yet in the late 1700s both were among Denmark's absolute top achievers in the field of natural history. Other important scientific figures outside the University of Copenhagen were physicians, officers, clergymen and pharmacists. Some had no professional qualifications at all, but were amateurs with their own financial means and no need to worry about tenure or salaries. These were the gentleman scientists, and although their breed was best known in England, Denmark also had its share. Generally speaking, the amateurs were indispensable, particulary as collectors of natural-history objects, and their activities earned them tremendous admiration, professionally and socially.

Two cultures

Around 1760, a new kind of formative ideal arose among Denmark's cultural elite that differed from previous ideals in its narrow focus on Latin literature and classical culture. The new French-inspired ideal was aesthetical and literary, often with a patriotic twist. This especially manifested itself in poetry and literary criticism written in Danish, as practised by such bodies as the Society for the Advancement of the Fine Arts and Useful Sciences, founded in 1759, which in its capacity as Denmark's most important literary society published the trend-setting journal *Forsøg i de Skiønne og Nyttige Videnskaber* (*Essays in the Fine Arts and Useful Sciences*). The aesthetic ideas of the Enlightenment were authoritatively formulated in a great work by the French philosopher Charles Batteux, which was translated into Danish in 1773–1774 as *Indledning til de Skiønne Konster og Videnskaber* (*Introduction to the Fine Arts and Sciences*). Batteux distinguished between the mechanical arts, meaning handicrafts that are useful or practical, and the arts whose object is enjoyment or *plaisir*, such as music, poetry, dance, sculpture and pictorial art. The new field of knowledge known as "the fine arts" was different from, and even opposite to, not only handicrafts but also the sciences. A good example of the importance attached to aesthetics and good taste was the creation in 1790 of an independent university chair in aesthetics, which went to the poet and literature professor Knud Lyne Rahbek.

Within this literary-aesthetic framework it could, at times, be difficult to find room for the natural sciences that in Denmark were so closely linked to practical utility and mundane agricultural economics. How on Earth could potato crops, mining, and malodorous laboratories ever be associated with

9.1 *Jens Schelderup Sneedorff was a professor at the Sorø Academy, as well as a prominent representative of the literary-aesthetical culture that did not identify with ongoing attempts to give science and technology a higher priority in education. Sneedorff's dispute with Jens Kraft on the issue of science and intellectual formation was an early example of the "dual culture" debate. Engraving from 1764.*

things sublime and noble? It was true that certain products of science could be pleasing to one's aesthetic sensibilities – say, the exquisite botanical encyclopaedia *Flora Danica*, or the elegant instruments made by Lorenz Spengler, royal ivory turner and manager of the delightful curiosities at the *Kunstkammer*.[141] But however enchanting, such things were only the tangible, external expressions of science and did not reflect any actual scientific content. A person could take great pleasure in beautiful botanical prints without having a clue about botany, or marvel at the gorgeous colours on a seashell without knowing a single thing about zoology.

An early example of the cultural gap detectable during that period is a literary dispute that broke out in 1761 between two professors at the Sorø Academy: the physicist Jens Kraft and the jurist and man of letters Jens Sneedorff.[142] Futile though it was, the heated debate between the two scholars deserves to be mentioned, as it clearly illustrates the cultural status of mathematical science during the period under review. Sneedorff, a literature lover and *bel esprit*, regarded man first and foremost as a sensitive and imaginative being, and he emphasized that the world in all its multiplicity cannot be understood in mathematical terms, as enthusiatic Wolffians claimed. Sneedorff was not against science, but in his journal *Den Patriotiske Tilskuer* (*The Patriotic Observer*) he denied that the exact methods Kraft praised so highly had any validity to the fine arts and sciences. He asserted that there were two ways to approach the world: one exact and scientific, the other artistic. A painter, he wrote, must admittedly understand so much geometry that he could present his picture in the proper perspective, "but who could abide a drawing of a meadow or wood where all lines were straight, and all were features in figures that could be dissolved according to the rules of mathematics." Kraft, a physicist and mathematician to the core, evidently perceived Sneedorff's article as an attack on the mathematical sciences and an attempt to belittle them in relation to the fine arts. Although Kraft was no more an "anti-aesthete" than Sneedorff was an "anti-scientist", he did not think of the fine arts and sciences – including what would later be known as "the human sciences" – as actually scientific. Their task was to evoke feelings, not to uncover truths, and they were obliged to rely on rhetoric in the place of reason and experiments. In his confrontational publication entitled *Kritiske Breve til Videnskabernes Fremvext og Smagens Forbedring* (*Critical Letters for the Promotion of Science and the Improvement of Taste*), Kraft expressed his scepticism: "The lovers of [good] taste know of a peculiar means – without reading, without exertion – to understand everything. Their fine sciences move through all hearts and all brains, without a body feeling it at all." Kraft was not distancing himself from good taste, but from the sort of superficial sophistication that became intoxicated with emotion and words without having any scientific foundation.

According to Kraft, mathematics and the natural sciences were not only useful; they were also central pillars upholding the culture of education, and were by no means the opposite of "good taste". But, he conceded, were he forced to choose, his preference would be clear. As he wrote: "We love good taste, but we revere the method of the thorough sciences above all else". He also believed that the natural sciences were more pious than humanistic and aesthetic studies, for as Kraft wrote in one reply to Sneedorff, "geometry and physics have rid the world of more, and of more dangerous, superstition than have all the writings of [good] taste." This was an often-used argument at the time, used for example in Kratzenstein's textbook on physics, in which he pointed out that physics was of value to society because it put paid to the idea of ghosts and other superstitious notions. Besides which it was, of course, extremely useful in a religious sense, since "against the atheists, by the world's arrangement and the laws of nature [one can] prove God's existence and supreme perfection and wisdom". The reason why Kraft, Kratzenstein and others could refer to science as a bulwark against superstition and magical beliefs was that such ideas were still very much alive in folk culture. The authorities found it important to counter any and every notion of trolls, witches and ghosts really existing, and this was where Kraft and his like-minded peers felt that science could be a useful ally.

In the course of his learned discussion with Sneedorff, Kraft also brought up the ideas of natural theology, which led him to claim that Divine Providence and God's creation of the world could "be proved with the same clarity as a mathematical truth". On the other hand, he did concede that truths that had been revealed (which he referred to as "the inner part of God") lay beyond the reach of the mathematical method. Kraft was not the first one who felt goaded into defending the physical-mathematical sciences. A fellow mathematician, Joachim Ramus from the University of Copenhagen, had served up a tirade in 1747 about "those people ... who imagine themselves to be so clever, and who know how to converse about all things, and who speak with contempt of the *Philosophia naturali*". Ramus legitimized the exact sciences with a reasoning built around utilitarianism and natural theology, but he did consider them intellectually formative.

The rhetorical dispute between Sneedorff the *bel esprit* and Kraft the mathematician can perhaps be perceived as a prologue to the discussion nearly two centuries later of what the British writer C. P. Snow would refer to as "the two cultures". In spite of Kraft's arguments it was clear that in this matter, too, he stood alone. The mathematical sciences did not belong in the culture of good taste.[143] With the arrival of Romanticism the issue was thrown into a new perspective, but the result remained largely the same and has since changed little, if at all.

In an age when science had not yet become specialized, it was still possible for a person to operate simultaneously within both cultural spheres; to remain active in literature even while working on scientific topics. Denmark offers examples of this, but only a handful or so, and they hardly epitomize the spirit of their time. The literary or aesthetical ambitions some scientists harboured would typically be revealed during their youth, and these ambitions usually did not persist into their scientific careers. As a young man Otto Müller, later to become Denmark's leading zoologist, would briefly nurture such ambitions. He translated a lengthy poem by the English poet Edward Young as well as writing a pastoral poem of his own, which appeared in 1767 as *Frodde den Fredegode* (*King Frode the Peacemaker*) in the fashionable journal *Essays in the*

Fine Arts and Useful Sciences. Although a review in the discerning *Kritisk Journal* praised Müller's historical eulogy, that was not the beginning of a literary career – its author being far too preoccupied with studying insects. Another prominent scientist, the veterinarian P. C. Abildgaard, was also briefly enamoured of aesthetics in his youth and contributed to several critical and literary journals. And at the age of twenty, H. C. Ørsted won an award in the university's academic competition on aesthetics with a paper on the relationship between prosaic and poetic language. In fact, his paper even appeared in *Minerva*, Denmark's most prominent literary journal at the time.

Another distinguished figure who was active in both the scientific and literary spheres was Oluf Christian Olufsen, who has earned a place in the Danish histories of agriculture and of literature. After completing his education as a surveyor, Olufsen wrote a light-hearted comedy in 1793 called *Gulddaasen* (*The Golden Canister*), which was extremely successful, and was followed a decade later by his second piece, *Rosenkjæderne* (*Chains of Roses*). At the same time he directed his scientific interests towards the economic tradition. After a study tour that included a stop in Göttingen with the German physicist Georg Christoph Lichtenberg and a visit with Goethe, Olufsen became a teacher at the short-lived Classen's Institute, inaugurated in Copenhagen in 1801, and later a professor of political economics at the University of Copenhagen, also achieving membership of the Royal Danish Academy of Sciences and Letters. His numerous professional literary contributions included a revised and translated version of Johann Erxleben's natural-science textbook in German, which was used at the university and at the Sorø Academy. Olufsen wrote a treatise in 1822 entitled *Om Menneskets Rolle i den Physiske Verden* (*On Man's Role in the Physical World*). This work, inspired by Lamarck, presented an evolutionary history of the Earth and the creatures that dwelt upon it.

Science for the masses?

Although many representatives of Denmark's literary culture were either sceptical of, or indifferent to, the natural sciences, generally the public was greatly interested in science and became increasingly captivated as the years went by. A popular science of sorts had emerged as early as the 1600s, but it was not until around 1760 that it solidified into a genre within the Danish professional literature. The many books and articles being published at the end of the eighteenth century indicate that there must indeed have been a large number

9.2 *Cartoon by the eminent painter Christoffer Wilhelm Eckersberg, c. 1806, showing a scientific presentation before a somewhat restless audience. In the early 1800s much of the citizenry was curious about science, and public lectures that included demonstrations and experiments enjoyed good attendance.*

of readers hungry for news of scientific advances and keen to learn more about the scientific world view.

Reading scientific presentations written for the general public was one way inquisitive citizens could satisfy their thirst for knowledge. They could also seek membership of Copenhagen's more scientifically oriented societies, or they could attend the private presentations or the numerous public (and therefore free) talks delivered by the likes of Kratzenstein and Ørsted at the university and other suitable venues. One year the citizens of Copenhagen were even fortunate enough to have the opportunity to attend a series of public "discourses on physics" adapted for the masses. After studying theology, the Norwegian-born scientist and educator Ulrik Green had gone to Sorø Academy and received instruction from Kraft. In 1765 Green announced his scientific talks for the general public, having received 400 rixdollars in financial support from the king's "fund for particularities".[144] Inspired by the English ideals of enlightenment, Green was convinced that "Nothing can more strongly promote a nation's happiness than knowledge of things natural and necessary", and his goal was to spread this good news to the common man. Above all he aimed to pass on knowledge of mathematics and chemistry, which he termed, respectively, "the Mother of all studies" and "their Father". More specifically he promised to give instruction

> on sympathy and antipathy, on the magnet, on quicksilver, on poison, on causes of illness and health, on the original substance of all things and the interstices in material things, on the nature and movements of the heavenly bodies, on the things that come into sight in the air, on winds and their nature, on hail, snow, and frost, on the earth, water, and bodies constituted thereof, both above the Earth and below it, on ebb and flow, on the sentient creatures, in particular Man and his immortal soul, and on the End of the World.

Quite a tall order to fill. Although some in the audience were satisfied, Green's ambitious undertaking – Denmark's first night school, as these discourses on physics have been called – was not a success. The instructive programme from 1765 was not repeated, and the only thing the idealistic Green achieved by his efforts was to sink himself into abject poverty.

The literature on scientific subjects that began to gush out of the publishing houses from the 1770s onwards was extremely diverse. Altogether it covered not only the entire scientific spectrum, but also the whole gamut of styles and genres, from easy-to-read children's books to academic-level professional literature. Some publications were meant to entertain rather than to enlighten, describing, for instance, intriguing experiments in physics and chemistry. A good example of such works in the "natural magic" genre is one published in multiple volumes in 1794–1797 and written by the surveyor Andreas Svendsen, whose title translates, literally, as *Natural Magic or the Secret and Wondrous Effects of Nature and Conjuring, Presented in Several Electrical, Chemical, Magnetical, Mechanical and Optical Tricks, for Practicality and Pleasure.* The second part of this work, *Economical and Technological Experiments and Experience*, is presumably the first book in Danish to use the word "technology". The concept was introduced and the term coined in 1777 by the German scientist and cameralist Johann Beckmann, whose grand oeuvre, *Anleitung zur technologie*, Svendsen translated into Danish in 1798 under a similar title.

Many of the books appearing in Danish were written by Danish authors, but even more were translations, chiefly from German but also from French, English and Swedish. For the first time, translators came to play an important role in disseminating scientific knowledge, so much more so as "a translation" in 1790 was very different from a modern-day translation. In the 1700s, translations of non-fictional works were not only free renditions in regard to language; usually they had also been revised and could even contain independent contributions, thereby differing significantly from the original. For instance, the third Danish edition of *Botanik for Fruentimmere* (*Botany for Housewives*) from 1803 was certainly based on Rousseau's original in French, but it also contained sections added by the translator Odin Wolff, who was himself a prolific popular-science author. All in all, the borderline between translating and publishing books under one's own name was indistinct, to say the least, and in many cases the latter were based on, or sometimes even copied from, works by foreign authors.

Other notable translators in the kingdom include a Norwegian lexicographer named Hans von Aphelen, who was very active and took it upon himself to publish a Danish edition in eight volumes of Jacques Valmont de Bomare's *Dictionnaire raisonné universel d'histoire naturelle*, which were released from 1767 to 1770. He followed this up in 1771 and 1772 with a three-volume Danish edition entitled *Chymisk Dictionnaire* of Pierre-Joseph Macquer's chemical encyclopaedia from 1766, but refrained from even mentioning Macquer's name. A review featured in *Kritisk Journal* emphasized the latter work as rising above and beyond a mere mechanical translation and commended Aphelen for having "rendered, both to science and to our Danish language, a new and real service ... he diligently seeks to contribute through his translations to enriching the [Danish] language".[145] His enrichment consisted, among other things, in the word "Brændvæsen" – roughly "firestuff" – to create a Danish name for the substance chemists normally called phlogiston. Then there was the military officer Christian Carl Pflueg, who also made a name for himself as a translator of more weighty works, including the brilliant mathematician Leonhard Euler's "letters to a princess in Germany" (as *Breve til en Prindsesse i Tyskland*) in 1792–1793, and the philosopher Immanuel Kant's cosmological treatise from 1755, which appeared in Danish in 1806 as *Almindelig Naturhistorie og Theorie over Himmelen* (*General Natural History and Theory of the Heavens*). The Danish translation is noteworthy given that, at the time of the translation, Kant's original German work (*Allgemeine Naturgeschichte und Theorie des Himmels*) was barely known at all, and gained no recognition until the 1840s.

The significance attributed to translations is underscored by the fact that authors and scientists were often translators and annotators themselves. Gregers Wad, a Copenhagen professor of mineralogy and zoology, translated

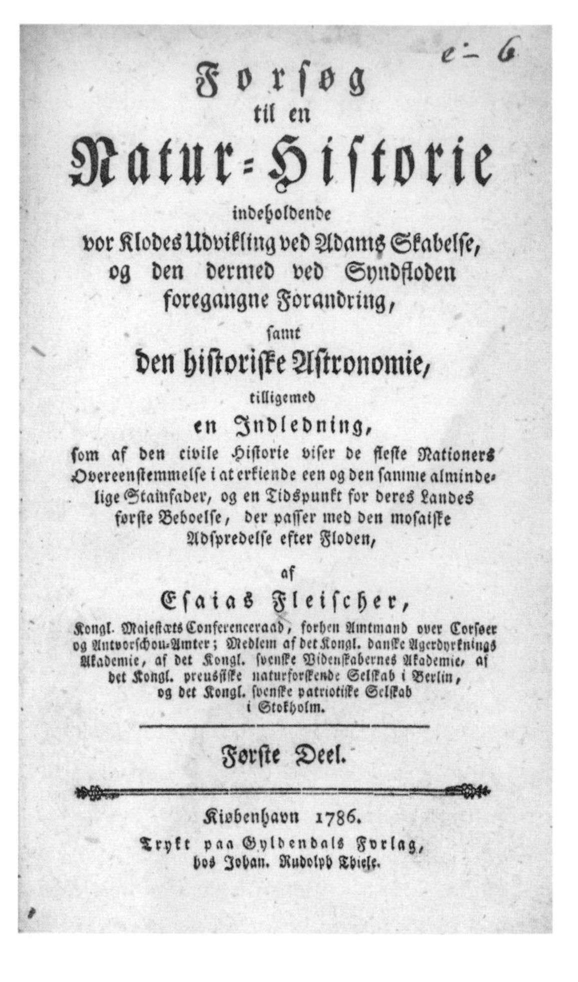

Forsøg
til en
Natur-Historie
indeholdende
vor Klodes Udvikling ved Adams Skabelse,
og den dermed ved Syndfloden
foregangne Forandring,
samt
den historiske Astronomie,
tilligemed
en Indledning,
som af den civile Historie viser de fleste Nationers
Overeenstemmelse i at erkiende een og den samme almindelige Stamfader, og en Tidspunkt for deres Landes
første Beboelse, der passer med den mosaiske
Adspredelse efter Floden,
af
Esaias Fleischer,
Kongl. Majestæts Conferenceraad, forhen Amtmand over Corsøer og Antvorschou-Amter; Medlem af det Kongl. danske Agerdyrknings Akademie, af det Kongl. svenske Videnskabernes Akademie, af det Kongl. preussiske naturforskende Selskab i Berlin, og det Kongl. svenske patriotiske Selskab i Stokholm.

Første Deel.

Kiøbenhavn 1786.
Trykt paa Gyldendals Forlag,
hos Johan. Rudolph Thiele.

9.3 *Cover from volume 1 of Esaias Fleischer's universal natural history – which is probably the largest single work in the scientific genre by any Danish writer. The 26 volumes that appeared from 1786 to 1804 were all in Danish, and largely modelled on similar works by non-Danish writers.*

both Cuvier's zoology and Saussure's geology (the latter, *Voyages dans les Alpes*, becoming literally a "handbook for travelling geologists" under the title *Haandbog for Rejsende Geologer*. For his lectures on zoology, Wad initially used the *Haandbog i Naturhistorien* (*Handbook of Natural History*) the Danish translation of a book in German by Johannes Friedrich Blumenbach done by the physician and chemist Ole J. Mynster in 1793. There was also the highly esteemed public administrator and man of letters Tyge Rothe, who was a member of the Royal Academy as well as a co-founder of the Society for the Advancement of the Fine and Useful Sciences. During his academic tour of Europe, Rothe had met the Swiss natural scientist Charles Bonnet, author of *Oeuvres d'histoire naturelle et de philosophie*. Rothe took on the formidable task of translating that work into Danish, producing *Naturen Betragtet efter Bonnets Maade* (*Nature Regarded after the Fashion of Bonnet*), a six-volume work in Danish appearing 1791–1797 and spanning nearly 2,000 pages. This work, too, was more than just a translation, and Rothe, like Bonnet, perceived nature as a continuous whole that had developed into the best it could be, in accordance with the Divine Creator's master plan. He imagined that the once-barren Earth had contained "Elements" and "Germs" that had, in time, produced life, and that God had minutely planned the entire development process. As Rothe described it, "God did not sit idly, pondering and choosing: He knew all from Eternity. He did not *see* what was there; rather, all possible things were one idea in His mind."[146]

One man earned an exceptional position in the broadly descriptive and generally oriented scientific literature of that age: the productive author and civil servant Esaias Fleischer, who wrote extensively on a number of subjects, including agricultural economics. Fleischer, who held a degree in theology, was a respected amateur scientist who was elected to be a foreign member of the Royal Swedish Academy of Sciences. In England, Fleischer became known by virtue of his friendship with the Welsh gentleman scientist Thomas Pennant, who travelled extensively to study natural history and supported the young Joseph Banks before he had made a name for himself. When Banks was preparing to visit Linnaeus in Sweden in 1767, Pennant provided him with a letter of introduction to Fleischer – which was never used, however, since Banks did not make it to Uppsala, or to Copenhagen.

None of Fleischer's works were notable for their originality, or valuable as independent research, including his colossal *Forsøg til en Almindelig Naturhistorie* (*Essays for a General Natural History*), which was published from 1786 to 1804 and took up a whopping 26 volumes with about 16,000 pages all told. That makes Fleischer's *Forsøg* one of the greatest works in the history of Danish scientific literature, at least in terms of quantity if not in terms of quality. This work was a critical compilation of the international scholarly literature, supplemented with Fleischer's own observations and occasionally spiced with interesting comments. The object was to give the reader an comprehensive knowledge of the results achieved in all fields of science. This naturally made the work extremely wide-ranging, and it treated topics ranging from cosmogony to anthropology. The two themes linking the numerous volumes were present in much of the period's literature. Firstly, Fleischer underscored the usefulness of natural history, in that it "liberates us not only from much damaging superstition, but also from unfounded and foolish fear, in which one has agonized in ignorance of these things, and yet in certain places agonizes still and tortures oneself" (vol. 2, p. xi). Secondly, the perspective of natural theology was all-pervasive, as Fleischer summed it up in his preface to volume 1 freely quoting Francis Bacon, in Danish, that "A small degree of philosophy and natural science leads away from God; but a more thorough consideration and a greater degree of these things leads ever more strongly back again to Him." As noted, Fleischer's *Forsøg* was no original work, but it is a superb source for anyone studying the history of science in the late eighteenth century.

The learned schools

Mathematical and scientific subjects played only a very modest role in the Danish grammar schools and *gymnasier* – academic preparatory schools – of the 1700s. The only reason these subjects were taught at all was that they were considered useful in two different ways. Firstly there was the die-hard faith in science as an important means of developing Danish agriculture and industry, and secondly there was the belief that scientific instruction was one way of bringing students to recognize the presence and omnipotence of the Lord God. This second motivation was important in the science that was taught, and the natural-theological approach continued to hold a place in the textbooks long after it had lost support from most scientists and theologians.

A national school ordinance issued in 1739 reduced the number of Denmark's grammar schools – the "Latin schools" – from 58 to 20, and it further stipulated that the schools were to teach not only mathematics, but also "natural philosophy", meaning physics and chemistry.[147] Very little is known of how these subjects were actually taught, or how often, and at what level. The scientific subjects were drastically trimmed in the grammar-school reform of 1775, which by and large transferred them to the university's reformed instruction for the *filosofikum* – the mandatory first-year exam in philosophy. At that point an education at grammar school offered only the most elementary introduction to mathematics and astronomy, if even that. Henrik Steffens, who had been a student at the grammar school Roskilde Katedralskole from 1785 to 1787, wrote in his memoirs that he received no instruction in mathematics, nor in the exact sciences. But despite the changes in the relationship between school and university, several of the learned schools still gave some basic instruction in the natural sciences. A debate arose in the 1790s about the form and content of the teaching at the learned schools, including the status of natural history. In 1795, as part of the ongoing effort to reform the grammar-school system and harmonize it with higher-level education, the university patron Frederik Christian, Duke of Augustenborg, proposed introducing mathematics, natural history and natural sciences in the intermediate level.

The duke believed that the instruction given in the two latter areas ought to be accompanied, whenever possible, by experiments and demonstrations. Reform efforts over the following years led to a new school ordinance, adopted in 1809, which was nothing if not anticlimactic as far as the scientific subjects were concerned: Though it did state that teaching in the natural sciences and natural history was to be given, this was only to apply "when and where this is possible". For all practical purposes this meant it was up to each individual school to decide whether its students would receive such instruction. The outcome was that at several Danish grammar schools the natural sciences disappeared completely.

Widespread debate arose in the 1830s over the ideal of intellectual formation that was universally agreed to be the very foundation of the instruction given at the grammar schools. One main issue was whether certain subjects (Latin in particular) were formative in the concrete sense that by mastering them a student could attain the level of general knowledge and "force of spirit" that must necessarily be the object of teaching. Where the natural sciences, or "exact" subjects, were concerned, the debate revolved around whether (or not) the value of such subjects lay not merely in their content, and hence in their utility, but also in their ability to promote students' concrete intellectual development on a par with Latin and mathematics. Beyond his own field, the professor of classical philology Johan Nicolai Madvig was keenly interested in, and fairly knowledgeable of, the natural sciences. In 1832 Madvig published a proposal that would place greater emphasis on mathematics in the grammar schools, and would introduce natural history, physics and chemistry into the curriculum. He did not argue from the angle of practicality or entertainment value, but stated that these were formative subjects that contributed to "the real education and elevated life of the mind".[148] Even though he affirmed that the classical languages must maintain their central position, they would have to cede some ground to make room for a more encyclopaedic distribution of subjects. And Madvig, a professor whose thoughts built on the work of the German philosopher and educator Johann Friedrich Herbart, staunchly maintained that the formative capacity of such subjects was inextricably linked to their factual content.

Not surprisingly, the two leading scientists in Denmark at the time – H. C. Ørsted and J. F. Schouw – agreed that science subjects ought to be strengthened in the schools. Schouw forcefully published his thoughts on the matter in *Dansk Ugeskrift*, his "Danish weekly", arguing that natural history, in particular, merited a place in the curriculum, and that time could be found for this by reducing the number of lessons in Latin and Greek. Naturally, such heretical views were not allowed to stand unchallenged, and so in the 1830s there arose a running polemical debate about the issue; a debate in which mainly theologians and philologists contested the intellectually formative value of the scientific subjects.

In 1850, the law governing Danish grammar schools and their relationship with the University of Copenhagen was amended. According to the new regulations, the eight-year education was to be an all-round, intellectually formative course of study structured as a cohesive whole and concluding with a battery of exams. This bolstered the scientific subjects, since the new rules meant that students would be examined in them. It also created the need for competent science teachers. Astronomy remained an ancillary part of mathematical instruction, whereas the new law stated that the instruction given in the natural sciences "must comprise the elements of mechanical and chemical physics, and is aimed not so much at a strict and predominantly mathematical development, but more as a clear and living demonstration of the chief phenomena and laws presentable by way of experiment, and of their connection." Even though the reform of 1850 did reinforce science in the lower and secondary schools, it still operated on classical, humanist terms, namely by defining the scientific subjects as "generally formative".

The somewhat abstract pedagogical and political debate going on among Denmark's intellectual and educational elite was one thing. The actual instruction in the debated subjects given at grammar schools across Denmark was quite another. Practices varied greatly from one school to the next, and the following account – from the grammar school at Søro – is just one example, and not representative of the general situation in the country.

Beginning in 1826, the educational institution at Sorø served a dual purpose, functioning as an upper-secondary-level academy as well as a lower-level grammar school. The teachers at the Sorø Academy held the title of *lektor* – "lecturer" – and also taught the oldest grammar school pupils. One of them, Carsten Hauch, was a lecturer in physics and zoology at Sorø from 1827 to 1846. As a young man Hauch had been a promising naturalist. He studied in Paris under Cuvier and was the first Dane to write a doctoral dissertation in zoology. But his interest in letters and literature continued to grow, gradually overshadowing his scientific pursuits, and in 1846 the University of Kiel offered him a professorship in Nordic languages and literature, which he gladly accepted. But during his years at Sorø, Hauch taught natural science to the academy students and to the younger pupils taking the "exact sciences curriculum" in a special programme established at the grammar school in 1837. This particular course of instruction had natural sciences and natural history as its predominance subject areas, accounting for 23 classes per week, whereas it included no Latin or Greek at all.

There were discussions in the 1840s of closing down Sorø Academy, and proposals were made to turn the academy complex into an agricultural institute, or into a *højskole* – a "folk high school" – inspired by a new movement that began in Denmark and was championed by the Danish scholar, pastor and poet N. S. F. Grundtvig. From the new humanist point of view, such proposals were atrocious, and they were forcefully rejected as being a threat to the truly humanist culture. The philologist and historian Christian Molbech protested against the attempt to give a higher priority to the technical-scientific subjects, at the expense of the classical-humanist culture. Although he conceded that technicians and scientists had promoted the Danish society's development with their knowledge and their inventions, that did not carry sufficient weight to convince a genuine humanist. The scientists were knowledgeable and skilled, but according to Molbech they utterly lacked the "divine strength, and divine spirit and facility" that characterized the humanists.

Journals and societies

In the early 1800s there were only very few Danish journals that were chiefly, or even partially, concerned with scientific topics. However, articles that dealt with popular scientific issues or topical debates were often found in literary and other journals that cannot otherwise be characterized as scientific or technical. It is quite noteworthy, for instance, that Ørsted's first real publication of his ground-breaking experiments with electromagnetism appeared in a small notice in the literary journal *Dansk Litteratur-Tidende*, presumably written on 15 July 1820 – six days before the appearance of Ørsted's much larger and far more famous treatise.[149] It was not until the mid-nineteenth century that Denmark saw the first examples of regular popular-science journals, such as the short-lived *Nordlyset* edited by the officer, translator and popularizer of science Christian Anders Schumacher (no close relation to the brothers previously described). *Nordlyset* (*The Northern Lights*) concentrated on the physical sciences, and readers of an introductory article in the journal were assured that they exerted a "beneficial influence ... upon the civic well-being, on the development of the intellect and the ennoblement of the heart."

A few of the new journals were more professional, most notably *Tidsskrift for Naturvidenskaberne* (*Journal of the Natural Sciences*). The influential threesome behind the new science journal consisted of the tireless Ørsted, the zoologist Johannes C.H. Reinhardt and the botanist Jens W. Hornemann. Published from 1822 to 1828 in five volumes encompassing 1,884 pages, the aim of this journal was not to bring translations or excerpts from dissertations published abroad, but to disseminate scientific information in a manner that appealed to a broad readership. The journal did not belong to the genre of popular science, but neither was it intended to serve as a medium for publishing original research. Another journal that was more important in a strictly scientific sense was *Naturhistorisk Tidsskrift* (*Journal of Natural History*), which the fish and crustacean expert Henrik Krøyer published in six large volumes from 1837 to 1849. This journal was very important for Danish zoology and botany, and it also became known internationally, as many of its articles were translated into German and appeared in *Isis*, the journal published by the German naturalist Lorenz Oken. From the very beginning, Krøyer's Danish journal was professionally oriented, targeting his fellow scientists rather than the public at large. This was underscored not only by the style of writing used in the contributions but also by their language, which was occasionally German or Latin. Krøyer's journal was succeeded in 1849 by *Videnskabelige Meddelelser fra Naturhistorisk Forening* (*Scientific Announcements from the Danish Natural History Society*), which accepted even lengthy treatises, and which became an important forum for Denmark's naturalists.

Several organizations were established during the first half of the 1800s with the goal of bringing scientific knowledge and news to a wider audience, particularly skilled craftsmen and small businesses. The most important, and the most long-lived, was the Danish Society for the Dissemination of Natural Science, which still exists today.[150] Ørsted apparently got the idea for such a society on a trip to Britain in 1823, finding himself greatly impressed with the many institutions for the promotion and dissemination of science and technology that he saw in the large cities. The most prominent was the Royal Institution of Great Britain, established in London in 1799. Its Danish counterpart, founded in 1824, soon gained a solid, paying membership base that included prominent citizens, university people, public servants and industrialists. It was involved in a wide range of outreach activities, of which its public lectures were the most important and the most visible. Although based in Copenhagen, from the very outset the society had specifically committed to spreading out its activities to include the larger towns around the country, and it trained a corps of "regional lecturers" for this specific purpose. Over a period of about 20 years, the Danish Society for the Dissemination of Natural Science held a total of 36 lecture programmes in 19 different towns around the country, reaching an estimated audience of three to four thousand.

9.4 *The journal* Naturhistorisk Tidsskrift *(*Journal of Natural History*) became an important Danish outlet for publishing on botany and, not least, on zoology. The first volume, from 1837, contained a report from the journal's originator, Henrik Krøyer, on a new genus of crustacean that he called* Geryon tridens. *This crustacean, which Krøyer had found in the waters of the southern Kattegat, was preserved in alcohol at the museum of the Danish Natural History Society.*

Even though natural history was mentioned in the Danish society's preamble, the field was never more than an appendix to the physical and chemical sciences. That was one reason why the Danish Natural History Society – also extant today – was established in 1833; a society that in many respects was an extension of the legacy Natural-History Society, described in an earlier chapter. The creators of the new society were Joachim Frederik Schouw, the eminent botanist, and Daniel Frederik Eschricht, who was a professor of zoology. These two, along with distinguished government official and patron of science Jonas Collin, were the directors of the Danish Natural History Society, which largely targeted the same circles as the Danish Society for the Dissemination of Natural Science: the higher echelons among the citizenry of Copenhagen. And like its older cousin, the new society of 1833 was initially a club for the social elite, its membership roster revealing a remarkable scarcity of young scientists. Many of them felt out of place in the new society and preferred to cultivate their interests elsewhere, which in this case meant the likes of another new organization, the Society of Natural History, which was created in 1844 and changed its name the following year to the Danish Botanical Society.

During the first half of the nineteenth century, throughout the process that strengthened science and transformed it into a powerful socio-economic force, the formation of national and international organizations played a crucial role. In 1822 the Society of German Researchers and Physicians met in Leipzig for its first gathering, and nine years later the British Association for the Advancement of Science was born. The huge and highly profiled congress of German scientists held in Berlin in 1828 included participants from all three Scandinavian countries, which helped to ripen the barely budding idea of creating a Nordic scientific organization along the same lines as the German body. The idea was discussed over the next few years, and although it gained a wide following it did not win universal support. The eminent Swedish chemist Jöns Jacob Berzelius spoke against the idea, doubting the scientific value of such meetings and finding it most likely that if they had any effect it would be to isolate the Nordic region from the world of international science.

The first convention of Scandinavian scientists was finally organized in Gothenburg in 1839, with Berzelius as a notable absentee – but there were equally notable attendees, including Ørsted, Schouw, Eschricht and Forchhammer, who all belonged to the Danish scientific elite. According to the bylaws of the Society of Scandinavian Natural Scientists, as the new organization was dubbed, plenary lectures and joint sessions would be open to the public, whereas voting and partcipating in sectional meetings were priveleges reserved for the professionals. The term "professionals" was construed very broadly, however, and the group consisted not only of active scientists and physicians, but also of people who simply held an academic degree in the relevant professional areas. The Gothenburg convention succeeded in building up a viable organization that for decades continued to gather at regular intervals in the Nordic capitals and came to serve as a vital forum for the entire region's scientists and doctors. Berzelius overcame his scepticism and from 1840 became a leading figure in the society, alongside Ørsted and the physicist Christopher Hansteen. At the Copenhagen convention in 1847, the number of participants had swelled to 472 in all – 338 of whom were Danes.

Theoretically at least, the meetings were equally open to women and men, but the first female scientist, Charlotte von Yhlen – a Swedish physician with professional qualifications that entitled her to full-fledged participation – did not actively take part until 1873. Women had also been present before her time, but had only accompanied their husbands or fathers as companions. Rather unlike the meetings in the German society, those in Scandinavia were not characterized by polemical, ideological disputes or personal antagonisms. It almost goes without saying that the keynote speakers did not always agree, and in the 1840s, Romantic natural philosophy was a watershed theme, although the debate was rarely conducted directly, or publicly. Great efforts were expended to achieve a basic consensus, and to form a seemingly unified front in the battle against the neo-humanist tendencies. The intellectually formative value of science was an essential theme, whereas its practical applications played only a negligible role. It was no coincidence that the wide variety of professional sections that made up the Scandinavian society did *not* include one for technical or applied science.

Ørsted and the Age of Romanticism

10

The influence that *Naturphilosophie* – the new philosophy of nature that arose in Germany – exerted on Denmark's intellectuals and academics was quite considerable. One of the foremost adherents of this line of thinking, referred to in Denmark as "Natural Romanticism", was the great physicist Hans Christian Ørsted, who for half a century was the pivotal figure around which these intellectual currents swirled. Even so, Ørsted did not foster a school of thought, and only very few of his contemporaries in Denmark made his ideas their own. What is more, Ørsted's brand of Romantic natural philosophy focused on the inorganic sciences, that is, chemistry and physics, whereas Danish naturalists in general avoided Natural Romanticism *à la* Ørsted.

One probable reason why thinking along the lines of genuine *Naturphilosophie* was blatantly absent from the Danish natural-history landscape is the high esteem in which the Danes held the French school. During the period under review here, the works of Buffon, Lamarck and Cuvier were mandatory reading for any well-educated Dane. Likewise, it was just as important for a young scientist on his *grand tour* to spend much of his stay in Paris at Cuvier's venerated feet as it had been during the previous age to study with Linnaeus in Sweden. In all events, most of the actual theories created by Danish naturalists during the first half of the 1800s were based on empirical-inductive principles, not inspired by speculation on the philosophy of nature.

There were a few exceptions, however, such as the geological works of Henrik Steffens, which will be described later in this chapter, and which most definitely belong to the school of Natural Romanticism. The Danish author Carsten Hauch, who today holds a prominent position in the history of Danish literature, is another notable exception. In his youth he studied philosophy and physics, and under the influence of Ørsted and others he was inspired by the new philosophy of nature. Having chosen zoology as his research field, Hauch developed a way to comprehensively and systematically arrange mammals in a system that was clearly indebted to Natural Romantic speculations. In a work dating from 1831 on the rudimentary organs of mammals, he expressed the idea of a principle of nature endowed with a soul; teleologically controlled evolution, which progressed from the imperfect towards

10.1 *Painting of Henrik Steffens from 1804. Steffens, who was born in Norway, was the earliest and most important representative of Romanticism in Denmark, and he never abandoned his ideals of a Romantic* Naturphilosophie. *His scientific contributions, which mainly lay in the field of mineralogy, were greatly valued by his contemporaries, but they lost their validity when the Romantic movement faded away.*

an ever-closer approximation of perfection. According to Hauch, even the human spirit was in a state of continuous progressive development, and was capable of eventually evolving into a more perfect spirit in the actual or divine world. Hauch often underscored the holistic dimension of science, and such thinking was by no means original. To him, the wholeness that the spirit impresses upon nature was an organism in which all elements were integrated into one indissoluble unity. Hauch's interest in natural history soon began to wane, however, and even when he took up a position as a lecturer at Sorø in 1827 he was already pushing aside the natural sciences in favour of his literary pursuits.

Romanticism and *Naturphilosophie*

In Denmark as in Germany, the road to Romanticism went by way of Kant.[151] The thought system of the great German philosopher played a significant role in the ongoing discussions among Danish philosophers in the 1790s and was also known to a handful of Danish scientists. The most important young scientist to study Kant and find inspiration in his dynamic *Naturphilosophie* was undoubtedly H. C. Ørsted, and Kant's ideas would later serve as a platform for his own Natural Romantic views. Even though Ørsted was familiar with Kant's greatest work *Kritik der reinen Venunft*, he mainly found his inspiration in another of Kant's works: *Metaphysische Anfangsgründe der Naturwissenschaft* from 1796, and thoughts expressed in the latter work had an extremely important place in the dissertation Ørsted published in 1799.

The substance of what Ørsted referred to as "the new natural metaphysics" was characterized by being dynamic; by possessing – indeed by consisting of – attracting and repelling forces. In keeping with Kant, Ørsted emphasized the aprioristic element; that the nature of the system was such that human thought could, in advance, arrive at the valid and necessary laws of nature. And likewise in keeping with the Königsberg philosopher, Ørsted rejected the mechanistic world view based on the ideas of the atom and vacuum, which was an image Ørsted claimed to be not only conceptually inconsistent, but also methodologically inadequate. Ørsted compared the dispute between the materialistic-atomistic system and the dynamic system of critical philosophy with the dispute between the chemical theories of Lavoisier and Stahl, and he did not doubt that the former dispute would end up in the same way as the latter had. Just as Lavoisier's theory had consigned the old theory of phlogiston to the wastebasket of history, so the dynamic theory of matter would make the atomistic theory a relic of the past.

In Ørsted's view, Kant was a step on the path leading towards Schelling, and it was under the influence of Schelling's ideas of natural philosophy, dating from the late 1790s, that Ørsted realized how heat, electricity, magnetism, light and chemical phenomena could all be embedded within a comprehensive dynamic programme. Schelling, and the other philosophers of nature Kant inspired, assumed there was a fundamental unity between mind and nature, such that one was unable to exist without the other. As mind was invisible nature, so nature was the visible mind. One could, of course, regard nature from without, as had traditionally been done in the empirical sciences, in which case nature would appear to be passive and devoid of mind. But nature also had an inner side to it, and by looking within one could divine much more of its true character, and as such, nature constituted an active and creative whole; and organism that cannot be understood by means of sensory input alone. And because this inner character of nature was inseparable from the human mind and spirit, it could be speculatively, or intuitively, recognized

10.2 *Portraits of Ørsted. The first drawing depicts a 23-year-old Ørsted in 1802–1803 as a student in Paris, while the second is from 1844, when the great physicist was 67. The daguerreotype from the late 1840s shows Ørsted as an elderly gentleman, complete with hairpiece.*

by particularly gifted thinkers – the geniuses of this world. Hence, a purely speculative physics was a possibility, and indeed the only possible path for those seeking to attain insights into nature at the most profound level.

Schelling and those who followed his thinking were not necessarily against experiments, but they had a different and more qualitative view of the role experiments played, a view in which measuring a thing's properties was unimportant. In certain cases natural philosophers went so far as to completely deny that sensing and experimenting could lead to any real insights into nature. The sort of nature that could be objectively sensed was regarded as a dull wrapping that contained and obscured the sort of real, non-objective nature that can only be recognized by taking the route of speculative physics. To the human senses, the objects of nature appeared to be fixed, material and permanent. Speculative insight would reveal, however, that in reality they are dynamic and eternally changing manifestations of opposite polarized forces in equilibrium; parts of a Heraclitic *panta rei.* In 1809, Ørsted, who greatly

appreciated the value of experimenting, gave a precise definition of the *Naturphilosophie* that he adhered to in many significant respects, despite his critical approach, explaining that

> The investigator of nature seeks to take command of the idea of the whole, so that from there he may take a larger view across the parts, the more perfectly to behold them in their context. This is what occurs in the *Naturphilosophie.* This science attempts to ascertain the nature of the World based on the necessary characteristic of things. It is the highest form of speculation and borrows nothing from experience, just as, conversely, experiential science should not be admixed with speculative sentences.[152]

Denmark was primarily introduced to the Romantic view of nature, humankind, and literature by the Norwegian geologist Henrik Steffens, whose adherence to the ideas of natural philosophy was more unconditional than Ørsted's.[153] Steffens had arrived at the University of Kiel in 1796, where he gave

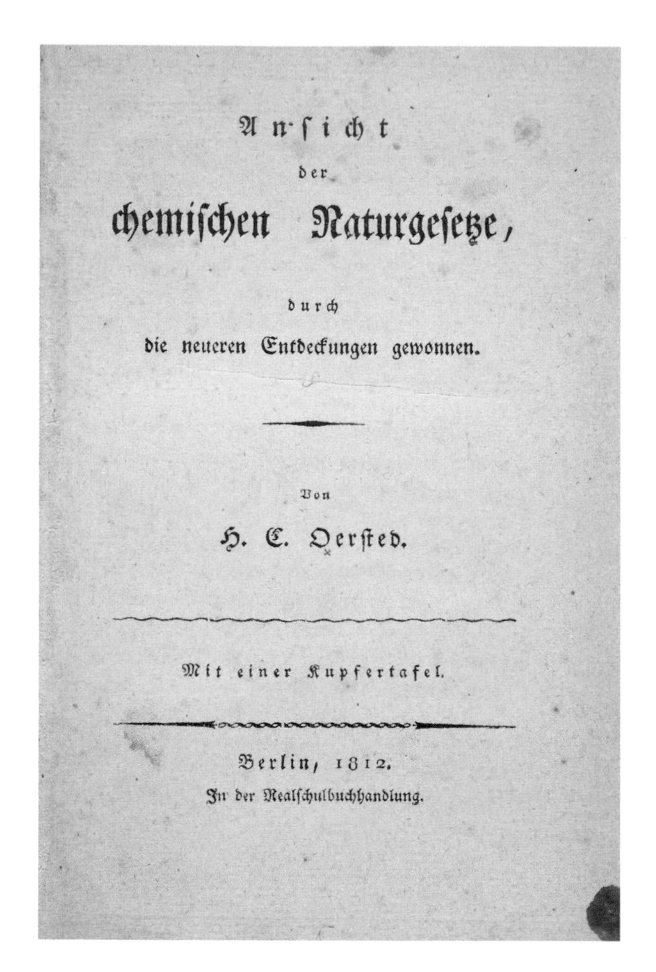

Ansicht
der
chemischen Naturgesetze,
durch
die neueren Entdeckungen gewonnen.

Von
H. C. Oersted.

Mit einer Kupfertafel.

Berlin, 1812.
In der Realschulbuchhandlung.

10.3 *The cover of Ørsted's* Ansicht der chemischen Naturgesetze *from 1812. In this important work Ørsted gave a meticulous, independent presentation of his dynamic theory of chemistry, making his textbook differ in many respects from those used in France, Britain and elsewhere. The ambitious work was known to leading chemist across Europe, but did not gain recognition.*

lectures while working on his doctoral dissertation in mineralogy. When he became acquainted with Schelling's main works two years later, he was profoundly impressed. With the aid of a travel stipend from "the royal fund for public benefit", Fonden ad usus publicos, Steffens travelled to Germany, and in Jena he met a number of the leading thinkers in the new Romantic tradition, including Goethe, Schelling, Fichte and Ritter. Steffens also made a natural-philosophy contribution of his own: an article on chemistry and geology that appeared in Schelling's *Zeitschrift für spekulative Physik.* In this article Steffens sought to present, in the grandiose Romantic style, the polarized basic forces of the universe, which including linking terrestrial magnetism, chemical oxidation processes, electricity and organic life processes.

After his stay with the renowned Abraham Werner at the mining school in Freiberg, Steffens wrote an important work in 1801 on geological *Natur-*

philosophie entitled *Beyträge zur innern Naturgeschichte der Erde*, dedicating the first volume to Goethe. In his *Beyträge* he attempted, in purely speculative terms, to present the basic forces as being contingent upon nitrogen and carbon, asserting (on the basis of flimsy analogous conclusions) that these substances "represent" the magnetism in chemical processes. Steffens further claimed that whereas carbon characterizes things of a vegetable nature, nitrogen characterizes things of an animal nature. Despite his studies with Werner, who was a sworn Neptunist, Steffens greatly emphasized the importance of volcanic activity, which he regarded as a sign of violent oxidation processes within the bowels of the Earth. In *Beyträge* Steffens laid out an evolutionary history for the Earth that was in accordance with Schelling's brand of natural philosophy. The true evolutionary path was unfathomable using only traditional natural history, and so it had to be contained in what Steffens called the Earth's "inner natural history". He was not alluding to something as mundane as the material inner substances of the Earth, but rather to an inner understanding of the Earth's dynamics. In Steffens's inner natural history, the inorganic was not distinct from the organic; the mineral kingdom not distinct from the animals and plant kingdoms. Like many of his contemporaries, Steffens supported the idea of *generatio aeqvivoca*, the spontaneous generation of primitive life forms out of inorganic material. His perception of the Earth's dynamic processes featured the new chemical theories in a prominent role, albeit in a decidedly speculative fashion.

Beyträge and other of Steffens's geological, chemical and mineralogical works were valued in their time, but when Romanticism waned they were quickly forgotten, and those who still remembered them saw them as nothing more than natural poetry devoid of scientific content. Changeability and impermanence were catchwords among the natural philosophers, who felt no sympathy for eternal, immutable laws of nature. The discomfiture the Romantics felt when forced to deal with concepts like constancy and eternity was often evident, as in the writings of Steffens, whose *Geognostisch-Geologische Aufsätzen* from 1810 rejected the law of the constancy of mass and even argued in favour of a sort of spontaneous creation of matter.

Steffens spent almost his entire career in Germany, but he kept up contacts in Denmark and Norway, where he was generally respected. In 1821, for instance, he became a member of the Royal Danish Academy of Sciences and Letters. In 1840 he participated in the second convention of Scandinavian scientists in Copenhagen, on which occasion he gave a talk "On the relationship between *Naturphilosophie* and empirical natural science" – which incidentally proved he was still as full-blooded a Natural Romantic as he had been in his

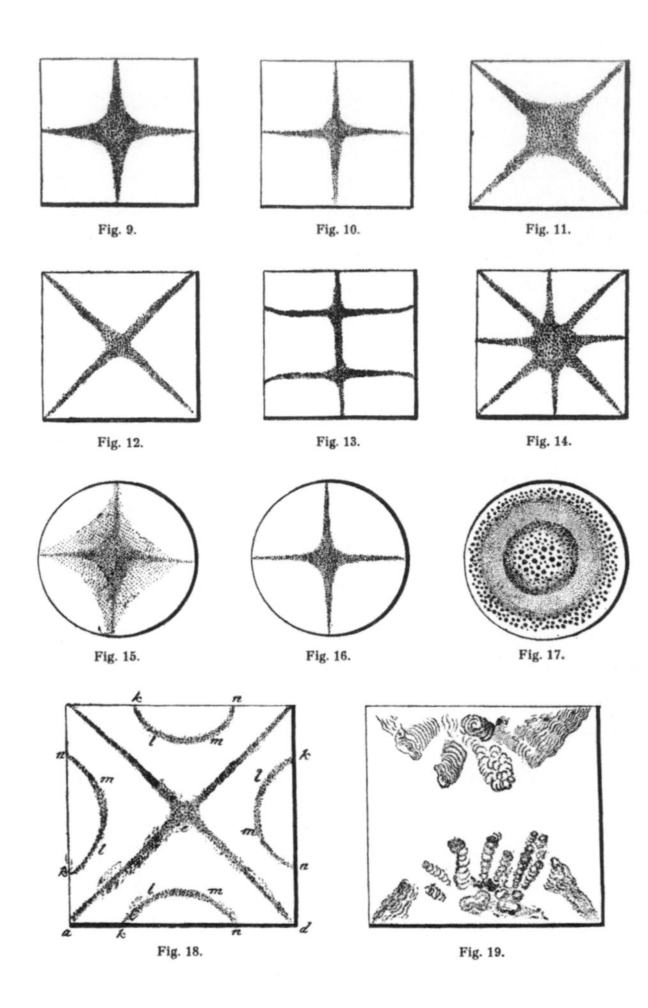

10.4 *Ørsted's experiments with sonorous figures. Drawings used in his own treatise from 1810. The German scientist Ernst Chladni had demonstrated that certain figures arise when fine-grained sand spread on a plate is affected by the sound waves from a violin string. Ørsted's curiosity was piqued, and he investigated the phenomenon further, erroneously believing it was the result of an electrification caused by the mechanical sound.*

youth. Steffens acknowledged that natural philosophy had been abandoned by most prominent scientists, who now considered it to be not only "a pointless fantastical preoccupation, but also harmful". Steffens himself had not lost his ideals, however, maintaining the essentiality of *Naturphilosophie* as a kind of alternative science, for as he expressed it "Natural science has reached an evolutionary epoch which necessarily forces the spirit to go beyond the binding limitations of the sensory [sphere] and observe its phenomena from a more elevated spiritual vantage point".[154] He was basically reiterating what he had said 40 years before, but in the meantime his thinking had taken a religious turn. At the Copenhagen science convention he argued that man was the very intention of nature. Nature in its entirety was the expression of God's own will, and hence *was* theology.

Chemistry without atoms

Ørsted's transformation from tentative Kantian to enthusiastic *Naturphilosoph* came to pass after a trip abroad that included stops in Berlin, Weimar, Jena and Paris. Upon his return to Copenhagen in early 1804, Ørsted received funds that would enable him to give public lectures in chemistry and physics, which he did with great success. Two years later, in 1806, Ørsted was appointed extraordinary professor of chemistry and retained that position until it was converted into an ordinary professorship 11 years later.

The most important source of inspiration for the young Ørsted was a German scientist named Johann Ritter, whom he had met during his stay in Germany. Ritter was an eccentric and original scientist who was adept at combining his remarkable talent for experimenting with his equally remarkable talent for speculative interpretation. His Romantic-holistic perception of nature was perfectly attuned to Ørsted's growing support for natural philosophy, so it was only natural that the two became fast friends, initially with Ørsted as the ardent admirer and Ritter as the revered oracle.[155] Although Ritter was speculative, he linked his speculations to experimental investigations; a practice to which Ørsted attached great importance.

While studying with Ritter, Ørsted happened upon a book by the somewhat obscure Hungarian chemist Jacob Joseph Winterl, and upon reading it believed that he was able to detect, in Winterl's views, the contours of a theory that fit the Romantic programme like hand in glove.[156] According to Ørsted and Ritter's reading of Winterl, their Hungarian colleague's work anticipated the Galvanism (a dualistically based chemistry) that provided a fertile alternative to Lavoisier's anti-phlogistic, "materialistic" theory. For one thing, Winterl perceived the properties acidic and alkaline as simply being two different states of a substance, and he explained the heat of neutralization as an electrical effect.

Moreover, according to Winterl water was an element, and when water was electrolytically split into two gases it was purely a matter of these two gases being compounds of water and demonstrating the principles of acid and alkali formation (corresponding to oxygen and hydrogen in the French chemistry). This interpretation was highly compatible with Ritter's ideas, the two principles merely having to be translated into the two types of electricity, and for a while Ørsted had no problem at all acknowledging the claim that water was an elemental, non-compound substance. Small wonder that some critics claimed the Natural Romantics had reinstated phlogiston and were merely using a new name.

Even though Ørsted remained sceptical of certain claims Winterl had made, he still believed that Winterl's system was, on the whole, worth defending, if only it could be presented in a more lucid form. Ørsted took it upon himself to do so, publishing his *Materialien zu einer Chemie des neunzehnten Jahrhunderts* in 1803. Nevertheless, the Winterl-Ørsted-Ritter version of an alternative chemistry was rejected by virtually all scientists of the day, not only in Britain and France but also in Germany, where Romanticism was thriving. *Materialien* was a complete and utter failure, as was Ørsted's subsequent attempt in Paris to promote Ritter and his views on electrochemistry – which Ørsted did by submitting some of Ritter's treatises in a competition for a sizeable academic prize that Napoleon had announced in his enthusiasm over Volta's invention of the battery. The French committee of judges were by no means disposed to acknowledge Ritter's greatness, and Ørsted did little to promote his cause when he failed dismally in his attempts to substantiate his friend's claims with experiments. Although Ørsted gradually distanced himself from Ritter's and Winterl's more concrete ideas, he kept his faith in the underlying spirit in their system. His continuing allegiance is reflected in his important work *Ansicht der chemischen Naturgesetze* from 1812. This book, in which Ørsted gave a detailed and independent presentation of his own dynamical-chemical theory, appeared in French the following year, but failed to win a following among chemists in France.

In Ørsted's terminology, traditional chemistry belonged to "the special physics", since it was a purely experiential science – and therefore according to Kant it was, strictly speaking, no true science at all. Ørsted's intention with his dynamic chemistry went to the heart of the issue, as he wished to alter the cognitive status of the field and have chemistry accepted as a general branch of physics, complete with an aprioristic foundation. Chemistry, he said, was not just the study of substances and their transmutations, but also comprised heat, light, electricity and magnetism; all phenomena that Ørsted believed could be understood as polarized forces in an ever-conflictual equilibrium. These phenomena were usually described in material terms as weightless substances or *imponderabilia*, but in Ørsted's system they were more in the nature of forces rather than substances. In an unpublished work written around 1812, Ørsted wrote that chemistry "now shows itself as a science of the laws governing the forces which bring about connections and separations". He also stated his belief that this new turn has made possible a chemistry that rested on a solid scientific foundation. Chemistry could now be based on rational laws of nature in the same was as mechanics, except that the precise nature of these laws had not yet been determined.[157] But it would be, he asserted, and after

10.5 *The chemist William Zeise was a student of Ørsted who went on to become Denmark's first professor of chemistry. He made his greatest contribution in the new discipline of organic chemistry, and today he is remembered as a pioneer in the study of so-called coordinate compounds.*

that it would become evident that the chemical and mechanical laws of nature could be merged into one comprehensive unitary theory.

It was a crucial aspect in Ørsted's natural philosophy that the substances one can sense were not the actual goal of natural cognition; the actual goal was to find the forces and laws that governed them. The enduring constituent was not the material objects themselves, but the forces that were manifested in, or that produce, material objects that our sensory faculties are able to perceive. Ørsted established this point in his *Første Indledning til den Almindelige Naturlære* (*First Introduction to the General Physics*) from 1811, stating that while the substances of which a body or object consists are constantly shifting, the laws that describe and define this shifting remain ever the same. Matter itself is nothing more than empty space filled by the forces of nature, he said, and it is the laws of nature that endow things with their various properties.

The laws of nature that held such a crucial place in Ørsted's conceptual world, and to which he so often referred, were laws of reason. They did not express the limited reason present in man's thinking, but a reason that was embodied in nature, which in turn was an expression of God's thinking, of which humankind was a part, and could partake. Commenting critically on some of Steffens's articles in 1830, Ørsted expressed it like this: "The laws of nature form a system of laws, and as this is a system of the laws of reason, it thence follows, in turn, that all of nature is an arrangement of reason, and that it is the business of naturalists to seek for reason in nature."[158] Ørsted's strong emphasis on a natural reason that man is able to access made his natural philosophy rather rationalistic. He believed, for instance, that the law of inertia was true *a priori*, being a "self-evident necessity of reason".

The atomic theory that celebrated such huge triumphs in the 1800s was unpalatable to Ørsted and his like-minded peers, especially as formulated by John Dalton around 1808. According to Dalton's theory, the atoms of each different element had specific weights that were expressed in units of the atomic weight of hydrogen, which underscored the materialistic basis of the atomic theory. Dalton's ideas of indivisibility and immutable particles of matter could be seen as an expression of the purest form of materialism, and must, for that reason alone, be abhorrent to anyone who was truly a *Naturphilosoph* at heart. Around 1812, however, Ørsted began to realize that the atomic theory could account, quite naturally, for a number of empirical laws of nature (such as the laws of simple and multiple proportions), yet even this was far from sufficient to win him over. On the other hand, by this time it had become amply clear to Ørsted that matter (or the forces of which it was made up) cannot continuously be universally present, but must have some sort of discrete structure.

Ørsted never came to accept the atomic theory – either in Dalton's or Berzelius's version – but he sought to formulate a compromise between atomism and dynamism, in which the discrete units or atoms had to be perceived more or less and as localized centres of force. Ørsted discussed such a dynamic atomic theory in 1829, on a strictly qualitative basis, in his correspondence with the German chemist and crystallographer Christian Weiss, but without being able to translate his ideas into an actual theory. As it happened, one of the strong points in the new atomic theory was that because of the atomic weights Dalton had introduced, the theory was quantitative, and Ørsted would therefore have to transfer these quantitative values to his brand of dynamic atomism. He suggested that the atomic weights – which he referred to as "the chemical numbers" or "elemental numbers" – in some way expressed the intensity of the chemical forces, although he was unable to explain his idea in any detail. Readers of a small compendium from 1820, meant for those attending his lectures, are told that "the chemical numbers surely show us the magnitude of the substances' chemical forces." He preferred to use the phrase "chemical units" to denote his discrete forces, rather than "atoms" – a word he intensely disliked.

What was important to Ørsted was being able to reproduce the explanations of the atomic theory without the use of material, physical atoms. That is why, in 1822, he spoke up in favour of a more phenomenalistic interpretation of the concept of the atom:

> Dalton calls the basic particles dealt with here *Atoms*, which causes this theory to seem like one that exclusively belongs to the atomistic *Naturphilosophie*; but nothing prevents us from assuming, in the dynamic system as well, [that there is] a basic size for each single force, such that this force cannot express itself in a smaller space. So here, as elsewhere in the experiential physics, one can leave the metaphysical disputes out of the game, and without disturbance observe the evolved laws of nature.[159]

Ørsted was not the only anti-atomist in the Danish chemistry community. William Christopher Zeise became Denmark's first professor of chemistry in 1822. As a young man Zeise had been Ørsted's assistant and had earned a degree in pharmacy in 1815. The study trip he made to Germany and France in 1818–1819 included visits with Christoph Heinrich Pfaff in Kiel and Friedrich Stromeyer in Göttingen, and while in Germany he was not hampered by being under the influence of Ørsted's dynamic, anti-atomic natural philosophy. No so in Paris, where his stay was marred when he discovered that the French chemists deeply disagreed with Ørsted's views. As he wrote in the account of his travels, he had been "raised in the German school, where the theory of atoms is now regarded as an absurdity", and there he was in Paris listening to his French peers hold forth on the atomic theory almost as if it were a matter of fact. His initial reaction was to reject "the opinion of the French", but he was soon obliged to admit "that atomistics also has its merits, and far more, I believe, than many a German scholar would be willing to admit, for there are actually several phenomena and experiences ... which are much easier to order and, to some degree, more readily lend themselves to explanation with the aid of the atomic hypothesis than with that of the dynamic."[160] Nevertheless, he was still only willing to accept the atomic theory as a useful hypothesis, not as an actual fact.

The discovery of electromagnetism

In his later years Ørsted was celebrated at conferences throughout Europe and received a variety of honorary titles and awards, including a prize from the Académie des Sciences in Paris and the prestigious Copley medal from the Royal Society in London. Ørsted's journey in 1846 through Germany, France and Britain developed into a veritable victory tour. Interestingly, Ørsted's renown did not spring from his tireless efforts to reform chemistry based on his ideas of dynamism, nor was it a result of his landmark achievement in isolating aluminium. It was purely a consequence of his discovering electromagnetism in 1820. When, in 1846, he participated as a guest of honour at a meeting of the British Association for the Advancement of Science, the renowned astronomer John Herschel introduced Ørsted as a great pioneer of science, referring to his famous experiments in which a magnetic needle was affected by a wire with a running electrical current.

It was well-known by the late 1700s that electrical discharges such as lightning bolts could affect magnetic needles and magnetize needles or nails made of steel, but beyond that no one had proved that a definite link existed between electricity and magnetism. On the contrary, the influential French physicists agreed that the two phenomena were quite different, and that there were sound theoretical reasons for refuting any ideas of kinship. Within the generally recognized research programme in France, which the great mathematician Pierre-Simon Laplace had played a key role in formulating, there was no cause to explore a potential connection – given the common knowledge that no such connection existed. The issue was rather different when seen from a natural-philosophical point of view, which inherently expected to find evidence of the unity of forces. And it is certainly understandable that the theme pops up in Ørsted's *Ansicht* from 1812, mentioned above; a work in which, among other things, the author outlined a dynamic explanation of the transmission of electricity in conductors and commented on magnetic force. After citing a series of well-known similarities or analogies between electricity

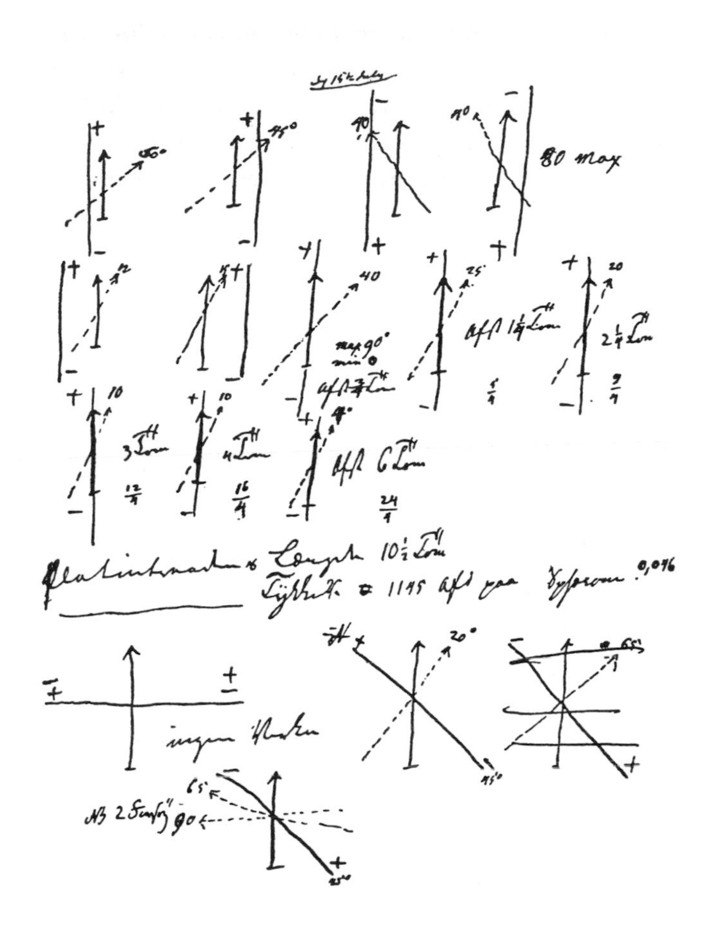

10.6 *Ørsted's notes from electromagnetic experiments he carried out on 15 July 1820. He had already suspected as early as 1812 that there was some connection between electrical and magnetic forces, which was to be expected in light of his natural-Romantic ideas. His 1820 experiments, which led to the discovery of magnetism, were motivated by his philosophical ideas of a basic unity in nature.*

and magnetism, Ørsted rounded off by recommending that investigations be made to determine whether electrical current might have some sort of magnetic effect.[161]

Ørsted himself believed that his musings in *Ansicht* might have precipitated his later discovery of electromagnetism and, more generally, that this discovery was a result of his Natural-Romantic belief in the unity of forces. Ørsted mentions in his Danish autobiography from 1828 that by the beginning of 1820, at the latest, he had reached the conclusion "that just as a body, through which courses a very strong electric current, will emit light and heat to all sides, so might it also be with the magnetic effect."[162] During a lecture Ørsted gave around that time, in the winter of 1819–1820, he sought to demonstrate the magnetic force in a wire carrying a current by showing its effect on a magnetic needle. Unfortunately the effect was too weak to allow the phenomenon to be investigated properly. The reason for this inadequacy was that, in keeping with his basic assumptions, Ørsted had expected the magnetic effect to correspond with the thermal and luminous effects and would therefore be most easily detectable in very thin lengths of wire (with high resistance and consequently low amperage).

He did not have time to look into the matter until July 1820 – but then proceeded to do so systematically, and with startling results. Monitored by witnesses of unquestionable integrity, Ørsted now detected a much stronger effect using a thick conducting wire. He also examined the direction of the deflections of the magnetic needle relative to the direction of the current, reaching the crucial conclusion that "the electric conflict is not confined to the conductor, but dispersed fairly widely in the circumjacent space". He also deduced that the effect came about in such a way that "this conflict performs circles".[163] Ørsted was aware that his discovery was a momentous one, and he realized that it must quickly be communicated to his fellow physicist. This may be why he chose a rather unusual mode of communication: a brief announcement dated 21 July 1820, printed as a four-page leaflet, paid for with private money, and sent to a number of Ørsted's professional peers abroad. The announcement had no illustrations and was, remarkably, written in Latin, a language that by 1820 had all but disappeared from the physics and chemistry literature. The announcement, *Experimenta circa effectum conflictus electrici in acum magneticam*, soon appeared in a Danish version, and in various international journals in French, German, English and Italian.

Ørsted's discovery was immediately greeted with great enthusiasm in Germany, where a translation of the Latin leaflet appeared in the country's two leading journals, which were edited by Johann Schweigger and Ludwig Gilbert, respectively. The latter stated in his commentary that Ørsted had chanced upon the discovery during a lecture, and this statement would later lead to a great deal of confusion as to the actual process of discovery. Gilbert's claim was later repeated, for instance by Pfaff, who in 1824 published the first comprehensive presentation of the history of electromagnetism, and who found it difficult to accept the discovery's natural-philosophical backdrop. The French physicists initially reacted with some scepticism, as they did not believe the phenomenon Ørsted had demonstrated could exist. Quite apart from that, they had also had unfortunate experiences with natural philosophers such as Ørsted – bear in mind his stay in Paris in 1803 – and they were therefore suspicious when confronted with what might prove to be "yet another German reverie", as the physicist Pierre Louis Dulong wrote in a letter to Berzelius.

As it turned out, the discovery was no reverie, but solid fact, as was demonstrated when the experiment was duplicated at the Académie des Sciences in

Paris. Within an astoundingly short time, physicists in France and elsewhere expanded upon the experimental insights Ørsted had achieved, and which would create the foundation of electrodynamics, a new field of knowledge chiefly defined by André-Marie Ampère. Ampère knew that Ørsted's electromagnetism really did not require any magnet at all, as it also manifested itself in an interaction between two current-carrying conductors. What is more, he interpreted this interaction very differently than Ørsted had. According to Ampère and his supporters, there was no "interactive battle" revolving around the conductor. What they proposed was an "action-at-a-distance" force working directly between the conductor's tiny particles. In the longer term, the reception Ørsted's discovery received in Britain was just as important as its reception in France, as it came to form the basis of Michael Faraday's law of induction from 1831, paving the way for the concept of electromagnetic fields that James C. Maxwell would later develop into a unified field theory applicable to electromagnetism and light.

Throughout all of this Ørsted himself remained on the sidelines, and his only significant contribution to electrophysics after 1820 was an experimental project he undertook in 1821 with the mathematical physicist Joseph Fourier in Paris to study Thomas Seebeck's new discovery, thermoelectricity. Ørsted's lack of interest in the further development of electromagnetism was even more pronounced when it came to its practical applications. Ørsted himself mainly saw the discovery as valuable by virtue of its ability to confirm his natural-philosophical views on the unity of forces, not because of its potential technological usefulness as a concrete scientific finding. And he was an equally passive observer of the developments taking place in the 1830s, aimed at putting electromagnetism to practical use. Within a decade, efforts led to the construction of the first electromagnetic telegraph systems – an innovation that was apparently of little interest to the Danish physicist with the Romantic inclinations.

Considering the immense and immediate importance attributed to the discovery of electromagnetism, it is quite understandable that Ørsted's priority has been questioned. There are no grounds for characterizing his discovery as serendipitous, and yet we must ask: Was Ørsted really the first person to notice, and document, an electromagnetic effect? Experiments into the possible magnetic effects of galvanism had been going on for years before Ørsted went down that road. One predecessor was the Italian amateur scientist Gian Romagnosi, who performed his experiments in 1802, and it has been suggested that Ørsted knew of them and at the very least ought to have referred to them in 1820. It is, however, highly unlikely that Ørsted's hand was guided by the early experiments of Romagnosi and others, even though he may have known of their existence.[164] Romagnosi never personally claimed to be the true discoverer of electromagnetism, and the effects he vaguely described in 1802 were probably electrostatic in nature. A review article from 1830 that Ørsted wrote on thermoelectricity for *The Edinburgh Encyclopædia* did actually refer to Romagnosi's experiment, and to its being briefly mentioned in a French-language textbook from 1804, though without stirring any interest. As Ørsted magnanimously wrote, had its importance been recognized, "it would have accelerated the discovery of electromagnetism by sixteen years".[165]

Colding and the unity of forces

Ørsted all but monopolized basic physics research in Denmark during the first half of the nineteenth century. However, there was another Danish contribution that has later been accorded its rightful place in the international history of science, although at the time it went virtually unnoticed.

The seminal concept of energy, with its wide-ranging significance, is intimately linked to the recognition that although energy can be transformed from one state into another, its size or presence remains unchanged. This realization was reached in the 1840s, although at that time scientists usually spoke of "force" rather than "energy", and most of the credit is due to the German physician Robert Mayer and the British amateur scientist James Prescott Joule. The applicable law was definitively and clearly formulated in 1847 by the German physician and physicist Hermann von Helmholtz in his classic paper *Ueber die Erhaltung der Kraft.* Just one year later and, it seems, independently of Mayer, Joule and Helmholtz, the Danish artillery captain Hermann Kauffmann of Copenhagen published a dissertation expressing the principle of the conservation of energy.[166]

Yet another Danish contribution to the process of discovery, and one that was even more significant, came from a young technological scientist and student of Ørsted named Ludvig August Colding, who is sometimes mentioned as an independent discoverer of the principle of energy conservation. After completing his training as a journeyman joiner, Colding passed the entry exams at the Polytechnical College, graduating in 1841 with a master's degree in mechanics. A few years later he embarked upon a successful career in engineering that in 1857 would make him Copenhagen's first *stadsingeniør*, or "city engineer". He began to ponder the perpetuality of the natural forces as early as 1839 or so, while still a student at the Polytechnical College.[167] As Colding himself stated on more than one occasion, this reflection was religiously

10.7 *Ludvig August Colding was a polytechnic graduate in mechanics and, in certain respects, a student of Ørsted. He earned a place in the history of physics for his early works concerning the "conservation of force", arguing theoretically and experimentally that there was some entity in nature – corresponding to what we now call "energy" – that is conserved in all events.*

founded, and he was convinced that the forces of nature expressed the very essence of God and must therefore be everlasting.

In 1840, Colding discussed his ideas with Ørsted, who was sceptical and wished to see these ideas substantiated by physical experiments. Thus spurred on to further investigate, Colding was able to report, three years later, to the Royal Danish Academy of Sciences and Letters on his ongoing experiments involving frictional heat. He believed his findings could verify that the amount of heat produced corresponded to the mechanical work used to generate the heat. Financial support, provided by the Danish Society for the Dissemination of Natural Science, enabled Colding to perform a more exhaustive array of experiments. As well as discussing his findings in treatises submitted to the Royal Academy in Copenhagen, he also gave an account of his work at the Scandinavian science convention held in 1847. Colding had not stated the numerical ratio of conversion between heat and mechanical work in 1843, but his subsequent experiments led him to find a reasonable value of about 3.7 J/cal (a figure 13% short of the correct ratio).

In the account from 1843, which remained unpublished, Colding proposed his first version of a conservation rule stating that "when a force disappears, it is merely subjected to an alteration of form and then becomes active in other forms". In other words, once the thing that Colding referred to as a mechanical force or an "activity" – later called "an amount of energy" – had transferred itself to another body, it must continue to reside there in some form or other. In 1850 he formulated the following claim, which was at least as rationally justified a priori as it was deduced from experiments, namely "That forces can never vanish in matter, and that it consequently must be a general law of nature that forces, without exception, merely undergo a change in form when they seem to vanish, whereupon they reappear as active sources of the same amount but under different forms."[168]

After becoming a member of the Royal Academy in 1856, Colding further discussed his theory in the Academy's series of writings under a heading worthy of Ørsted: "Naturvidenskabelige Betragtninger mellem det aandelige Livs Virksomheder og de almindelige Naturkræfter" ("Scientific Observations between the Activities of Intellectual Life and the Ordinary Forces of Nature"). In this piece he expressed the strongly teleological–religious belief that had motivated his research, and which he still perceived as the basis of his system of natural cognition – and which was, incidentally, *passé* among contemporary physicists. He was firmly convinced "that God, from the very beginning of the world, … has provided all the forces by means of which the cosmic evolution will reach fulfilment". Colding held that the eternally preserved forces included not only the natural forces such as heat, chemical processes, electricity and mechanical work, but intellectual activity as well:

> It is my contention not only that life in general demands its nourishment [from forces], but in particular that intellectual activity – the act of thinking – may also be viewed as work demanding nourishment, and I do not believe I am mistaken in expressing the view that it is the forces of nature, in their many forms, which serve as support for the intellect, and that intellectual activity evolves at the expense of these. … As the intellectual life evolves at a rapid pace, the abundance of forces of nature must be in a continuous decline, because the sum of all these forces is that invariable quantity originally created by God![169]

Colding was, however, obliged to admit that he could not experimentally substantiate his claim that intellectual activity was a form of energy. His speculations concerning a link between the physical forces of nature and mankind's intellectual life was not inspired by Ørsted alone, but also by the thinking of the Danish physiology professor Daniel Frederik Eschricht, who in his book from 1850 had intimated the permanence of things intellectual and spiritual, at the expense of the transience of things bodily.

Colding's works concerning a kind of energy conservation did not earn much attention among scientists in his native Denmark, where not even Ørsted understood them, let alone attributed them any value. And because they were written in Danish, they at first remained unknown abroad, although Helmholtz did mention Colding's early contributions in a lecture he gave in 1854. Because the principle of energy conservation soon gained such great significance, Colding believed he was justified in claiming priority over others, most notably over Robert Mayer – and he actually did so, first writing in Danish in 1856, and then in an English article that *Philosophical Magazine* printed in 1863. In his "Scientific Observations" paper from 1856, where he had presented the human intellect as a new and acutely refined force of nature, Colding chided the materialists for denying the existence of man's intellect, or at least for their linking it only to man's transient flesh. It was unavoidable that his views would become part of the religious and cultural battle of wills that was emerging in Denmark as a result of impulses coming from Germany. One response to the Christian–spiritual position expressed in Colding's paper came from the writer and journalist Rudolph Varberg, who represented a materialistic, agnostic position. Varberg was a professed atheist, having written a defence in 1851 of Ludwig Feuerbach's controversial atheistic perception of religion. Far from denying the existence of the soul, he held that it existed but was connected with man's transient physical body. Varberg also pointed out in his polemical discussion that Colding's world view was, in reality, deistic, as God had no influence on the workings of the physical world.

Although today Colding holds a place in the history of the law of energy conservation, he was not justified in claiming priority. Furthermore, it is quite reasonable to question whether the "forces" and "activities" Colding spoke of express the same energy concept that Hermann von Helmholtz, Rudolf Clausius and William Thomson established around 1850.[170] At any rate, in 1843 Colding was not using any sort of general unitary concept of a conserved entity – whether in the form of an "activity", a "force" or "energy" – and he always used the plural when speaking of the perpetuality or "immortality" of forces. In addition to his concept of forces being very vague, it also contradicted Ørsted's concept of forces on many essential points, which contributed to Ørsted's lack of enthusiasm and support. Although Colding was strongly influenced by Ørsted in many respects, it was not so in all respects. He certainly shared the natural metaphysics of his mentor, but he was no adherent of Natural Romanticism in the sense that Ørsted was, and one view he did not share was Ørsted's dynamic perception of matter and heat. Colding supported a kinetic heat theory, according to which heat is a macroscopic manifestation of atomic or molecular movements. He accepted the existence of material atoms, even going so far as to speculate on whether they might have an internal structure.[171]

Colding deviated from Ørsted in another area as well, namely in his view of the significance of mathematics to physics. Ørsted's knowledge of mathematics was very limited, and he was critical, even sceptical, of the mathematization of physics that began around 1820. Colding, on the other hand – speculative natural metaphysician that he was – felt quite comfortable with mathematical physics and was, presumably, the first Danish scientist after Jens Kraft to make any high-level contributions to that field. As early as 1850 he had already developed his thoughts on the conservation of "forces", complete with mathematical analyses of physical phenomena, and he later presented theoretical contributions on thermodynamics and hydrodynamics that proved worthy of international attention. These efforts reached their culmination in 1853 with the appearance of Colding's fully developed, mathematically formulated analysis of the theory accounting for the efficiency of the steam engine, a theory based on earlier works by Sadi Carnot and Émile Clayperon, the two French pioneers of thermodynamics. Colding's scientific work fully lived up to the highest international standards, ultimately earning him a reputation well beyond Denmark's borders.

Contributions to chemistry and natural history

II

The reason this chapter is dedicated to certain areas of natural history and chemical research in Denmark is not that they specifically enjoyed more attention than so many other areas, nor that they attracted especially gifted scientists. They do have a special merit, however, which is that when seen together they illustrate Danish science at its most typical, in addition to which they offer examples of interesting research at various times throughout the period under review. At the same time these areas can give the reader an idea of the shifting conditions for, and attitudes towards, scientific research during this period – one shift being the gradual professionalization that took place during the first half of the nineteenth century.

Whereas the motive of practical utility was of paramount importance to science in the 1700s, particularly where Denmark's agricultural economy was concerned, utility as a criterion became less important throughout the next half-century, as through new insights science gradually came to be viewed as legitimate. The utility motive was not generally replaced by a spiritual or intellectual ideal in Romantic terms, but rather by an empirical ideal that regarded science as valuable in and of itself. Traces of this shift away from technical and economic utilitarianism and metaphysical system-based thinking can be found, for instance, in the medical research taking place around 1840, and in the work of prominent scientists like J. G. Forchhammer and J. F. Schouw.

Two innovative zoologists of the 1700s

Zoology played no significant role in Danish science until the last few decades of the eighteenth century; decades during which the situation changed quite dramatically as figures such as Otto Friedrich Müller and the somewhat younger Johann Christian Fabricius made their mark in the international zoological community. Even in the late 1700s it was rare to find a clear-cut distinction made between zoology and botany: as a scientist one would cultivate the field of natural history, and both Müller and Fabricius certainly contributed all round. And yet the most impressive and most important work of both men lay within the realm of zoology, and it makes perfect sense to refer to them as "zoologists", even though using the term in that context may be slightly anachronistic. Besides their nationality and the general time period of their active careers, Müller and Fabricius also shared a penchant for investigating nature's less conspicuous creatures – from protozoa to insects – and for conducting their research independently. In Müller's case this meant staying completely outside the University of Copenhagen, and in Fabricius's case it meant having only a slight affiliation with the old institution.

Otto Müller was an amateur of science, in the truest and noblest sense of the word. He had an academic background in theology and law, but never received any formal training in the natural-history subjects in which he would later do such excellent research. The autodidact Müller was able to use his scientific reputation to climb the social ladder, doing so with great efficiency and singularity of mind. Known for his ambitious, self-promoting, pretentious and polemical personality, the socially successful Müller was an extremely prolific and internationally oriented writer. In fact, he also sought, with equal success, to make a name for himself through learned academies and societies across Europe, achieving no fewer than a dozen such memberships and becoming a corresponding fellow of the exclusive Académie des Sciences in Paris. Müller's inclusion in Cuvier's *Histoire des sciences naturelles* from 1845 – in which a prominently positioned article in volume 5 emphasized his studies of the class of animals known as *infusoria* – illustrates the international recognition he enjoyed.

Many of the organisms Müller found especially intriguing were microscopic, and existed in the borderland between the plant and animal kingdoms that had yet to be fully explored. He was deeply fascinated by the world his microscope revealed, believing that it confirmed his natural-theological views. A treatise on worms eloquently expressed his feelings: "All bears witness to the finger of God, transporting the observer into a state of reverent jubilation". Müller was able to find beauty and divine Providence even while "scrutinizing ... the bowels and entrails of animals, and becoming acquainted with the invisible denizens of this world."[172] Throughout his life he remained captivated by the beauty of nature, and by the modes of literary expression he could employ to describe it.

Müller did most of his early work in the field of botany, closely following the Linnaean tradition and mainly concerning himself with what Linnaeus called "cryptogams", a poorly defined group of primitive, non-flowering plants including fungi, ferns, algae and mosses. In 1764, Müller published his *Fauna insectorum Friedrichsdalina*, a list of nearly 900 insects, arachnids and

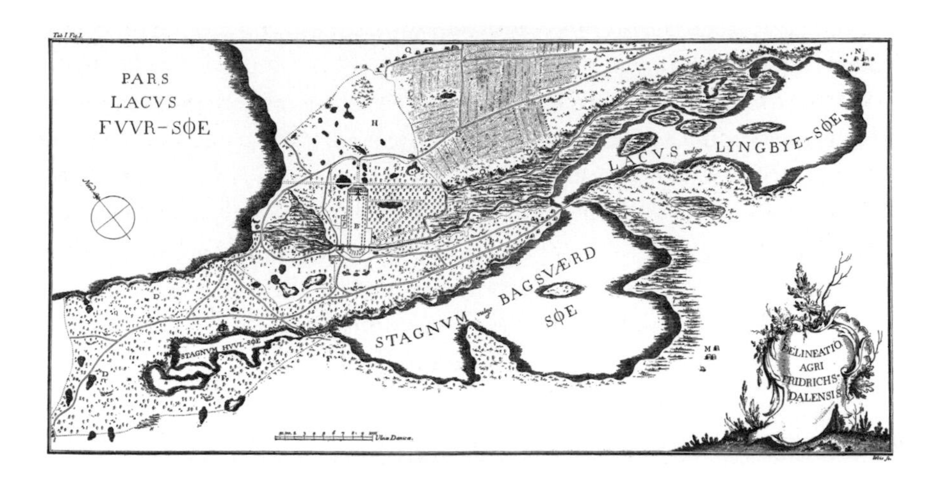

II.1 *The area north of Copenhagen where Otto Müller carried out his first botanical investigations. From his* Flora Friedrichsdalina, *1767. Müller's interest in science was awakened while he was serving as a private tutor at Frederiksdal Castle north of Copenhagen. Here, inspired by Linnaeus, he began his extensive studies of the area's microscopic fauna.*

similar creatures he had found in the habitats around Frederiksdal in northern Zealand, and three years later came his *Flora Friedrichsdalina*, a corresponding volume about the plant life in that area. Müller spent much of his early years as a scientist travelling, which contributed to his international outlook and excellent connections with scholars around Europe. While in Switzerland he met with Albrecht von Haller, Horace Bénedict de Saussure and Charles Bonnet, and was even introduced to the aging Voltaire. He also met Abraham Trembley, whose experiments with coelenterates (corals and jellyfish) were considered quite sensational, having proved that these animals were able to fully regenerate from a piece of severed tissue.

Even though Müller made significant contributions to botany, it was not his botanical achievements that made his reputation as one of Europe's most prominent naturalists. This he mainly owed to his studies of small animals like worms, water mites, tiny crustaceans and microscopic rotifers. His great two-volume work on worms and molluscs, *Vermium terrestrium et fluviatilium*, appeared in 1773–1774. Among other things, Müller described the genera *Cyclops* and *Daphnia*, actually proposing both genus names, and he also pioneered the scientific research on *Ciliata*, which he himself referred to as "*infusoria*". He first described these unicellular organisms in 1773, but did so most thoroughly in *Animalcula infusoria fluviatilia et marina*, published posthumously in 1786 by fellow zoologist Otto Fabricius.

Müller's interest in the reproduction of primitive organisms led him to speculate on the fundamentals of the reproductive process. He proposed that all livings beings were filled with what he called "molecules" which "contain an infinitesimally small *drawing*, partly of the animal in its entirety and partly of its constituent elements, and ... [which, when it expands] produces an embryo that in all things resembles the body in which it previously lay concealed". Müller held that these elemental germs of life were present everywhere, in huge numbers, and that they constituted the very essence of each being:

> To my mind, these *monads* ... give all organized creatures life and movement. ... They are imperishable, and are the basic schemata of the creatures... They lose life, then once again become living. They gather themselves into a lifeless mass, remaining thus constrained for an indefinite time, perhaps for centuries, until they are released from one another and each again regains its life and freedom.[173]

Elsewhere in his work he speaks of these basic schemata or governing rules embedded in the monads, as "embryons". But Müller's theory of reproduction was not quite as original as he made it out to be. It was actually a variation on the theory of preformation, which was popular at the time and was supported by such prominent physiologists as Bonnet and Haller. Unlike them, however, Müller did not believe the basic schemata were found only in animal ova; he believed they were present in every kind of life germ, even down to the most primitive organisms.

Denmark's other innovative zoologist was Johann Christian Fabricius, a native of the Schleswig region who had spent two blissful years in Sweden, studying with Linnaeus in Uppsala. These years not only made him an unwavering Linnaean, but also provided him with the foundation for his life's work: reforming entomology – the study of insects – in accordance with the principles laid down by his mentor. After his stay with "the immortal Linné", Fabricius went on another scientific journey that would last five years, three in Germany, the Netherlands and Britain, and the last two in France, Italy and Germany. Fabricius returned to Copenhagen in early 1770 and spent a few years as a professor at the university there before taking up a chair in 1775 in Kiel, where he taught natural history, economics and cameral science. Due to the university's small size and meagre finances, it could only offer humble working conditions, and so Fabricius arranged his life to work away from the university for the better part of every year, either in Copenhagen or in Paris –

which was the absolute centre of natural history at the time. Apart from residing in Kiel during the winter to teach, Fabricius was constantly on the go, and his frequent journeys set him among the most widely travelled Danish scientists of the eighteenth century.

One of the scientific attractions that often drew Fabricius for long periods of study was the collection created by Joseph Banks, the powerful president of the Royal Society in London. Fabricius and Banks were close friends, and the Dane was intimately acquainted with Banks's collection in Soho Square. Fabricius also spent much of his time in Copenhagen engrossed in similar studies, for instance at the Sehested–Lund entomological collection, eponymously named after two prominent public administrators. It ranked among the finest of its kind in Europe, thanks, in part, to the many species sent back to Denmark from the colonies.

The superordinate principle behind Fabricius's entomological system was to divide insects according to the shape and arrangement of their mouth parts. He found the mouth parts to be an ideal tool for classification because by reflecting the animals' feeding habits and diet, they also described its biological conditions and habitat. Fabricius's work, mapping out and classifying thousands of insects, led him to write a series of comprehensive and highly erudite works, beginning in 1775 with *Systema entomologiae*, covering all of 800 pages. Later, in 1792–1798, he published the five-volume work *Entomologia systematica emendata*. Fabricius was, overall, tremendously loyal to Linnaeus, whom he profoundly admired. This is quite obvious in his theoretical grand oeuvre *Philosophia entomologica* from 1778, which in form and structure is virtually a replica of the Linnaeus's *Philosophia botanica*. In Fabricius's own words, "I studied ... the systems and rules of the botanists, but most especially of my great and excellent teacher Linné, that I might see and know the solid foundations for scientific certainty and apply them to entomology."[174]

In his *Resultate natur-historischer Vorlesungen* from 1804, Fabricius mainly set out his thoughts on evolution – even though most had already been expressed in 1781 in his *Betrachtungen über die allgemeinen Einrichtungen in der Natur*. In the high style of French authors, he launched into his cosmic evolutionary scenario by describing how air and water joined, in a manner only vaguely described, to form minerals and thus the solid Earth, still in its primordial phase and as yet devoid of life. Then, somehow, the first germs of life came into existence, enabling stones to transmogrify into plants, and plants into animals. Through years innumerable and generations untold, changes in the climate and soil combined with cross-breeding ("bastardization") to yield new varieties of plants and animals that slowly settled into a mosaic of new species. Mankind, too – the master of all Creation – was the result of such an evolution, and descended from the apes. As for the Negroid race, Fabricius could only recognize them as the half-brother of mankind, since he believed they resulted from a cross between the Caucasian race and the great apes. "Can new species arise in nature within the animal, plant and mineral kingdoms?" His reply to this question, which he himself posed in the beginning of *Resultate*, was this:

> Admittedly, I do not see the earliest beginnings of the arising creatures, but I do see, quite clearly, their slow and gradual development on the basis of this primal germ through a long line of elapsing millennia. In the expression of Moses, which is most certainly taken from the Egyptians, even that last and most splendid work in the Creation known to us – namely Man – arose from a clod of earth. Undoubtedly, [this being] must have developed through an almost infinite number of intermediate links until finally, long, long after, it assumed the shape of human perfection.[175]

Expressing himself more bluntly, he asserted that "Man is wholly an animal, as far as his body is concerned." Unlike Buffon and certain other colleagues, Fabricius was convinced that the great ladder of evolution would lead upwards to ever more perfect species, and that even mankind's place in nature's household would one day be taken over by an even more perfect being.

It may seem that Fabricius, with his evolutionary conviction, had moved away from the Christian faith, but that was not how he saw it. Like other of his contemporaries he was inspired by natural theology and believed that his explanations helped to clarify God's existence. He considered the laws of nature, and their handiwork, to be "the most visible and strongest evidence of His wisdom, goodness and magnificence". And yet his ideas certainly disagreed with a literal reading of the Bible, and he was, perhaps, the first Danish scientist to clearly and radically distance himself from the accepted biblical chronology. Some historians of science have pointed out so many similarities between Fabricius and Lamarck that they find it possible the Frenchman in Paris was inspired by the Dane, implying that the latter was, in reality, "the father of Lamarckism".[176] This is, without doubt, a disproportionate interpretation of Fabricius's work from 1804, since the author himself did not consider it very important, but seems to have regarded it as merely a series of musing arising out of the prevailing atmosphere of popular science. *Resultate* belonged to that part of Fabricius's writings he himself described as "decked out in a great mass of words". In spite of his interesting thoughts he was no Lamarck, and certainly no precursor of Darwinism.

A peaceful chemical revolution

The 1780s witnessed the breakthrough of what was later called "the chemical revolution", a decade during which the traditional idea of explaining combustion processes based on the hypothetical substance *phlogiston* was replaced by Lavoisier's new chemical system based on oxygen. The discussion about the basis of chemistry that raged among the major European nations did not completely bypass Denmark, where the literature at least hints at the ongoing battle between the two schools. C. G. Kratzenstein's textbook from 1787, for instance, did not mention water as an element, but conceded that it was probably a compound consisting of two gases. On the other hand, Kratzenstein described these gases in accordance with his phlogistic views, using Joseph Priestley's names for them: "dephlogisticized" and "combustible" air.

Naturally it was not necessary for Danish scientists, any more than it was for those in other countries, to declare themselves adherents or opponents of the new chemistry. They could anticipate further developments, or acknowledge certain aspects of Lavoisier's system while criticizing others. One interesting example was the Danish apothecary Nicolai Tychsen, who in 1784 published a chemical handbook entitled, simply, *Chemisk Haandbog*, providing the first exhaustive account of chemistry in Danish. While this first edition was based on the phlogiston theory, the second edition from 1794 was decidedly eclectic. Tychsen made it quite clear in his 1794 preface that he wished to treat both theories, and also that he did not perceive them as directly conflicting:

> I have in all parts sought to explain things in accordance with the newest basic tenets. I have therefore, in most places, where various things were explained according to the old Stahlian system, additionally, like Hermstaedt [*sic*], explained things according to be basic tenets of the antiphlogistic theory, by which I thought to be of use to adherents of both these systems. I have also, here and there, remarked upon various things that defied the new antiphlogistic theory. ... Perhaps by this means, I thought, the conflicting parties and the old and the new theory could be united together, but perhaps I have erred as well, though as to that, time will tell.[177]

One of those who in the 1790s clearly preferred Lavoisier's new chemistry was the officer and amateur scientist Adam Wilhelm Hauch, a wealthy patrician who conducted many and varied experiments and his well-equipped private laboratory in Copenhagen.[178] Hauch was an avid reader of scientific literature, and while travelling in 1788–1789 he met a number of the most prominent

11.2 *Lord Chamberlain Adam W. Hauch, an officer and a gentleman – as well as a prominent public administrator, collector, nobleman and scientist. He was an influential member in the Royal Academy in Copenhagen, and his physical and chemical experiments attempting to clarify the composition of water helped to introduce Denmark to the new chemistry. He was also acknowledged abroad as an accomplished scientist in the empirical tradition.*

European scientists of his day, including Martin H. Klaproth in Germany, and Joseph Priestley, Henry Cavendish and Joseph Banks in Britain. Hauch was strongly preoccupied with the composition of water, which he considered to be the pivotal question in the controversy between the two chemical systems.[179] Certainly, Lavoisier had demonstrated that hydrogen could be derived from steam, and that hydrogen and oxygen could combine to form water, but as phlogistic chemists argued, this did not necessarily imply acceptance of the claim that water actually *consisted* of the two gases. Hauch therefore decided to duplicate, refine and vary experiments involving the analysis and synthesis of water, sometimes using a gasometer of his own construction. Like Lavoisier he concluded that the release of hydrogen when steam reacts with hot metal had only one possible explanation: that the metal has reduced the water, and that consequently the gas must be a constituent of the water. His experiments were not particularly innovative, but their quality was so outstanding that they won recognition all over Europe and helped the new chemistry gain acceptance among German scientists.

Hauch's most significant influence on Danish science came through his *Begyndelses-Grunde til Naturlæren* (*Introduction to the Physical Sciences*), a Danish work that was published in 1794 – the very year Hauch was appointed lord chamberlain. *Begyndelses-Grunde* rapidly became the Danish standard work on physics and chemistry, was translated into German, and served as a textbook not only in Denmark, but in Germany and Norway as well. Its sections on chemistry faithfully followed Lavoisier's system. He did mention phlogiston as the essence of "the elder chemistry", but only as a prelude to rejecting it.

Another contribution to the chemical revolution in Denmark came from a small group of young scientifically interested citizens in Copenhagen who began publishing a new journal in 1794 with the ambitious title *Physicalsk, Oeconomisk og Medicochirurgisk Bibliotek for Danmark og Norge* (*Physical, Oeconomic and Medico-Surgical Library for Denmark and Norway*). This journal was the mouthpiece of the new chemistry, and from the very start its position was consistently and aggressively anti-phlogistic.[180] The group behind the new journal included Henrik Steffens, who in 1794 wrote an extensive exposition on the "calcination" of metals, in which he consistently supported Lavoisier's oxidation theory. Incidentally, a later article by Steffens, which appeared in the journal in 1799, has been called the first Romantic manifesto in Denmark.[181] According to Steffens, the changes our planet had undergone were the result of an "immense phlogistic process", and he proposed that earthquakes were caused by electrical processes within the Earth.

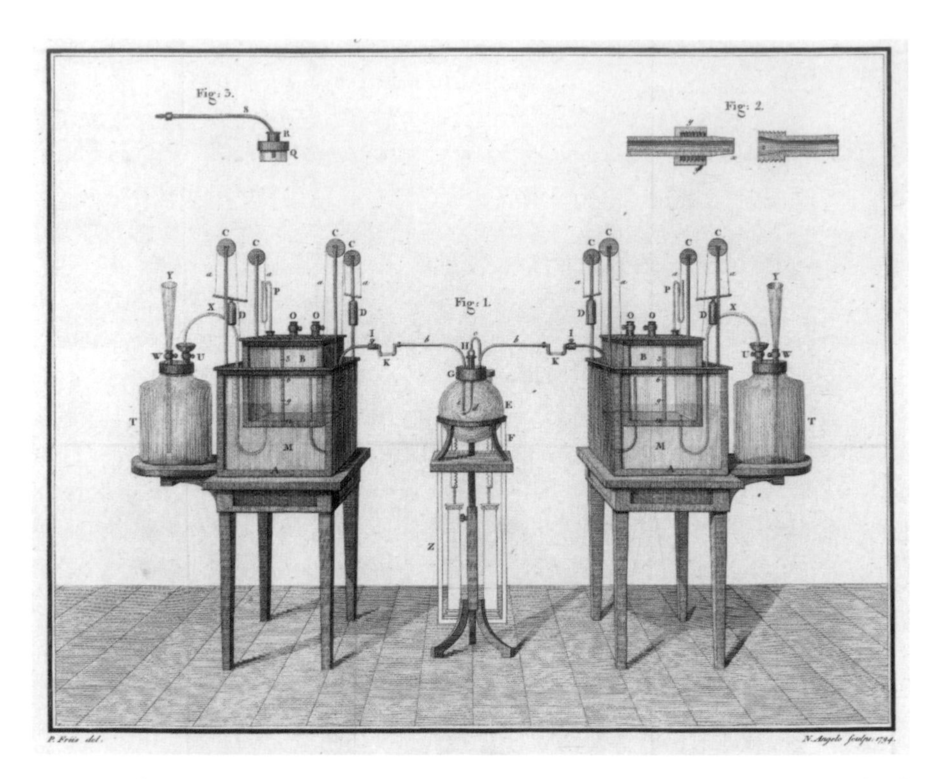

11.3 *Hauch's gasometer from 1793 for examining the composition of water. The two smaller cubical containers, open at the bottom and floating in larger containers with water, are filled, respectively, with oxygen and hydrogen led in from the large bottles at either side. At the centre sits the round beaker that houses the combustion process resulting in water. As the gases are consumed, the containers will sink in the water, allowing the gas consumption to be read using the scale on one side of the glass container.*

With the 1794 contributions of Hauch and Steffens, Denmark, too, was largely liberated from the old idea of phlogiston. Nevertheless, as exemplified in the account of Ørsted given in the previous chapter, even this did not result in the uncritical acceptance of Lavoisier's French chemistry.

Forchhammer and the cycles of the elements

The eminent Johan G. Forchhammer was Denmark's first professor of mineralogy and geology, and although he had never taken any formal degree in geology, he was well-versed in the chemical sciences. Once an apothecary's apprentice, he had gone on to study at the University of Kiel under Pfaff and specialized in analytical chemistry. His doctoral dissertation, completed in 1820 and entitled *De mangano*, examined different coloured compounds. Forchhammer was actually the first to demonstrate that all of them were compounds of manganese with different "oxidation numbers" (to use the later terminology). Sweden's eminent Jöns Jacob Berzelius found his conclusion dubious, but it was confirmed by the German chemist Eilhard Mitscherlich. The following year, in 1821, Forchhammer published a brief account of his findings "On the preparation of pure salts of manganese, and on the composition of its oxides" in Thomas Thomson's *Annals of Philosophy*.

Forchhammer's early geological works included a study of Møns Klint, a series of striking white cliffs facing the Baltic on the Danish island of Møn. This study caused quite a stir among geologists in Denmark and abroad, who were hesitant to accept his claim that the chalk deposits on the eastern coast of Møn were from the Tertiary period. As it happened, Charles Lyell – the most celebrated geologist of his day – became involved in the controversy while visiting Denmark for a few days in the spring of 1834.[182] He accompanied Forchhammer on an inspection tour of Møns Klint and subsequently described his investigations in an article published in *Transactions of the Geological Society of London*, which demonstrated that Forchhammer's geological dating of Denmark's white cliffs was incorrect.

In spite of the professional disagreement it elicited, that first meeting in 1834 between Forchhammer and Lyell would be the beginning of a lasting friendship. The two geologists corresponded regularly, and Forchhammer visited Lyell on several of his trips to Britain. Forchhammer was generally quite well-connected with prominent figures in the geological community, not least in Britain. His contacts there included Roderick Murchison, who introduced the Silurian as a geological period in the Earth's early history, and whom Forchhammer had met during the Scandinavian science convention in Christiania (today's Oslo) in 1844. Forchhammer also corresponded with Charles Darwin after the great scientist invited him, in 1849, to participate in the annual meeting of the British Association for the Advancement of Science. Owing to the ongoing rebellion and ensuing hostilities in the Schleswig-Holstein region, Forchhammer initially felt obliged to decline – "the state of our country and its affairs is so unsettled" – but in the end he decided to risk the journey and was able to give two talks at the association's meeting in Birmingham.[183]

Forchhammer's widely used textbook in chemistry, *Lærebog i Stoffernes Almindelige Chemie* (*Textbook on the General Chemistry of Substances*) from

1842, was descriptive and experimental. In addition it contained what is probably the first defence written in Danish of the so-called Proutian hypothesis, a theory proposed in 1815 by the English physician and chemist William Prout in the wake of Dalton's atomic theory. The essence of Prout's controversial hypothesis was that all atomic weights could be expressed as multiples of the weight of hydrogen, the reason being that all elemental atoms ostensibly consisted of a primordial substance in the form of the hydrogen atom. Around 1840, when Berzelius and other leading European chemists believed they had refuted the Proutian hypothesis, there was Forchhammer supporting it – although that did not mean he was an advocate of composite atoms. He found it "highly probable that one of the laws by which the atomic weights of the elements are connected is this: that hydrogen, the substance in which the weight of the basic constituents is the smallest, is a *communis divisor* for all of the others." Forchhammer clearly did not share Ørsted's dislike of atoms, considering his use of the Ørstedian phrase "the weight of the basic constituents" in the same sentence as the non-Ørstedian expression "atomic weight".

Geochemistry as a scientific discipline barely existed in Forchhammer's day, and he himself must be counted as one of the pioneers of that field. The very word "geochemistry" – a word Forchhammer never even used – was not introduced until 1838, by the German chemist Christian Schönbein. Forchhammer's many notable contributions included his development of precise analytical methods to map the occurrence and natural cycle of trace minerals. One effort, namely his extensive research programme from 1844 to 1859 that aimed to determine the constituents of seawater and their geographical variations, put him at the cutting edge of oceanographic chemistry. He was able to inform Berzelius in 1845 that "I have, over the past year, been occupied with analyses of seawater, and have now finished with the Kattegat, the North Sea, and the northern part of the Atlantic Ocean ... I yet await, this autumn, water from the Baffin Bay and from a voyage between St Thomas in the West Indies and the Channel."[184] Forchhammer assembled these and subsequent results in a lengthy dissertation in Danish from 1859, which appeared in translation six years later in the *Philosophical Transactions of the Royal Society*. The multitude of samples he analysed included 27 supplied to him by the ambitious Danish expedition travelling on the corvette Galathea.

Around 1850, Forchhammer became interested in different issues relating to biospheric chemistry, such as the cycles of the elements and the interplay between mankind and nature. As he wrote when introducing a series of lectures, "The basic idea is that the elements circulate between nature's various kingdoms, but that these cycles are never fully contained, that something goes

11.4 *Johan Georg Forchhammer held degrees in chemistry and pharmacy, but despite his lack of formal education in geology he became Denmark's leading professor in this field. He was an gifted geochemist with a visionary outlook, whose work included pioneering measurements to determine the content of various elements in the great oceans. His geological mapping of Denmark was a first step in reaching an understanding of the country's geology.*

missing in the course of the cycle, and that this is partly replaced by the effect of nature – that is, through an even larger cycle – and partly by man's meddling with nature."[185] For the most part he would present his thoughts on such issues in a practical agricultural context, lectures at agricultural associations and the like, but he never published them internationally. Generally there was a great interest at the time in agricultural chemistry, particularly after the appearance in 1840 of Justus von Liebig's famous work in German propounding the mineral theory of plant nutrition rather than the humus theory. This work appeared in a Danish translation in 1846 under the title *Agricultur-Chemie.* Whereas Liebig rather one-sidedly pointed to mineral substances as the only source of plant sustenance, Forchhammer held that one must also add nitrogen to the soil, for example in the form of manure. The question of whether plants were capable of biologically fixating or binding nitrogen from atmospheric air was still unresolved, but Liebig and most other authorities rejected the idea. Forchhammer did not agree with them, although he was unable to explain what sort of mechanism might bring about such a process.

The far-flung Kingdom of Denmark played host to yet another pioneering effort in the field of geochemistry. The person behind that effort was no Dane, however, but the brilliant German chemist Robert Bunsen, who in 1846 participated in an expedition to Iceland sponsored by Christian VIII of Denmark.[186] Bunsen performed crucial studies of volcanic gases and proved that the water in geysers is rainwater, not seawater, as many scientists believed. Bunsen, in a collaborative effort with the French mineralogist Alfred Descloizeaux, measured the temperature in geyser tubes, and based on their findings he set out a theory of geyser eruption.

One last effort worth mentioning here in the context of geochemistry and atmospheric chemistry was led by Baruch Levy, a Jewish chemist and natural scientist of Danish-French descent.[187] After completing a degree in pharmacy Levy worked under Forchhammer and his prominent colleague William Zeise, then travelled abroad and earned his doctorate in Berlin in 1839 after defending a dissertation on arsenic. While in Paris he had become acquainted with another pair of renowned chemists, Jean-Baptiste Dumas and Jean-Baptiste J. D. Boussingault. This led him to become involved in measuring the composition of atmospheric air; a project that would keep him occupied for a year back at the Polytechnical College in Copenhagen and elsewhere. Following a three-year research journey in Colombia for the Académie des Sciences, Levy continued his work in Paris studying and investigating mineralogy and plant physiology, but never completely severed his ties with Denmark.

Schouw and plant geography

One of Denmark's leading scientists during the first half of the nineteenth century was Joachim Frederik Schouw, who has already been mentioned earlier in this work. Like Ørsted, Schouw was involved in many diverse activities, influencing not only the world of science, but culture and education as well.[188] In 1821, at the age of 32, Schouw had already been appointed to a professorship in botany, and soon afterwards he began publishing the work that would make his reputation: *Grundtræk til en Almindelig Plantegeographie* (*Basic Features for a General Plant Geography*) and the companion volume of illustrations *Plantegeographisk Atlas* (*Atlas of Plant Geography*). This work appeared in German in 1823 as *Grundzüge einer allgemeinen Pflanzengeographie.*

By the 1820s it was already widely known that many plants, and for that matter many animals, are only indigenous to a very limited area, and possibly to similar habitats elsewhere on the globe. The basic ideas of plant geography had been laid down by a German named Karl Ludwig Willdenow and his student Alexander von Humboldt, whom many regard as its founders. However, it was not until Schouw had systematized and organized the field that the

11.5 *Drawing of Joachim Frederik Schouw, 1838. He and Ørsted were the two most prominent Danes in the Romantically influenced science of the first half of the nineteenth century. His scientific work concentrated on plant geography, where he carried on and developed the work of Humboldt and others. Another cause that Schouw championed was strengthening science in schools and among the general population.*

skeleton was fleshed out and plant geography took on the shape of an actual science, often referred to as phytogeography.[189] As early as 1818, Schouw had written a long article in *Jahrbücher der Gewächskunde* demonstrating a stance towards the great Humboldt's works on plant geography that was exceedingly critical. Actually, Schouw chose to publish the article anonymously, which was probably very sensible, or at least completely understandable. It was mainly the category of plants known as "Alpine growths" that motivated naturalists to begin drawing lines to connect the dots. As it happens, alpine growths are not merely plants that grow in the Alps, but plants that are found to occur naturally in areas reminiscent of the Alps; areas like the Andes, where Humboldt had studied the vegetation. Schouw sought to find a universal explanation based on climate and soil analyses, combining this with his understanding of the actual historical development of the Earth's crust.

Schouw was very deliberate in his rather narrow definition of plant geography, sharply distinguishing between it and the historical evolution of plants, and preferring not to deal with the latter issue at all. In his *Grundtræk* he

11.6 *The vegetation zones in the western hemisphere, as represented in Schouw's* Plantegeographiske Atlas *(*Atlas of Plant Geography*). Schouw was the first scientist to systematize and organize the many data about the distribution of plants, and only through his efforts did the field begin to crystallize into an actual science: phytogeography.*

defined plant geography as "that science which presents the *current* incidence, growth area and distribution of plants, as well as the differences in the Earth's *current* vegetation, all with regard to the external factors affecting the plants." Essentially, this is what would later come to be known as "ecological plant geography". Schouw's somewhat ahistorical approach did save him from much evolutionary speculation, but it also imposed a variety of arbitrary limitations on his project. By defining himself out of the historical dimension, so to speak, it was difficult for Schouw to explain the connection between conditions such as plant distribution areas and climate. Several of his works nevertheless dealt with the issue of the origin of species. Schouw subscribed to the idea of the spontaneous generation of primitive life (*generatio aequivoca*) and believed that the higher-order organisms had evolved under the influence of a physical "cosmic force", but he did not believe this force was capable of transforming one species into another.

Schouw continued to support Cuvier's brand of catastrophism (or "revolutionism"), meaning he believed that no momentous climatic changes had taken place within historical times – that is, after the most recent natural cataclysm, thought to have occurred before the dawn of mankind. This opinion disagreed not only with the ice-age theory, but also with the conclusions that the young zoologist Japetus Steenstrup reached after his investigations of peat bogs. What is more, Schouw largely ignored palaeontology, and his catastrophistic views were diametrically opposed to the uniformitarianism that with Lyell's support would soon become the prevailing dogma in the geosciences. Speaking at the fifth Scandinavian convention of scientists in 1847, Schouw argued against the hypothesis that all plant species shared a common ancestor, cautiously concluding that several original plant ancestors had existed. These had gradually evolved into the forms now known, he asserted, but he did not believe that new plant species would arise in the future. He raised the question of whether today's flora included any species that had originated in the "prior world", using that expression to indicate the Earth as it had been before its last cataclysm. Since no one had any idea of the nature of this catastrophe or when it had occurred, Schouw felt unable to make any decisive pronouncements on the matter. Even so, he did believe he could supply an answer, stating that there is "no proof that they [the plants] survived the natural revolutions that took place immediately before our current age; for if one assumes that the same species might have arisen at the same time in different locations, then it might likewise have arisen at different times."

Schouw distanced himself from geographical or naturally conditioned determinism, perceiving nature instead as a thing that could be affected and altered by human intervention. As previously mentioned, Forchhammer also supported such thinking, although he arrived at that point by a very different route than Schouw, whose reasoning ran as follows: "Man is a part of nature, is affected by it, and is subject to its laws; but man also, in some sense, stands outside nature and can therefore … exert an influence upon it, alter it, and to a degree even rule it and give it laws."[190] Although only indirectly, these thoughts would prove to be quite significant for the Danish Heath Society, an association founded in 1866 by a small group including Enrico Dalgas – who was Schouw's nephew and was strongly influenced by his uncle. The fact of the matter was, it was important for the Danish Heath Society to establish that Jutland's extensive tracts of heath were not merely the result of processes conditioned by nature, but had largely resulted from societally conditioned changes to Jutland's physical environment.[191]

Besides his career as a scientist and professor, Schouw kept busy with a multitude of other activities. He was the founder and editor of *Dansk Ugeskrift* (*Danish Weekly*) from 1832 to 1836, producing a magazine that was typical of the period for an audience of generally educated readers. Publications of this sort would mix culture and science with political commentary, and during those years such magazines, directed at a much wider audience, ran a surprisingly large number of important natural-history studies that would later appear in specialist journals. Schouw's recognition of the close link between plant incidence and geographical conditions reinforced his interest in physical geography, and in 1832 he published the first actual textbook in that field, entitled *Europa. En Letfattelig Naturskildring* (*Europe. A Straightforward Depiction of Nature*). This book, while aimed at the same broad group, was also a part of Schouw's endeavours to gain a place for the natural sciences in the general education given at Denmark's grammar schools, and to put science on a par with the classical subjects. Despite the dedicated efforts of Schouw and many others who championed the same cause in speech and writing, many years would pass before physics, chemistry, geography and natural history found a place in Denmark's secondary-level curriculum on an equal footing with Greek and Latin.

Travel and exploration 12

Compared with the century of 1580–1680, Danish science in the period 1730–1850 showed little inclination towards cosmopolitan or international thinking. Prominent Danish scientists certainly continued to visit colleagues and universities abroad, to study and find inspiration there, but for the most part they still published their findings in Danish and considered the reaction within the national scientific community to be at least as important as reactions abroad. On top of that, the small number of students going abroad to study meant that a much larger proportion of graduates had been educated exclusively at the University of Copenhagen. So in spite of the increased level of activity that characterized Danish science from the late 1700s, it was considerably more local – often provincial - during that period than it had been during the Latinized 1600s. One major difference between the two periods was that during most of 1730–1850 there existed a scientific literature and a wide range of scientific journals in Danish, meaning that most Danish scientists published their work in a language that was incomprehensible outside Scandinavia. Hence they were not under the same pressure as before to publish internationally, or to live up to international standards.

Of course Danish scientists were not isolated from the rest of the world, nor were they indifferent to broadening their horizons professionally and geographically. One way they were able to expand their outlook was by going on expeditions whose aims were partly scientific. The two largest and most highly profiled were the promising and much-publicized journey to Arabia in the early 1760s and the ambitious Galathea expedition in the 1840s – both funded by the royal coffers. To a certain extent, the waters and lands of the North Atlantic also became objects of scientific exploration, but the relatively limited activities there were just a foretaste of things to come in the 1850s and beyond.

Foreign travellers in Denmark-Norway

The Kingdom of Denmark was not the most attractive place for foreign scholars and scientists, yet some did choose to visit, either on brief stays or to study natural phenomena, especially those in Iceland and Norway. One foreign visitor was the Austro-Hungarian Jesuit priest and astronomer Maximilian Hell, director of the observatory in Vienna. Hell arrived in Copenhagen with his assistant Johann Sainovics and a servant in 1768, continuing from there to Trondheim in Norway. The Danish government had invited Hell to observe the transit of Venus expected to take place on 3 June 1769, a celestial event that captured the interest, and coordinated the efforts, of astronomers and scholarly societies all over Europe.[192] Even though the Danish professor of astronomy Christian Horrebow had previously witnessed the planet's passage in 1761, his career was all but over in 1769. The observatory in Copenhagen had fallen into neglect, and Horrebow's international reputation was tarnished. The observations he made of Venus during the summer of 1761 were of such poor quality that the European astronomical community simply chose to ignore them.[193] Instead the Danes called in assistance from abroad, in the embodiment of Hell, to observe and measure the 1769 transit from the island of Vardø in northern Norway – an invitation which, needless to say, was a complete disavowal of Danish astronomy.

After many trials and tribulations Hell at last reached his northerly destination in October 1768, describing Vardø as "the end of the world, where only few people live, and where the sky is covered by a black shroud". Hell's was the only one of the northbound expeditions that succeeded in observing the transit of Venus in all of its phases, which made his data particularly valuable to the European project. The two Hungarians spent their time on other scientific pursuits as well, being fortunate enough to witness a solar eclipse on 4 June – just one day after Venus had completed its passage of the Sun. They also had ample opportunity to observe the northern lights during their stay, which Hell described in a letter: "This magnificent spectacle we had during our entire journey, every single night, provided the weather was clear, and in such a manifold variety of shapes! Mairan and all of his devotees are quite mistaken to believe this to be caused by the atmosphere of the Sun."[194]

On the return trip, Hell and Sainovics, his assistant, stopped in Trondheim, where they were duly honoured and became the first foreigners to be granted membership of the brand-new Trondheim Learned Society. Back in Copenhagen, both Hell and Sainovics gave talks at the Royal Danish Academy of Sciences and Letters, which also accepted them as fellows. Finally, after presenting the Danish king with a copy of his report, Hell was able to return to his home in Vienna in august 1770. Considering the sheer cost of the Danish-funded expedition – a tidy sum of 6,398 rixdollars – it is quite remarkable that no Danish astronomers participated in it, not to mention paying attention to its progress and results. The expedition would, incidentally, have certain unpleasant repercussions: The renowned French astronomer Joseph-Jérome de Lalande insinuated, in less than subtle terms, that Hell had revised, perhaps even forged, his observational data. Only much later did it become clear that Hell's scientific integrity was irreproachable.

Some years later, Fransisco de Miranda, an envoy from Venezuela, stayed in Denmark for a time, and his travel journal gives a good impression of the limited interest in science he experienced in Copenhagen. After visiting the university there in February 1788, Miranda wrote how "Greek, Latin, natural law, history &c, those are the subjects one studies, whereas neither physics nor mathematics holds any significance". And remarking upon a brief visit at Borchs Kollegium, the college donated by polymath and chemist Ole Borch in 1691: "They have their own library &c, in order that they may study chemistry, but great was my surprise upon observing that instead of this science, one cultivates the writings &c of Luther to educate clerics."[195] Miranda had a personal interest in science and took the opportunity while in Denmark to carry out electrical experiments with the German-Danish general Wilhelm Theodor Wegener, who had studied mathematics. In his laboratory at the Army Cadet Academy, where he taught, Wegener kept a collection of instruments that included an "electrification machine", which he and Miranda used in their experiments.

The British writer and feminist Mary Wollstonecraft spent a few months in Denmark and Norway in 1795, finding that the kingdom had "some respectable men of science, but few literary characters, and fewer artists. They want encouragement".[196] Denmark proper was rarely considered sufficiently interesting as an object of study, whereas other parts of the kingdom – notably the mountains of Norway, spilling over into Sweden, and the Arctic and North Atlantic were considered worthy of investigation. One traveller in the Scandinavian region was the British scholar Edward Daniel Clarke, who was particularly interested in mineralogy and chemistry, and who arrived in Hamburg accompanied by several colleagues from Cambridge University in June of 1799. Clarke's party included the famous British economist Thomas Malthus, who had just published his controversial demographic work *Essay on the Principles of Population.*[197] Clarke's tour of the Nordic region was only one minor leg of a much longer journey, which he later described in six substantial volumes appearing between 1810 and 1823. Unimpressed with the level of the scientific and cultural activities he observed in the Danish capital, Clarke wrote:

> To our eyes, it seemed, indeed, that a journey from *London* to *Copenhagen* might exhibit the retrocession of a century; every thing being found, in the latter city, as it existed in the former a hundred years before. … In literature, neither zeal nor industry is wanted: but compared with the rest of *Europe*, the *Danes* are always behind in the progress of science. This is the case, also, with respect to the Fine Arts.[198]

Clarke met several of the Copenhagen professors, including the naturalist Gregers Wad, but only in the quality of the zoological and mineralogical collections did he find any reason for a more positive assessment of the city's scientific state. And when the prominent German geologist Leopold von Buch visited Copenhagen seven years later, he too toured the city with Wad as his guide, likewise praising the mineralogical collections. In truth, Buch was only passing through Denmark en route to his actual destination, Norway, where he would spend most of 1806–1808 on detailed studies of the country's geography and geology. After returning to Berlin, he completed his great work *Reise durch Norwegen und Lappland*, which was published in 1810 and gave an account of his observations and scientific findings.

The British chemist James Johnston visited Copenhagen in the summer of 1829, describing in detail the city's academic and scientific life in an article run in *The Edinburgh Journal of Science* the following year. Johnston pointed out that the city was home to several excellent scientists, but that the students and the population at large showed little interest in science. The Polytechnical College, the city's college of advanced technology, was newly opened, and Johnston gave an account of its arrangement, construction, curriculum and other details. As for the university, he was quite critical of the instruction it provided, and of the buildings in which this took place: "They have nothing in their exterior … to awaken any of those undefined, yet peculiar feelings with which we pace the halls and courts of those seats of grave and venerable learning founded by the liberality of our forefathers". He also commented on the general shape of Danish science, observing that

> The state of natural science in Copenhagen is not very high, though the University reckons several celebrated men among its scientific professors. Chemistry, botany and zoology, are the favourite studies. Even these, however, can boast of comparatively few cultivators. … There are few who cultivate science for its own sake, few amateurs, those who do give themselves up to it having generally a view to the professions. Among a population of less than two millions, by no means rich, many independent cultivators of science are hardly to be looked for, and yet, even here, they contrive always to have a very respectable amount of talent and industrious perseverance collected together in the professional body.[199]

Exotic expeditions

In Denmark, the mid-1800s were the golden age of scientific exploration, although previous centuries had also seen a couple of noteworthy journeys that included at least some scientific research. In 1737 a royal expedition was outfitted and departed from Copenhagen, its original aim to forge trade connections with the Emperor of Ethiopia.[200] Heading the expedition was Frederik Ludvig Norden, a naval officer and graduate of Naval Cadet Academy. Norden had been thoroughly educated in the naval sciences in the Netherlands, France and Italy, where he also developed an interest in Antique art and attained membership in the academy in Florence.

Norden was the obvious choice as leader of King Frederik IV's 1737–1738 expedition to the African continent. While travelling through Turkish-controlled Egypt, he drew the first reliable maps of the lower Nile, meticulously noting down and drawing everything he saw, including the great pyramids and the ruins of Thebes and Aswan. Upon Norden's return – without having reached the fabled kingdom of Ethiopia – the king ordered him to write an account of his travels. The results garnered attention from scientific circles across Europe, not least in Britain, where Norden had published a sample leaflet in 1741 with drawings from the intended work. In 1742, Norden was elected a foreign fellow of the Royal Society, but died that very year, a mere 33 years old. The arduous task of transforming Norden's notes and drawings into an impressive printed publication was handed over to the newly formed Royal Danish Academy of Sciences and Letters. In 1750 the Royal Academy could present the first volume of the work, entitled *Voyage de l'Egypte et de Nubie*, completing the second volume in 1755. An English translation appeared two years later, and a German translation in 1779. Note that unlike the Royal Academy's other publications, *Voyage* was a splendidly produced work that was internationally oriented and written in the major scientific languages of the period, serving Denmark as a worthy monument to the Crown's interest in scholarship and exploration.

The second and rather more famous Danish expedition to *Arabia felix* - the blessed land of Arabia, or Yemen – was also based on romantic, hazy ideas of the world beyond Denmark's borders, although in this case the enterprise was not motivated by commercial interests. A well-known professor of his day, Johann David Michaëlis of Gottingen, suggested in 1756 that mounting an expedition to western Arabia would result in a better understanding of the original Hebrew language, thereby enabling a truer interpretation of the messages in the Bible. Michaëlis was able to convince the Danish foreign minister,

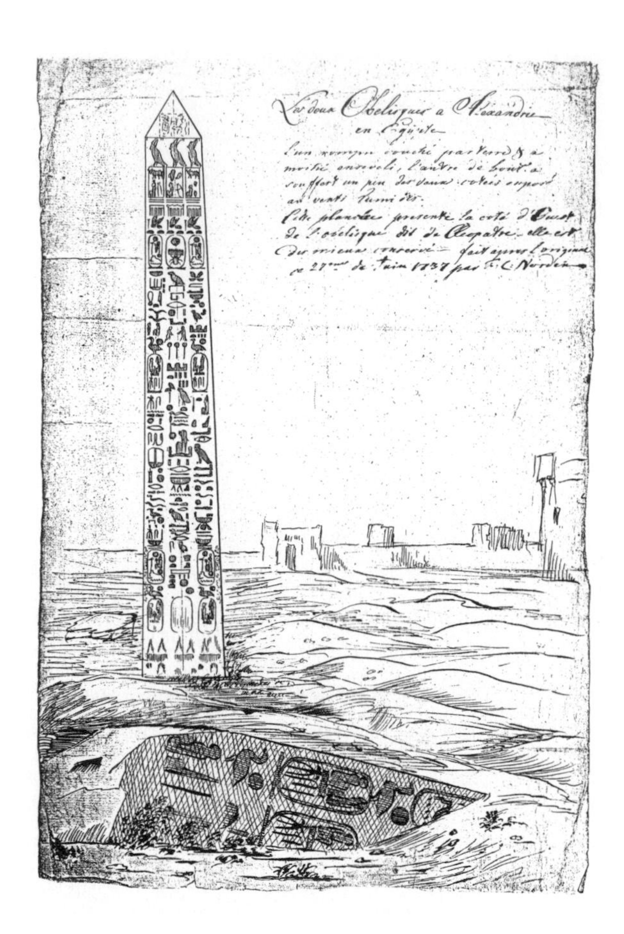

12.1 *Drawing from Alexandria by Frederik Ludvig Norden, 1737. Norden, a naval officer, and his companions explored the lower reaches of the Nile in 1737–1738. Contemporary readers were quite taken with his descriptions of this cradle of ancient culture, especially as presented in the pair of magnificent volumes that made up Norden's* Voyage de l'Egypte et de Nubie *(1750, 1755).*

J. H. E. Bernstorff, of the plan's value and viability, and when King Frederik V agreed to finance the expedition the stage was set. It was agreed that beyond seeking enlightenment on philological and Biblical issues, the expedition would also use the opportunity to gather samples and learn more about the region's natural history. The elaborate royal instructions issued in 1760 ordered that the participants in the expedition must "ever bear in mind this principal objective, presented by Our good grace, that they must make as many discoveries for science as they prove able to make".[201] Furthermore, it was of the utmost importance to the king and his advisers that the journey be publicized as a *Danish* contribution to international science – the expedition being meant to reinforce the king's own reputation as a patron of the sciences.

The task of planning the expedition's scientific content largely fell to Christian G. Kratzenstein, who has already been mentioned in previous chapters. He discussed in detail the various methods of determining longitude, and

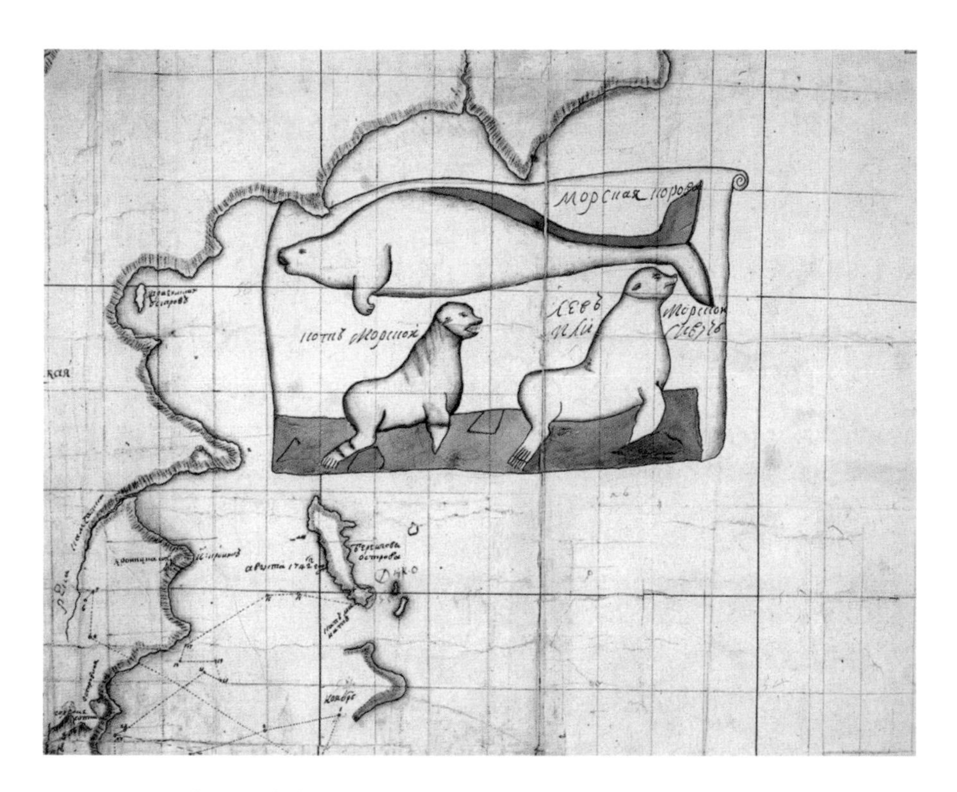

12.2 *Segment of a map of Siberia's eastern coastline. The boxed-in illustration refers to the animals (sea cow, sea bear and sea lion) that Bering and his men discovered in the Arctic Ocean off Siberia. Although Bering's expeditions to Kamchatka and eastern Siberia mainly had geographical and political aims, they ended up collecting large amounts of scientific information about this little-known region.*

also (among other things) suggested systematic marine-biological studies, even going so far as to have a special net made for collecting samples of small marine life. In early 1761 the warship *Grönland* left Copenhagen and sailed to Smyrna (modern-day Izmir), where the six expedition members boarded a Turkish ship bound for Alexandria by way of Constantinople (Istanbul).

Two of the expedition's six members had been specifically selected to handle its scientific mission. One was the German-born Carsten Niebuhr, who was studying as a surveyor in Göttingen when he was encouraged to serve as the expedition's astronomer and geodesist. Having accepted the proposition, he spent the interim studying astronomy in Göttingen and Copenhagen with a special view to longitude determination.

The other scientific member of the expedition was one of Linnaeus's most promising students, a Finnish-born Swede named Per Forsskål, who in addition to being a brilliant naturalist was also knowledgeable in Oriental philology – the latter indeed being his actual profession, which he had studied under Michaëlis in Göttingen. Possessing this dual qualification as Biblical philologist and naturalist, it was only logical that Forsskål be charged with "devoting particularly scrupulous attention to the *naturalia* that appear in the Bible, and which in part remain unknown".[202] The great importance the Danish state attributed to its Arabian expedition is reflected in Niebuhr's appointment as lieutenant engineer at the age of 27, and Forsskål's appointment as professor at the age of 28. The Danes often refer to the entire undertaking as "Niebuhr's expedition", but it actually had no leader, the royal instructions being that all five scholars (the sixth member being a servant) were to share equal status.

Sidestepping a detailed account of the dramatic events and tragedies that befell the ill-starred expedition,[203] suffice it to say that two participants died in Yemen, and Niebuhr returned alone to Copenhagen in the autumn of 1767. Despite his harrowing ordeal, Niebuhr was able to bring back a number of valuable maps, drawings and manuscripts and, not least importantly, to preserve the better part of Forsskål's collections of fish and plant specimens. There has been considerable controversy over the expedition's scientific results, but if not in all areas, then at least in some its findings were significant and were perceived as such by contemporary scholars and scientists. This was mainly due to the merits of Forsskål's work. Not only had he assembled a herbarium with a wide variety of plants previously unknown in Europe. He had also gathered valuable zoological specimens and observations, and was perhaps the first to account for the migration of birds and to describe a large number of fish, corals and conchs.

The main fruits of Niebuhr's labour were his maps over areas stretching from Suez to Mocha, his copies of cuneiform inscriptions from the ancient city of Persepolis, and more generally his rich and accurate descriptions of ethnographic and Orientalistic features he observed. Upon his return he was extolled as one of the great heroes of science and immediately set to work on the account of his travels he had pledged to write. It initially appeared in 1772 as *Beschreibung von Arabien*, and the German version was soon translated into Dutch and French. A few years later, in 1774–1778, Niebuhr published his comprehensive *Reisebeschreibung nach Arabien und andern umliegenden Ländern* – now also available in Danish, with a delay of some 230 years.[204] With the aid of botanists, Niebuhr additionally oversaw the posthumous publication of Forsskål's natural-history works, which appeared in Latin in 1775–1776 as *Descriptiones animalium*; *Flora Ægyptiaco-Arabica*; *Icones rerum naturalium.*

Back around the time Norden was travelling to Egypt, another expedition with a prominent Danish participant was setting off for a very different and much colder part of the globe. That Dane was Vitus Bering, famous for his pioneering expeditions in the waters between north-eastern Siberia and Alaska in what is now known as the Bering Sea – though in all fairness, his exploits are hardly part of the Danish history of science. Vitus Bering was certainly born in Horsens, a market town on the east coast of Jutland, but after entering Russian military service at the tender age of 22 he remained a subject of the tsar until the day he died.[205] The aims of the elaborate and immensely expensive expeditions Bering led to Kamchatka in 1725–1730 and 1733–1741 were almost exclusively geographical, commercial and political. While they greatly expanded the geographical world, they made only modest contributions to science, although at an analytical level it can be difficult to distinguish between politics and geographical discoveries.

Besides Bering, its leader, the Second Kamchatka Expedition also counted the French astronomer Louis Delisle de la Croyère and the German naturalist Johann Georg Gmelin, both of whom were also in Russian service, and travelling with their own missions that were independent of Bering's. Croyère was mainly concerned with determining longitudes, while one of Gmelin's primary tasks was gathering plant specimens, his efforts ultimately resulting in the *Flora Sibirica* (1747–1769). Another German scientists, Georg Wilhelm Steller, took part in the final stage of the ill-fated expedition, which he – unlike Bering – was fortunate enough to survive. Steller wrote of many things, including the large, eponymously named sea cow that is unknown today, having been hunted into extinction just a few decades after its first sighting.[206] Bering's work was monumental and led, among other things, to Alaska's incorporation into the Russian empire, but his Danish ancestry is, after all, a coincidence and in no way makes his contributions Danish. Small wonder that Russians consider Bering a Russian, and that almost all of the literature on the great explorer is written in Russian.

Greenland and Iceland

There was little, if any, interaction between the kingdom's sparsely populated possessions – the Faeroe Islands, Iceland and Greenland – and the Danish science community in Denmark proper, though the far-flung lands were rich in resources and served as objects of study. There was a certain interest in exploring the majestic landscapes and nature of these islands, and if possible in exploiting the raw materials they held, but hardly any independent research was done in the more remote reaches of the kingdom. In 1750s, the Royal

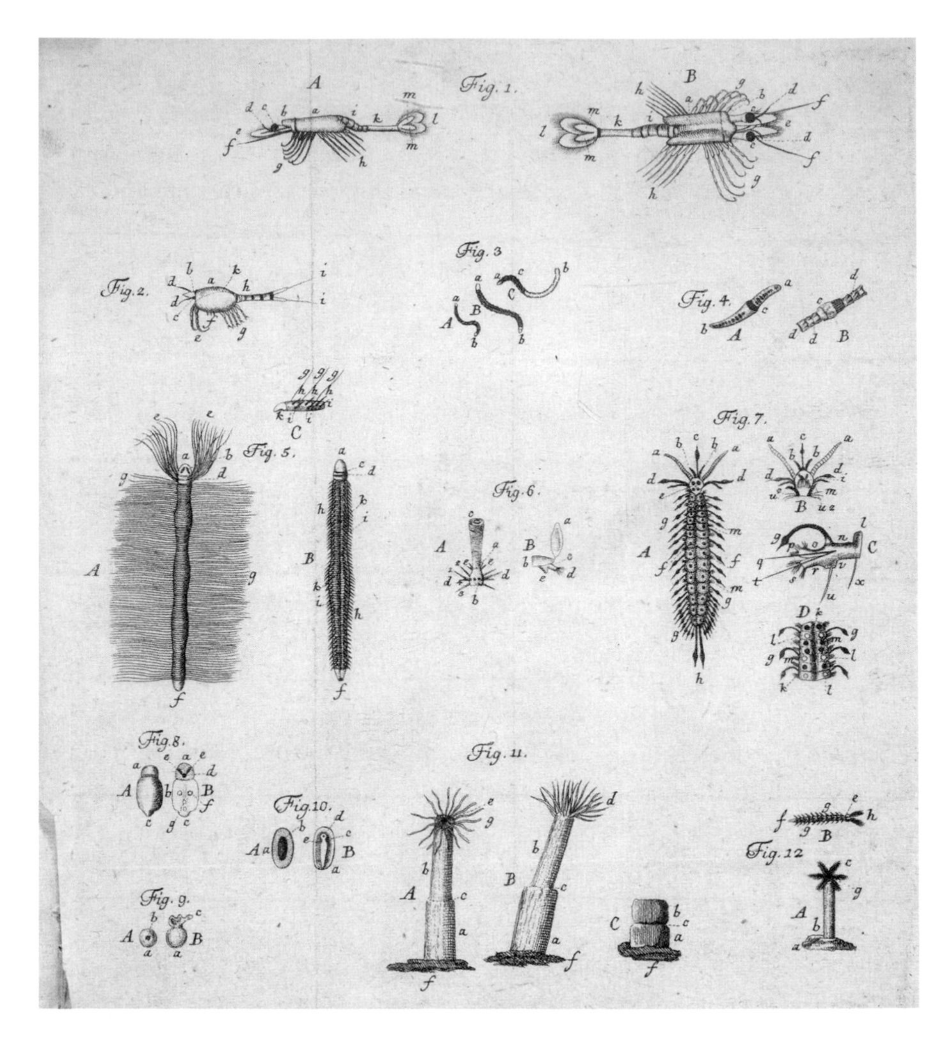

12.3 *Plate from Otto Fabricius's* Fauna Groenlandica. *Fabricius, a Lutheran minister and a naturalist, pioneered the study of Greenland's natural environment and its animals. Many of the species described in* Fauna *were new to European science, which had not yet taken an interest in the island's natural history. Besides the flora and fauna, Fabricius also wrote about the daily life, customs and culture of the native Greenlanders.*

Academy in Copenhagen sent two Icelandic students on a scientific journey to the island of their birth. One was Eggert Ólafsson, who later completed an account of their journey (published in 1772), which included descriptions of native Icelandic vegetation.[207] While it was true that in geographical terms Iceland clung to the very fringes of Europe, the island still had an active scholarly community interested in the same sort of things as the Danes were – natural history, topography and economics. A good example was the printer Olaus Olavius, who published economics studies and natural-history bulletins in the transactions of the Icelandic literature society. In 1772 the eminent British botanist Joseph Banks – one of the few foreigners of that era to do so – visited Iceland on a scientific voyage that would also take him to the Hebrides and the Orkney Islands.

Huge stretches of Greenland's coastline were still uncharted in the late 1700s, a fact that did not seem to cause much concern in Copenhagen. One of the few Danish expeditions to the high North Atlantic set out in 1786, after the naval officer Poul Løvenørn had outfitted two ships with the support of the public benefit fund known as Fonden ad usus publicos. Part of his mission was to investigate Greenland's east coast, but his primary instructions directed him elsewhere – this being an "expedition to Iceland to make geographical and astronomical observations and to seek opportunity to discover ancient Greenland or [the lost medieval Nordic settlement of] Østerbygd".[208] Although drift ice prevented the ships from sailing close enough to Greenland's coast to do any surveying, the expedition did clarify that Frobisher Strait was not located in Greenland, as many contemporary maps indicated, but off North America. This "strait", named after the British explorer Martin Frobisher, actually turned out to be a bay in the southern part of Baffin Island, and is located almost 1,000 km away from Greenland. Generally speaking, the scientific exploration in Greenland during the 1700s was very modest, and quite random, and understandably enough the scientific activities taking place in Greenland itself were even more limited. In keeping with the spirit of Enlightenment prevalent at the time, a few of the missionaries attempted to instruct the native Greenlanders on matters other than the salvation of their souls, and tried to acquaint them with science. But these few and often feeble attempts made no impression whatsoever and were regarded with suspicion by most Danes – and presumably with incredulity by the people of Greenland. As one inspector with the Royal Greenland Trade Department put it in 1793: "What to do the Greenlanders care for the routes of the stars, or whether one or the other is called Venus, Mercury, Jupiter or something else."[209]

Undoubtedly, the most valuable scientific contribution on Greenland in the 1700s came from Otto Fabricius, one of the great Danish naturalists of his age, but also a well-known figure in Danish church and mission history.[210] His main work, the internationally recognized *Fauna Groenlandica* from 1780, was based on observations and collections gathered in Greenland, and it earned him a fellowship with the Royal Academy in Copenhagen. In this work

Fabricius described 473 animal species, ranging from baleen whales to marine sponges, many of them new to European science. He described not only their appearance, but also biological conditions such as their distribution, food, reproduction and habits, as well as the uses to which the Greenlanders put them. *Fauna Groenlandica* first dealt with mammals, which Fabricius arranged in the following remarkable order: 1. Greenlander (*Homo groenlandus*), 2. Walrus, 3. Sea cow, 4. Sea bear, 5. Hooded seal, and so on. Describing the first of his mammals, Fabricius wrote that Greenland man was

> A human being that is active during the day, dusky red in colour with black, straight, thick head hair, sparse of beard ... They have wide nostrils, small eyes and large lips, a large mouth, strong forehead, broad shoulders, short feet, a choleric-phlegmatic temperament, and are of slight build, quick, stubborn, outspoken, content, fearsome, superstitious, and of little fecundity.[211]

Fabricius continued his scientific activities after returning to serve as a minister in Denmark. He published a treatise on Greenland's drift ice in 1788, and a few years later described Greenland's seals in a work that long remained the most comprehensive and precise treatment of that topic.

One of the few Danes to carry out actual scientific investigations of Greenland was the widely travelled botanist Morten Wormskiold, who in 1812–1814 went on an expedition to the great island. Wormskiold's journey brought him into close contact with the prominent British mineralogist and naturalist Robert Jameson, and he ended up with a sizeable collection of indigenous fauna.[212] On a later voyage with a Russian ship that circumnavigated the globe, Wormskiold returned to Denmark with an extensive collection of plant specimens gathered during his 1816–1818 stay in Kamchatka, and elsewhere – subsequently leaving the field of botany for other pursuits.

Commercial vessels returning from Greenland would often bring back stones and minerals thought to be of potential interest to museums and collectors. In the late eighteenth century the Danish scientist H. C. F. Schumacher examined a whitish stone, which he described in 1795 for the Natural-History Society as a hitherto unknown mineral. Following a qualitative chemical analysis performed by Peter C. Abildgaard in 1800, the mineral was dubbed cryolite – "cold stone" or "ice stone" – and its composition immediately sparked an interest among chemists and mineralogists all over Europe. Martin Klaproth in Berlin and Louis Nicholas Vauquelin in Paris found the new mineral to be composed of soda, alum earth and hydrofluoric acid, but it was not until 1823 that the great Swedish chemist Jöns Jacob Berzelius determined the mineral's stoichiometric composition, expressed in the formula Na_3AlF_6. In the early 1800s cryolite was still regarded as an extremely rare mineral that could, at best, be of interest to museums and the world of science.[213]

Only in the 1820s did scientists realize that cryolite was found in a very limited geographical area around the Arsuk Fjord, near the mining town of Ivittuut, where, on the other hand, it was abundantly present. This realization originated with the German collector and mineralogist Karl Ludwig Giesecke, one of the more colourful characters in the Danish history of science, and also a man of many talents.[214] Giesecke already had experience hunting for mineral deposits in Iceland and Norway when he was sent to Greenland in 1806, where he was obliged to remain for more than seven years because of ongoing hostilities between Denmark-Norway and Britain. He took advantage of his unintentional stay to work on extensive mineralogical, natural and ethnographic studies of the country, most of which remained unpublished for many years after his death.[215] Later, in 1822, as a professor of mineralogy in Dublin (under the name and title of Sir Charles Lewis Giesecke) he described the rich cryolite fields of the Arsuk Fjord. Initially this did not change the mineral's status, and it was still considered a rarity worthy of little more than a place on a museum shelf. During the 1850s it finally became clear that cryolite could serve as a useful technical-chemistry resource in the production of soda.

As for Iceland, the government in Copenhagen had no significant political or economic interest in the country, which, for all intents and purposes, lay outside Danish control in the years around 1810. Many Danish intellectuals in the Romantic tradition nevertheless continued to find Iceland intriguing because of its unique history and perhaps more than anything because of its language. Urged and aided by the Danish linguist Rasmus Rask, Iceland founded its own literary society in 1816, Hiðíslanska bókmenntafélag, which had a branch in Reykjavik and one in Copenhagen. This society was involved in one way or another with most of the island's intellectual life, and published works on Icelandic history and literature, as well as descriptions of nature. The learned school at Bessastaðir near Reykjavik was an important seat of culture and learning. The teaching staff included the mathematically gifted Björn Gunnlaugsson, who had handled geodetic assignments for H. C. Schumacher in Altona before joining the school as a teacher in 1822. Gunnlaugsson wrote a textbook in arithmetic, and at the request of the Icelandic literary society he also completed a huge project surveying the island. The result was an improved map of Iceland, which was published with the support of the Royal Academy in Denmark and appeared in the late 1840s, along with a description of the island's natural environment.

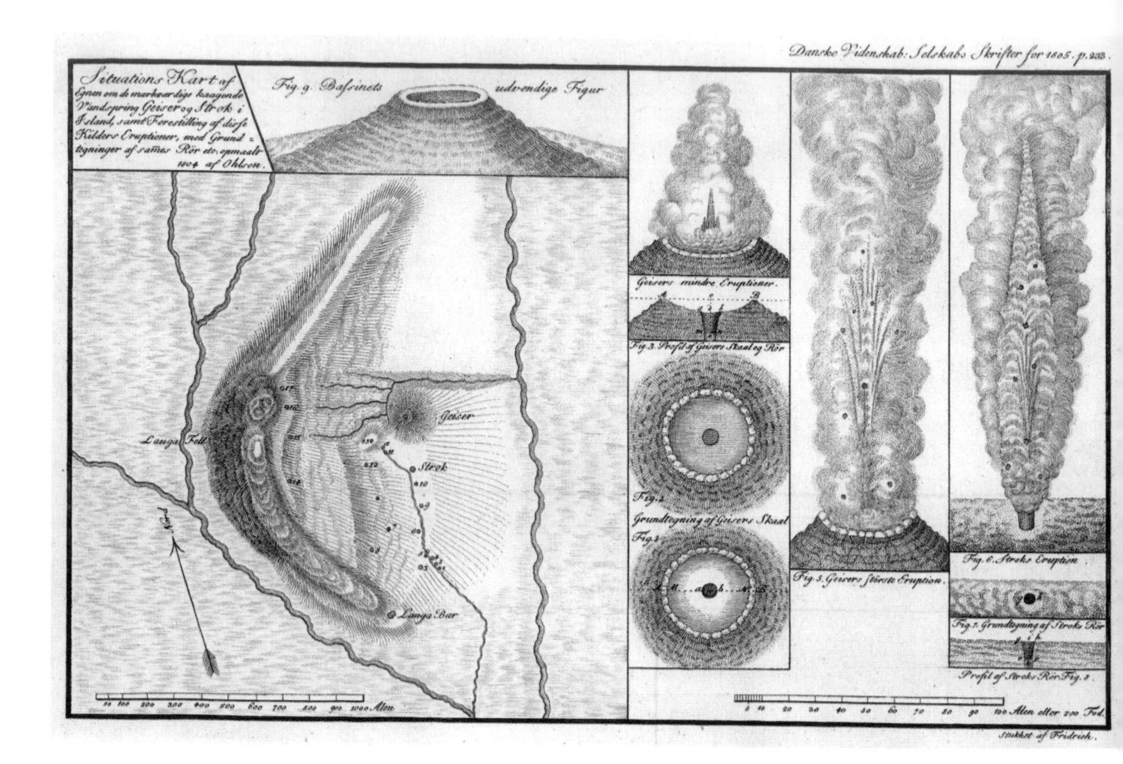

12.4 *Eruptions of the Great Geysir and Strokkur, as reproduced in lieutenant Ole Ohlsen's article from 1805. During the early to mid-1800s a number of scientists travelled to Iceland to investigate the geysers and study the country's geology in general. One foreign scientist drawn to the geysers was the eminent German chemist Robert Bunsen, who spent a part of 1846 on the island.*

One of the pupils at Bessastaðir learned school was Jónas Hallgrímsson, who after studying in Copenhagen returned to Iceland on scientific journeys in 1837 and 1839–1842. He collected copious amounts of material, intending to compose a comprehensive natural description of Iceland after he returned to Denmark; an intention he was unable to carry out. But Hallgrímsson was a poet, a man of letters, and a natural scientist in the Romantic tradition, and his evocative descriptions of the nature and history of his native Iceland played a valuable role in the country's budding patriotism. In 1835, Hallgrímsson was active in publishing the influential journal *Fjölnir*, to which he also contributed articles on fossilized fish, the Earth's evolutionary history and other scientific topics. He also made contributions to Denmark's scientific literature, including an article in *Naturhistorisk Tidsskrift* (*Journal of Natural History*) about Iceland's geysers. Hallgrímsson was very good friends with the zoologist

Japetus Steenstrup, who had made a scientific journey to Iceland in 1839–1840, in the company of the polytechnic graduate and geologist Jørgen Christian Schythe. So close was their friendship that Steenstrup even invited Hallgrímsson to come and live in his home at Sorø.[216]

Scientists were mainly drawn to Iceland because of its distinctive natural settings, which lured many geologists and other naturalists to the island. One visitor was the Danish lieutenant Ole Ohlsen, who in 1840 while surveying Iceland's southern coast took the opportunity to examine the great Geysir itself, and its brother Strokkur. In a prize-winning paper submitted to the Royal Academy, Ohlsen described the eruptions, detailing how they could reach up to 212 feet, and how the water at the base of the water column was nearly 100°C. One of the foreign scientists to visit Iceland in the first half of the nineteenth century was Scottish mineralogist George Mackenzie. He arrived in 1810 and after returning to Scotland published his *Travels in the Island of Iceland*, an account of his findings on mineralogy and a wide variety of other topics. Another visitor from Britain was the reverend Ebenezer Henderson, who in 1814–1815 spent more than a year on the island and wrote the comprehensive two-volume work *Iceland*, which appeared in 1818.[217] Some years later the German geologist Otto Krug von Nidda went on an expedition to Iceland, which in 1834 resulted in an account of its complex geology. Later still, a French mineralogist and explorer by the name of Eugène Robert also came to investigate the island's stones and minerals, an experience he described in a book from 1840 entitled *Voyage en Islande*.

The circumnavigation on the *Galathea**

The years 1845 to 1847 witnessed the achievement of the most ambitious expedition Denmark had yet mounted. Its aim was partly scientific, and posterity would refer to it as "the first Galathea expedition" – with the second taking place in 1950–1952 and the third in 2006–2007.[218] This prestigious and extremely expensive project was financed by the Crown, and Christian VIII, quite the naturalist himself, followed its progress with keen interest. The king decided that the Royal Academy would handle the scientific aspects of the expedition and appoint the scientists who were to undertake the long voyage,

---

* This section is based on Michael Sterll's contribution on the Galathea Expedition included in *Fra Middelalderlærdom til Den Nye Videnskab – Dansk Naturvidenskabs Historie*, vol. 2.

in addition to which the king himself would appoint the head zoologist. His choice fell on Wilhelm Behn, a professor at the university in Kiel, who was to be accompanied by the two Danish zoologists Carl Emil Kiellerup and Johannes Theodor Reinhardt. The expedition's botanist was Bernhard Kamphövener, who was forced to leave the expedition because of tuberculosis, and whose duties were taken over by Ferdinand Didrichsen, who was actually one of the expedition's physicians.

Looking back today, choosing the corvette *Galathea* for an expedition on the seven seas may seem a rather odd decision. In those days a corvette was a three-masted, fully rigged warship; effectively a miniature frigate with an open upper gun deck. The *Galathea* was outfitted with no less than 36 cannons, which sounds like an incredibly heavy armament for a ship on a scientific mission. But given the precarious political and social situation in many of the places the expedition planned to visit, the vessel's impressive firepower was as much for business as it was for show. The ship's greatest advantage was her short keel, enabling her to penetrate deep into shallow estuaries, and prior to departure her rigging was altered to facilitate navigation in narrow waters. The greatest disadvantage was that the *Galathea* was by no means equipped with the sort of technical gear associated with modern marine exploration and research, nor even with the type of fishing gear that one of the most advanced contemporary fishing vessels could have offered.

In the 1840s it was still assumed that life could not exist at great depths in the ocean, so deep-sea exploration was not a part of the Galathea expedition's mission. Another scientific task the expedition neglected, seen from a modern point of view, was investigating the plankton present in pelagic waters. Firstly, scientists had not yet understood that such a proliferation of microscopic life existed, and secondly they had no idea that the ocean's animal life was entirely dependent on its plant life. Around that time science was just beginning to understand this complex interaction thanks, not least, to Anders Sandøe Ørsted. He was a botanist, and a nephew of the physicist H. C. Ørsted, and in 1849 he provided evidence of the role that microscopic plants – phytoplankton – play in the ocean's nutritional cycle. His idea was not readily accepted, however, and most of the scientific community brushed A. S. Ørsted's crucial research aside as mere natural-philosophical speculation.

The Galathea expedition's route largely followed the well-known trade routes, departing from Copenhagen and sailing to the Atlantic, then north around the Shetland Islands. Then came the great Atlantic passage and the southbound journey hugging the coast of Africa, around the Cape of Good Hope and eastwards to the Bengal coast. Here they had two principal tasks:

12.5 *Golden-Age master C. W. Eckersberg's painting of the* Galathea *from 1839, six years before the ship set off on its voyage around the world. The expedition, partly scientific, would later be known as "the first Galathea expedition" – a designation that was obviously not used at the time.*

transferring a number of minor Danish possessions to Britain, more precisely to the East India Company, and investigating the Nicobar Islands with a view to long-term colonization. The Danish trading posts in India had been costing the Danish state money for years, and the Nicobar Islands produced nothing of commercial value. The Galathea expedition remained in the area for a few months thoroughly studying and surveying the islands, and a group led by the Danish geologist Hinrich Johannes Rink stayed there to establish a permanent port and settlement. In 1847, Rink was able to publish *Die Nikobarischen Inseln*, which included a geological description of the Nicobar Islands. Meanwhile, the attempt to settle the islands proved fruitless, and a few years later the small band of would-be colonists were brought back home.

Passing through the Chinese archipelago, the *Galathea* managed the long Pacific crossing and reached Valpareiso in Chile, sailing from there north along the coast to Bolivia and then on to the south, rounding Cape Horn. At this point the expedition collected a valuable array of marine-life samples from the Antarctic region. After visiting Montevideo and Buenos Aires, the ship finally arrived in Rio de Janeiro. Here the Danish scientists met a German-born botanist named Luiz Riedel, who had previously travelled through much of Brazil in the company of the Danish zoologist and palaeontologist Peter Wilhelm Lund.[219] Riedel had been appointed director of the emperor's botanical gardens and his vast collection of natural specimens, both of which he showed to the Danish scientists. Here, for the first time, they saw actual proof of South America's extinct megafauna, and they left the gardens weighed down with specimens of their own.

The *Galathea* returned to Copenhagen on 28 August 1847 and received a welcome befitting its achievements – but that was more or less the end of it. The king, who had made sure in advance that the expedition's results would be published in monumental style, died just a few months later. And because the Three Years' War, beginning in 1848 and lasting until 1850, had so heavily drained the country's resources, Denmark had neither the means nor the will for any sort of monumental publication. As a consequence, a substantial part of the material from the first Galathea expedition was used only sporadically, and even today much of it still has not been treated or published. Scientifically the expedition was, in many respects, a great success, but the follow-up was nothing short of scandalous, the bulk of the material remaining pigeon-holed – untouched and unheeded. To fully appreciate the bizarre nature of this situation, one must also consider its financial background: The total cost of the expedition on the *Galathea* was a mind-boggling 462,000 rixdollars, making it by far the most expensive scientific enterprise the Danish state had ever underwritten.

PART III

Light over the land

1850–1920

Peter C. Kjærgaard

Science and democracy 13

From 1850 to 1920 the world known to man was growing, and with it the world of science. Scientists scrutinized the manifold facets of nature, from the smallest building blocks to the most distant nebulas, and explorers mapped the Earth's last uncharted regions, from the humid interior of Africa to the icy vastness of the Arctic. New disoveries revealed secrets about the birth of our planet and the beginnings of life hundreds of millions of years ago, and about the ultimate, inevitable demise of our solar system – even while more and more people were enjoying modern conveniences like trains and telegraphs, indoor plumbing and electric lighting. The world was constantly changing, with science and new technology profoundly affecting the daily lives of the Danes. There were some who paused, standing back to nostalgically lament the rapid pace of change, but most people enthusiastically embraced the new. The nineteenth century was proclaimed "the century of science", and the trend continued with renewed vigour after the turn of the century. This third part of our historical narrative outlines the impact of these developments.

The age of science

"Our century is the age of science", the chemistry professor Julius Thomsen exclaimed in 1884 during the annual banquet at the University of Copenhagen. Although the banquet that year was to commemorate the reformation of the Protestant church, Thomsen had chosen to talk "On Molecules and Atoms". Its tone was almost triumphant. Here was the voice of science, speaking from society's and culture's most prominent podium previously monopolized by the Chuch. He did not mention the Reformation once, not even in a respectful tribute to the past. His was an effervescent speech, brimming with energy and confidence in the present, a day and age so full of potential that it almost took one's breath away. Firmly convinced of mankind's fair future, Thomsen pointed to the wondrous scientific achievements already witnessed in the nineteenth century.[220]

The science he spoke of was reshaping everything from household chores

13.1 *The many new discoveries and inventions fostered great expectations to science, and to the future, and they let the imagination soar. Here, a dramatic situation from Vilhelm Bergsøe's futuristic tale of a steam-propelled flying machine –* "Prometheus *the Flying Fish" – which ran as a serial in the popular weekly magazine* Illustreret Tidende *in 1870. Like several other contemporary novelists, Bergsøe had an academic background in science.*

to outer space, and studying everything from the ingredients in soap to the chemical composition of the stars. It was a world of science which, in Thomsen's own words, bordered on the incredible – even on the supernatural – while at the same time it was becoming such an integral part of daily life that few people gave it a second thought. The ongoing scientific investigation of nature held an amazing potential, he said, and yet there was an inherent risk in the development of science the world had witnessed during the 1800s: There had been so many discoveries, and their influence upon society and daily life had been so overwhelming and so revolutionizing that people had gradually become accustomed to the unexpected, being now so used to incredible things that they had begun to regard them with a certain indifference. The sheer number of surprises had a dulling effect upon the perception of their significance, Thomsen believed, and accordingly scientific discoveries and inventions no longer invoked the admiration and recognition they had previously enjoyed. In the midst of all Thomsen's enthusiasm lurked the anxiety of what would happen if attention turned into indifference.

But the nineteenth century was not the right century to worry about that. The world of science was getting attention like never before, while its significans to people's everyday lives increased. Science had entered the private sphere to a hitherto unseen extent. Scientific discoveries and scientists were all over the newspapers, reaching thousands of readers in the capital and across the country. Popular science found a broad audience in weekly magazines, periodicals and journals, in book series and public lectures. The general public enjoyed close encounters with science at exhibitions and museums, and welcomed their scientific heroes back from long, dangerous expeditions. Contrary to the scenario painted by Thomsen, the fact that people were beginning to get used to scientific progress neither diminished its significance nor reduced its pace.

Denmark had become a democracy, to some extent representative, and there was seemingly no end to the advantages that could be anticipated from science and technology. During the period 1850–1920 a general perception of science and research arose that is familiar to modern Danes. Science became more professional as it specialized and solidified into many of the disciplines still found at universities today, embedded within a structure we have become so accustomed to that it seems perfectly natural. The academic world changed considerably over those seven decades, and scientists began to behave largely as they have since then, operating in a competitive international scientific environment where only the best was good enough.

During this period there were also clear expectations to the practitioners of science, and to science itself. Addressing the founding assembly of Denmark's Society for the Natural Sciences in 1911, the chemist Martin Knudsen proclaimed that scientists were first and foremost interested in pursuing science for its own sake and in the same breath defined scientific progress as increased knowledge and mastery of nature.[221] Such progress opened a world of possibilities for non-scientists as well, and it certainly did not take long to realize and exploit the potential of the new discoveries and inventions that followed in its wake. Agriculture and industry were particularly alert. Initial efforts were driven by the desire to make agricultural practices more rational and scientific, ensuring greater yields and more uniform produce, which in turn would increase Danish export earnings. Later in the period, however, industrial players also grew more proactive, until it became increasingly clear, in the early twentieth century, that the industrial exploitation and processing of Denmark's agricultural resources, which built upon a scientific and technological foundation, was the best way forward. From 1850 to 1920 Denmark witnessed its own transition from a traditional agricultural society into a modern industrial nation, undergoing a process where science played a pivotal role.

The philanthropist and founder of the Carlsberg breweries, Jacob Christian Jacobsen, invested enormous personal commitment and funds in scientific research, founding the Carlsberg Laboratory 1875. Meanwhile, professional scientists did not shy away from exploiting the potential that new scientific breakthroughs offered – not working *solely* for the sake of science itself, but for their own benefit as well. One of these enterprising scientists was the same Julius Thomsen who had waxed poetic about the "century of science". He made the most of his chemistry expertise to establish the factory "Øresund", where Greenland's natural deposits of cryolite were used to manufacture soda applying a method Thomsen himself had devised and patented.

New career opportunities arose for university graduates holding a degree in the natural sciences and it was easier than ever before to make money on inventions and scientific methods. Following the French Revolution in 1789, the "right of invention" was introduced into the international human rights standards. It was thought to serve as a continuation of the right to own property and the inalienable rights of the individual, and during the nineteenth century patent regulations were introduced in most western countries. Even though patent handling was not given a legal framework in Denmark until 1894, a prior system had been in place for granting monopolies or "sole rights" that were valid for five years at a time. Such grants were frequently sought. For the most part, the new professional class of scientists and engineers had no qualms about profiting from their inventions.

Denmark may have been a small country, but it had great ambitions. Its scientific research had to be comparable in quality to the research being done abroad. The most important aim of science may have been to serve mankind, promoting public health and improving or embellishing people's lives – which, at the time, was often declared to be the calling of any true scientist – but at the same time there was definitely an awareness that science could also serve other purposes, bringing glory to the nation while generating a substantial financial profit for its supporters.

Science was in a strong position. Confidence was running high, as were the scale and scope of the country's scientific ambitions. Science and technology merged in the minds of their practitioners during this period, creating a rather different picture of their mutual relationship, evolving during the twentiety century.[222] Yet precisely these elements of interrelatedness, optimism and indomitable vitality are important in understanding society's perception of science, and the scientific community's perception of itself. Both metaphorically and quite literally, it was full steam ahead.

Making a small country larger

During the nineteenth century Denmark became smaller in one very concrete sense and larger in another. Geographically the country had shrunk greatly. As a result of the long-simmering rebellion in the border regions south of Jutland, an actual border war erupted. Denmark suffered a devastating defeat to Prussia in 1864, losing the duchies of Schleswig, Holstein and Lauenburg – as well as its military confidence. This also meant the loss of the university in Kiel, and of what had actually been the country's most technologically advanced and most progressive province. Denmark still had its overseas dependencies: the Faeroe Islands, Greenland, Iceland and the Danish West Indies (today the three largest of the US Virgin Islands). These territories enabled the country to maintain its geographical confidence as a former global power, and some of their unique geological and climatic conditions had a distinct impact on the development of certain sciences in Denmark. Even so, they were never able to fill the position of developmental dynamo that the lost north-German duchies had left vacant in 1864.

There was no prospect of Denmark ever recouping its external losses. If the nation should grow anew, it would have to do so within the borders of the existing realm. To that end the Danish Heath Society was set up as a commercial foundation in 1866. It was headed by the energetic Enrico Dalgas under the motto "What is outward lost must be inward won". Dalgas was a former military officer and veteran of 1864 and a highly experienced road construcion engineer. He was also a man of progress and tirelessly led Denmark's great heath reclamation project, thereby "inwardly" expanding the amount of arable land available to Danish farmers. At the pan-Nordic industrial and art exposition held in Copenhagen in 1872, the words of Dalgas were echoed as the rallying cry to Denmark's new industrialists and engines of growth. King Christian IX personally decorated over 300 exhibitors with medals, each bearing the full two-line inscription of Danish poet Hans Peter Holst's rendition of

Dalgas's enduring words: "For every suffered loss, restitution comes anon; and what is outward lost must yet be inward won". The medal's other side bore a cloverleaf signifying the three merits "diligence, ingenuity and taste". Thus was outlined the future of the Danish nation – and the plan worked. Denmark had certainly become smaller in size, but it was a country in the throes of rapid growth. Heathland was reclaimed, agriculture rationalized, yields maximized, industry mechanized, production standardized, and exports procedures optimized. And as expected, the desired results materialized. Despite significant emigration Denmark's population grew dramatically, particularly in the capital city of Copenhagen. The economy improved and wages rose, while education became better, the standard work-week shorter and average life expectancy longer. Denmark was a growing nation, but also a nation in transition.

The world of the Danes was deeply influenced by two great wars in 1864 and 1914–1918, and by the introduction of democracy and the women's liberation movement. It was not as if a truly representative government was suddenly put into place, or women were suddenly given the same privileges as men. It took time for the democratic system to function smoothly and settle into the well-established multiparty framework Denmark has today, but the absolute monarchy had come to an end. On 5 June 1849 the country enacted its new democratic constitution, and on 30 January 1850 the newly elected Danish parliament, the Rigsdag, assembled for the first time. In the beginning, actual democracy was quite limited, and although Danish electoral law stipulated that everyone over the age of 30 had the right to vote, that was not strictly true. Farm hands, domestic servants and others who did not live under their own roof could not vote, nor could paupers, convicted criminals or "fools" – citizens with physical or mental disabilities. Moreover, the authors of the new Danish constitution had, as a matter of course, precluded half of the adult population from participating in the parliamentary elections – the half who were not born male. But that too would change. The growing awareness of the inequality of the sexes found a strong voice when the Danish Women's Society was founded in 1871 with the express purpose of making women fully active participants in society on an equal footing with men. Progress was made, one laborious step at a time. Danish women gained access to the university in 1875 and won suffrage in 1915, but not until 1921 were they entrusted to hold public positions – and even then a woman was excluded from becoming a minister or joining the military forces.

Although the years between 1850 and 1920 held both failures and successes, good times and bad for the various industries and, crucially, the defeat in 1864 at the hands of the Prussian empire that demoralized the entire nation, most

Danes had their eyes firmly fixed on progress. There were more mouths to feed, but also new and more efficient ways to do so. Poverty and destitution were well-known phenomena, but for ordinary Danes the general standard of living had nonetheless improved significantly. Agriculture was still the country's largest sector, but industry was playing an ever more important role. In the mid-nineteenth century Denmark was a small, underdeveloped agricultural society on the very edge of Europe. By 1920 it had become a growing industrial society, and a nation which, thanks to its strong merchant navy, had overcome its disadvantage – a lack of natural resources – in the race for industrial breakthroughs. Agriculture was still going strong, and like all other commercial activities it had experienced growth during the period, though no more than a good 10%. In striking contrast, the growth rates for trade, transport and the retail business had exceeded an impressive 200%, surpassed only by the incontestable winner, industry, which had grown by over 300% during the same period. Agriculture still carried the weight of Denmark's economic progress, but it was industry that propelled the country forwards. Industry also offered employment opportunities to Denmark's increasingly large number of high-level graduates in engineering, chemistry and other scientific fields.

An age of invention

From 1850 to 1920 the citizens of the world, Denmark included, witnessed a revolution in communication and transport, as they had seen in so many areas already. Spectacular changes had taken place in the span of a single lifetime. Things once regarded as the most fanciful flights of the imagination had become tangible, everyday realities in Denmark's towns and cities. The rest of the world moved closer, and travelling beyond Denmark's borders was no longer a major effort. It became possible to telegraph across the Atlantic, and eventually to telephone. Trans-Atlantic travelling became possible first by steamship and from 1919 by aeroplane.

The railway network's steel skeleton united Denmark, assisted by a number of new ferries and bridges creating links to the country's numerous islands. Because the timetables had to be precisely timed, Denmark was effectively synchronized and the astronomical time difference was equalized. The sunrise in Copenhagen occurred 16 minutes earlier than in the new harbour town of Esbjerg in the south-western corner of Jutland, a fact that would obviously have to be disregarded by a country in need of cohesion. The first timetable

13.2 *The opening of a new railway line was always a joyful occasion, often attended by members of the royal family. This scene from 1869 shows the festively decorated train arriving in Aalborg after its maiden run north on the new stretch from Randers.*

between Copenhagen and Viborg remarked that "railway time" was approximately 10 minutes earlier than "Viborg time".[223] This discrepancy was already changed in the next timetable, and all of Denmark's clocks and watches were set to "Copenhagen time". By 1875 most major stretches of the Danish rail network had been completed. A good 30 years after the opening of the first stretch – known as "King Christian VIII's Baltic Railway" – running between Altona and Kiel in what, at that time, was still a part of the Danish kingdom, the country was now connected from east to west, and from north to south.

The invention of the internal combustion engine paved the way for the arrival of the motor car. In 1900, Denmark had 50 car owners. Barely thirty years later the number of cars in the country had mushroomed to 90,000. The path leading from invention to mass-production became ever shorter. The internal combustion engine was also the basis of another spectacular contraption: the aeroplane. In 1852 the Danish author Hans Christian Andersen – who, although best known for his fairytales, was a genuine science and technology enthusiast – had predicted in his brief tale "Thousands of Years from Now" how young Americans would cross the Atlantic in steam-powered airships, taking in all the memorable sights of ancient Europe on an eight-day tour. Andersen's time-frame may have been some distant, unknown future, but within 50 years after he wrote this scenario the Wright brothers successfully took a "heavier-than-air", motorized flying machine off the ground. In 1910 the young Danish reporter Carl Th. Dreyer – later to become a film director of international fame – proclaimed that "The aeroplane is no longer a possibility of the future. It is a fact."[224]

The first passenger flight in Denmark took place on 26 June 1909, when a Mrs Erna Lange took a seat behind the legendary French aviator Georges Legagneux during an air show held at Klampenborg racecourse north of Copenhagen – in a plane that climbed to the altitude of 10 metres. As early as 1917, first lieutenant "Cowboy" Kofoed-Jensen beat the Scandinavian altitude record, passing the 5,000 metre mark in a Danish-built fighter plane from the aeroplane department of manufacturers Nielsen & Winther.[225] The national operator Danish Airlines (the Danish component of the later amalgamation Scandinavian Airline System – SAS) was founded in 1918, and two years later a regular service was set up between Copenhagen, Malmø and Warnemünde.

People were exhilarated about the "great age of invention" and its seemingly endless wonders. Progress was moving at a speed that defied even the imagination of the technophile Hans Christian Andersen. Although the average Dane would hardly have been able to sentiently experience the whole seven decades from 1850 to 1920, it is still true that society as a whole and each individual citizen witnessed a series of momentous changes throughout the entire period. It was as if the two ends of the period represented two different worlds, one a rural agricultural society, the other a highly developed society whose growth and lifestyle were founded on science and technology. Chemistry, electricity and Denmark's new technological infrastructure with trains and telegraph wires profoundly changed Denmark's landscape and the Danish way of life.

There were so many new things to grow accustomed to. Not just the steady flow of new technology and new knowledge, but also a new system of weights and measures based on the decimal system – with its litres, metres and grammes, and the prefixes deci-, centi- and milli-, deca-, hecto- and kilo-. The process of switching to "metric measures" was completed rather more slowly in Denmark that in its neighbouring countries, however. Germany managed the changeover in 1868–1871, Norway in 1875, and Sweden in 1878. The system of centimetres, seconds and grammes may have been used by Danish scientists in the nineteenth century, but not by the country's general population. The scientific advantages of using standardized measures was obvious. A scientist might work anywhere on the globe, but the units would be identical. As long as the three basic units of length, mass and time were defined, they could be used to derive all other units in the system. This became possible in Denmark in 1889, when the country received a standard metre and a standard kilo weight from the Bureau International des Poids et Mesures. It was not until 1907, following several parliamentary debates in the Rigsdag, that the metric system officially became the measuring system of the Danish people. Finally on 1 April 1916, after a transitional period of six years, the metric system became the only legally applicable system in Denmark. Once again society had to learn new steps as the measuring stick of science tapped out the beat.

Looking back on 1850 from a 1920 point of view the world had changed to an extent that was almost beyond belief, in Denmark and elsewhere. Science had explored and revealed nature and the universe, in the minutest detail and on the largest conceivable scale. There was much to know, and much to learn – and this was just the beginning.

The rise of a new profession 14

A professional science

The dream of one day establishing a single comprehensive scientific explanation of the world and the universe had been reinforced through the success of the mechanistic world view in the seventeenth century and the centuries that followed. The great breakthrough came with Isaac Newton's *Principia Mathematica Philosophiae Naturalis* from 1687, which dealt with the mathematical principles of natural philosophy. Newton demonstrated in *Principia* that the same laws of nature applied to Earth and to the universe, thereby creating the underpinnings for the idea that everything in the world and beyond could be regarded as a mechanical system that acted, and could be described, according to the laws of mechanics. This provided scientists with a common outlook and a comprehensive foundation on which to base their scientific work – regardless of field or discipline. During the Romantic period in the early nineteenth century the idea of unity was also expressed as a sort of mutual interconnectedness, implying that everything in nature was somehow linked. Indeed, this was also the backbone of the great Danish physicist H. C. Ørsted's natural philosophy and his idea of "the spirit in nature".[226]

As the nineteenth century moved on and the number of fields and subdisciplines grew, it became increasingly difficult to hang on to this idea. Paradoxically, the unity of science was a dream that appeared more and distant with every step science made. Scientific development was closely linked to the explicit specialization of various branches of science. In order to create results and contribute to the pool of scientific knowledge, scientists required less general knowledge and more specific knowledge about their own particular discipline: its theoretical and historical foundation, empirical material, objects of study, techniques, measuring equipment, laboratory facilities and scientific practices. This further meant that the system had to begin educating people to acquire particular skills needed in the various scientific fields. There was no longer place for generalists in the Danish educational system. Students of science had to quickly choose a specific direction, then specialize even further in a bid to meet the growing demands of an ever-expanding industrial sector, and to blaze new trails and expand the existing fields.

In the 1850s the dream did not seem so distant. A sort of wistful sadness at the increasing degree of specialized segregation was discernible in many scientists still influenced by the Romantic ideals. But naturally there remained common goals that scientists could seek to achieve together, even though their disciplines, their work and their theories might seem far apart. A good example is Julius Thomsen, who as a young chemist in 1856 asserted in his popular-science book *Vandringer på Naturvidenskabens Gebeet* (*Wanderings in the Field of Natural Science*) that there was no way around scientific specialization. Anyone attempting to encompass all the fields of natural science would soon find his capacity exhausted. One must, instead, concentrate on closely related phenomena, ignoring the rest of nature's diversity. Every scientist had to make a choice and decide upon a certain direction: "he elects a field in which he will work, and over which he will endeavour to shed light."

Thomsen kept his optimism, however, maintaining that for all their divisions and subdivision, the sciences continued to work towards the same goal.

> The great task that natural science has to solve is to find a true perception of nature itself: In each individual branch of science work is being done to solve this task, and gradually as the discrete parts of science move closer to this common goal, they also enter into a more intimate contact with one another. There will come a time when all individual branches of science will once again have merged into one strong current which, flowing quickly, seeks to reach the goal, infinitely distant even now.[227]

Years later, at the inauguration of the Polytechnical College's new buildings in 1890, an ageing Thomsen spoke as the director of the sextagenarian institution, sharing experiences from his long life in science. The goal was still to come closer to a true perception of nature itself, but attaining a level of complete understanding, he now realized, was an impossible task.

> In mathematical terms, science's relationship to the enigma of existence is a hyperbolic function; for just as the hyperbole on its rule-bound path moves ever closer and closer to its asymptote without ever being able to reach it, so the development of science brings us ever closer to the recognition of truth, but without it ever being possible to fully reach its source.[228]

And so, by the end of the nineteenth century, after an almost endless succession of scientific triumphs unpretentious reflection was aired that neither belittled the scientific results hitherto accomplished nor lessened the expectations to future achievements. It still spoke of an optimistic, goal-oriented sci-

14.1 *A group of Danish naturalists on an outing in 1914; professors and students, men and women.*

ence. "Science strives ever forwards and permits no respite", as Thomsen put it. Nevertheless, the subdivision of subjects, the specialization and the necessary disciplining of students, teachers and topics were all unavoidable elements in the professionalization of science and its practitioners. And with professionalization now an incontrovertible fact of life, the unity of science was no longer a viable aspiration.

Specialization brought more and faster results from the various fields of science. Consequently, the fields gradually became estranged from one another, creating a variety of field-specific cultures with their own journals, associations and meetings. This in turn meant fewer common platforms where scientists could meet, making it more difficult to understand those from other professional fields. In theoretical and practical terms "the unity of science" came to be regarded as an echo of the past. This view was also reflected in discussions about the relative positioning of the various sciences, for instance in Denmark's educational system.[229]

Whereas a scientist of the 1850s had benefited from a broad scientific education, the type of scientist in demand in the 1920s, whether in research, commerce or industry, was the expert. Education programmes, research and application had all become professionalized. Science had become more efficient, and it delivered what it promised. But scientists from the different fields increasingly spoke their own separate languages. In a sense, science had become a tower of Babel.

Making an academic career

Professionalization fundamentally altered the general status and behaviour of scientists: from natural philosophers in 1850 to scientists in 1920, from male generalists then, to male or female specialists seven decades later. Earlier natural philosophers addressed a broadly educated scientific audience and were part of the scientific community at large. Later scientists addressed a select group of professional peers and were part of an international research community narrowly defined by the limits of their discipline.[230] This also applied to Danish scientists, who followed the European and North American pattern for scientific education, research areas and career paths.

In the mid-nineteenth century many a scientific career still relied, at least to a certain degree, on coincidence, personal contacts and a broad, general scientific education. This was very closely linked to the fact that the university courses and degrees in science had not been fully institutionalized – a process that did not take place until the Faculty of Science was established at the University of Copenhagen in 1850. As early as 1848, however, students could earn a master's degree in natural history. Before that time it was not possible to take an actual scientific degree in the field in which one specialized at the university. Although the establishment of the Polytechnical College in 1829 (today the Technical University of Denmark) remedied this situation somewhat, the existence of the new college also impeded the development of the scientific subjects at the university.

One scientist who took a non-institutionalized path to an academic career was Japetus Steenstrup, who as a zoology professor at the University of Copenhagen became one of the period's most distinctive figures – for better and for worse. Steenstrup, who had set out to study medicine, but eventually came to spend most of his time on natural history, never graduated. In the mid-nineteenth century that in itself was no obstacle to a future career at the university. A mere 32 years old and with no science degree, Steenstrup was appointed to a chair in zoology; an appointment that would have been absolutely unthinkable fifty years later. Steenstrup held his position for the better part of a lifetime, controlling scientific careers in Denmark, manipulating many people and frustrating even more.[231] Through his powerful professorship and concomitant responsibility for the zoological museum, Steenstup wielded great influence on new academic appointments and on determining zoology's profile in terms of research and education. He became a fellow of the Royal Danish Academy of Sciences and Letters at the age of 29, later serving as its secretary for 13 years. He gave a large number of scientific and popular-science lectures at the Danish Natural History Society and also attended the Scandinavian science conventions and international archeology conferences. In addition, by virtue of his friendship with J. C. Jacobsen, the prominent brewer and patron of the arts and sciences, Steenstrup assisted in establishing the Carlsberg Foundation in 1876 and was a member of its first board of directors.[232] Creating this sort of career without a scientific degree, and publishing almost entirely in his native tongue, with little concern for the international community's ability to understand Danish, would be an impossibility for the next generation.[233]

It had become usual for scientists to go on research trips abroad, for shorter or longer periods of time, and then to return to Denmark bringing back new knowledge of laboratory construction and experimental practices, as well as new techniques and equipment. At the same time there was a growing awareness among scientists of the importance of sharing their knowledge in scientific and technical publications targeting their international peers, and also in more popular publications targeting a wider audience. In 1850 there was no obvious difference between the two, but by 1920 there was a clear-cut distinction between scientific publications and works of popular science.

In contrast to Steenstrup another prominent scientist, Carl Wesenberg-Lund, is an excellent example of a scientific career path towards the end of the period. Like Steenstrup, he too was a zoologist and ended up a professor, but the road he took and the scientific behaviour he exhibited were very different and, quite consciously, far more internationally oriented. The first marked difference lies in the two men's education. Wesenberg-Lund earned his master's degree in 1893 and became an assistant at the Royal Veterinary and Agricultural College. Like Steenstrup, Wesenberg-Lund also won a gold medal from the university, and he also went on a northern expedition. Whereas Steenstrup had gone to Iceland, Wesenberg-Lund was part of the Ingolf expedition that sought to explore the depths of the North Atlantic. Upon his return Wesenberg-Lund began work on his doctoral dissertation on rotifers, which he defended in 1899. Even before this, however – in 1897 – he had received support from the Carlsberg Foundation to set up a freshwater biological laboratory on Furesø, a large lake northwest of Copenhagen. Leaving his position at the Veterinary College, for quite a few years he lived solely on his funding from the foundation, supplemented by the income he earned writing works of popular science.

Wesenberg-Lund was in close contact with scientists abroad, and visited research centres in Germany, Austria, Switzerland and Scotland. He was also a co-founder of *Internationale Revue der gesamten Hydrobiologie*, the first international journal on freshwater and marine biology. Not until 1911, at the age of 44, was he offered a permanent position at the University of Copenhagen and began working there as a lecturer. Until then he had run his laboratory and pursued his scientific career on external funding alone. In supporting him the

Carlsberg Foundation had played a decisive role in ensuring that Danish freshwater biological research not only could survive, but actually thrive in the international community. The university appointed Wesenberg-Lund extraordinary professor of zoology in 1922. By then he was 55 and quite a contrast to the young Steenstrup, who assumed his chair of zoology at the age of 32.[234]

Societies and associations

Even as Danish scientists began to travel more and publish their written work in non-Danish journals, a growing number of specific professional journals and associations began to appear within Denmark itself. Still central was the

14.2 *A host of specialized scientific societies arose in the nineteenth century, but the Royal Danish Academy of Sciences and Letters continued to play a pivotal role in Denmark's science community. One main reason was its close links to the Carlsberg Foundation, which became an important factor in the funding of Danish research. This meeting at the Royal Academy, painted by P. S. Krøyer in 1897, shows a preponderance of elderly gentlemen and a conspicuous absence of young men – and women.*

Royal Danish Academy of Sciences and Letters, founded in 1742, which provided a general framework for the many different branches of science. But the Royal Academy was an elitist, closed forum, formally organized and not ready to open its doors to everyone. Instead scientists were obliged to go elsewhere, gradually grouping together into various professional associations whose members, despite internal differences, concentrated on a particular field or discipline.

Groups of scientists met at lectures and excursions, and most societies and associations had their own journals, transactions or proceedings. These channels were used to disseminate information consisting of everything from brief presentations of scientific results to lengthy articles, as well as announcements of lectures, meetings and job opportunities. At the beginning of the period the mutual contact among Scandinavian scientists had already been institutionalized in the form of Scandinavian science conventions, which had been held at regular intervals since 1839. Furthermore, international contacts forged through foreign societies and meetings gradually increased throughout the entire period.[235]

It is characteristic that during the last half of the nineteenth century there was a drastic increase in the number of new, specialized scientific societies (see table 1), which followed the general trend elsewhere in Europe and in North America. However, because nineteenth-century Denmark still had only one significant intellectual and academic centre, many of the meetings and other activities were accessible only to those living in or near Copenhagen. This tendency was, to some extent, offset by new journals, study groups and local alliances, but those who wished to meet scientists from abroad, invited by Danish colleagues, had no choice but to travel to the capital.

Many of these meetings received extensive attention in the media. The international archaeology conference in Copenhagen in 1869, for instance, enjoyed high-level coverage and bolstered Denmark's national pride. Many prominent foreigners were interviewed and thoroughly profiled and their activities reported in great detail. The farewell banquet, hosted by the king, was abundantly illustrated in the country's newspapers and magazines.[236] The same was true of the Eighth International Medical Conference, held in the Danish capital in 1884. At this event the media were particularly eager to report on Louis Pasteur and Joseph Lister, who were both international scientific stars at the time.[237] The intense public interest in such international meetings was not just a thing of the 1800s. In 1911, when Copenhagen hosted the Second Scandinavian Mathematics Conference, the Danish radical-liberal daily *Politiken* ran the opening ceremony as its front-page story, complete with portrayals of several leading mathematicians. This conference was also the object of extensive coverage that featured illustrations, interviews and serious reporting on a number of lectures.[238]

Table 14.1 Scientific societies and associations from 1850 to 1920

Original name and English translation		*Founded*
Det Kongelige Danske Videnskabernes Selskab The Royal Danish Academy of Sciences and Letters	# •	1742
Landhusholdningsselskabet The Royal Danish Agricultural Society	# •	1769
Det Classenske Litteraturselskab for Læger The Classen Literary Society for Physicians	#	1802
Selskabet for Naturlærens Udbredelse The Danish Society for the Dissemination of Natural Science	•	1824
Naturhistorisk Forening The Danish Natural History Society	# •	1833
De Skandinaviske Naturforskeres Selskab The Society of Scandinavian Natural Scientists		1839
Den Botaniske Forening The Danish Botanical Society	# •	1840
Danmarks Apotekerforening The Danish Pharmaceutical Association	# •	1844
Polyteknisk Forening The Polytechnic Associations	•	1846
Entomologisk Forening The Entomological Society	# •	1868

Det krigsvidenskabelige Selskab The Royal Danish Military Society	# •	1871
Det medicinske Selskab i Kjøbenhavn The Medical Society of Copenhagen	# •	1872
Nationaløkonomisk Forening The Danish Economic Society	# •	1872
Matematisk Forening The Danish Mathematical Society	# •	1873
Det kongelige danske geografiske Selskab The Royal Danish Geographical Society	# •	1876
Den tekniske Forening The Technical Society	#	1877
Kemisk Forening The Danish Chemical Society	# •	1879
Dansk Tandlægeforening The Danish Dental Association	•	1881
Dansk Ingeniørforening The Danish Society of Engineers	•	1892
Dansk Geologisk Forening The Geological Society of Denmark	# •	1893
Biologisk Selskab The Danish Biological Society	# •	1896
Jydsk Forening for Naturvidenskab The Jutland Society for Science	•	1903
Dansk Vind Elektricitets Selskab The Danish Wind-Electricity Society		1903

Fysisk Forening The Danish Physical Society	#	1908
Danmarks Naturvidenskabelige Samfund Denmark's Society for the Natural Sciences	#	1911
Astronomisk Selskab The Astronomical Society	•	1917
Dansk Medicinsk Selskab The Danish Medical Society	•	1919

List of scientific societies and associations active during the period 1850–1920, with # indicating a related journal, proceedings or other publishing activity relevant to the society or its work, and • indicating that the society is still in existence.

Science branched out and sprouted a multitude of new subdisciplines during the nineteenth century. This was also reflected in the education of scientists and engineers, who now belonged not merely to a broad group generically referred to as "scientists", but to various specialist groups each with their own particular field. Both in Europe and the United States such scientific specialists became society's new professionals, complete with a formal education, membership of exclusive clubs, subscriptions to specialized journals and access to professional gatherings that enabled them to meet with their peers and exchange news and views.[239]

Despite the growing number of specialist associations and societies, not all of the generalist societies were phased out. A number of them survived (some up to the present day) and were even joined by a few new organizations. One of these, established in 1911, was especially important in bridging the gap between the worlds of academia and industry. Actually, Denmark had lacked neither industrially oriented scientists – such as Julius Thomsen, a professor of chemistry who also ran his own chemical plant – nor scientifically oriented industrialists – most prominently J. C. Jacobsen, founder of the Carlsberg Breweries.[240] But apart from the Carlsberg Foundation, a combined holding company and charitable foundation that can hardly be defined as a society, science and industry had no common ground.

Then, on 14 December 1911, Denmark's Society for the Natural Sciences was founded, and several lengthy addresses were made to mark the occasion.

The Danish physicist Martin Knudsen praised the attempt to bring men of science and industry together. He expressed his hope that the industrialists would not be driven away by the scientists' dedication to science for its own sake. It was therefore imperative, Knudsen argued, that the new society consist of men who appreciated the gravity and importance of their common cause and the society's main purpose: the advancement of science.

One way to advance science was to bring Danish and foreign scientists together. And the society did so by inviting scientific guests to lecture in Denmark, supporting and organizing international and Scandinavian conferences and meetings, and also arranging visits to laboratories in Denmark and abroad. By using such events to bring together men of science and industry, the society united academic and practical interests in its quest to advance science. Scientists could learn much by looking beyond their laboratories, and industrialists could gain a better understanding of the nature of scientific research, they hoped.

In his 1911 speech Martin Knudsen also stressed how important it was that members of the society took pride in supporting *Danish* science, and scientific results produced in *Denmark*, with the clear expectation that this would help Denmark compare favourably with other nations. "Let the group consist of men who would be proud to see Danish scientific achievements match other nations", Knudsen said, "and let it always be an honour for merchants or industrialists to be accepted in the group". Knudsen's remarks were embodied in a statement expressing the essential goals of the new society: "for the advancement of science, for the benefit of mankind, for the glory of the nation".[241] The members of Denmark's Society for the Natural Sciences concurred.

National identity was indeed a matter of concern to those discussing the advancement of Danish science in the early twentieth century. Clearly the situation called for action and would not improve of its own accord. This was a great challenge for the Danes – who had been forced to redefine themselves and adjust to their new status after Denmark's geographical and military losses in 1864. As early as the 1870s, a group of Copenhagen intellectuals had derided the Danes for their complacency, the most prominent and outspoken being the literary critic Georg Brandes, who bluntly declared that "the Danish people have always been very disinclined to do anything to make changes in the present state of affairs, and dearly they love their own indolence".[242] The members of the newly founded Denmark's Society for the Natural Sciences had a plan for "putting on a fresh shirt in the morning", as Brandes had advised his fellow countrymen to do, urging them to direct their attention to the world beyond the Danish duck pond – so scathingly portrayed in Hans

Christian Andersen's allegorical tale "The Ugly Duckling" from 1843.

And thus, industrialists and scientists were united in the endeavour to compare favourably with other nations. For a small country like Denmark it was essential that its engineers, scientists and businessmen had the best knowledge available, in order to profit from cutting-edge science and technologies in their various fields. After the turn of the century these fields coalesced, resulting in the emergence of new applied sciences and a technical-scientific understanding of culture that seriously challenged the traditional liberal education and the classical virtues of the humanities. In the new international world of science, Denmark and other small nations had every opportunity to grow. Seen against this backdrop, personal, commercial and national interests converged when the physicist Martin Knudsen and G. A. Hagemann, an industrialist who was also the director of the Polytechnical College, joined forces with Denmark's Society for the Natural Sciences.[243] Much was at stake. But the important thing was to promote science as a useful enterprise, for the benefit of mankind *and* for the glory the nation – whilst earning an additional income.

Many other professional scientific societies and associations also organized activities and contributed to the increasingly international approach among the new generation of scientists and engineers. However, as far as the pronounced dialectic of the national–international relationship is concerned, it is Denmark's Society for the Natural Sciences that particularly draws interest. While on one hand this society conceded that circumstances obliged Danes to operate in an international scientific environment, on the other hand it found that the specific situation also provided an opportunity to bolster Denmark's national identity; an opportunity the new society exploited with considerable success.

A new framework for science

At the University of Copenhagen, science eventually found a place of its own marked by the establishment in September 1850 of the university's Faculty of Science – the first of its kind in Scandinavia.[244]

Only France had been quicker than Copenhagen in splitting off the scientific disciplines from the faculty of philosophy. The University of Christiania (now Oslo) was quick to follow Denmark's example, setting up its independent faculty of science in 1860. The Swedish universities in Uppsala and Lund chose to split their Faculties of Philosophy in 1876. Of the 22 universities in

Germany, only three had faculties of science at the turn of the century. The pace of this development in Britain was similar. In both countries the entire process of institutionalizing science, with all of the professorial chairs and academic degrees this entailed was late in coming compared to France and Denmark. In other words, it was not necessity that drove the effort to found an independent faculty to support and sustain scientific development. The Danish decision to set up a faculty of science can be seen as a strongly symbolic action reflecting the advance of science in the academic world. And the decision's symbolic nature is important to the role science came to play in Denmark, although the symbolism was not initially evident, nor was it the basic motivation for deciding to disaffiliate science from other academic fields.

The growing significance and strength of science at the university naturally played a role in the formation of the new faculty. However, as indicated above it was more a matter of internal university politics than of urgent professional necessity. Even though Hans Christian Ørsted had already in 1813 expressed his hope to see the sciences enjoying the same privileges at the university as they would in an independent faculty. Nonetheless, the time was not ripe, and his wishes came to nothing at the time. At that point the problem was that only the ordinary professors of mathematics and astronomy had the right of representation in the *konsistorium*, the university's governing body. In 1815 the Faculty of Philosophy had a total of six science professors, two thirds of whom were kept outside any political influence by the university's own rules.

It became clear that the university's politics were untenable when H. C. Schumacher, then 35, was appointed professor of astronomy, thereby gaining access to its governing body, while Hans Christian Ørsted, who at that time had been the university's professor of physics for nine years, was still barred from entering. In short, the university's Charter – which outlined its fundamental rules – was unable to accommodate the new subjects that were branching off the academic tree of learning. The process of specialization, which continued throughout the nineteenth century, was nurturing the growing number of scientific disciplines. It became increasingly unacceptable that a growing group of the university's professors were cut off from many of the privileges accompanying membership of the *konsistorium* – such as rent-free accommodation, rent allowance and potential allocation of university grants. A solution would have to be found.

The first remedy was a decision in 1817 to expand the governing body from 14 to 16 members. Although the idea of setting up a fifth faculty at the university was voiced, it was rejected with reference to the centuries-old tradition of

14.3 *The Zoological Museum was inaugurated in Copenhagen in 1870. Its layout and lofty central atrium with its glass ceiling won admiration and were soon copied at museums in Hamburg, Leiden and Paris.*

a four-faculty university, and to the fact that none of the recently reformed universities had found such a step necessary. Those citing the latter as an example were thinking of the universities in Germany, while conveniently forgetting that the French universities had already introduced a structure including a science faculty in 1808.[245] And even the expanded *konsistorium* did not yield any additional seats to science members, at least not at first.

The battle over university politics continued in the decades that followed, fuelled by a mixture of tradition and power-hunger in equal measure. Finally the internal pressure became too great to ignore and at last, in 1849, a decision was made to split the Faculty of Philosophy in Copenhagen into two, thereby creating an independent faculty of science. This equipped the university with a strong scientific foundation on which to build a framework capable of handling the changes and challenges that awaited. These challenges included field differentiation and increased specialization, as well as new and growing demands to scientific research and to the education of science graduates. In these endeavours the university did not stand alone.

The increasing specialization and professionalization of science was mirrored in an increased institutionalization. Societies, associations and professional journals were all contributory factors, but the trend also resulted in other manifestations that were public, and far more visible: new buildings and scientific institutions. From 1850 to 1920, science was given more buildings, education programmes, teachers, researchers, laboratories and, as an essential prerequisite for this development, more funding. Science gained more room to manoeuvre physically, intellectually and financially. These gains could be seen and felt, not only in relation to the effects of science on society, but also in relation to Denmark's rapidly expanding industrial sector.

One of the most significant state-run institutions was, of course, the university's Faculty of Science, established at the beginning of the period. In addition the University of Copenhagen had a zoological museum, a mineralogical museum, an astronomical observatory and a botanical garden with an affiliated botanical laboratory. But science also had other homes outside the university. The institutions of higher learning located in and around the capital included the Polytechnical College (1829), The Royal Veterinary and Agricultural College (1858), the School of Dentistry (1888) and the Pharmaceutical College (1892). These were joined by a number of other state-run institutions, such as the Bureau of Statistics (1850), the Danish Meteorological Institute (1872) and the State Serum Institute (1902). But even beyond that, the Danish state was not the only player in the scientific arena. The country's greatest philanthropists backed some of the most distinguished scientific institutions of their time. In 1875 J. C. Jacobsen founded the Carlsberg Laboratory as a privately run research facility with two independent laboratory departments: one for chemistry and one for physiology.[246] Another independent effort was headed by the physician Niels Ryberg Finsen, who convinced the industrialists G. A. Hagemann and V. Jørgensen to back a venture involving medical treatments with light. The Finsen Medical Light Institute was founded in 1896 and expanded in 1902, rapidly making Finsen a scientific celebrity receiving the Nobel Prize in 1903.[247]

With the educations professionalized, the new scientific institutions in place, and the associations, meetings and journals creating professional communities and identities, an abundance of new career paths and opportunities

14.4 *Science was now a natural part of the public debate. Twenty years after the opening of the Zoological Museum, a satirical magazine offered this suggestion for increasing the number of visitors.*

became available in a country that was becoming increasingly dependent on the efficient, targeted efforts of scientific specialists. Danish engineers and chemists tendered their expert knowledge and skills to a society that gradually came to rely on them, supporting a new social order based on science. As G. A. Hagemann said about "the engineer" upon retirement as director of the Polytechnical College:

> Science, to him, shall not be like the cultivated rose, whose exquisite colour, shape and scent win our admiration; it must be like the wild dog rose, or like the apple blossom. No less lovely to behold; but it must bear fruit, and preferably such a fruit to be enjoyed by all Mankind.[248]

From 1850 to 1920, Denmark constructed a new framework in which the entire gamut of scientific activities could unfold. The new institutions bore witness to the growing need for scientifically educated Danes who could propel their old agricultural society into the industrial age. Science was useful to society. It was well worth the investment and moreover it was essential if a small country was to compete in an international market.

Denmark's public administration

The state bureaucracy in Denmark grew throughout the nineteenth century, as did the number of scientifically educated civil servants. Society in general, and the monitoring and control of the population, grew more and more scientific in the quest to calculate and enumerate human activity. This quantitative and objective ideal became an overarching ambition for society as a whole, and for the individual. In the writer and poet Holger Drachmann's novel *En Overkomplet* (*A Supernumerary*) from 1876, Erik Dahl, the son of an academic, spends his spare time on his "favourite studies, the best inheritance after my father, statistics and economics, as these sciences had developed during the momentous world events in recent years". Erik's educational journey demonstrates how scientific thinking pervaded all aspects of life, from the intimate personal sphere to the broadest social sphere, as the young Erik listens to an old painter's unsentimental reflection on the times: "In our day, the 'humane' is understood as providing food for the poor and hygienic statistics, suffrage and sewers".[249]

Statistical information became an important official business for the new Danish democracy, which set up the new Bureau of Statistics (today Statistics Denmark) to handle these tasks in 1850. Its first director was A. F. Bergsøe, a professor of statistics at the University of Copenhagen and the author of *Den danske Stats Statistik* (*Statistics of the Danish State*), a four-volume work published from 1844 to 1853. Bergsøe died just a few years after his appointment and did not greatly influence the bureau's activities and ideological responsibilities to Danish society. This changed with C. G. N. David, who had a background as a professor in economics at the university, but was chiefly known for his political work. Statistics was still a relatively new social science, characterized by a distinctly international trend towards development and standardization. This was the golden age of international statistics conferences, which greatly benefited the new science as statisticians took an international perspective and agreed upon which tasks to take on, thereby ensuring the comparability of statistics from different countries. This was not fully in keeping with the statistical ideology at the Danish Bureau of Statistics near the turn of the century, at which time the bureau's director Marcus Rubin pronounced that in previous years one had easily forgotten "that what was important, first and foremost, was to have each individual country's economic and social life as thoroughly elucidated as possible, and that a statistical practice of that particular nature was less easy to reconcile with international uniformity".[250]

One of the bureau's most important tasks was to coordinate Danish census activities. It was an arduous undertaking involving the cooperative efforts of various parties, including the municipal statistical offices spread across the country. The statistical information, which was published every five years until 1860 and then once every decade until 1900, gave a detailed quantitative picture of daily life in each Danish town and parish, specifying the population by sex, age, weddings, births, deaths and population movements. Rural and maritime areas received particular attention in statistics for agriculture, trade and shipping. Certain other statistics fell outside the bureau's scope of activity, such as information on rail transport of goods and passengers, and post and telegraph information, which were handled by the local administrations. As industry became increasingly important to Denmark's economy and employment, there was a growing pressure to add industrial statistics as well. When this was done, it produced a picture of a country in change, and with a solid growth in the number of factories and new products. At the same time it also became increasingly clear that it was necessary to elaborate a set of specific social statistics in order to comprehensively summarize the state of the nation. Such social statistics would need to include information on the financial situation of urban and rural workers, on their relation to their employers, their health insurance, their receipt of old-age pensions, and their drinking habits. Further expansions added education statistics, crime statistics and financial statistics, thus providing numerical data on virtually every activity in the country, whether related to the citizens in Denmark's rural municipalities, the number of detainees in correctional institutions, or agricultural exports. Reams of information were now available about Denmark, all backed by numbers.

There are, in particular, two statistical publications that merit special mention. The first is *Danmarks Statistik* (*Denmark's Statistics*), published from 1878 to 1891 and written by V. A. Falbe-Hansen and William Scharling, both professors of economics in Copenhagen. *Danmarks Statistik* was an impressive collection of geological, geographical and climatic information, accompanied by a status on Denmark's population, trade and industry, and public institu-

14.5 *All aspects of Danish daily life were named and numbered. Those were busy years at the Department of Statistics, seen here around 1918.*

tions. The volumes treated a wide range of topics and included lengthy chapters by prominent scientists like J. F. Johnstrup, the university's professor of mineralogy.

The second distinctive statistical publication during this period was a work written by the cabinet secretary J. P. Trap and entitled *Statistisk-Topographisk Beskrivelse af Danmark* (*Statistical-Topographical Description of Denmark*). The first edition, dating from 1858 to 1860, was superseded by a second edition available from 1872 to 1879, and by a third edition appearing from 1898 to 1906 under the new title *Kongeriget Danmark* (*The Kingdom of Denmark*). The author cooperated closely on these volumes with the Danish Bureau of Statistics, and his work gave an extensive and richly illustrated portrayal of the country's natural make-up and landscapes, as well as describing the entire country's public and private institutions – including scientific and educational institutions, associations, buildings, and trade and industry, retailing, transport and the financial sector, plus the country's social legislation, constitution, public administration and economic conditions. Trap's *Description of Denmark* does much more than simply convey statistical information. It paints a uniquely comprehensive picture of Denmark's nature, culture and science, and the lives of its citizens and business community; an excellent example of the way science was used to describe the world of the Danes, and their daily lives.

The meticulous number-crunching at the Danish Bureau of Statistics certainly benefited the country's civil servants and scientists, but it was also supposed to tangibly benefit the country's citizens. That is why, from 1869, the bureau began to compile brief summaries of key statistics covering the main fields of activity, along with excerpts of rail, post, telegraph and bank statistics and exchange rates. As the bureau director C. G. N. David explained in his preface to the first summary:

> The information on conditions in Denmark, which the statistical bureau collects and publishes or which is made public in other ways, will hardly gain entry into the consciousness of the general public or be used to advantage in such a way as would be considered desirable. One of the reasons for this might perhaps be that the degree of detail in this information is, most often, so copious and overwhelming that it dampens or eradicates one's inclination to fix the main findings in one's mind. It is therefore believed that it would be useful, as is done in England, from time to time to summarize the statistical information and to present such summaries to the public.[251]

Over the following decades the statistical summaries appeared at irregular intervals. However, because they proved quite useful, because the information was put to practical use, and – a very weighty argument – because it was common practice in other countries, it was decided that David's initiative would become a permanent arrangement. And so the first *Statistisk Aarbog* (*Statistical Yearbook*) appeared in 1896, as it has continued to do every year since.

The information from the Danish Bureau of Statistics revealed an increasingly differentiated population whose world was constantly shifting and changing, with migrations from rural to urban communities and a multitude of new professions, not to mention the completely new area of leisure activities and hobbies. Nevertheless, despite the many different directions the Danish citizens' lives took, they still had one thing in common: the weather.

Naturally, in an old agricultural and seafaring nation it had always been important to stay abreast of developing weather conditions. The Danes had been using barometers and thermometers to predict weather conditions and read temperatures since the seventeenth century, but no systematic, reliable observations had been made until the nineteenth century. The only weather information available was a weekly newspaper notice, published from January 1786 onwards, initially prepared by the Copenhagen astronomy professor Thomas Bugge listing the barometer pressure and wind, weather and temperature in Copenhagen.

Hans Christian Ørsted had instituted a practice in 1820 that the Royal Danish Academy of Sciences and Letters would publish precise meteorological observations from as many weather stations as possible across the entire kingdom, including Denmark's possessions in the tropics and the Arctic region. In 1827 the Royal Academy also set up a meteorological committee that was responsible for coordinating observations and gathering data. There was, quite clearly, a contemporary scientific interest in meteorology, and its practical utility was obvious. Even so, science still lacked the necessary technology to allow meteorology to be a part of daily life in Denmark and to become what it is today: the most widespread, regular and continuous sharing of scientific knowledge with the public at large.

The new communication infrastructure with trains and telegraph lines enabled a rapid dissemination of information. This allowed daily weather reporting and facilitated the gathering of far more detailed descriptions of the precipitation, temperatures and wind speeds observed across the Danish kingdom. Around the world the collection, statistical treatment and publication of weather data was coordinated by national meteorological institutes. The Netherlands were the first to set up their meteorological institute in 1854. France established a permanent weather service in 1856, as did Britain and Norway in 1866. Denmark followed suit in 1872, going along with the general trend of nationally run meteorological reporting and security services based on weather telegrams sent at set times each day.

Initially the Danish Meteorological Institute had three tasks: obtaining observational data, passing its findings on to the public, and developing a scientific meteorological practice. It soon became clear that the institute held an important position, and it rapidly became a great success. One result of the institute's extensive data-collection activities was the preparation of so-called "synoptic weather maps" in an effort headed by the institute's first manager, Niels Hoffmeyer. Such maps consequently became known as "Hoffmeyer maps", and within the international meteorological community they came to play an important role in developing the field. The many observations also laid the foundation for establishing Denmark's gale warning service.

The University of Copenhagen did not wish to draw the institute into its sphere of activity, being worried that the institute's practical work would take first priority to the detriment of its scientific work. As the university's governing body wrote to the Danish ministry of marine affairs: "even though the scientific goal must be taken into account, a meteorological institute will, in the first instance, direct its enterprise towards satisfying the multifarious practical duties that are, essentially, of no concern to the university, or which are more removed from it, examples being the interests of the shipping trade and agriculture".[252] Nevertheless, in scientific terms Denmark's Meteorological Institute fully equalled analogous institutions abroad, and Danish meteorology made its mark as one of the country's most thoroughly international sciences in the late nineteenth century.

The long and winding road to equality in science

The democratic ideals of freedom and equality for all were gradually extended to encompass women as well, but even getting that far was a long, slow process, and the degree of equality that was achieved was by no means complete. From 1850 to 1920, women more than ever before became public figures with privileges that, at least in part, were similar to those that men enjoyed. The most significant breakthrough came when women's suffrage became a reality, ensuring equal democratic rights. The first step towards the actual democratization of Danish society came when it was decided in 1903 that women would have the right to vote at parish council elections, and in municipal elections in 1908. However, it was only with an amendment of the Danish constitution, enacted on 5 June 1915, that suffrage was legally consolidated as an unconditional and equal right, to quote the 1915 constitution's article 30: "The right to vote in parliamentary elections belongs to every man and woman who has [Danish] citizenship, has attained the age of 25, and is permanently living in this country." At the same time women, like men, became eligible to run for office in the national parliament. After the 1915 enactment Klaus Berntsen, a member for the liberal party Venstre and one of the most ardent advocates of constitutional reform, had proclaimed from the rostrum of the parliament that

> We men of all parties have willingly and voluntarily shared power in the future with the women of Denmark. We cordially welcome them to participate in the collective efforts towards the future of our nation. Women are now so much better equipped to step into public life than men were when they received the right to vote in 1849.[253]

This was the climax of a long, gradual process, in the course of which women's emancipation had been debated back and forth, here, there and everywhere. Women did, however, experience other significant breakthroughs during the period.

Besides suffrage there was, of course, the right to work, which was an absolutely fundamental element in promoting women's emancipation and equality.[254] In 1857 unmarried women, widows and divorcees were granted a limited right to run their own business. On the other hand they were not allowed to hold public office until 1921 – and even after that the armed forces and the Danish church were still exclusively male. Nevertheless, there were a few exceptions that allowed women to be taken into public employment. The Danish Bureau of Statistics and the Danish National Archives, for instance, had female employees as early as the 1890s, although they earned far less than their male colleagues. Women could become assistants, but not public officials.[255]

The women's movement in Denmark became officially organized in 1871, the year the Danish Women's Society was founded.[256] Mathilde Bajer, the society's first chairwoman, had worked with vigour and determination to improve the conditions of women in Danish society, assisted by her husband Frederik Bajer, who would later receive the Nobel Peace Prize. The new Danish Women's Society gave women a voice in the public debate, and a number of other women's associations followed in its wake. These included the Women's Reading Society (1872), the Free Women's Society (1873), the Women's Progress Society (1886) and the National Danish Association for Women's Suffrage (1907). This development was accompanied by the appearance of various journals that focused on women's rights, notably *Kvinden og Samfundet* (*Woman and Society*) published by the Danish Women's Society (1885 to the present) and *Hvad vi vil* (*What We Want*, 1888–1894). Political issues were at the top of the agenda, and several suffrage organizations backed the *Kvindestemmerets-Bladet* (*The Women's Suffrage Magazine*, 1907–1913), *Kvindevalgret* (*Women's Suffrage*, 1908–1915), *Kristeligt Kvindeblad* (*Christian Women's Magazine*, 1910–1913) and *Dansk Kvindeblad* (*Danish Women's Magazine*, 1913–1918). But women's influence on the journalistic scene was not confined to the area of women's rights. There were also figures like the physicist Kirstine Meyer, who founded *Fysisk Tidsskrift* (*Journal of Physics*) in 1902 and served as its first editor until 1913.

The story of Danish women in science closely mirrors the history of Danish education. In 1859 it became possible for women to take a teacher's examination that authorized them to teach at girls' schools. Before the end of the

14.6 *The number of women in science gradually increased. Here a mixed group of students at work in the Polytechnical College's chemical-analytical laboratory.*

next decade the authorization to female teachers was expanded to include all subjects taught in the Danish primary school system. In 1872 the Commercial School for Women opened, and in 1875 women were allowed to sit for exams at upper-secondary level, to train as nurses at the country's hospitals and, finally, to matriculate at the university. The technical schools did not open their doors to women until 1907. Prior to that women had already gained better education opportunities with the advent of the domestic-science schools – the first of which opened in Sorø in 1895, followed in 1902 by the domestic-science institute Ankerhus Husholdingsseminarium. This latter type of school was not an emancipating element, as it was more focused on keeping women in the traditional family roles and occupational patterns.[257] Nonetheless, it was through these schools that many women first became acquainted with science, especially in the guise of domestic chemistry. The trend gave rise to a string of scientific publications on housekeeping from the 1890s onwards, such as *En lille letfattelig Husholdningskemi* (*A Little Book of Basic Household Chemistry*) by C. C. Larsen, released in 1899.[258]

Although only a tiny minority of women made their way through the university system, they had an enormous symbolic significance in the fight for women's emancipation, rights and opportunities. There was one very corporeal reason for the University of Copenhagen to begin admitting female students. Her name was Nielsine Nielsen. A 24-year-old school teacher, Nielsen had read in 1873 about how American women had begun to study medicine. As she later described the experience: "Never before or since in my life have I experienced such internal emotion. It was as though suddenly I had gone from being blind to being sighted."[259] Backed by the physician and politician Carl Emil Fenger, who had previously supported Mathilde Fibiger in becoming Denmark's first female telegraphist, Nielsen submitted an application in 1874 to the Danish Ministry of Church Affairs and Education, requesting that she be allowed to first take her exams at upper-secondary level and subsequently be admitted to the university to study medicine. The ministry passed her application on to the university's governing body, where the faculty of law was asked to prepare a written statement responding to Nielsen's request.

At the time women did not have the right to demand admission to the university. On the other hand, nothing in the university's charter prevented their admission. As the university's statement concluded: "Neither directly nor indirectly has the law anywhere stipulated a prohibition against women's access to the university." The recommendation was that women should be allowed access, as this was regarded as "a further development of the university's general goal of promoting and propagating the cultivation of science, in such a fashion as the *Zeitgeist* might demand".[260]

The faculty of medicine also looked favourably on Nielsen's application, and the following year she was able to begin her medical studies. Graduating in 1885, Nielsen became the first woman to obtain a master's degree from the University of Copenhagen. Some were less thrilled about her achievement, however, expressing worries at the implications of women's access to the field of medical study. One sceptic was the professor Mathias Saxtorph, who openly expressed his aversion to the new idea, declaring to the medical faculty that

> A woman who is able, to such an extent, to set aside every sense of decency and modesty that she might wish to attend, along with the students, lectures upon the anatomy and diseases of the male body, and that she might wish to examine and treat the surgical and syphilitic diseases in the various hospital wards, can only be looked upon with revulsion and loathing by the students.

So although the *Zeitgeist* might demand more extensive rights and privileges for women, in Saxtorph's view there ought to be a limit to their emancipation. It was bad enough that the state recognized, permitted and kept an eye on "the prostitution services", thereby taking indecency under its wing – although, he conceded, one probably had to accept prostitution as a necessary evil. Female doctors, on the other hand, were an utterly unnecessary evil "from which every Danish man must surely wish to see his country spared."[261]

Saxtorph may have been expressing what many others kept to themselves, but he was certainly outnumbered at the university. So it was that on 25 June 1875, King Christian IX signed an ordnance concerning women's access to "acquiring academic civil rights" at the University of Copenhagen, which stated the following in Section 3:

> The female students matriculated at the university shall have the same access as the other students to pursue the study of the subjects selected by them, and to present themselves for the standard examinations and academic degrees arranged by the University, when in every respect they fulfil the same requirements made of the male students with respect to previously passed examinations and to preparatory studies, and to the actual concluding faculty examinations and dissertations.[262]

Women had gained access to the university at last. The opposition had clearly had an impact, however, and they were still far from equal to their male counterparts. This was evident in relation to the upper-secondary exams they must pass prior to entering university: female students were obliged to take private lessons, and initially they were only granted the right to sit for the upper-secondary exams, not to attend the preparatory schooling. In 1877, N. Zahles Skole – a Copenhagen girls' school founded in 1852 by Nathalie Zahle – began offering an upper-secondary-level course for women. Still, it was not until 1886 that N. Zahles Skole was authorized to examine and graduate students, making it easier for young women to take an upper-secondary education. Meanwhile, they were not granted access to the state-run *gymnasium* system until 1903.

Female students were also at a disadvantage vis-à-vis their male peers when it came to privileges like study grants and residence halls. A few grants were specifically set up for women during the final decades of the nineteenth century, but no semblance of equality between the sexes was achieved until 1920. The student organizations also initially denied women access, but when Studentersamfundet – the "Students' Society" – was founded in 1882 by liberal academics, who aspired more to educate the underprivileged classes than to promote academic interests, they at once admitted women as members and seized every opportunity to emphasize how welcome they were. The old academic club Studenterforeningen – the "Students' Association", against which the "Students' Society" was basically a reaction – stolidly maintained, until 1898, that membership was a male privilege. No actual women's club existed until 1922, when the interdisciplinary Danish Association of University Women was founded.[263]

The exact number of university students during the first part of the period is unknown, but up to 1880 or so the total annual number of students at the faculty of mathematics and science would hardly have exceeded 20. Gender data is only available from 1913, at which time women accounted for about one-sixth of the scientific student body. About 20 women completed their studies with a master's degree from the early 1890s to 1920.[264] The Polytechnical College also followed suit, with more than 30 women completing degrees in advanced technology from the late 1890s to 1920, most of them as industrial engineers.

During the period stretching from the mid-nineteenth to the early twentieth centuries, Danish science underwent a professionalization that followed the pattern of the large European nations and the United States. There was still a long, hard struggle to be fought against the privileges and nepotism of former times, which proved hard to eradicate, not only at the university but at other higher-level institutions as well. Nevertheless, scientific merits came to play an increasingly decisive role in appointing individuals to academic positions. At the same time, the number of scientific graduates was growing, and they were spreading throughout society and the various trades and industries. This gave the concept of "career" a whole new meaning, and a new appeal among the growing class of science professionals in Denmark. Over the years career opportunities would also gradually open up to women, although they still had to fight for every bit of progress made towards the goal of gender-blind equality.

A small country in an international world of science

15

A changing world view

From a scientific perspective, the period offered a multitude of new theories, findings, breakthroughs and discoveries that in many respects broke with existing thinking and led to an entirely different perception of the world. By 1920 biology had a new view on man and the development of life, geology had a new view on the Earth itself, chemistry had a new view on the constituents of matter, and physics had a new view on the workings of the natural universe.[265]

In the nineteenth century physicists still endorsed a classical mechanistic world view securely rooted in Newtonian mechanics. Much had been added and there was much to discuss, but the foundation remained the same. The world obeyed the classical laws of motion. Scientists applied the fundamental concept of absolute space and time and relied on the assumptions of atomic indivisibility and immutability. Electromagnetic waves, light waves and heat radiation were propagated through an ether. Matter and ether were integrated, and energy obeyed the laws of thermodynamics. The theories of relativity and quantum mechanics in the twentieth century utterly transformed the scientific perception of the physical world.

Even before that time, a number of factors had already begun to destabilize classical physics. In 1887 the American physicists Albert Michelson and Edward Morley set up an experiment to confirm the existence of the ether – and failed. But while Michelson and Morley, and many others with them, continued to believe in the ether, their experiment provided the empirical underpinnings that enabled Albert Einstein to write, in 1905, that the ether was no longer a necessary hypothesis. As for the building blocks of nature, scientists discovered near the end of the nineteenth century that atoms were anything but indivisible. The background for this discovery – studies on cathode radiation – had taken place in the 1870s, but the pace quickened when Wilhelm Conrad Röntgen discovered his new rays in 1895, closely followed by Henri Becquerel's observations of uranium affecting a photographic plate in 1896. Just one year later, J. J. Thomson proved the existence of the electron and determined its size to be 1/2000 that of a hydrogen atom. In 1898 Marie Curie referred to the phenomenon that Becquerel had discovered as radioactivity.

At the dawn of the new century scientists demonstrated through radioactive decay that atoms were not immutable, even as Max Planck was introducing his theory of light quanta, claiming that light could be described as tiny packets of energy. In 1897, J. J. Thomson had likened the atom to "a raisin pudding", but in 1911 Ernest Rutherford declared that atoms mainly consisted of empty space. In 1913 Niels Bohr introduced his atomic model, which resembled a solar system that had a nucleus of protons with electrons moving around it in "shells". Bohr's work paved the way for modern quantum mechanics and the understanding of light as a wave–particle phenomenon.

Einstein formulated his special theory of relativity in 1905 asserting that physical systems could not be regarded from a privileged viewpoint; that all observations depended on the position and speed of the observer. It was also in this context that Einstein introduced what has become the world's most famous equation – $E=mc^2$ – declaring that Newton's laws were merely approximations correct at "low velocities". In 1916 Einstein presented his general theory of relativity stating, that light was deflected because of gravity. It was a huge triumph for the general theory of relativity when this was empirically verified during the solar eclipse of 1919. However, a far more controversial aspect of the theory was that outer space was not regarded as a regular Euclidian space, but instead was seen as a four-dimensional space-time continuum. By means of the non-Euclidian geometry, gravity was perceived as a curvature of space, which was bent by heavy objects. Thus, the planets moved around the Sun not because of gravity, but because they were obliged to follow the shortest path through a curved space.

Great changes had taken place in the physical world view since nineteenth-century scientists had sought to build it on the basis of Newtonian mechanics. With the advent of thermodynamics, the dream of a completely reversible mechanical universe crumbled. In the 1840s, along with James Prescott Joule, Julius Robert Mayer and Hermann von Helmholtz, the physicist Ludvig August Colding made a Danish contribution to the theory behind the constancy of energy and the first law of thermodynamics. Soon after, William Thomson and Rudolf Clausius presented the second law of thermodynamics, asserting that a thermodynamic system will always tend to equilibrium, which was formulated in 1867 as an increase of entropy. The theoretical foundation of electrodynamics, building on the discoveries of Hans Christian Ørsted and Michael Faraday, had been reinforced by the British physicist James Clerk Maxwell. The strong theoretical basis of physics led Maxwell to predict the connection between electromagnetic waves and light in 1861. This connection

15.1 *Albert Einstein's visit to Copenhagen in 1920 received extensive exposure in Danish newspapers and magazines. The caption of this satirical drawing by the popular cartoonist Storm P. reads: "Professor Einstein, defier of the law of gravity, visiting a wine bar in Copenhagen".*

was confirmed in 1888 with the German physicist Heinrich Hertz's discovery of radio waves – a discovery that launched the most dramatic change in human communication since Gutenberg perfected the art of printing in the fifteenth century. A few years later, in 1894, the Italian electrical engineer Guglielmo Marconi conducted his first experiments with Hertz's "invisible rays" at his father's plantation, and by December 1901 he was transmitting messages in Morse code 3,200 kilometres across the Atlantic.

Within a few centuries, the Earth had become older and older. Whereas terrestrial history was counted in thousands of years in the 1700s, and in millions of years in the 1800s, by the 1900s scientists were counting the Earth's age in billions of years. Theoretical physics and geological observations were constantly pushing the birth of the planet back in time. Even though the German geologist Alfred Wegener's theory of continental drift, first presented in 1915, was initially too controversial for his peers, it was clear that the Earth's geology had been shaped by massive forces at work for millions of years. What is more, the biological theory of evolution – most famously marked by the publication of Charles Darwin's *On the Origin of Species* in 1859 – required an exceedingly long time span to account for the development of life on Earth.

Other sciences also underwent similar drastic developments from the mid-nineteenth century until around the beginning of World War I. Fundamental theories were verified and reinforced, the empirical approach to science was strengthened, and its practical utility was highlighted. From mathematics to medicine, and from chemistry to genetics, the sciences evolved into distinctly profiled fields with professionally organized special disciplines. Science was advancing fast, in an increasingly international world where only the best and brightest were able to keep up. Danish scientists felt it too, and while some achieved fame in the international arena, others made it big in their own small country. Some were renowned for their scientific prowess; others became famous for doing the right thing at the right time. Whatever the case, they all belong to the history of Danish science from 1850 to 1920.

A distinctly Danish science

Historians of science have long discussed whether it makes sense to talk about national styles in the sciences.[266] One could argue, for instance, that the strong influence German philosophical idealism had on Danish scientists in the mid-nineteenth century – not least through H. C. Ørsted's Romantic inclinations and powerful personality – made at least some of them adhere to a special Danish style in their scientific practices. On the other hand, many reacted against Ørsted's influence, which had a particularly long-lasting impact on physics and zoology. There was, however, another way in which a particular science was specific to Denmark.

In a Scandinavian context, Arctic science played this role. Moreover, the question of science and national identity was closely related to Arctic exploration.[267] The creation of a public understanding of science was successful in the sense that in the eyes of the Scandinavian peoples, Arctic exploration was indeed *scientific* exploration. The message to the public was that polar research brought glory to the nation.[268] This understanding of Arctic adventures was eagerly supported by the explorers themselves, although at times the scientific content of their expeditions was neither very obvious nor very impressive. Despite its tragic end, the issue of national identity is evident in the very name of the illustrious *Danmark-Ekspeditionen* – The Denmark Expedition – in the years 1906 to 1908. The initiative for this high-profile expedition to the coast of north-east Greenland came from the journalist Ludvig Mylius-Erichsen.

One of its principal objectives was to confirm or disprove the existence of the so-called Peary Canal. In the process Mylius-Erichsen and two of his men disappeared. Only one body was ever recovered.[269]

In a narrower, more national sense, the importance of farming and agricultural productivity placed a natural emphasis on scientific experiments to increase crop yields using fertilizers and pesticides. These became central research topics for many plant physiologists, chemists and entomologists. Furthermore, the emphasis on maximizing agricultural efficiency even gave rise to an attempt, albeit short-lived, to formulate a particular Danish agrarian science known as "rational farming".[270]

At yet another level, Danish naturalists were influenced in their choice of topics by the country's lowland geology. Additionally, the former colonies in the West-Indies, the Faeroe Islands, Iceland and Greenland all had an impact on Danish natural history, providing a major part of the empirical data collected in the nineteenth century, with Greenland playing an especially influential role for Danish geologists. Nonetheless, many naturalists preferred to stay at home, cultivating Denmark as their "scientific garden".

One example was the pioneering research that led to an understanding of *kjøkkenmøddinger* – "kitchen middens". Briefly told, a kitchen midden is a heap of waste left by a prehistoric settlement or, more specifically, a deposit consisting primarily of shells and related cultural material found in coastal locations. The cultural artefacts in a prehistoric kitchen midden can be dated to tell us about the settlement's daily life, while animal remains can be analysed to reveal information about hunting patterns, animal domestication, seasons and climate.

The science of kitchen middens was founded when the Royal Danish Academy of Sciences and Letters in 1848 commissioned a report on certain prominent mounds of marine shells found along the coast of Zealand. The archaeologist Jens Jacob Asmussen Worsaae, geology professor Johan Georg Forchhammer and zoology professor Japetus Steenstrup conducted the investigation. They concluded that the mounds were of human origin and were, in fact, piles of domestic refuse, which they dubbed "kitchen middens". Interestingly, the Danish term *kjøkkenmødding*, as originally used by Steenstrup, worked its way as the official scientific term into other languages such as English, French, Italian and Spanish. The Danish scientists discovered that the mounds were of considerable age, dating back to the country's earliest inhabitants.[271]

It was not until the mid-1860s and early 1870s that references to kitchen middens began to crop up in English and American journals, such as the *Jour-*

15.2 *Kitchen middens were an important part of Danish science in the 1800s. This drawing shows the participants of the International Archaeology Congress in 1869 on an excursion to the Sølager kitchen midden near Roskilde Fjord. Each participant was allowed to take home a souvenir from the excavation.*

nal of the Anthropological Society of London, *Journal of the Ethnological Society of London*, *Proceedings of the Royal Geographical Society of London*, *The Journal of the Anthropological Institute of Great Britain and Ireland*, *The Journal of Anthropology* and *The American Naturalist*. An early example is an article by Charles H. Chambers in *Anthropological Review* recalling a visit to the ethnographic collections in Copenhagen.[272] The Danish term was even preferred over the English translation in texts referring to midden sites in Brazil, Britain, Northern Europe and North American – in orthographic variations like "kjoekken-moedding", "kjökken-mödding" and "kjokken-modding".[273] By the 1880s, the *kjøkkenmødding* had become a household concept among anthropologists and archaeologists throughout the English-speaking world. It also found its way to the pages of journals like *Science, American Anthropologist* and *The American Journal of Archaeology and of the History of Fine Arts.* In the 1850s and early 1860s, however, research on kitchen middens was published exclusively in Danish.

As a case in point, kitchen middens exemplify a field that existed not merely as a science in Denmark, but as a specifically Danish science – at least for a short while, until scientists in other countries began working on their own local kitchen middens in Cork Harbour, Pembrokeshire and San Francisco Bay, and at sites in South Africa, New Zealand and Japan.[274]

By the turn of the century the situation had changed completely, and in a progressively internationalized scientific climate it was hardly reasonable to talk about specifically Danish sciences – with the notable exception of "polar research", which was still framed in the public consciousness as specifically Scandinavian. Naturally this conception of Artic science was not accurate, but in the early twentieth century that was how the scandinavian public perceived it. In Scandinavia, polar research was regarded as a Scandinavian science – a bit more Danish in Denmark, and a lot more Norwegian in Norway.

A new pattern of publication

A transnational academic world was not an invention of the nineteenth or the twentieth century. Long before the rise of the nation state in the late eighteenth and nineteenth centuries, academics, scholars and learned men, and just a few women, had travelled far and wide, disregarding borders and communicating in Latin, the lingua franca of the pre-industrial world. The development taking place in the late nineteenth century was of a different nature, and resulted from the foundation of many new and highly specialized centres for the education of scientists and engineers. This gave rise to a new class of academic professionals, creating new career patterns at universities and research institutions, and in public and private corporations. The new scientific elite consisted of academic specialists and industrious entrepreneurs – a remarkable contrast to the older generation of mid-nineteenth-century scientists who still, to a large extent, allowed themselves to remain scientific generalists.

A comparison of the publication records of typical nineteenth-century generalists with those of turn-of-the-century specialists reveals several significant differences. First, the major publications of the generalists are in Danish, whereas a change in publication patterns is evident for the specialists, whose major publications mostly appeared in either German or English. This clearly indicates that the audience had changed. The scientific writers were no longer primarily addressing their fellow countrymen: The audience for Danish men of science had become international. Second, while generalists would publish important scientific findings both as monographs and in Danish journals with a more general outlook, the new Danish specialists – male and, increasingly, female – preferred to publish their results in specific professional journals in English, German or French. Gradually during the twentieth century, the English language came to dominate the publication pattern in most academic fields. Danish scientists followed the general trend along to publish in specialized English languaged journals.

Chapter 14 outlined the careers of two prominent Danish zoologists: Japetus Steenstrup, a generalist of the mid-nineteenth century, and Carl Wesenberg-Lund, a specialist of the early twentieth century. Although Steenstrup was a prolific writer he never earned a reputation as a profound thinker, and he had a tendency to venture "rather far out in the fantastic", as one commentator put it.[275] He was nonetheless instrumental to the progress of natural history as an up-and-coming discipline in Denmark. He was a much-loved lecturer, a driving force behind the university museums – particularly the new Zoological Museum – and an active participant at meetings and in societies. In short, his discipline benefited significantly from his contributions, and he was well known internationally. He corresponded with Charles Darwin and arranged for specimens to be sent from Copenhagen to Down House, Darwin's country home in England. Steenstrup was even on the exclusive list of people chosen to receive a signed copy of *On the Origin of Species* upon its publication in 1859. Yet when we look at Steenstrup's publication record, his major findings appeared in Danish.[276] Communicating with his international peers was not his main concern. Such indifference to an international audience was a luxury the new generation of scientists could ill afford.

Two examples demonstrate how drastically the publication patterns of the next generation of Danish scientist changed. The zoologist Wesenberg-Lund and the physiologist August Krogh were near-contemporaries, Wesenberg-Lund being born in 1867 and Krogh in 1874. Of Wesenberg-Lund's 15 scientific publications prior to 1900, only 1 was not in Danish.[277] From 1900 to 1909 he had 27 scientific publications, 9 of which were in Danish, besides 11 in German, 5 in English and 2 in French. From 1910 to 1920 he had 26 scientific publications, 8 of which were in Danish, 15 in German and 3 in English. His last publication in German appeared in 1914. During World War I he published very little, and in Danish, but from the 1920s most of his numerous publications were in English, with a few in French and some – especially those

15.3 *Japetus Steenstrup typified his generation of scientists, with broad interests and a primary orientation towards a Danish audience. This cover of a commemorative album from former students – in their own words, his "grateful disciples" – bears symbols of Steenstrup's scientific work in zoology and archaeology. But many gifted young naturalists got a raw deal as a result of his manifestly nepotistic dominance over the scientific community, which detractors dubbed "the Steenstrupiate".*

on wildlife preservation – in Danish. An interesting feature of Wesenberg-Lund's publication record is that he worked hard as a scientific specialist, making his findings available to an international scientific community while at the same time contributing significantly to the popular science literature designed for a lay audience. In that respect he was a model specialist, segregating his professional and popular activities through a strict publication strategy and earning an extra income from his popular writings.[278]

August Krogh received the Nobel Prize in 1920 for his discovery of the capillary motor-regulating mechanism.[279] Krogh published nothing prior to 1900, but from 1901 to 1909 he had 33 scientific publications, 15 of which were in Danish, 10 in German and 8 in English. From 1910 to 1919 he turned out 59 scientific publications: 39 in English, 7 in German (5 during World War I) and 13 in Danish. As in Wesenberg-Lund's case, most of Krogh's scientific publications are in English from the 1920s onwards, with some in French, a limited number still in German and some in Danish.[280]

The turn of the century marked a significant change in the behaviour of Danish scientists as they became more oriented towards an international world of science. In the days of the mid-nineteenth-century generalist it was still possible to be a *Danish* scientist. At the dawn of the new century one was merely a scientist in Denmark. The scientific audience had changed, as had the publication venues and the behaviour of scientists.

From kitchen middens to genetics, between nepotism and Darwinism*

Natural history is an eminent example of the professionalization, disciplinization and specialization that took place throughout the world of science from 1850 to 1920.[281] Biology, zoology, geology and geography all became independent fields with corresponding academic degrees from institutions of higher learning. Gradually Danish scientists, too, were able to make the difficult transition from generalist to specialist, and some even won international acclaim. During this period, Danish natural history achieved a number of remarkable results thanks to the innovative efforts of Eugen Warming in ecology, Carl Wesenberg-Lund in freshwater biology and Wilhelm Johannsen in genetics, culminating with August Krogh's Nobel Prize in physiology/medicine in 1920. But it was also a period marked by stagnation and nepotism, particularly in the early years, leading to a long and bitter feud.

The link between science and application was particularly important when it came to gaining support for, and expanding, the natural-history disciplines.

---

* This paragraph is partly based on Kaj Sand-Jensen's chapter on natural history in Peter C. Kjærgaard (ed.), *Lys over landet – Dansk Naturvidenskabs Historie*, vol. 3.

In addition there was an acute awareness of how education, research, knowledge and technology could be used in the service of the nation. This was expressed most tangibly in the founding of the Polytechnical College and also clearly evidenced in the inauguration of the Royal Veterinary and Agricultural College in 1858. However, there were also other interested parties such as the Royal Danish Agricultural Society, founded in 1769, and the University of Copenhagen sponsoring a series of academic prizes on topics like the calorific value of bog peat. A number of Danish naturalists from the period were frequent travellers, visiting destinations in Denmark as well as Iceland, Greenland and the Faeroe Islands. Often their mission was to evaluate the presence, quantity and quality of naturally occurring resources, to drill wells, to optimize seed-growing programmes, or to establish what food sources were available to marine life in Danish waters. Such efforts typically involved two or more, or even all, of the natural-history disciplines, and in a sense they helped to hone the professional profiles of each one. What is more, they were carried out in close cooperation with the state administration, resulting, among other things, in the establishment of the Commission for the Geological and Geographical Investigations of Greenland (1879), the Geological Survey of Denmark (1888) and the Agricultural Research Laboratory (1883).

Three institutions were particularly important in professionalizing the disciplines in the field of natural history: the University of Copenhagen, the Royal Veterinary and Agricultural College, and the privately financed Carlsberg Laboratory. Botany and zoology had already been established disciplines at the university for a couple of centuries and had their own chairs, but not until 1848 did it become possible to take a *magisterkonferens*, a master's degree, in natural history specializing in botany, zoology or mineralogy. J. G. Forchhammer became Denmark's first professor of geology when the field gained an independent chair in 1831, but it would be a good half-century before the university established a master's degree in teaching natural-history subjects at secondary-school level. This happened in 1883 when examinations were set up in zoology, botany, geology and geography, with physics and chemistry as ancillary subjects. In the early twentieth century, the academic institutionalization of natural history was a reality. The specialization was evident in the biological sciences: plant pathology, forest botany, forest zoology and bacteriology became firmly established at the Veterinary College, while plant physiology, animal physiology and genetics were established at the university. Scientists flowed quite freely among the two high-level institutions and the Carlsberg Laboratory.[282]

J. C. Jacobsen, founder of the Carlsberg Brewery, had a keen interest in sci-

15.4 *The Carlsberg Laboratory.*

ence and was deeply committed, personally and financially, to promoting scientific pursuits. He was one of the most important philanthropists of the period. Jacobsen had developed a natural interest in topics like yeast physiology, seeking to optimize his brewing processes, but his interests would eventually extend far beyond his own business activities, resulting in the Carlsberg Laboratory and the Carlsberg Foundation, set up in 1875 and 1876, respectively. The driving force behind these initiatives was to promote the general utility of science, and consequently experimental results from the Carlsberg Laboratory were made available to the public immediately. Jacobsen was adamant about publishing the laboratory's findings. This resulted in the publication series *Meddelelser fra Carlsberg Laboratoriet.* In order to reach an international scientific audience, each issue had a supplementary summary in French entitled *Comptes Rendus des Travaux du Laboratoire Carlsberg*, which over the years evolved into an independent parallel journal featuring articles in French, German and English.[283] The laboratory was split into two departments, one for chemistry and one for physiology. Several of the period's most eminent laboratory scientists carried out some, or even most, of their professional work there, among them the naturalists Emil Christian Hansen and Wilhelm Johannsen, and the chemists Johan Kjeldahl and S. P. L. Sørensen.

The Danish naturalists of the period were frequent travellers. Many made brief trips, often to Germany, but also to Sweden, France, Italy and the United Kingdom. Some also went on more lengthy expeditions to destinations like Greenland, Iceland, the Faeroe Islands, Brazil, Mexico, North America and the Danish West Indies. Many of Denmark's leading zoologists and botanists went to Germany or to southern Europe on prolonged stays as resident researchers. In other words, Danish scientists had excellent opportunities to travel and create contacts abroad. In time these activities were also reflected in the publication patterns, but as mentioned earlier, they were not necessarily

15.5 *J. C. Jacobsen, founder of the Carlsberg Brewery, was one of the greatest patrons of his time. In Denmark he came to epitomize a productive relationship between industry and science.*

part of that change. Even into the early twentieth century, scientists could still find themselves criticized for publishing work in German and for fraternizing with German scientists. One of those subject to such disapproval was the freshwater biologist Carl Wesenberg-Lund, who was quick to retort and vehemently defend sending young Danish scientists out into the world:

> All science is, in the profoundest sense, international; in the world of science no political boundaries are known, nor should any be recognized by its practitioners. Should a small nation seek to maintain them, all that it will achieve is confinement and stagnation. (...) It is only beyond Denmark's borders that our young people must themselves make an effort to obtain the stamp of first-class work that the levelling tendencies prevent us from using here in this country.[284]

All the same, this disagreement was a mere tiff compared to the conflict that festered in the Danish natural-history community from the mid-1840s until the turn of the century.[285] The conflict was rooted in an intense mutual, and personal, dislike between two of the period's most notable zoologists: the professor Japetus Steenstrup and the curator Jørgen Christian Schiødte. Ever since his youth Steenstrup, aided by good family connections, had successfully allied himself with the scientific power base consisting of the Ørsted dynasty, the Jacobsen brewery family and the geology professor Johan Georg Forchhammer. By contrast Schiødte, who came from a family of modest means and had no personal network, had to fight much harder than Steenstrup to make it in the world of science, and he was long obliged to work as a teacher outside the university. But gradually Schiødte was also able to win allies within Denmark's political circles so that he, too, managed to make a strong position for himself. The actual feud began when Steenstrup, backed by Forchhammer, proposed merging the royal natural-history collections with those of the university. Schiødte and the other royal curators were strongly opposed to being brought under Steenstrup's control, not least because Schiødte was embittered at the considerable power Steenstrup wielded in light of the insubstantiality of his scientific research. It would be 17 years before Steenstrup's plan was actually carried out. By then Schiødte was able to retain his independence by virtue of his appointment as titular professor. Even so, the inflexibility on both sides lingered on. The animosity was so palpable that as a young student in the 1860s and 1870s, the eminent fermentation physiologist Emil Christian Hansen felt he had landed in "the evil world of science".[286] Hansen, who was initially a student of Schiødte, was nevertheless later recommended by Steenstrup for his position at the Carlsberg Laboratory.

Steenstrup exploited his high standing to position his own disciples favourably when vacancies came up. Schiødte's supporters stood outside the established circles and were often out of poor families. This caused competition among the young Danish scientists eager to climb the career ladder; a competition that was largely defined by their ability to praise Steenstrup or, better still, to badger his enemies. While Schiødte was terminally ill with stomach cancer in 1884, he was the target of a scathing attack from Frederik Meinert, an erstwhile student of Schiødte himself who had now gone over to the Steenstrup faction – the "Steenstrupiate", as some contemporaries called it. Meinert's efforts bore fruit, and he was appointed as the new manager of the Zoological Museum, prevailing over H. J. Hansen, who was Schiødte's preferred candidate. Hansen did not take his defeat lightly and was one of those who perpetuated the irreconcilable atmosphere between the two factions. He also volubly expressed his bitterness, for instance in the book he wrote in 1901 entitled *Danmarks Stilling og Tilstand* (*The Position and State of Denmark*), which attacked head-on the morality of past events and the university's scientific standards. His was not the only book to debate these issues, and so the feud lived on, even after Steenstrup's death in 1897.

In 1908 the talented zoologist William Sørensen, whose career had been sidetracked, published his *Fromme Sjæles Gode Gjerninger* (*The Good Deeds of Pious Souls*). Sørensen had previously argued polemically that rather than Steenstrup, the archaeologist J. J. A. Worsaae should have been credited with discovering the Stone-Age kitchen middens, and he now applied his sharp pen against Steenstrup's successors, who were allegedly just as power-hungry and nepotistic as their predecessor had been. Emil Christian Hansen, although eventually coming to an understanding with Steenstrup, never felt quite comfortable about him: "The worst, the most intolerable and, I might add, the most dangerous thing about this old scoundrel is that he perpetually inclines his head piously to one side and swathes himself in a mild benevolence glazed over with a churchly varnish".[287]

Professionally, the position of Danish natural history was quite strong, despite the hostile, career-seeking atmosphere that reigned. One achievement was the confirmation of the Ice-Age hypothesis by the geology professor J. Frederik Johnstrup, who travelled throughout the country digging up more and more empirical evidence for the wanderings and workings of the ice. By the early twentieth century, Danish geologists had not only described the country's geology, but also provided a scientific explanation of how it had been formed. Other noteworthy figures were Eugen Warming and Christen Raunkiær, who both worked in the field of plant ecology. In 1895, Warming published his *Plantesamfund, Grundtræk af den økologiske Plantegeografi*, which is generally considered the first major work on ecology. This work was translated into German and Russian and also appeared in English as *Oecology of Plants: An Introduction to the Study of Plant Communities*.

Wilhelm Johannsen was Denmark's first professor of genetics. With his ground-breaking work on what he called "pure lines" as expressed in the self-pollinating pea bean, Johannsen demonstrated that every individual is a product of both heredity and environment. In his first important contribution to the new science – *Om Arvelighed i Samfund og i rene Linier* (*On Heredity in Communities and in Pure Lines*) from 1903, which appeared in German that same year – Johannsen introduced the concepts of "genotype" and "phenotype". His reasoning and concepts were simultaneously presented in a more popular form as an attack on Darwinism in the popular journal *Tilskueren* (*The Spectator*).[288] Johannsen's genetic research was presented more fully in *Arvelighedslærens Elementer* (*Elements of Heredity*), published in 1905 and made accessible to an international audience in 1909 with the appearance of the German version entitled *Elemente der exakten Erblichkeitslehre* – instrumental in making "gene" a household word.

Johannsen's reaction against Darwinism was based on the genetic theory of heredity, in opposition to Francis Galton's Darwinian heredity. Johannsen was only the last in a series of Danish naturalist since the 1860s who were hesitant to accept the theory of evolution. Because of Johannsen's professorship in genetics, his words carried a weight that prolonged this tendency among certain Danish naturalists for several decades. The caution Danish scientists exhibited did not reflect a lack of familiarity with Darwin's theories, although initially his ideas had a greater influence in public than in scientific circles.[289]

"How I wish that you believed in evolution!" Thus wrote Charles Darwin in his last letter from 1881 to Steenstrup.[290] Their correspondence began in 1849 when Darwin had some questions about the barnacles at the Zoological Museum in Copenhagen. Over the years Steenstrup provided Darwin with various specimens from the collections. Darwin acknowledged Steenstrup's assistance in the introduction to his *Monograph on the Fossil Lepadididæ* and reiterated his gratitude in the *Monograph on the Sub-Class Cirripedia*, both published in 1851. In *Cirripedia* he even quoted some of Steenstrup's publications from the original Danish, "translated to me by the kindness of a friend", as he put it.[291] Darwin continued to refer to Steenstrup's work in *On the Origin of Species* (1859), *The Variation of Animals and Plants under Domestication* (1868) and *The Descent of Man* (1871). Nonetheless, Steenstrup was never truly convinced about Darwin's ideas on evolution and natural selection. In general

15.6 *Despite the bitter controversies raging within the Danish natural-history community some productive research environments did emerge, for instance around the botanist Eugen Warming, who played a central role in founding the burgeoning field of ecology.*

chemie in 1911. Two years later, in 1913, the young Niels Bohr published his article series in *Philosophical Magazine*, setting the course of Danish physics for many years to come.

Having defended his doctoral dissertation on the electron theory of metals, Bohr went to England to work first with J. J. Thomson in Cambridge, then with Ernest Rutherford in Manchester.[301] Based on his experiments with the scattering of alpha particles on thin metal foils, Rutherford had already concluded in 1911 that the mass of the atom was concentrated in a tiny, positively charged nucleus surrounded by light-weight electrons, but a significant element was still missing: a stable model for the structure of the atom. According to classical physics, Rutherford's atomic model would be unstable both mechanically and in terms of radiation. Bohr resolved this dilemma by suspending certain aspects of classical physics, claiming that in nuclear physics a system can only exist in "steady states". Describing the hydrogen atom, Bohr explained how its electrons were only able to occupy certain orbits around the nucleus. Despite initial scepticism his theory soon gained general recognition and was further developed by Bohr himself and other leading physicists, including Arnold Sommerfeld. One major reason why the theory was so successful is that many of its predictions were experimentally confirmed. And although it was proved to be inadequate as early as the 1920s, the theory came to play a vital role in the history of physics, not only as the first successful quantitative theory to explain the structure of the atom, but also as a foundation for the further development of quantum mechanics.[302]

In the field of astronomy, a new observatory was built to replace the venerable but no longer workable historic Round Tower in Copenhagen. The aim was to help Denmark regain its former status as a key player. A prominent figure in this endeavour was the Prussian astronomer Heinrich Louis d'Arrest, who was summoned to take up a chair in astronomy at the University of Copenhagen in 1857. When the new observatory was commissioned in 1861 d'Arrest embarked upon the task of mapping nebulas. His 1867 catalogue *Siderum Nebulosum Observationes Havniensis* mapped 1,942 celestial objects, 215 of which had not previously been observed. Although this publication did not generate the same international attention as the English astronomer John Herschel's *Catalogue of Nebulae and Clusters of Stars*, which had appeared three years earlier, it was included in his Danish-Irish colleague Johan Dreyer's *New General Catalogue of Nebulae and Clusters of Stars*, published in 1888.

Amateur astronomers played an interesting role. There were already several amateur observatories in Denmark, and the founders of the Astronomical Society in 1916 were amateurs. The Swedish-Danish astronomer Elis Ström-

15.8 *Ejnar Hertzsprung was one of the most renowned astronomers of his time. His studies at the Polytechnical College were the beginning of a long international career. Seen here while in Leiden (standing at far left) in a party including Arthur Eddington, Paul Ehrenfest and Albert Einstein.*

gren, who was appointed a professor of astronomy in Copenhagen in 1907, was initially opposed to the new society and did not become a member until 1919. What is more, the university-based opposition to the amateur-controlled society meant that Strömgren and his associates ignored its journal *Nordisk Astronomisk Tidsskrift (Nordic Astronomy Review).* When Strömgren finally chose to become a member, however, the change was remarkable, and immediate. He at once took charge as chairman, additionally ensuring that his assistant Julie Vinter Hansen became the journal's Danish editor. Even the amateurs' initiative to launch a society and a journal was ultimately professionalized.

The physical world-view had changed dramatically from 1850 to 1920. Even though certain theoretical advances were appropriated relatively late by some Danish physicists, the process of change made its mark on Danish physicists and astronomers. Change was evident in the topics they discussed and the methods they used, and in their precision in experimentation, education, organization and career paths. There was still room for amateurs, but both physics and astronomy were increasingly becoming narrowly defined scientific fields cultivated by educated specialists.

Beer, bacteria and light*

The chemical and medical sciences had a long-standing tradition for coherence and cooperation. As the period wore on, however, they too underwent a process of professionalization and specialization. Institutionally, this set chemistry free from medicine, resulting in several independent research and development areas and infusing medicine with a new need for scientific legitimacy.[303]

Chemistry in Denmark was intimately linked with agriculture and medicine, in relation to which it served as an ancillary subject at university level. This was also reflected in the presence of chemistry at the Royal Veterinary and Agricultural College and at the Pharmaceutical College. Furthermore, chemistry had played a prominent role at the Polytechnical College since its inception. All in all, this meant that during the nineteenth century there were a growing number of scientifically educated graduates with a background in chemistry, or at least a decent working knowledge of the field. These graduates went on to careers in the educational sector and in Denmark's growing industrial sector. This new group of chemists and chemically interested professionals organized in The Danish Chemical Society, founded in 1879 and provided a social and professional framework for a group that was gradually growing more and more diverse. The field of medicine did not have to fight to justify its existence at the university, but the rapid development of the natural sciences during the nineteenth century created a growing demand for revitalization, and for medicine to develop a more well-defined scientific foundation. One result of this was the establishment of research laboratories and the formulation of an experimental medical science.

Danish chemical research during the period had two particular profiles: Julius Thomsen and Sophus Mads Jørgensen, both professors at the University

---

* This section is mainly based on Anita Kildebæk Nielsen's chapter on the chemical sciences and Morten A. Skydsgaard's chapter on medical science in Peter C. Kjærgaard (ed.), *Lys over landet – Dansk Naturvidenskabs Historie*, vol. 3.

of Copenhagen. There were also many gifted chemists at the Carlsberg Laboratory, such as Johan Kjeldahl, who was mainly known for the method he and Emil Christian Hansen devised to determine the nitrogen content in organic compounds. Hansen, who translated Darwin's *Voyage of the Beagle*, also vitally enhanced beer production thanks to his research on yeast cells. Finally there was the Carlsberg Laboratory chemist Søren P. L. Sørensen, who in 1909 introduced the concept of "pH value".[304]

Julius Thomsen was an indefatigable promoter of science. During his many years as a chemistry professor, factory manager, journal editor and university administrator he made his voice heard in public debates and in the research community. Unlike Ørsted, Thomsen firmly believed that all material was made up of atoms. His self-appointed mission was to map the internal structure of molecules, thereby revealing the forces that bound the atoms together. He sought to fulfil this mission by means of thermochemical investigations, which involved studying the heat conditions relating to chemical processes. Thomsen's meticulous experiments resulted in his publication in the 1880s of the four-volume work *Thermochemische Untersuchungen*, in which he had gathered the results of more than 3,000 experiments.[305] Sophus Mads Jørgensen, his professorial colleague, had earned an international reputation for his meticulous and systematic work on inorganic complex compounds. Nevertheless, Jørgensen's attempt to create a theoretical system for complex chemistry did not have the impact he had hoped for, colliding as it did with the coordination theory developed by the young Swiss chemist Alfred Werner.[306]

The notable figures in the Danish medical-science community included Peter Ludvig Panum, Carl Julius Salomonsen and the Nobel Prize winner Niels Ryberg Finsen. Panum became known as Denmark's first "laboratory doctor" after Carl Emil Fenger – one of the country's most zealous advocates for scientific medicine – had long expressed his hopes of seeing a "Newton of medicine", who could bring clarity to medical science and win back what medicine had lost through its lack of exact scientific methods and results. Fenger continued to hope in vain, but Panum made a valiant effort to bring science to the field of medicine in Denmark. This involved the vivisection of animals as well as chemical and physiological studies. One of Panum's institutional achievements was setting up the first modern biological laboratory at the University of Copenhagen.[307]

The tradition of laboratory research was further strengthened by the bacteriologist Carl Julius Salomonsen, who visited the German bacteriologist Robert Koch just before he published his studies on tuberculosis. Writing

15.9 *After the physician Niels Finsen received the Nobel Prize in 1903, the Danish public came to see him as the quintessential scientist working in the service of humanity.*

about his visit to Koch's laboratory, Salomonsen described how "One finds it quite chilling, seeing these hundreds of dishes with this malicious little fiend growing inside".[308] But frightening or not, the meeting was an inspiration for Salomonsen, who later succeeded in establishing a chair in bacteriology, complete with a laboratory, at the University of Copenhagen – the first of its kind in Scandinavia. Bacteriology's usefulness to society soon became abundantly clear. Louis Pasteur was celebrated as a hero during his visit to Copenhagen in 1884, having himself become a living legend and a symbol of how medical research visibly and successfully worked to the benefit of mankind. In the wake of Pasteur's discoveries serum therapy, developed in the 1890s, came to serve the cause of medical research and the Danes well with the founding of the State Serum Institute in 1902. Even before that, however, as early as 1894, Salomonsen was actively producing serum at the university with the assistance of Thorvald Madsen.

Another example of how modern science-based medicine experienced immense popular support and success was the "light treatment" conceived, practised and refined by Niels Finsen. Finsen, however, was conspicuously unlike most of his colleagues, and both professionally and institutionally he stood outside the established medical-science community. The Finsen Medical Light Institute, which treated skin tuberculosis (lupus vulgaris) with great success, was established with private funding provided by some of Denmark's most prominent industrial patrons.[309] What ensured Finsen his international attention was a high-profile stand at the Paris World's Fair in 1900. The official catalogue reporting on the fair described the exhibit from the Finsen Medical Light Institute as taking up "one-third of the space in that Section, and forming the actual basis of it".[310] The exhibit certainly had the desired effect, and the institute received the fair's highest honour: the *Grand Prix*. The following year Finsen was nominated for the Nobel Prize and received it in 1903, having been chosen over Robert Koch. Finsen's focus on treating without first researching to determine the medical cause was later criticized by his peers, including Salomonsen, who had been one of his most ardent supporters from the early 1890s on.[311] Nevertheless, this in no way detracted from Finsen's status as a national Danish hero of science. In Finsen the nation had found its first Nobel Prize winner. The fact that he was quickly forgotten outside his native country was evidently of little concern.

To the Danes, their own country still felt larger than the world, but science had already moved beyond that stage. The scientific culture had changed, and Danish scientists of both sexes now turned to face the world of international science.

Science and the city 16

Science comes to town

By the mid-nineteenth century, Denmark's major towns were in the process of installing gas lines. Odense, located on the island of Funen, with more than 10,000 inhabitants, was now the second-largest city in Denmark. Technologically progressive, Odense was the first to install gas in 1853 and also had a central waterworks up and running that year. Copenhagen came in second, setting up its gasworks in 1857. In terms of new technology Denmark was highly dependent on other countries. The 2,200 new gaslights ordered to replace the capital's 1,800 "train-oil" (blubber-oil) lamps were installed by the British firm Cochrane & Co. When the new lights were lit on 4 December, Copenhageners took to the streets to admire the city's new illuminated look. One spectator exclaimed: "Before, I could see nothing for sheer darkness. Now I can see nothing for sheer light."[312]

The literary critic Georg Brandes also recalled the transition in his young days as an event that, quite literally, threw the Danish capital into a new and different light. "It was a great day when, upon stepping out through the school gate, one saw all the new gas lamps erected, and a great evening when, for the first time, these lamps were lit."[313] Homes were equipped with indoor plumbing, complete with running water, washbasins and flush toilets – though naturally this was a gradual process. In Denmark, it was the provincial town of Nakskov on Lolland that could boast of having the nation's first water closet, installed in 1884. Still, nine years later only 200 of the town's 900 houses had one. Although latrines were still the most widely used solution, Danish families had begun to recognize the sanitary benefits that science and engineering made available.

Following the Copenhagen cholera epidemic of 1853, which cost almost 5,000 lives, the authorities began to seriously address the sewage issue, which posed a growing problem in a city whose population now totalled 163,000 – about the same as all of Denmark's major towns put together. The sewers of Copenhagen led into the city's canals and harbour. Odense used its small river for the same purpose until 1907. The population of Copenhagen had more than doubled since the mid-nineteenth century, and still its sewage simply drained into its open waterways. By the early 1890s, with more than 100,000 barrels of latrine waste stored in locations around the capital, it had become abundantly clear that the problem was rapidly getting out of hand. Engineers, doctors and bacteriologists were consulted. Many found water closets too expensive, pointing instead to other technical solutions that would permit the composting of bodily waste. In addition, the issue of fertilizer value weighed heavily in a country still largely dependent upon agriculture. The Association of Farmers' Unions in Jutland protested vigorously against the water closet after the Municipality of Århus tabled an initiative to reform waste management in Århus and environs. The farmers' unions on Zealand agreed with the Jutland organizations, demanding that latrine waste from Copenhagen be transported out of the city by rail at reduced rates to avoid squandering its valuable fertilizer potential. Nevertheless, the epidemic issue carried a lot of weight with the authorities, as did the ever-growing bills from private waste-handling contractors, and both factors reinforced their desire to create an independent waste-management system.

The solution to the capital's pressing problem was presented by Julius Thomsen, who, in addition to being a professor of chemistry and the managing director of the chemical plant "Øresund", was a member of the Copenhagen city council. Professor Thomsen's proposal was that the city's new sewer system must be linked to a pumping station that could push the wastewater out into the strait of Øresund. In addition, the city would construct a storage facility to hold three days' worth of latrine waste, which could then be transported by rail to agricultural areas outside the city, for farmers to use. If they did not, that too would be pumped into the strait. This was a cheap solution that made the municipality of Copenhagen independent while at the same time satisfying the farmers. After the plan was realized in 1903 the waters of Øresund were no longer the same – neither for the animals that had lived there, nor for the people who used to swim there.[314]

Around the turn of the century water closets were still quite expensive. A connection to the new sewer system cost 100 Danish kroner. Given that a skilled worker's daily wages amounted to 3 or 3.50 kroner, the water closet was still a luxury reserved for the higher segments of society. Those who could not enjoy the privilege of using the new sanitary technology themselves could read about it in the 18-volume illustrated encyclopaedia *Salmonsens Store Illustrerede Konversationsleksikon* – the most ambitious work of its kind in Denmark. Appearing in 1893–1907, the encyclopaedia contained an article for "Klosét" spanning a full four columns, including illustrations. No less would suffice to explain its features and functions. Around 1900, reading about the water closet was still the closest most people would get. In *Salmonsens Konversationsleksikon* the engineer Frederik Valdemar Meyer, who had just transferred from a position with the Copenhagen Gas and Electricity Services to the Copenhagen Road and Sewerage Services, provided readers with detailed technical explanations as well as instructions on how to construct and comfortably arrange one's water closet. However, the engineer did not stop at that. For those who feared that the new contraption – what was conceived as the ultimate luxury of modern civilization – would be too comfortable for those frequenting it, Meyer was able to offer prudent and practical advice: "If, on the other hand, one intentionally wishes a visit to the C. to be unpleasant, thus preventing the stay from becoming unnecessarily protracted, one can arrange for a jet of steam to pass beneath the seat at regular intervals." In other words, there was virtually no limit to the detailed information, practical instructions and guidance available from "a marvellous new world of invention", and no shortage of engineers with bright ideas.

The world's first public power plant was built in 1882. From 4 September, the Pearl Street Station on lower Manhattan provided 82 consumers in New York with enough electricity to operate 400 lamps. It would be almost a decade before Denmark followed the American example of building public power plants. Odense and Køge installed electrical street lighting in 1891, with Copenhagen following suit a year later. Similar delays in comparison with the United States were seen in many European cities, where the authorities had invested heavily in gas infrastructure and were averse to inviting competition from a new lighting system. The first private application for a permit to build a central electricity plant in Copenhagen was submitted in 1884 – and rejected by the municipal authorities, who simultaneously formulated the rules that would apply to all future applicants. The conditions they set were so rigorous that private applicants were unable to comply.[315]

This did not mean, however, that Copenhageners had no opportunity to see the new "magnificent electrical light".[316] Thomas Edison introduced his filament bulb in 1879, but more than two decades before, in 1857, the Danish fireworks specialist Chief War Commissioner Høegh-Guldberg demonstrated "a peculiar physical phenomenon belonging to our age", by suspending an electrical light from a lamp-post at the riding grounds behind Christiansborg Castle.[317] Høegh-Guldberg's light was not much brighter than one of the city's gas lamps, but the invention of the filament bulb brought on rapid development. Although the municipal authorities obstructed the building of large-scale private electricity plants, small generators were not prohibited, as long as their purpose was to supply electricity to a single company or private property.

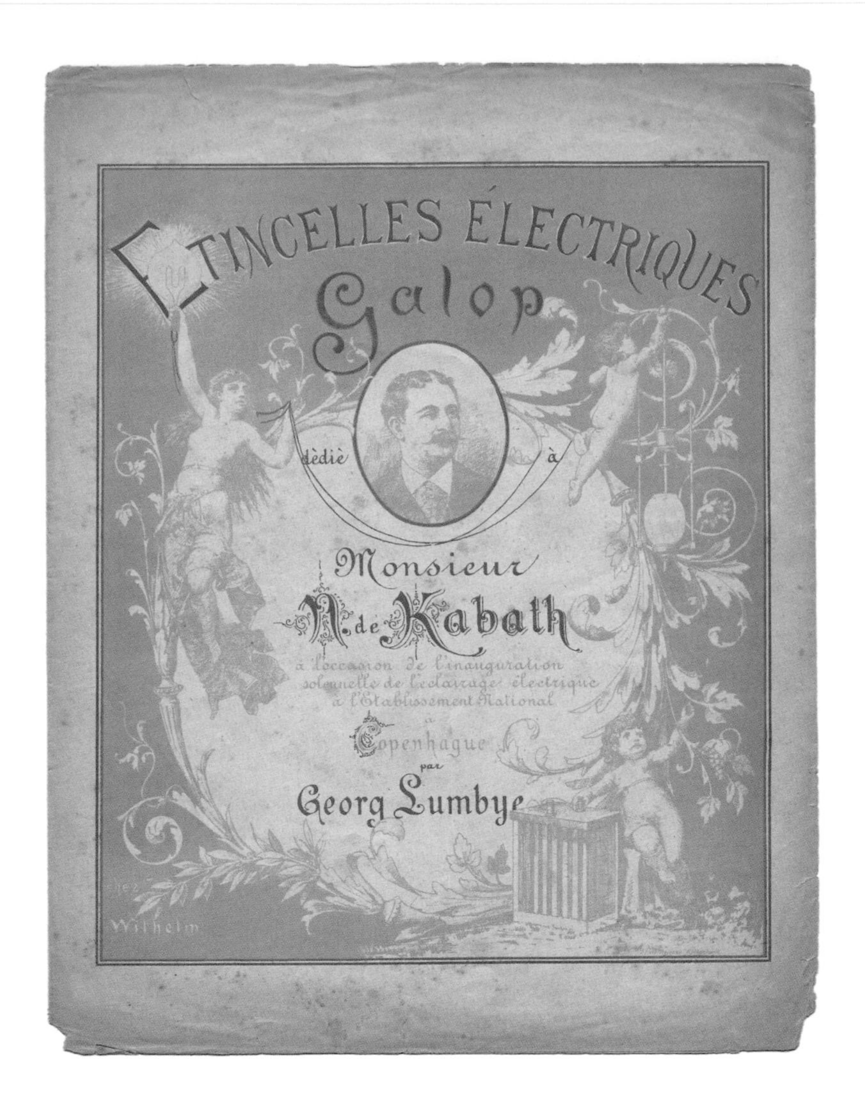

16.1 *The new electrical light created considerable public attention. "The National" – a fashionable venue in Copenhagen – even announced an "electrical extravaganza" to celebrate its new lighting installations. To mark the event, the Danish composer Georg Lumbye presented a galop aptly entitled* Etincelles électriques *– "electrical sparks".*

In 1879 the shipbuilders Burmeister & Wain were the first to introduce electric lighting. Three years later Carlsberg installed two light machines complete with arc lamps and filament-bulbs lamps.

But ordinary citizens, too, were soon able to enjoy the electric light. In March 1882, *Illustreret Tidende*, a conservative weekly targeting the upper middle class, informed their readers that they could expect something quite remarkable at the newly opened entertainment venue "The National", which had already impressed Copenhageners and out-of-town visitors arriving by train: "Despite the owner's having installed fully adequate lighting with numerous gas brackets, he has already concluded an agreement with the *U.S. Electric Lighting Company*, which, from the month of April, will arrange for the illumination of the entire establishment with 200 Maxim's filament lamps, arc lamps, and a projector."[318] On 25 July, The National opened its doors to host a spectacular "electrical extravaganza" where the atmosphere was reported to have been almost "Oriental". From the ceiling the light from of the projector beamed down upon the festive crowd while the popular composer Georg Lumbye conducted the orchestra to the strains of his *Etincelles électriques* – "an energetic *galop* bursting with electrical sparks, composed for the occasion". There had been much talk about the number and types of lamps installed, and about how powerful the generator was. Nevertheless, the reporter from *Illustreret Tidende* was evidently more interested in conveying his own impression of the experience – no doubt influenced by the generous provisions his host had supplied. "It was a clever fellow who came up with the idea of bottling champagne: May his memory live on, in the blaze of the strongest electric light".[319] The citizens of the capital were accordingly swept off their feet in their electrical intoxication, or at least that was the image embraced by the popular press.

Yet the ongoing development worried some of Copenhagen's progress-oriented politicians, who feared that too many private generators might make it impossible to set up a rational municipal electrical system. In 1889, the city therefore decided to send gasworks engineer Ib Windfeld-Hansen on a four-week European study tour that would enable him to report on the advantages and disadvantages of various techniques and administrative solutions. Windfeld-Hansen's report was a decisive factor in the developments that would follow in subsequent years. It gave the Danish politicians what they wanted: good arguments that electrical lighting would not damage the sale of gas, combined with good arguments for constructing a system in which municipal direct-current plants would make up the backbone of Copenhagen's electricity supply. The reaction was quick in coming. In 1889 the municipal authority submitted its own application to the city council, requesting permission to establish a municipal electricity supply that would initially consist of one central electricity plant. The application was approved, resulting in the 1892 opening of the electricity plant in Gothersgade, where Windfeld-Hansen was employed as works manager.

Electricity quickly gained ground and made its mark on urban life in Denmark. By 1920 the capital city of Copenhagen – which in 1850 had been malodorous and filthy, and whose inhabitants, crammed together within the confines of its old ramparts, were easy prey for cholera and other epidemic diseases – had evolved into a modern, bustling metropolis with bicycles, motorcars and electric tramcars. The Danes quickly took the new technologies to heart, although they were not blind to the problems such progress entailed. Newspapers and magazines offered scathing satire, and the realities of the new traffic situation were frequently, and comically, lampooned.

Progress, with all its inventions and changes to city life, also ricocheted back, impacting science in unexpected ways. One report from the Danish Meteorological Institute, featured in *Illustreret Tidende*, described how the new tramlines so strongly affected observations of terrestrial magnetism with their "vagrant currents" that the scientific results were completely off kilter. But as the journalist wryly remarked, "about this, nothing can be done; science must yield to the demands of this day and age for rapid, easy and comfortable transport".[320] The magnetic observatory was later permanently removed to more tranquil surroundings in the forest of Rude Skov, north of the capital.

A city in change

The new scientific institutions in and around Copenhagen greatly affected the face of the city and were the object of considerable public attention.[321] This also made them recurring topics in daily newspapers and weekly magazines. *Illustreret Tidende*, for one, was always pleased to inform its readers about the plans, arrangements and end results in the nation's proud new edifices of science, and to report on the men behind them. The importance of these institutions was communicated to the Danish public through an abundance of impressive illustrations showing façades, laboratories, and serious gentlemen of science – almost all of whom were depicted, around the turn of the century, in white coats and engaged in scientific pursuits meant to benefit the nation and humanity at large. Besides being useful, these institutions were also necessary if Denmark was to follow the path of progress and ensure the quality of the scientific work being done – which would, inevitably, affect the nation and its people. These sentiments are echoed in the following quote from the inauguration of the physician Niels Ryberg Finsen's new Medical Light Institute in 1902:

> We have moved, particularly over the last few years, from a sceptical and quite infertile period in medical science into a very fertile but rather uncritical age in which theories and medicines that 20 years ago were banned are now once again being noticed. But precisely for that reason, we need scientific institutes where new things can be critically and dispassionately tested, and in that respect one can gladly look to the Medical Light Institute.[322]

Denmark's national pride went hand in hand with international fame, thanks to Finsen's "wonderful chemical rays" and their almost miraculous ability to cure the disfiguring disease lupus vulgaris, popularly known as skin tuberculosis. Every Dane was familiar with the image of Finsen in his white coat, and knew of his institute and the cause he championed. Another source of national pride was the Carlsberg Laboratory, which to the Danish public was a symbol of what Danish traditions and values could amount to when allowed to shine on the scientific firmament. The laboratory was known for its altruistic and industrious work, which "already at this stage has borne fruit in theory and in practice, the value of which the entire world has recognized, and benefited from."[323]

Because the scientific institutions were visible monuments to Denmark's progress and prosperity, their aesthetic significance and appearance was the focus of intense interest. As a case in point, comments to that effect were liberally tossed around in public at the inauguration of the new botanical laboratory in 1890. Going against expert advice, the original structure dating from the 1870s had been built "as cheaply and as frugally as possible", resulting in a "shoddy-looking construction". The new building was contemporary, suitable for the scientific activities it was meant to house, and capable of accommodating "all future requirements for expansion". All in all, the appearance of the new building was found to be satisfactory, a high point being its "red brickwork with ornamental use of lighter varieties of stone", which gave "a light and lively – almost over-lively – impression. It is reasonable to assume, however, that this will become somewhat less prominent when the Copenhagen air has deposited its usual black coating of grime upon our porous domestic stone".[324] Air pollution was already a well-known problem in the late nineteenth century.

Generally speaking, the scientific institutions were noted for their architecture, which underscored pure, elevated science; the work of mind and body complementing each other. The main building of the new State Serum Institute, for example, was emphasized as being "of extremely attractive construction in the palatial style", while the Royal Veterinary and Agricultural College had an "exceptionally lovely location" and buildings that were "beau-

16.2 *At the time of its inauguration in 1859, the Royal Veterinary and Agricultural College was located in park-like surroundings, despite its proximity to central Copenhagen.*

tifully grouped in relation to one another". The Finsen Medical Light Institute was similarly admired for its "monumental stone edifice".[325]

These and other large building projects completed during the period gave science a highly visible presence in and around the capital. It was a remarkable upgrading effort that required massive investments, but which also yielded enormous benefits in terms of research results and international prestige. In the decades from 1850 to 1920, science mentally and physically came to hold sway in the public and private sphere. One encountered science in the city's prominent façades and in its most private closets. The people of Copenhagen no longer lived in a little city that was clogged, cramped and congested, but in a modern, spacious metropolitan area blessed with sewers, electrical trams and scientific institutions, as befitted the capital of a small nation.[326]

A new alliance

Although many scientists had begun to appreciate the commercial potential of their field, there were still many who were more gifted in science than they were in business. One of Denmark's most original inventors, Poul la Cour, had a master's degree in physics and excellent technical skills. La Cour submitted a number of patent applications in Denmark and abroad.[327] In 1875 he sought to patent an invention that would solve one of the most irksome problems telegraphy was facing at the time, namely the ability to send many messages simultaneously on one telegraph wire. His patent was accepted in Denmark, but ran into priority issues abroad. In this case, and in others, la Cour was obliged to go through a number of exhausting patent lawsuits. He was, however, highly recognized in his native Denmark. His telegraphy invention, for instance, earned him a gold medal from the Royal Danish Academy of Sciences and Letters. His recommendation for membership of the academy, written by some of the country's leading scientists – the physicists N. J. Fjord, C. Holten and L. A. Colding and the agricultural economist B. S. Jørgensen – stated that, provided certain technical and practical difficulties could be overcome, "then the significance of the discovery cannot be overestimated, and it must be classified – not only scientifically, but also from a purely financial standpoint – alongside the most important inventions to have appeared over the last 50 years in the field of electromagnetism".[328]

Nevertheless, la Cour found it too difficult to take his invention to an international market and to capitalize science on his own, and so he formed an alliance with the powerful entrepreneur C. F. Tietgen, who in 1866 had set up the tremendously successful "Store nordiske Telegrafselskab" – Great Nordic Telegraphy. In 1871, Tietgen's company had completed the wire running across Siberia and all the way to the Pacific, and in just a few years the enterprise grew to be almost as large as its counterparts in America. In other words, Tietgen had a flair for science and technology as well as a nose for business, and he readily identified the weak point in la Cour's endeavours: la Cour was still too much the scientist, and not enough of a businessman. This motivated Tietgen to write to la Cour's brother: "It does not come naturally to your brother, you know, to make the thing practically usable, and that is what lies at the heart of it". As Tietgen went on to explain,

> he feels especially called to continue his scientific research, convinced as it were that the practical significance of the thing will be evident some day. But that is wrong. What would it have availed, had Columbus sailed out, and after having become certain himself that there lay a South America, and then changed course in order to arrive at a personal certainty that there also lay a North America, or an Australia, though without having obtained fully valid evidence of the sea route to the first location, but relying only on his subjective conviction about it. No, it is about the practical result, and as soon as this has been achieved we can use his apparatus on the Nordic telegraph company's wires.[329]

The scientist sorely needed the businessman if he was to step out of the laboratory and into the new world controlled by finances. Then again, the businessman needed the scientist's professional skills to develop new products he could sell.

The practical uses of scientific knowledge had moved to the very top of the agenda in the early twentieth century. The common interests of the two different worlds became clear when, working side by side, men of science and men of industry founded Denmark's Society for the Natural Sciences. Although the physicist Martin Knudsen was the initiator and a driving force behind the society, he had the assistance of the industrialist G. A. Hagemann, a great patron of science and a solid support. Hagemann had no doubts, as he expressed in a letter to Knudsen in September 1913: "It seems to me that connecting the purely scientific works with practical goals must be the aim of all science." Although he did not intend for scientists to think about the utility of their results each time they entered the laboratory, Hagemann still declared that "science, in progressing, has found many uses, and these uses have in turn had a stimulating and supportive effect upon science".[330] Electrotechnology, dye chemistry and light technology were all examples of practical uses that in their turn had positively affected scientific development. The cooperation and the reciprocal influence of the two communities had been a precondition for this development, and had contributed to bringing Denmark up to the level it had reached just before the outbreak of World War I, worth maintaining and reinforcing.

The period was characterized not only by the rise of the professional scientist, but also by the rise of the industrialist. A new, strong alliance was forged between two of contemporary Denmark's most distinctive groups – one in white laboratory coats, the other in dark suits and top hats.

Science in a world of business*

The slow, gradual change of an old agricultural country into a modern industrial nation meant changes in the landscape and in the workforce. Factories appeared, and with them a new working class, new neighbourhoods and new

---

* This section is based on Henrik Knudsen and Peter C. Kjærgaard's chapter on private and industrial research in Peter C. Kjærgaard (ed.), *Lys over landet – Dansk Naturvidenskabs Historie*, vol. 3.

social movements arose. But there was also room for a new type of employee in the labour market: the scientific or technical specialist.

The mass production of textiles, soap and beer satisfied an ever-growing demand for material goods in step with the new patterns of consumption, and along with mass production came stricter requirements for product standardization. Quality was, above all, a guarantee that consumers could rely on product standards. This became one of the most significant driving forces on the domestic market and in exports. Lur Brand butter, standardized margarine colour and beer brewed using controlled fermentation became symbols of the new industrial age. To the Danish population such uniformity signified stable production, and on export markets it signified an exclusive Danish-grade quality that consumers could trust.

In Denmark's delayed industrial revolution, which lagged behind countries like the United Kingdom and Germany, science came to play a crucial role in the new business culture as a driving force and a hallmark of quality.[331] The Polytechnical College, teaching advanced technology to a new generation of engineers, was a central factor, but so were groups like the pharmaceutical graduates, who were influential as industrial processes increasingly came to use, and indeed to depend upon, scientific methods. Just as academic science had many faces, industrial science, too, was diverse. It was evident at the family dining table and in household chemistry, but it was equally obvious in the large manufacturing halls, the processing of raw materials and employee routines, not to mention the managing director's ambitions.

In this context the metaphorical light of science shone brightly. By founding the Carlsberg Laboratory – Denmark's first real major industrial research laboratory – in 1875 J. C. Jacobsen had already made his scientific ambitions clear. The same thinking preceded his establishment of the Carlsberg Foundation, a charitable endowment that would benefit science and the arts. As Jacobsen wrote in a letter addressed to the Royal Academy in 1876:

> ... given that such an institute, intended for special studies, can only thrive when it is upheld by the spirit and permeated by the light that issues from the sciences and letters at large, and given that this light has been, for me, a source of joy and contentment, it is my heartfelt wish that in partial repayment of my debt I might also make a contribution to the general promotion of the sciences, mainly in those directions in which, it appears to me, the state has not so far used, nor will with any probable likelihood in the future be able to allocate, all of the necessary funds.[332]

Through the efforts of the Carlsberg Laboratory and Carlsberg Foundation, the light of science was intended to shine selflessly over Denmark. These bodies enabled the achievement of many important scientific findings that won considerable international recognition, not least because of the civic spirit and freedom of mind that made the Carlsberg Laboratory's results immediately and readily accessible to the public. There was no close guarding of industrial interests. Scientific research was considered one of the pillars of modern society to be shared by all, thus benefiting as many as possible.[333]

The financial capital that the business community had at its disposal also contributed considerably to basic scientific research in Denmark. It was, for instance, with the support of the Carlsberg Foundation that Carl Wesenberg-Lund was able to set up and run Denmark's first freshwater biological laboratory independently of the university in Copenhagen (see chapter 14). As Jacobsen himself expressed it, it was "for the benefit of science and the glory of Denmark".[334]

There was a very different and far more practical approach towards science – in fact an urgent need for it – in the highly technology-dependent manufacturing companies. For them, the industrial laboratory was at the very heart of developing new products and processes. Around 1900, a handful of the largest Danish companies had set up in-house works laboratories that not only verified quality and operations, but also experimented to improve products and techniques.[335]

The cryolite-processing plant "Øresund" was already running a small testing laboratory as early as the 1850s, but the Carlsberg brewery was the first example in Denmark of a company that systematically conducted scientific research on a large scale in connection with its production. The first brewery laboratory at the Carlsberg complex, located in Valby in the western part of Copenhagen, was set up in 1871, but just four years later a new and larger laboratory was commissioned as scientific activities were intensified. Even by international standards, Jacobsen the brewer was leading the way in strengthening the ties between science, technology and industry.[336]

As far as laboratory activities were concerned, the Danish agricultural sector was ahead of industry. In the 1880s the Danish state and the Royal Veterinary and Agricultural College joined forces and set up the Agricultural Research Laboratory, which carried out advanced experiments relating to agricultural production. A number of private agricultural businesses were also quick to begin innovative work based on systematic technical experimentation. Industrial food manufacturers like "De danske Sukkerfabrikker" (sugar), "Christian Hansens Osteløbefabrik" (rennet), "Aarhus Oliefabrik" (vegetable

16.3 *Around 1900, the world's largest production of margarine was located in Århus. Scientific knowledge was converted into financial wealth as the Jutlanders made good in the capital, conveniently aided by modern, high-speed transportation.*

oil) and "Otto Mønsteds Margarinefabrik" (margarine) had laboratories of their own in the 1890s.

Business and science went hand in hand. If the large Danish industries were to hold their own in the growing global market, their technological development would have to support basic scientific research as well as easily applied knowledge. This research increasingly took place behind factory walls, and behind the closed doors of private laboratories that were keeping their results secret for commercial reasons. However, the business community also generously supported scientific research elsewhere, and clearly it was becoming more dependent on graduates with degrees in science and advanced technology. Especially after the turn of the century, the frequency and intensity of contact between scientists and the captains of industry grew, gradually creating the modern Danish society's most powerful alliance. Small companies also benefited from the dissemination of scientific and technological expertise that was far removed from the university, but which still found a concrete and tangible expression in the plethora of new products coming from the development of scientific processes.

This development was a result of contributions from a variety of institutions, companies and individuals. The Polytechnical College was a key factor in at least two respects: first, in its education of graduates who could readily fulfil much of Danish industry's crucial demand for employees, and second, in its practical, utility-oriented approach that stood firm in an otherwise academic scientific culture. Speaking at the college's 75th anniversary in 1904, the prominent industrialist G. A. Hagemann expressed the following sentiments:

> As true as it is that the Polytechnical College has, so far, been the home of chemistry, physics and the technical sciences in this country, it is equally true that our entire production today is based on these sciences, and equally true that the nation's meadows have yielded a rich crop. From this place, teachers have been educated for chairs at the University as well as the technical schools. From this place came a Thomsen, a Fjord, a Segelcke and a whole series of names. But first and last; from this place came the spirit of science that has been infused into the entire people of Denmark. Scientific perspectives are now taken by the smallholder as much as by the landowner and the industrialist. Systematic measurements and investigations are the lever that is thrust under each unresolved agronomic and industrial question.[337]

The spirit of science flowed through everyone and everything. During the years from 1850 to 1920, Danish industry developed a dependency upon scientific knowledge and skills that became an elemental part of people's daily lives, permeating everything from the food they ate to the beer they drank and the soap they used for washing.

Trade and industry exhibitions*

One distinctive trend that emerged worldwide during the nineteenth century was a new exhibition culture. Accordingly, many Danes first encountered science and science-related products at fairs and expositions. Within trade, industry and agriculture people recognized this new avenue as an excellent way to approach a wider customer base. They could directly reach and affect potential buyers, partly by offering a personal experience and partly by making the events increasingly spectacular to attract the attention of the media. Denmark was no exception.

---

* This section is based on Rikke Schmidt Kjærgaard's chapter on fairs, expositions and museums in Peter C. Kjærgaard (ed.), *Lys over landet – Dansk Naturvidenskabs Historie*, vol. 3.

16.4 *The Danish Technological Institute, founded in 1906, became an important component in the integration of science, technology and business. The institute's main building, shown here in 1918 at the centre of the photo, was located in the heart of Copenhagen.*

The manifestation of science in the public sphere gave the Danes many new opportunities. This was true for exhibitors, inventors and scientists, but also for regular citizens, upper class and working class alike, to whom science and its topics became accessible in a genuinely new and more palpable sense. It was true that science in the public sphere was often far removed from science in the scientific sphere, but it was nevertheless visible, and it was reaching a wider audience than ever before. At the same time newspapers and the illustrated press were making the events and the objects on display accessible to a captivated and growing audience.[338]

Denmark never hosted a world's fair, but between 1850 and 1920 the country did host a number of expositions that concentrated specifically on industry, or on certain regions or professions. Early in the period, such events were still held exclusively in the Copenhagen area, but gradually, as the trend became more established, they began to appear in provincial towns as well. After 1900 provincial Denmark even gained ground in relation to the large industrial fairs Copenhagen had monopolized during the preceding century. Actually, the very first industrial exposition in Denmark was held in 1810 and featured exhibits from 66 companies. In 1831 the Society for the Dissemination of Natural Science, collaborating with the Royal Danish Agricultural Society and the Royal Danish Academy of Fine Arts, arranged a fair in Copenhagen for domestic products, and also headed up two similar though smaller events in 1834 and 1836.

A few years later the Confederation of Industries in Copenhagen, founded in 1838, took over the administration of national and international industrial fairs, and took industrial exhibitions to a whole new level. The organization was already able to open its first event at Charlottenborg palace in central Copenhagen in 1840, aiming to arrange such an exhibition every four years. Ambitions also ran high elsewhere in Denmark, and several major provincial towns would later host specific, small-scale trade shows in connection with meetings, rallies and conferences.

All of these activities were inspired by the many international industrial shows held during the first half of the nineteenth century, most often with Denmark as a participant. The first major Danish exposition to achieve financial support from the Danish state was the industrial fair held in Copenhagen at the riding grounds behind Christiansborg Castle – just one year after the 1851 World's Fair in London. The approximately 750 stands, which were not exclusively Danish, were seen by about 100,000 visitors during the fair's three-month run. This was a breakthrough event for all those involved – exhibitors, planners and visitors – and even included the publication of a magazine called *Industri-Udstillings-Tidende (Industrial Fair Times).* Although not previously seen in Denmark, such publications were later a natural part of several similar events.

During the second half of the nineteenth century Denmark hosted two large pan-Nordic art and industrial shows, both located in Copenhagen. The first took place in 1872 and coincided with the inauguration of the Confederation of Industries' new rooms next to Tivoli Gardens. This show also had its own magazine *Industri- og Kunst- Udstillings-Tidende.* This publication was free and featured brief articles on various aspects of the exhibition itself, supplemented with descriptions of other locations and events worth seeing in the Copenhagen area in 1872. There were also advertisements for assorted shops, information on opening hours for museums and gardens, listings of steamship and railway information, a list of Copenhagen's banks and financial institutions, a briefing on the Danish postal, telegraph and customs services, a list of public baths, rates for hansom cabs, listings of exchange rates for Danish

16.5 *The main building of the Nordic Industrial, Agricultural and Art Exhibition in Copenhagen, held in 1888.*

kroner in relation to Swedish, Norwegian and Prussian currencies, a map of Copenhagen, notices of major events taking place during the exposition, and much, much more – in short, everything a visitor might conceivably need to know.

The second major show was held in 1888 as a centennial celebration to commemorate Denmark's abolition of adscription – bonded tenure or, effectively, villeinage. This show was called *Den Nordiske Industri-, Landbrugs- og Kunstudstilling i Kjøbenhavn,* the name reflecting its location and wider scope, which included industry, agriculture and the arts. The fairgrounds spread over what is today the City Hall Square, Tivoli Gardens and an extensive area stretching all the way to the waterfront. This show aimed at uniting the Nordic region, science and progress in one common cause. A wave of patriotism rode on the back of science and technology, which were listed among the most crucial ideological resources for a future Denmark that would be strong and closely linked to the rest of the Nordic region. The 1888 show was a remarkable success. Never before had such large sums been invested in a Danish exposition, never before had so many things been on show or the exhibits been seen by so many people, and never before had any show received so much exposure. For its entire duration, the illustrated press ran lengthy articles about inventions, items, activities, business deals, technology and science, spiced with reports on fashion, high-society life and must-see events. The two largest weeklies, *Illustreret Familie-Journal* and *Illustreret Tidende*, vied for reader attention with pages of richly illustrated reporting.

The Regional Fair in Århus in 1909 was the first in a series of similar fairs that could give the great expositions in Copenhagen serious competition. A commemorative publication, *Festskrift til Landsudstillingen i Aarhus*, contained articles about the town of Århus, which by then had grown into a modern provincial capital, just like Hans Christian Andersen's ugly duckling had grown into a beautiful swan. Great pains were taken to create and maintain a highly professional atmosphere focusing on science, technology and the objects on display.

Critical voices had previously complained that such events were degenerating into mere fun fairs, with more entertainers and scams than legitimate content. The organizers of the Regional Fair in Århus had been very conscious of this, and had therefore kept amusements and diversions at a minimum in order to keep visitors more focused on the objects and installations. Their efforts evidently bore fruit, and the press coverage of the opening in the regional newspapers of Funen and Jutland was effusive: "Such a work can only be created by a master" *(Fyens Stiftstidende),* "A beautiful portent of our nation's progress" *(Bogense Avis),* "The regional exposition surpasses the World's Fair in Stockholm" (*Ringkøbing Amts Dagblad*), and "A more magnificent exposition has never been seen in Denmark" *(Esbjerg-Posten).*[339]

At the national and regional expositions Danes experienced science put into practice. Industry and agriculture were prominently represented with a focus on electricity, household chemistry and the mechanization of society. White coats and laboratories were far away, but these events gave them a means to reach the public. The expositions, fairs and shows gave science a face that citizens could easily recognize and identify with, thereby greatly facilitating the integration of science into the daily lives of the Danish people.

White coats and wellingtons 17

From tailcoat to lab coat

A great celebration was held on 7 November 1850 in honour of the physicist Hans Christian Ørsted's fiftieth anniversary at the University of Copenhagen. With due pomp and circumstance, Ørsted received a life tenancy at Fasangården, a picturesque residence nestled in the extensive gardens of Frederiksberg Have. The geology professor Johan Georg Forchhammer and the dramatist Johan Ludvig Heiberg praised him in speech and song, the university's rector gave him a diamond-studded doctoral ring, and the king appointed him Titular Privy Councillor. He was applauded by students young and old, and the Students' Association made him an honorary member. That evening there was a torchlight procession in his honour. When he died just a few months later, following a brief illness, Denmark mourned a much-loved scientific hero.[340]

More than half a century later, in 1908, the Danish Society for the Dissemination of Natural Science instituted a gold medal, H. C. Ørsted Medaljen, in memory of the society's founder. This medal was first awarded in 1909, to S. P. L. Sørensen, head of the chemistry department at the Carlsberg Laboratory. There was no doubt of the honour being bestowed upon Sørensen. The illustrated weekly *Illustreret Tidende* reported on the event, exclaiming that "it will bring about general concurrence and pleasure that this modest and agreeable man of science should be the first to hold this distinction."[341]

Although the honour accorded to Sørensen bears no comparison with the celebration of Ørsted's anniversary and the degree of national grief at his death, there is a very significant difference in the way these two scientists were portrayed. Ørsted was always impeccably dressed and often surrounded by the symbols of science, and he exuded an aura of serenity and calm authority, fitting for a national patriarch of science. Sørensen, on the other hand, was always just taking a short break in his busy day, surrounded by glass bulbs and beakers in his white laboratory coat, and eagerly awaiting a chance to resume his work. This is illustrative of a major presentational shift. By the end of the period, the model scientist first portrayed as an unflappable gentleman in an immaculate tailcoat had evolved into a specialist on the go, clad in his trusty

17.1 *The white lab coat became the new symbol of science, as seen in this painting of the chemistry professor S. P. L. Sørensen surrounded by his research assistants, both male and female. Impeccably dressed, the professor could now leave the dirty work to them.*

white lab coat. While the former had to be drawn or painted, the latter might just as well be captured on a photograph, dashing from one place to another. Despite these differences, one thing Ørsted and Sørensen did have in common was their status as objects of national adulation and pride.

During the period 1850–1920, a new role was conferred upon the practitioners of science. It was the role of national heroes; icons who symbolized progress and a better life for all. Their new celebrity status was apparent in many ways.[342] The increased visibility of science in academia, in towns and cities, and in the everyday lives of the Danes also made the scientists themselves more visible. The individuals behind the great discoveries and inventions, behind the professional expertise, behind the science used in everyday living, were sought out by newspapers and popular magazines alike. The public enthusiastically kept up with the latest developments, even as the practitioners of science increasingly hung up their tailcoats and pulled on their lab coats. They became heroes, and were therefore inherently interesting, which also made them the object of considerable attention.

Life in the laboratory

Public interest did not stop at the representatives of science. It also extended to their surroundings and the places where they lived and worked. People had discovered that the world of science was inhabited, and now they also wanted to know what it looked like, and what went on there. Scientists in action were followed with the keenest interest, and their work was carefully chronicled, whether they were in the chemistry lab in their white coats, or in the marshland in their trusty wellingtons.[343]

The Danish laboratories were a popular topic, regularly praised in weekly magazines for their pure aesthetic beauty. The laboratory offered readers an opportunity to meet science and the men behind it – as well as the women who, as yet, largely continued to participate only as students and assistants. It was here, in the laboratory, that the nation's future wealth and glory would be ensured. Danish laboratories were fully on a par with those found abroad – and then, according to the national press, they had that special something that, put them in a league of their own. As *Illustreret Tidende* put it when describing one of the Polytechnical College's treasures: "As concerns equipment and size, our electrotechnical laboratory can measure up to similar laboratories abroad, and as concerns tasteful appearance, I hardly think it is surpassed by any other".[344]

The State Serum Institute similarly exemplified the sort of scientific style and classic design ideals that exuded scientific respectability: proper scientists doing proper science in proper surroundings. Tradition combined with modern convenience to reinforce the institution's almost moral atmosphere of scientific propriety, so obvious in its exterior and interior design. The tour concluded at the very top of the institute's hierarchy:

> At the eastern end is a beautiful chemical laboratory, a weighing room, and an attractively furnished room for the Director, Prof. Salomonsen. It almost goes without saying that all of the rooms have electrical lighting, produced by an in-house dynamo driven by gas, as well as central heating and telephones connecting all of the various rooms. Nowadays a well-equipped laboratory can hardly do without such amenities.[345]

17.2 *The electronic laboratory at the Polytechnical College was regarded worldwide as one of the most modern facilities of its kind – and, in Denmark, as one of the most beautiful.*

However, it was not only the aesthetics of such rooms that was fascinating. The activities that went on among Bunsen burners and bell jars also held a special attraction. Reading detailed accounts of the white-coated scientists' everyday lives was a favourite diversion, and at the same time it gave people a picture of the practical considerations that went into constructing and equipping a scientific laboratory. Passages like this description of a bacteriological laboratory transported readers right into the lab, allowing them to vividly imagine its daily routines:

> Here beneath the windows is a worktable, along the length of which run four pipes with numerous taps and valves. The one pipe contains cold water, the second hot, and the third pipe leads to an air pump and can thus be used for suction. Finally, the fourth is a run-off pipe to the sink, supplied in each workplace with a funnel. This arrangement, which is repeated throughout the laboratory, enables much time to be saved in fetching and carrying water and wastewater.[346]

Such laboratories ensured the health and wealth of Denmark and its people. They were the country's very foundation. It was quite natural that the Danes wanted to know what they were like, and to feel secure in the knowledge that their own future and that of the nation were in the best possible hands, and had the best possible conditions. The laboratory was the little-known and rarely-seen place that made the modern world go around, with white-coated scientists as the inconspicuous champions of their age. The implicit argument ran that any non-believer would be convinced after such a visit by proxy to Stein's Laboratory, which was the independent analytical laboratory serving Denmark's agricultural sector:

> Upon walking through the laboratory one will find, in flasks and retorts, in test tubes and crucibles, under the microscope or in the photographic dark room, everything that we in Denmark live and build upon, burn, fertilize, live off, and suffer from. There are samples of coal and kerosene, the most up-to-date treatment of fertilizers and feedstuffs, the tuberculosis bacterium innocently resting under the microscope, magnified handwriting samples – fraudulence exposed – the matter-of-fact solving of blood dramas, ... not to mention the analytical treatment of our good Danish butter or the examination of other foods.[347]

Science was everywhere in the lives of the Danes, who hungrily devoured news from the laboratories and now knew whom to thank for the improvements they enjoyed on a daily basis. The contemporary coverage in newspapers, magazines and journals bears witness to the massive public admiration and reverence accorded to science and its practitioners.

Scientific heroes

With the coming of electricity, the steam locomotive, the telegraph, the telephone, the radio, the motorcar and the aeroplane, science became firmly established on the social and cultural agenda. It appeared in practically every conceivable context, accompanied by an unprecedented growth in new technology. It brought enthusiasm and optimism, giving rise to a new social class with its own national heroes. The public consciousness no longer distinguished sharply between science and technology. The borders were blurred by the abundant inventions that fuelled visions of a cleaner, better and more efficient society whose amazed and elated citizens would benefit from an endless succession of new marvels. These visions had the engineer, the chemist, the bacteriologist and the inventor standing side by side. Together, they fulfilled a

brud (Breaking Through) from 1888 has one character, Dr. Plesner, who in an opening speech to the Students' Society says: "The science that sits in splendour on Olympus has lost its significance. The ear of the people is open; you must go out to the people. The scientist's most sacred calling is to use his science to build a bridge between the segments of society."[351] This gave Jørgen, the novel's protagonist, a good reason to study the natural sciences.

But the people also had to enter the domain of the scientist. In 1906 the notorious debater and literary critic Georg Brandes spoke to the university's new first-year students about his visit to the inventor Valdemar Poulsen's laboratory. He had gone there to see Poulsen's new "arc transmitter" – a system that could transmit continuous high-frequency electromagnetic waves from a sender to a receiver, and was, in fact, the foundation of wireless telephony. It was not the technical details that captivated Brandes, however, but the very act of observing the scientist at work. The opportunity to meet him in his natural setting, the laboratory, and gain insight into his daily world and work.

> For the layman who is led into Poulsen's laboratory, it is amusing and exciting to see what forces of nature are set in motion here by such diminutive machines; to see the fire, whizzing like music, dart out of the apparatus, shaping itself like a multiforked trunk of blue fire; and then, when two arcs unite, to see the bluish, crackling flame become a yellow blaze that lets sparks fly with a noise like a waterfall's thunder.[352]

"If only Denmark had a few more like him!" an impressed Georg Brandes exclaimed.

Even though such encounters enabled reporters to visit the practitioners of science in their own world, in their element, there was no doubt as to the formidable stature and importance of the hosts. *Illustreret Tidende* invited its readers on such a visit to the Meteorological Institute for a talk with another Poulsen: the institute's director, Adam Poulsen, who with "masterly calm and never a hint of bitterness" remarked that people only remembered weather forecasts when they were wrong. And since no one ever noticed when the forecasts were right, his was a highly ungrateful post. But this did not make the scientist's hands tremble. An unassuming man, Adam Poulsen still knew his own scientific place and his own scientific worth, and he courageously took up the heavy burden of ensuring reliable weather reports for those who depended on them. The impact upon the reporter of meeting the composed and sharp-witted director clearly shone through:

> And indeed, it is a large kingdom that Mr. Adam Poulsen rules from his office, where the walls are ablaze with northern lights. Like Tycho Brahe, he has the heavens all around him, knowing like Noah that never, as long as the Earth exists, shall frost and heat, summer and winter, day and night – in short, the meteorological phenomena – ever die down.[353]

The meteorologist was always on the spot, the readers were assured – in the service of the people and the nation.

Such effusive hagiographical portrayals of men of science were a common standard into the twentieth century. Another example from the Meteorological Institute demonstrates that its practitioners were recognized for more than their role in the weather service. In 1909 the return of Halley's Comet caused a public panic as the news spread that the comet's tail contained cyanide gases, which were known to be lethal. "The prussic-acid comet", as it was popularly called, was a major issue in Denmark and elsewhere, and newspapers hotly debated the pressing question of whether the tail's potential contact with the Earth's atmosphere might be lethal to human beings, or was completely harmless. Enterprising quacks successfully sold large numbers of "comet pills", claiming they would protect users against the toxic gases from outer space. And the authorities in Boston even planned to issue a general alert if the comet was visible on Wednesday, 18 May 1909 – the day the comet's tail would skirt the Earth over a period of six hours.

Even this could not unnerve "the brave" scientists at the Meteorological Institute. That very evening, Carl H. Ryder (the institute's director), Dan la Cour (head of the weather service) and Georg Krebs (skipper and engineer) intrepidly rose towards the heavens, lifting off from Valby gasworks in a hot-air balloon loaded with scientific instruments. Like the rest of humanity, they came through the incident unscathed – but returned, alas, without any significant mesurements.

The Meteorological Institute had also been at the centre of momentous events the previous year after the American explorer Frederick A. Cook turned up at the Nordre Toldbod customs-house wharf in Copenhagen, virtually on the institute's doorstep, and was given a hero's welcome as the first person to reach the North Pole. Just days later, Cook's compatriot Robert E. Peary announced that *he* had been the first to reach the North Pole, which made it necessary to investigate the matter in depth. Director Ryder of the Meteorological Institute chose to regard Cook's claim as credible until disproven. A committee was set up to resolve the dispute with Dan la Cour and the astronomy professor Elis S. Strömgren among its members. The committee's report, made public a few months later, established that based on meteorological and

17.4 *First flights were enormously prestigious and the pioneers of aviation were highly competitive. The Danish inventor Jacob Christian Ellehammer was one of many who sought recognition as the first airborne European. Although others beat him to that title, he soon became a hero in his native Denmark. This is the first picture immortalizing Ellehammer in the air – though not the first flight on European soil.*

astronomical criteria, Cook's account was highly dubious. Peary subsequently received the honour and retained it for most of the twentieth century, although modern-day scientists seriously doubt that either of them ever made it to the North Pole at all. But at least initially Cook was celebrated in Denmark, until science came along and took away his laurels. Being a hero of science was not always easy.

A scientist's heroic standing was no shield against satire either, as clearly exemplified in the public coverage of the versatile inventor Jacob Ellehammer, the first Dane to become airborne. Ellehammer dreamed of being the first to fly in Europe, but was beaten by the Brazilian aviator Alberto Santos-Dumont, who received official recognition for the first motorized flight on European soil after flying 61 metres in Paris on 23 October 1906. In 1908, however, Ellehammer won a major aviation competition in Kiel, along with 5,000 Deutschmarks and some of the honour that had previously eluded him. According to Ellehammer himself, it was the easiest money he had ever made: lasting 11 seconds, the flight had earned him a good 450 Deutschmarks per second. He received a diploma in 1909 at the aeronautical exhibition in Paris in recognition of his pioneering efforts. Nevertheless, his contribution to the history of flying was soon overshadowed by the many new records achieved in the years that followed. And yet in his native Denmark, Ellehammer's limited international recognition had no effect at all on his status as a scientific hero – or a favourite butt of public jokes. A darling of the nation, he was respected and admired for his scientific talents and perseverance, and at the same time fondly teased by his fellow Danes.[354]

As the period wore on, a number of scientific archetypes crystallized: the tireless servant of humanity, the absent-minded professor, the white-coated lab researcher, and the inventive genius with the quirky ideas. All of these types were images reflecting the public perception of the scientific practitioner. People now knew what a scientist looked like, where he worked, what he did, and why. From 1850 to 1920, the scientist came to be seen as a recognizable type. He was a hero, sometimes an eccentric, but most often a friend of the people, a son of the nation.

Even so, the scientist – be he ever so much a hero, an icon, and archetype, a servant of humanity – was also something else. Behind his public image, a new view of science and of human beings was gradually emerging. In the words of the poet and writer Sophus Claussen, as expressed in his manifesto "Ny Aand" ("New Spirit") from 1925: "The men of science are people, just like the rest of us".[355] The scientist still carried the responsibility for the future on his shoulders, but the spell was broken. The superhuman hero had become a mortal man.

The most popular and widespread archetypes among men of science fell into three categories: the engineer, the Arctic explorer and the agricultural consultant. And at long last the woman of science also entered the scene.

The engineer*

The advancement of the industrial society, coupled with the intense cultivation of spectacular technical machinery, and combined with a distinctly progress-oriented optimism, made the engineer a model hero of the new age.

* This section is based on Michael F. Wagner's chapter on engineers in Peter C. Kjærgaard (ed.), *Lys over landet – Dansk Naturvidenskabs Historie*, vol. 3.

During this period the engineer was celebrated as few other professionals, and he was willingly and prominently drawn into the public arena. Georg Brandes wrote of progress in 1883 that it was "a monstrous, ever-restless, never-ending procession; humanity's eternal quest towards an ever more ideally perceived freedom; the ceaseless triumphal procession of civilization and progress".[356] The cultivation of scientific advances, the veneration of new technology, and the heroizing of the engineer made people feel comfortable and confident about progress. When things were moving so quickly, the Danes wanted to rest assured that they were in good hands, and only a few critical voices were raised.

The engineer was also used in the efforts to morally bolster the Danish public. During the 1870s some of the tireless popularizers of science such as André Lütken and Herman Trier translated a number of Samuel Smiles's puritanical writings. They made works like *Egen Kraft* (originally *Self-Help* from 1859) and *Pligt* (*Duty*, 1880) and edifying engineer biographies accessible to a Danish audience. Lütken went on to adapt what finally became *Opfindelsernes Bog* (*The Book of Inventions*), a thoroughly edited Danish version of the German work *Das Buch der Erfindungen.* And Trier threw his energies into the educational–pedagogical journal *Vor Ungdom* (*Our Youth*), which subsequently came to play a significant role in the country's pedagogical development and debate.[357] The Victorian railway engineer George Stephenson gave readers a British example they could emulate, showing how to negotiate the rocky road to success using will-power, vigour and perseverance.[358] The Smilesean heroes "succeed because they possess determination to the highest degree".[359] These engineer portraits were done in a realistic style that matter-of-factly presented the social conditions and the long, hard road towards their goal, which the hero attained solely by the force of his own strength of mind and ability to overcome the obstacles in his way. The engineer became the consummate modern hero, who not only quite literally was helping to build the modern world, but who was also able to serve as an example and promote the personal development of others.

Engineers were also celebrated in the weekly magazines. *Illustreret Tidende*, for instance, ran a series of engineer portraits with elaborate engravings and extensive eulogies that helped to consolidate the picture of engineers as society's faithful servants and the modern world's finest men. Both in form and style this series was strongly reminiscent of *Nature*'s famous "Scientific Worthies", but there were certain differences. To qualify for inclusion in the Danish series one preferably had to be not only a "polytechnician" – a graduate of advanced technology from the Polytechnical College – but dead as well.

17.5 *Many novels, like Otto Rung's* Den hvide Yacht *(*The White Yacht*) from 1906, typecast the engineer as a heroic protagonist.*

The series was launched in 1866 with an obituary of Edward August Scharling, a faithful supporter of Ørsted, and there was no limit to the scientific and personal virtues attributed to the deceased: "On no occasion and in no place would he ever compromise on the truth, just as Ørsted never would. He has, therefore, in his scientific chemical works, upheld [Denmark's first chemistry professor William C.] Zeise's great reputation for an unusual degree of cleanliness, accuracy and order, and every work that bears Scharling's name inherently also vouches for its great reliability. The same fundamental view also applied to his personal comportment".[360]

Sharling's obituary was written by another polytechnician, Eugen Ibsen, who had been one of Scharling's students. In this case it did not matter that Scharling was a chemist. As a professor at the Polytechnical College he had been part of the old-boy network, and his was the first of many portrayals that were strongly idealized, compared to other available descriptions of these figures. Virtually all of these portraits followed Samuel Smiles's pattern of the self-made man; the selfless and hard-working specialist with the high moral standards, who put the interests of society and others far above his own.

The engineer was also elevated to a literary figure during this period, in Denmark and in other countries. This literature featured an increasing number of engineers and other technical professionals, often casting them as protagonists rather than mere supporting characters. One of the best known and most enduring examples in this genre was written by Henrik Pontoppidan, who shared the 1917 Nobel Prize in literature with another Danish writer, the Darwinist Karl Gjellerup. As a youth in the Jutland market town of Randers, Pontoppidan had observed first-hand the building of the Danish railways. After a meeting with British railway engineers decided to become a man of the modern age himself. Thus motivated, Pontoppidan took off to Copenhagen to study at the Polytechnical College. This was the story behind his eight-volume odyssey *Lykke-Per*, which appeared between 1898 and 1904. Like the author himself, Lykke-Per – "Lucky Peter" – is a parson's son who travels to Copenhagen to become an engineer and seek his fortune. His dream is to make Denmark an industrial nation, and to exploit the energy potential hidden in the ocean's waves. His high-flying ideas find little support at the college of advanced technology, but a wealthy entrepreneur with an eligible daughter finances Per's technical educational journey through Europe. Just when life seems successfully lined up for this "man of the twentieth century", Per's parents die and he experiences an intense spirituality that he thought he had abandoned long ago. *Lykke-Per* becomes a story of how technical triumphs, faith in the future and optimism all combine and then, just when everything seems to fall perfectly into place, how they leave the characters with an unmistakable sense of loss.

An altogether different sort of book is the German journalist and writer Bernhard Kellermann's action-packed futuristic novel *The Tunnel*. His original work *Der Tunnel* dates from 1913 and was an instant success, with three impressions in its first year alone and subsequent translations into 25 languages, including Danish. *The Tunnel* follows the trials and tribulations of MacAllan, a brave engineer fighting valiantly for global progress. Modelled on Isambard Brunel's visionary tunnel proposals, MacAllan's dream is to link the old world and the new by drilling a gigantic railway tunnel beneath the bed of the Atlantic Ocean. The tunnel itself becomes a symbol of the great project of civilization, intended to link all of humanity and transcend historical boundaries. The novel's uncritical heroizing of the engineer as the virtuous defender and saviour of the future was expressed in Kellermann's novel as the literary conviction that technology's superiority would ultimately be decisive for the global development of civilization.[361] And this called for men who were cast in the right mould. Fortunately, such men were easy to find in the weekly magazines and contemporary literature; characters as convincing as they were one-dimensional. The Danish readers could feel secure about leaving their future with the engineers. Pontoppidan had initially felt the same, until as a young student of advanced technology he was surprised to find that such dreams were made of a different stuff than the tangible realities at the Polytechnical College.

The Arctic explorer*

Denmark's great Arctic adventures unfolded in the late nineteenth and early twentieth centuries. The "polar explorers" of the north were not just a part of the scientific world, notably in the fields of geology and geography. They also played a very prominent public role. Arctic heroes were popular, and the Scandinavians had become quite used to extolling their achievements. The Norwegians applauded Fridtjof Nansen and Roald Amundsen, while the Swedes celebrated Alfred Gabriel Nathorst and Nils A. E. Nordenskiöld, and the Danes revered Knud Rasmussen and Lauge Koch. One did not have to be a Scandinavian to receive the admiration of the people. When the American explorer Frederick A. Cook arrived in Copenhagen in 1909 and declared himself the first to reach the North Pole, thousands gathered on the waterfront by the Meteorological Institute. Everyone was eager to see the new "polar hero", and neither the Copenhageners nor the newspaper and magazine readers were disappointed. The event was front-page news, and Cook's purposeful countenance could be studied in the extensive photographic reports provided in special supplements. Of course, all of this was forgotten when Cook was stripped of the honour a few months later.

There is no doubt that the Scandinavian peoples' Viking legacy played a role in their glorification of Arctic explorers. The earliest Scandinavian settlements in Greenland had been described in the Icelandic sagas, and had been internalized as part of the Nordic heritage. The efforts made in the early seventeenh century to explore, describe and map Greenland and its indigenous

---

* This section is based on Christopher Jacob Ries's chapter on polar explorers in Peter C. Kjærgaard (ed.), *Lys over landet – Dansk Naturvidenskabs Historie*, vol. 3.

17.6 *The classic heroic Arctic explorer with dogsled and whip was gradually superseded by a new polar research culture. In 1922, the Danish explorer Lauge Koch made one of the earliest attempts to employ modern technology in the Arctic. In fact, just moments after this photograph was taken, the caterpillar tractor got stuck in the snow and was not used again.*

peoples were indeed partly founded on a wish to search for early Scandinavian settlers. Another issue that affected the Danes' own perception of Arctic exploration and research was Greenland's status as a Danish colony for several centuries.

The many Arctic expeditions taking place from the early twentieth century onwards enjoyed massive attention, thanks not least to the American explorer Robert E. Peary. His country was interested in securing areas that enemies might use to conduct an overland invasion of North America. Alaska was purchased from the Russians in 1867, which meant the United States only needed to buy Denmark's possessions in the West Indies and Greenland – but they were not for sale. In 1892, Peary had discovered the much-debated Peary Canal, ostensibly establishing the north-eastern part of Greenland as an independent island that was given the name Peary Land. The Danes were clearly interested in keeping Greenland in one piece, so to speak, and thereby maintaining Danish sovereignty over the entire region. In other words, the exploration of Greenland was intimately linked to large-scale geopolitical considerations, and to Denmark's national pride and integrity, which is why the findings of the Arctic explorers carried an additional weight that lay beyond the realm of science. This, and the fact that the expeditions were bound for one of the most dangerous regions known on the planet, a region that already had a long history of heroic struggles and tragic deaths, had the makings of legendary tales.

The numerous Danish expeditions to the Arctic gave rise to two great figures, each with his own iconic profile. One was Knud Rasmussen, who personified the spirit of Arctic conquest combined with romantic primitivism and ethnic reconciliation and brotherhood.[362] Rasmussen was of mixed Greenlandic-Danish descent and had been raised in Greenland. This endowed him with an outstanding knowledge of Greenlandic language and culture and made him familiar with the local population's own practices and techniques for moving around in their vast country. The Danish writer, Nobel Prize winner and advocate of Darwinian theory Johannes V. Jensen enthusiastically described Rasmussen as the perfect mix of Greenlandic and Danish features: "What organs that man must have had! And if he inherited these from an ancient people, then we have here the key to explain how humanity was able, before the advent of technology, to subjugate the entire Earth, all the way from the Arctic Ocean to the Tierra del Fuego".[363]

To the Danes, Rasmussen was great in life, but even greater in death. As the Danish newspaper *Nationaltidende* wrote in 1935: "He was personal charm from head to heels. He was ever surrounded by an aura of light and festivity, and one got the impression of a man who lived his life free of all complexity. Easily and playfully he conquered, the youth gave him torchlight processions, the university made him an honorary doctor, the entire country called him by his first name".[364] The great Knud was everyone's friend – and everyone's hero. He knew Greenland and the Greenlanders, and he linked history with the present, forging ties between the two countries and their peoples that the Danes had longed for.

The other great figure was Lauge Koch, who was Rasmussen's diametrical opposite and the personification of scientifically based, rational Arctic exploration. Koch was extremely diligent in ensuring the scientific usability of his results, and he augmented the field with every modern scientific gadget he could think of. The classic "polar hero" with dogsled and whip was gradually challenged by a new technological culture. In his attempts to modernize Arctic research, Koch brought a tractor with caterpillar belts on his 1922 expedition. It was to be used for transporting large geological samples over short distances, but in practice the idea was a total failure. The powerful machine was bogged down in snow during its initial test run and was never brought into use again. Clearly the dogsled had not outlived its usefulness. As it happened, just before the breakdown of technology in the Arctic, Koch was photographed on his indomitable tractor – with an air of the modern scientific hero and master of the wilderness.

Koch's personality made him much less of a beloved hero than Rasmussen was. As the same newspaper that had glorified Rasmussen explained: "It never occurred to anyone to call Lauge Koch by his first name. As for his doctoral title, he had to earn that himself on a geological dissertation. Heavily and stubbornly he worked his way to the name he wished to make for himself as an explorer". After 1920 it had become more difficult to be a scientific hero. Apparently, it could even be seen as an impediment to have worked to acquire one's own scientific title. Rasmussen, with his honorary degree, held a much stronger position in the hearts and minds of the Danes than Koch ever would, despite his numerous and hard-earned scientific merits.[365]

The nature of Arctic exploration changed. New technologies appeared in the shape of field stations, aeroplanes and radios, replacing the traditional tents, dogsleds and handwritten accounts left in stone cairns for others to find. Denmark's mental image of the lonely "polar adventurer" was replaced by international specialist teams of Arctic scientists. Koch moved on, down the path of science. Rasmussen stood still – but also stood out crisp and clear in the collective consciousness of his fellow Greenlanders and Danes.

The agricultural consultant*

The economy of Denmark, an old agricultural nation, was still chiefly based on things that grew in the fields or grazed in the meadows, as trade was becoming a more important factor in maintaining the country's economic growth. It is hardly surprising that the Danes were interested in making agriculture and animal husbandry as efficient as possible, and here science had much to offer.

---

* This section is based on Anita Kildebæk Nielsen's chapter on agricultural consultants and Casper Andersen and Hans Henrik Hjermitslev's chapter on rural science in Peter C. Kjærgaard (ed.), *Lys over landet – Dansk Naturvidenskabs Historie*, vol. 3.

17.7 *Against the backdrop of the natural alliance between agriculture and science in Denmark – an old agrarian nation – a completely new type of hero was born: the agricultural consultant. With his white coat and portable laboratory, the agricultural consultant became the personification of hope and progress in rural communities across Denmark.*

The various institutions of higher learning, particularly the Polytechnical College and the Royal Veterinary and Agricultural College, were involved in numerous cooperative networks among scientists, educators and agriculturalists.

What farmers increasingly realized during the nineteenth century was that science and its practitioners would be excellent allies in their endeavours to improve resource utilization and optimize yields. Such improvements were necessary, not only in light of Denmark's rising trade abroad, but also in light of the country's growing population. Merely expanding the amount of arable land would not be enough to fill export orders and feed the growing number af Danes, nor was there much room left for expansion once the heaths of Jutland had been put under the plough. Intensification was the only way for agriculture to expand. The response to these needs was the development and dissemination of "rational farming", which was largely based on scientific insights. This theory included improved breeding methods, industrially manufactured equipment and machinery, professional training of crop and livestock farmers, and the establishment of an agricultural-science environment that could convey to the public the advantages, and the necessity, of rational farming.[366]

Obviously, institutionalized science as it existed at the places of higher learning did its part, but it was clear that science now had to make its way into the fields. From 1850 to 1920 a knowledge-based infrastructure developed that handled the education of farmers at folk high schools and agricultural schools. In addition to this, Denmark established systems of test farms that worked closely with scientifically trained staff, and it was this context that made the agricultural consultant a scientific hero in his own right, personifying qualified science rolled out to each farm. It was the consultant who would help agriculture ride the wave of industrialization; who would transform the old smallholding into an efficient production facility, allowing the farmer himself to evolve from peasant into operations manager – and to run his agricultural undertaking much like a factory manager would run an industrial enterprise.

The concrete link between research and practical farming, between the Agricultural College and the farms, came in 1882 when the Agricultural Research Laboratory was set up in Copenhagen. The first head of the new laboratory was N. J. Fjord, who had taught physics, meteorology and mathematics at the Agricultural College since its founding in 1858. Actually, Fjord had begun a systematic agricultural testing programme many years earlier, initially financed by the Royal Danish Agricultural Society. Gradually the programme had began to receive support from the Danish state, understanding how important it was to promote the Danish economy as a whole. Although the new Agricultural Research Laboratory began very modestly with just a chemistry laboratory and a small staff, it was soon expanded with departments for bacteriology and animal physiology.

Although based at the laboratory, the agricultural consultants would visit individual farms, arriving and setting up their portable laboratories if a farmer was having production problems that required further investigation, for instance, an inexplicable aftertaste. One such case led Denmark, as the first country in the world, to pass legislation in 1898 ordering the pasteurisation of all skimmed milk destined for use as calf feed. Another requirement, dating from 1911, called for pasteurization to be used at all dairies wishing to export butter under the Lur Brand. This distinctive new brand, introduced in 1906, was a mandatory quality stamp required for all Danish butter sold abroad, and it quickly became an international symbol guaranteeing consumers a wholesome, clean, uniform product.[367] Thanks to this new and closer relationship between agriculture and research, between problem and solution, Danish farmers developed a confidence in science, and in the scientific institutions and their representatives.

The number of agricultural consultants grew during the final decades of the nineteenth century, but it was still a small profession. Around 1900 the total number of state-employed national agricultural consultants was 18. Twenty years later there were 22. Meanwhile, during the same period the number of consultants employed by local agricultural associations had swelled from 40 to around 150. Most of these local consultants were graduates from the Agricultural College, and their activities were co-financed by the Danish state, which paid half of their salary.[368]

Knowledge was a necessity for the modern farmer. He could educate himself, read reports, and attend public lectures. But when things went wrong and the farmer, despite his best efforts, was at his wits' end, he could call on the agricultural consultant. White-coated and with his portable bacteriological laboratory he became a celebrated hero of his time.

The women of science

Even the world of science did not consist of men alone. That is not to say it was easy for women to find a place for themselves alongside the men of science. Gaining access to the university and securing the right to sit for exams was just the first of many steps on what would turn out to be a very long journey. One of the several reasons for this was that the unequal treatment of men and women reflected a pronounced and robust perception of biological dissimilarity between the sexes. This perception was supported by various scientific hypotheses, which made it all the more difficult to eradicate. Such views were chiefly propagated in male-dominated academic circles.

As late as in 1925 the Danish geneticist Wilhelm Johannsen wrote that women were not well suited as scientists. It was one thing, he said, to acquire existing knowledge, but quite another to do original research and create new understanding. In this area women were not biologically equipped with the energy and fortitude that such efforts required. Women did, however, make superb assistants. The more useful biological qualities of the human female included conscientiousness, patience, diligence, and an aptitude for repetitive

17.8 *Only after the 1903 reform were women granted access to Denmark's publicly run secondary schools. Prior to this, they were obliged to take instruction at special schools for girls, as here in a mathematics class at N. Zahles Skole.*

work. The genetics professor therefore applauded the presence of women at the university, even going so far as to say that female laboratory assistants were superior to male assistants. In Johannsen's view, "qualified female assistance is virtually indispensable", whereas "in independent research, I believe that women, on average, stand far weaker than men."[369]

Johannsen was not alone in these views. Lis Jacobsen, a linguist receiving her doctorate in 1910, agreed that women had not contributed significantly to research. The main reason, in her opinion, was that women lacked the intellectual power and the capacity for concentration that men possessed. The greatness of men was determined by their ruthlessness, a character trait to which women were not predisposed. The incontrovertible biological reason for the difference was, Jacobsen said, that a woman's emotional instincts were greater than her cognitive powers. A woman's true calling was not science, but motherhood.[370] These views were not universally shared, but they did account for a great deal of the resistance afflicting the women's movement during this period.

The actual courses of study at the university's faculty of mathematics and science were no different for female students than for male. No distinction was made, even though individually the professors and male students showed discreet consideration for the fact that women were now present in the lecture halls and in the labs. The physicist Kirstine Meyer, writing about her years as a student, recalled how her geology professor, Johannes Frederik Johnstrup, felt rather nervous at having only one female student among all the young men. And so, for her sake, "he was always extremely punctual in turning up for his classes, in order that she could make the time she was obliged to spend alone with her male peers quite brief." When arranging an excursion to the rocky Baltic island of Bornholm, Johnstrup arranged for a female acquaintance to accompany them, so that his one female student would not again find herself alone and surrounded by men. This was, as Kirstine Meyer put it, "very considerate – but quite unnecessary".[371]

Julius Thomsen, the professor of chemistry at the University af Copenhagen, was the only one who was displeased with women entering the university. As Meyer expressed it, Thomsen preferred to ignore their existence altogether. In 1888, for instance, when Thomsen was the director of the Polytechnical College and new buildings were being furnished, he opposed any special arrangements in the toilets and changing rooms that might anticipate the presence of women among the student body. By contrast, the chemistry professor Sophus Mads Jørgensen outfitted a temporary changing cubicle for women in a supply room near the laboratory.

Men still clearly dominated the scientific arena when it came to education and employment, yet gradually more women began to appear and to make their mark on Danish society. The first Danish woman to obtain a degree in sciences from the University of Copenhagen was Sofie Rostrup. Having graduated from upper secondary school in 1884 as a "private" pupil, Rostrup concentrated on natural history at the university and successfully defended her master's thesis, obtaining her *magisterkonferens* in 1889. Following a suggestion from her thesis supervisor Jørgen Christian Schiødte (see chapter 15), Rostrup had chosen insects as her topic. Taking additional inspiration from her father-in-law, the plant pathologist Emil Rostrup, she further specialized in the interaction between insects and plants. This gave her a solid foundation on which to firmly establish herself in a new and useful field: pests and how to control them. This field would prove to be immensely important to Denmark's economy. With financial support from the government, she and Emil Rostrup went on a series of journeys into different parts of Denmark from 1896 onwards. Their mission was to investigate pest attacks and assess possible measures to fight them. This led to the publication in 1900 of her handbook *Vort Landbrugs Skadedyr blandt Insekter og andre lavere Dyr (Our Country's Pests among Insects and Other Lower Animals).*

Even before this, however, Sofie Rostrup had gained notice as a serious plant pathologist with her *Danske Zoocecidier (Danish Zoocecidia)* from 1896 and *Grønlandske Phytoptider (Greenlandic Phytoptidae)* from 1900. These two books had earned Rostrup a good scientific reputation, and when the Cooperation of Danish Farmers' Unions established phytopathological testing activities in 1907, Rostrup was an obvious choice as its associated zoologist. The testing station became the National Research Laboratory for Plant Pathology in 1913, and in 1919 Rostrup was appointed manager of its zoology department. A position at the university was out of the question, but that did not prevent Rostrup from pursuing a professional career in science. However, because her numerous publications only appeared in Danish, they reached only a Nordic audience, at least until her handbook on agricultural pests was finally translated into German in 1928.

Like Sofie Rostrup, Denmark's first female university graduates in physics – Hanna Adler and Kirstine Meyer – were also highly distinguished figures. Both of them graduated from upper secondary school in 1885 with "private" degrees after taking courses at Nathalie Zahle's school for girls. By 1892 both Adler and Meyer had achieved a master's degree in physics from the university. Since women were still barred from holding public office, both of them were forced to follow other career paths than their male counterparts.

After completing a study tour in the United States, Adler founded Denmark's first co-educational school in 1893, where girls and boys had equal access and were educated and nurtured side by side from kindergarten to secondary-school graduation. H. Adler's Mixed School was donated to the Municipality of Copenhagen in 1918, and from that point on Adler served as the school's headmistress.

Meyer had passed her own teaching exams at Nathalie Zahle's school in 1882, and while still a university student she was signed on to teach the natural sciences at her alma mater. From 1900 onwards she worked at Adler's school, where she recieved the title of lecturer in 1920. Meyer maintained a keen interest in educational issues. Even in her early years as a teacher at N. Zahle's School she had introduced the practice of having her physics students carry out experiments themselves. In fact, she succeeded in getting this idea included in the Danish school legislation enacted in 1903. She held a variety of different positions in the Danish educational system, and in 1910 she successfully applied for a post as education assistant to the government's education inspec-

17.9 *Women had now become part of daily life in the world of higher scientific learning, and yet the sexes were still discreetly segregated. Here Professor N. Steenberg flanked by his female students at the Polytechnical College in 1913.*

tor for upper-secondary schools – after being passed over in 1903 for a similar post because of her gender. In 1902, Meyer published her *Lille Naturlære (Science Primer),* a textbook for students in the 12 to 14 age group.

Meyer's commitment to education did not curtail her academic activities. In 1899 she received the gold medal of the Royal Danish Academy of Sciences and Letters for her article "Om overensstemmende Tilstande hos Stofferne" ("On Concordant States in the Elements"), and in 1909 she became the first woman to earn a scientific doctorate with her dissertation *Temperaturbegrebets Udvikling gennem Tiderne (The Development of the Concept of Temperature*

17.10 *Caroline Rosenberg was one of the most remarkable Danish amateur botanists of her time and highly esteemed by her professional male colleagues.*

throughout the Ages). In 1920, Meyer edited and published H. C. Ørsted's *Naturvidenskabelige Skrifter (Scientific Writings),* as well as two independent works on Ørsted. Meyer, too, found all routes to employment at the university blocked, but that did not prevent her from making important research contributions, both in physics and in history of science.

Not all Denmark's women of science had an academic background. Caroline Rosenberg had been taken in as the foster daughter of Charlotte and Niels

Hofman Bang at the country manor Hofmansgave on Funen, a place well known for its progressive and proactive agricultural practices. Niels Hofman Bang, who had himself studied natural history in Germany and France, encouraged her to take up independent studies of botany. Inspired by the botanist Hans Christian Lyngbye, sometime tutor at Hofmansgave, Rosenberg threw herself into her studies, concentrating particularly on algae. She soon accumulated a sizeable collection of samples, which she gathered on her many journeys around Denmark and the rest of Scandinavia. In this respect Rosenberg contributed significantly to Hofmansgave's unique herbarium of Scandinavian flora.

Rosenberg corresponded with several naturalists of her day, including the prominent Danish botanist Eugen Warming, who held her contributions to mapping and understanding the Danish flora in high esteem. Moreover, Rosenberg's list of algae genera occurring in Denmark was included in the botanist Johan Lange's *Haandbog i den danske Flora* (*Handbook of the Danish Flora*) published in 1851. Lange also used her findings and specimen collections in his ongoing work as the tenth and final editor of the ambitious *Flora Danica* begun by Georg Christian Oeder in 1753. Incidentally, this ambitious undertaking, described in chapter 8, took 123 years to complete and ultimately ran to more than 50 fascicles, and well over 3,000 plates. Rosenberg was selected in 1866 for honorary membership of the Danish Botanical Society. She was a woman, an autodidact botanist and an amateur who – from her position at Hofmansgave manor, and without any opportunity to create an actual scientific career – still achieved widespread appreciation and the scientific respect of professional male botanists across Scandinavia.

There were also women with no academic background who nevertheless used science as a means to create a professional career. Eline Begtrup, for instance, had passed her standard school exams, but she never went on to take a higher education. Instead she expanded her scientific knowledge and skills by means of self-study and specific courses. In 1886 she and her brother Holger Begtrup, later a prominent public figure himself, moved to the village of Askov in southern Jutland. They were both attracted by the *højskole* located there, which was an energetic hub in the folk high school movement. She became head of the village's independent primary school Askov Free School; a position that also enabled her to teach several subjects, including zoology, at Askov Folk High School. She was never permanently employed there, however, partly because she explicitly stated that she wished to teach male students. Following a brief stint as a natural-history teacher at Ry Folk High School in central Jutland, Begtrup herself founded a folk high school in the region in 1897, in

Levring. In addition to the usual staple subjects such as history and literature, her curriculum also comprised subjects like science and women's emancipation. Begtrup was an active participant in the public debate through *Højskolebladet (The Folk High School Magazine)* and *Kvinden og Samfundet (Woman and Society)*. She was also the author of several books, including *Naturen og vi (Nature and Us)* from 1911 and *Carl v. Linné* from 1914.

There were very few Danish women in science around 1850, and those who were active in the field often worked unseen, promoting their husband's scientific career, for example. But as new opportunities arose for women's education at upper-secondary and higher academic levels, women became more visible – and more voluble. They demanded the right to work, like their male peers, and in many cases they found or created careers that required tenacity and originality.

Johanne Krebs, a Danish painter and champion of women's emancipation, offered an explanation in 1885 of why women were now seeking employment in areas that were formerly the preserve of men: Women had been pushed out of their traditional tasks in the "domestic industry", including baking, brewing, dairy production, spinning, weaving and dyeing. Now that men and machines had largely taken over such tasks, rendering women's labour superfluous in many areas, it was only natural that women had begun to look towards the areas and careers previously dominated by men. As Krebs succinctly put it: "That they could not be ousted from one spot without this obliging them to find another spot is in accordance with the order of nature, and must be easily comprehensible in this age of natural science".[372]

And so it was a natural and understandable necessity that women had now begun to make their way in the spheres of education and employment. Generally speaking, equality between men and women was still not a reality in Denmark. And as for equality in the world of science, by 1920 although the situation had improved immensely, there was still a long way to go.

Popular science and public culture

18

A new public culture

From 1850 to 1920 the world beyond Denmark drew closer to the Danes. Copenhageners and citizens throughout the country were briefed on events from near and far with increasing speed. New technologies like the rotary press were already revolutionizing the media, and in 1849 the *Kölner Zeitung* became the first newspaper to use telegraphic reporting. After that, there was no looking back. The advent of news agencies changed reporting and made it, quite literally, "news". The Danish news agency Ritzau's Bureau was established in 1866, primarily to provide the Danish dailies with cable news from around the world. As newspapers became cheaper, local editions began to appear across the country.[373] The new practice of advertising rapidly spread, reducing the price of newspapers even further. In 1889 the periodical press in Denmark (newspapers, journals and weekly magazines) comprised about 250 different publications. By 1913 this figure had grown to about 1,450. Concurrently there was a corresponding growth in cheap and therefore easily accessible works of fact and fiction. In short, the Danes now had a huge range of reading material at their disposal.[374]

Thanks to a greatly improved educational system, more and more Danes were actually able to read. The rising standards of living, the falling prices, and the general and growing trend of enlightenment among townspeople, rural populations and the working classes all combined to create a much broader literary public domain, extending far beyond the privileged bourgeoisie. Knowledge and entertainment came in all shapes and sizes and were now liberally disseminated to the public at large. Some material was political, some of lasting local interest, and some so fleeting that it was forgotten the next day. Various channels – written, oral, and event-based – were used to convey information, much of which found its way into the higher, broader, and lower levels of society in Denmark's sitting rooms, workplaces, folk high schools and private homes. And science was a part of it all.

The museums opened their doors, the university extension announced public meetings, and the periodical press soon took science to heart. People

18.1 *Danish artists were closely associated with the scientific community. A large number of scientific portraits were commissioned and many artists were inspired by scientific topics and events. One was J. F. Willumsen and his painting* A Mother's Vision: Two Boys Floating in the Air *from 1910.*

wished to partake of the abundant scientific knowledge available, and there were more and more ways in which they could do so. Artists and authors found inspiration and allowed scientific themes to reflect the times in which they lived. In "our busy age of thundering machines", Hans Christian Andersen had the Muse of the new century arriving "on a dragon of a locomotive, rushing through tunnels and over viaducts". Such "primaeval forces" may have given the great swan's cloak of the imagination its strength and magnificence, but the weaver of that cloak was science.[375]

Similar sentiments were stirring in the Danish painter Jens Ferdinand Willumsen, who captured the attention of crowds and critics in 1910 at the annual art exhibition at Den Frie Udstilling in Copenhagen with *A Mother's Vision: Two Boys Floating in the Air*. The painting was acclaimed in *Illustreret Tidende* as the show's most significant picture, a work surpassed only by Joakim Skovgaard's altarpiece. And yet the reporter had not fully grasped the painting's scientific background, for although he was able to identify the realistic inspiration from studies of bathing boys, he had failed to catch any cosmic references apart from the stars scattered across the dark sky.[376] The two boys were actually gliding through space on a comet.That was no coincidence: In 1909–1910 two different comets appeared in astrophysical space as well as in public space.[377]

Willumsen used science both as an inspiration and as an artistic motif. Explaining his decoration on the unique frame he had built around his painting *Jotunheim*, Willumsen said: "The infinitely large is represented by a nebula, the infinitely small by some microbes. The figure at the bottom feels inspired, the figure at the top feels confident of his research."[378] Here, science was something elevated, serious, an opening into the depths of human existence. But in satirical magazines and comic strips, science had quite the opposite face and was used for caustic commentary on current affairs. One example from the Danish humorist Robert Storm Petersen had members the Society for the Beautification of Copenhagen serving a "refreshment" to some red-nosed vagrants in the working-class neighbourhood of Christianshavn. The tramps, a favourite character type of Storm P., thirstily eye what seems to be a tray of familiar green beer bottles. Little do they know that instead of the usual frothy brew, the bottles contain radium. Yet another problem solved by the Beautification Society's bastions of propriety.[379]

Science came to cover the whole gamut of human activities. As part of the common referential universe, science was a well from which art, literature and social debate could draw inspiration and sustenance. The democratic Danes were both able and willing to learn and experience new things. The opportunities to meet science first-hand were many: One could sit down with a newspaper, magazine or journal, spend time on literature and art that was in vogue, or visit exhibits, museums and Copenhagen's botanical or zoological gardens. One could even sign up for a course at one of the folk high schools springing up across the country. The new public domain was spacious enough to accommodate the bourgeoisie, the working classes and the rural population.

The rise of popular science*

Professional science had a counterpart in popular science that was now able to reach a far larger group than ever before. New technologies brought faster and cheaper printing. Journals and magazines were now able to offer illustrations – eventually in colour – to readers beyond the small, exclusive circles of the well-established elite. For this purpose, science was a popular subject. At the community hall and the folk high school, in the newspaper and the weekly journal; everywhere the captivated audiences and popularizers of science were balancing utility and entertainment.

Public lectures had long been a distinct feature of Danish cultural life. Speaking to the Students' Society in 1913 on the topic of general scientific learning, the bacteriologist Carl Julius Salomonsen emphasized two phenomena that he regarded as being of particular importance to the Danes: the popular public lecture and the rural culture. The first of these, the public science lecture, was by no means a Danish invention, but was firmly established in other countries by bodies like the Royal Institution in London and the Sorbonne in Paris. Nonetheless, Salomonsen claimed, "Denmark was surely one of the countries at the forefront in the area of popularization, and one cannot very well speak of disseminating general scientific learning in Denmark without mentioning three names such as Ørsted, Schouw and Eschricht."

The other Danish phenomenon, the rural culture, had to do with the significance of science to Denmark's agricultural sector. "Hardly anywhere else is the cultivation of crops and breeding of cattle better substantiated with scientific insight than in Denmark," Salomonsen continued, concluding that "as regards the dissemination and thriving of general scientific education, such cooperation between farmer and scientist is of great value."[380]

But there were many others besides Ørsted, Schouw and Eschricht who disseminated scientific knowledge and education to the public, and they catered to a much wider audience than farmers. Men, and progressively women, of science actively entered the market for popular science. Here was an attentive audience and money to be made. Many were attracted to activities popularizing science, and soon their ranks included schoolteachers, journalists, authors, and various other professionals. Farmers needed popularized sci-

---

* The next two sections are based on Hans Henrik Hjermitslev, Casper Andersen and Peter C. Kjærgaard's chapter on popular science in Peter C. Kjærgaard (ed.), *Lys over landet – Dansk Naturvidenskabs Historie*, vol. 3.

ence for practical purposes, civil servants wished to be kept abreast of new developments, the lady of the house enjoyed a pleasurable diversion, and the children wanted entertainment – preferably of the edifying sort, if the parents had any say in it – and science faithfully fit the bill.

Popular science is defined by its content: scientific knowledge communicated to people who are non-specialists in the field. However, this simple definition covers a wide spectrum of intentions, from education to entertainment, and of recipients, from researchers to factory workers. This, combined with the enormous range of topics, authors and target groups, makes the genre almost boundless, which is also reflected in the popular-science media. Some were generally oriented while others specialized in, for example, natural history. Because the Danish market was small compared to Britain, France and Germany, much of the material dealt with topics of a general scientific nature. On the other hand, the audience for popular science was very diverse. In 1850 there were already a number of journals aimed at everything from the youth at folk high schools and the new working classes to the upper middle class. Regardless of age and social status the people of Denmark were curious, and the new media placed the world of science at their fingertips. Lethal comet rays, home-made volcanoes and primitive man all became part of modern everyday life.

Illustrated news

"Events quickly follow one another. It is the task of the daily papers, aided by the inventions that human perspicacity has devised and made the property of all, to relate these events as speedily as possible."[381]

Such was the message issued Sunday, 2 October 1859, to the readers of Denmark's first real weekly magazine, *Illustreret Tidende.* It was concieved as the dawn of a new age and the magazine presented the newest fashions and the most recent astronomical discoveries side by side. The concept of "news", of current events quickly recounted, was invented along with the rotary press, the telegraph and the steam locomotive. Readers of daily papers and journals had previously been obliged to accept a certain time lag when learning of events, but new inventions and technological infrastructure made it possible to bring stories from reporters on the spot with little delay. In addition, the development of more efficient printing methods, cheaper paper and more advanced image-reproduction technologies positively revolutionized the printed medium in the nineteenth century.

Better educational standards and rapid population growth continuously increased the potential readership, swelling the markets for newspapers, magazines, journals and books. This development peaked in Britain with more than 100,000 different journals during the nineteenth century. Though many had a low circulation and short lives, several of these journals survived for decades or more and in some cases inspired counterparts in Denmark and elsewhere. One notable example is the satirical magazine *Punch, Or the London Charivari* (1841–2002), whose Danish equivalent *Punch* was launched in 1873 (1873–1895). Another is the *Illustrated London News* (1842–), which inspired several European magazines such as *L'Illustration* in France (1843–1944), *Illustrierte Zeitung* in Germany (1843–1944) and *Illustreret Tidende* in Denmark (1859–1924). A cheaper competitor to *Illustreret Tidende, Illustreret Familie-Journal (Illustrated Family Journal),* was launched in 1877. Along with *Nutiden i Billeder og Text (The Present in Image and Text,* 1876–1890), *Illustreret Familie-Journal* (later *Familiejournalen)* also emulated some of the most successful magazines abroad.

Appearing just once a week, *Illustreret Tidende* had to have something different to offer readers than the dailies had. The weekly format meant that news stories could be presented in greater detail and be richly illustrated. "For it is only now, after the pictorial presentation has united with the descriptive presentation, that a literature has arisen which, accessible to anyone, has meant that no event of general interest, no prominent figure, no important discovery within human knowledge, will easily escape the attention of the general public – as long as one knows how to awaken the feeling for it in the appropriate manner", as the editors reassured the magazine's readers. Here there was a place for science. Presenting its content, the magazine informed readers that it would indeed cover "the things of general interest taking place within the areas of science, literature and art".[382]

Illustreret Tidende kept its promise, although there was not much pure science in the earliest issues. The "prominent personality" of the first week was the British Arctic explorer John Franklin, plus a few brief notices under the headings "Statistics" and "Technical Matters". The latter enlightened readers about "The use of compressed air" in an item introducing the visionary idea of using pressurized air as a new and cheap source of energy for towns and industries, which, unlike steam power, would be completely harmless. Although the reporter acknowledged that the idea still sounded somewhat fabulous, "yet it does not seem undoable, particularly not when weighed up against such difficulties as the culture of our century has been able to surmount."

Science had come to the people, and vice versa. Popular science was easy to

18.2 *For the Danish version of the satirical magazine* Punch, *the transits of Venus in 1874 and 1882 had considerable entertainment value. Astronomy and celestial events generally received much public attention, ranking high among the scientific topics treated in journals and newspapers.*

understand, and although it may have been far from life in the laboratory, it carried with it an optimism and a confidence in the improvements science would bring throughout the decades to come; a confidence that built on the things people had already seen. Science had become a central element in the public consciousness and through new popular periodicals it was reaching a larger audience than ever before. This was a gradual development in the weekly *Illustreret Tidende*, and the competitors that appeared over the years, such as *Nordstjernen (The North Star)* and *Familiejournalen*, ran regular columns with "News from the world of science and invention".

Internationally, a number of science journals and magazines provided models taken up by others. However, neither the monthly *Scientific American* (1845–), for instance, nor the weekly *Nature* (1869–) had a viable Danish counterpart, even though several spirited attempts were made and many popular-science journals saw the light of day between 1850 and 1920. Developments in technology and distribution channels, acting in conjunction with a growing market and generally heightened level of knowledge, made popular science a genre as well as a profession. Previously, announcements to the general public concerning scientific topics had known no systematic form outside the strictly scientific literature. That changed, however, with the resurgence of popular-science journals, series of books and lectures, and the various other broadly targeted instructive activities taking place in Denmark during this period.

Scientific edification

"Man is born a scientist." This was the opening sentence in the Danish physics professor Heinrich Oskar Günther Ellinger's book *Naturen og dens Kræfter (Nature and its Forces)* from 1898. It was an exemplary work of popular science, presenting a cornucopia of experiments that most people could understand and carry out at home with just a little effort and equipment. Scientific enthusiasm generally ran high around the turn of the century, but Ellinger was particularly dedicated to physics. In his opinion, physics was the driving force behind modern science, its applications, and even the development of humankind.

> If we then look further, and in our minds go through the long progression of the physical discoveries and inventions that are attributable to the human mind, and which increase every year – it suffices to mention the steam engine, the gas machine, the railway, the telegraph, the photograph, the telescope and microscope, light houses, lightning rods, electrical lighting, et cetera – we at once realize what enormous significance this study of nature and its forces has held for humanity and its development. Besides this, there are a number of other sciences and enterprises that build, to a greater or lesser extent, upon what physics has taught us. This is true of agriculture, of many technical subjects, of medical science, et cetera.

There was, however, more than merely scientific and material development at stake. Physics also had something special to offer for the individual, something that imparted knowledge and promoted one's personal development, and something that could benefit everyone. In a word, physics was "edifying". "By studying physics," Ellinger went on, "one accomplishes a proficiency and becomes well acquainted with a method that raises one's gaze to the things taking place all around us in our daily lives. It moreover trains a person in distinguishing the threads from one another; in discerning what is decisive, what is significant; and in deducing, from the observable effect, the cause that gave rise to said effect."[383]

During the nineteenth century science became more integrated into the

RANDERS DAMPBRÆNDERIS NY NATIONAL KOMMEN AQUAVIT ORIGINAL AFTAPNING Pris 65 Øre pr. Fl.

KUND SKAB

FREM

: ER : MAGT

28. FEBRUAR 1909

12. AARGANG NR. 22

REDAKTION: JULIUS SCHIØTT OG HELGE HOLST

PATENTER
VIGGO C. EBERTH og MARCUS-MØLLER
NYGADE 6 KJØBENHAVN
Varemærker og Mønstre registreres

GENNEM HELE VERDEN MED
LUKSUSTOGET
EN TUR 25 ØRE
33 · AMAGERTORV · 33

SUDAN SAFT
tilberedes af Urter fra Afrika og er anvendt med udmærket Held imod Mave- og Underlivssygdomme. Sudan Saft er fortrinlig anbefalet baade af Læger og private. Prøv 1 Flaske, og De vil ikke fortryde Forsøget. Pris 1.65 og 3.00 franco fra Generaldepotet Thy-Apothek Hvidbjerg St.

DET DANSKE MÆLKE COMP
DMC

CHARLES DARWIN, f. 12. Febr. 1809, død 19. April 1882

VEGA PALMIN
„Kokkepige"
„Solsikke"

Uld Musselin
flere Tusind Alen prima heluldne
50 Øre
Værdi 75 à 90 Øre. Prøver franko.
Vaskestof til Værdi 40 à 85 Øre sælges under Udsalget fra 25 Øre.
Udsalg i alle Afdelinger.
MESSEN

GEORG BESTLES VINFORRETNING
SKINDERGADE 45 & 43

ANKER HEEGAARDS
• Emaillegods •
ANKER HEEGAARD EMAILLE KJØBENHAVN
• Faas overalt •
Bemærk ovenstaaende Etikette i rødt Tryk paa sort Grund.

Pris 25 Øre. KASINOL-SÆBE Anbefalet af ca. 40 Læger. A/S N. N. Binmensaadts Fabriker. Odense.

18.3 *"Knowledge is power" – 'kundskab er magt" – was a widely used catchphrase, serving for instance as the motto of the popular-science journal* Frem. *This front page from the issue dated 28 February 1909, celebrating the 50th anniversary of Charles Darwin's book on the origin of species and the 100th anniversary of his birth, featured advertisements showing science put to daily use in new, industrially manufactured products.*

debates about the nature of a liberal education. The classical subjects, including Greek, Latin, and a traditional literary canon, held a strong position at the educational institutions and in Denmark's public intellectual life. But due to the growing significance of science in the second half of the nineteenth century, voices from several sides demanded that science be incorporated into the general learning ideals. A well-rounded individual of the modern age had to be equipped to deal with the modern world – and that purpose could not be fulfilled by Homer, Cicero and Luther's Catechism. Many felt that the natural sciences could no longer be ignored.

Still, changing the situation was easier said than done. For one thing, opposition was staunch among the prophets of traditional edification, who harboured an ingrown scepticism against science entering the realm of learning and the mind. For another thing, many believed that the raw material at the heart of the issue – human beings – might not be as receptive as Ellinger hoped. In an article about the H. G. Wells' novel *The First Men in the Moon*, first published in 1901, the Danish historian Christian Villads Christensen characterized the human race and its limitations.

> Cavor, the hero in this narrative, is the interplanetary representative of our Earth. Cavor is the very essence of the diversely amalgamated human race, the race with the warm hearts and the strong emotions that never learns to conform to the cold calculations of the reasoning mind. The hard-headed race whose rigid skulls put a limit onto the development of its intelligence, but also stand up formidably to a punch.[384]

Humankind had to be equipped to deal with the future. To Danish educators and popularizers of science the best weapon was knowledge. Unyielding skulls might well be curtailing evolutionary potential, but there was still much the human race could learn, and had to learn. "Knowledge is power" Francis Bacon wrote in 1597, thus supplying a suitable motto for the new sciences emerging in the seventeenth century. By the end of the nineteenth century, his aphorism had become the rallying cry of all forces striving for a new enlightenment in Denmark, rendered in Danish as "*kundskab er magt*". This slogan appeared in many contexts, including the banner of the Danish Electricians' Association and the front page of the periodical *Frem*.

A sample issue of *Frem*, sent out in 1897 to attracts readers to subscribe, stated that this was precisely the key for the youth who would be conquering the new century in the ever-more intense struggle to survive – in which scientific and technical prowess was absolutely crucial.

> In this day and age everyone ought to know how steam is exploited, how electricity is generated, how telegrams are sent through a thin wire that stretches for miles and miles, how the elements are broken down and once again combined. Everyone ought to know the construction of their body and its life processes, know the cosmos that overarches the Earth and its development throughout the immeasurable time-span of creation.[385]

The editor even promised that this self-development and instruction would actually be interesting. As long as they were presented well, the lives and deeds of Columbus, Pasteur and Edison were just as exciting as any good novel, and the evolutionary history of planet Earth and the human race just as captivating as any drama. Not only was this new learning necessary and useful. It was also entertaining.

The new wave of enlightenment that washed over Denmark was heavily influenced by the Cultural Leftism movement and the circle around the outspoken and controversial literary critic Georg Brandes. Ardent defenders of realism, Brandes and his supporters shared an antimetaphysical world view that passionately relied on science and freethinking. In his *Forklaring og Forsvar (Explanation and Defense)* – written in 1872 to counter the criticism of his hotly debated lectures on major trends in nineteenth-century literature, delivered at the University of Copenhagen in 1871 – Brandes wrote that the sole and single force driving him was his conviction that free science was justified. In this context he levelled a fierce blow at Danish culture, which he saw as cultivating everything *but* free thought. "The idea of free science has mighty allies in all countries and in all sciences, except in the countries of Denmark and Norway, and in the sciences of astrology, theology and alchemy," Brandes thundered, asserting that as far as free science was concerned, "Every good book that is brought here across the border from Germany, France, Britain, the Netherlands or Italy brings free science with it, conveyed by all of the greatest names in science and literature."[386] Knowledge was international, and thoughts were free. Brandes intended to shake some life into a country afflicted with intellectual lethargy, and his ideas found a large following.

One of his supporters was a young biology student, Jens Peter Jacobsen. Besides translating the writings of Charles Darwin into Danish, J. P. Jacobsen also emerged as one of the most influential Danish writers of the period. In one of the shortest poems in Danish history, from 1884, J. P. Jacobsen concisely synthesized the period's eagerness for public enlightenment, its faith in science and its optimism for the future: "Light over the land – That is our aim."[387] This became a catchphrase of his time, science and literature walking hand in hand; the catchphrase of the free mind, of the Danish people looking towards the dawn of a new century.

In the old, rural heartlands of Denmark, the folk high schools were traditionally at the centre of the general population's formative education and learning after their basic schooling had ended. Characterized by a joyful and optimistic outlook on life, and by alleged equality between the sexes, the folk high school culture in Denmark provided an alternative to the university and other places of higher learning in and around the capital of Copenhagen. Among the three leading folk high schools – in the villages of Askov and Testrup in Jutland, and Vallekilde on Zealand – the importance of science and technical training was most pronounced at Askov Folk High School, thanks chiefly to the school's employment in 1878 of the physicist, meteorologist and inventor Poul la Cour. His promotion of modern science is something of a paradox in light of his clearly stated anti-Darwinist position. But for la Cour there was nothing contradictory about being faithful to science while rejecting

18.4 *Science was an important part of the educational and formative ideology of Askov Højskole, one of Denmark's most influential folk high schools. Poul la Cour (middle row, far left), a physicist and later a teacher at Askov, constructed an experimental "electrical windmill", generating power and a good deal of public attention. The photograph shows the participants in Askov's first course for electricians, who took an exam and received a diploma to document their qualifications.*

the theory of biological evolution. His meteorological experiments helped to recast the folk high schools as a new and more broadly accessible seat of science and technical expertise. Here, science was popularized and made available to the people, but never sensationalized or capitalized on to turn an easy profit. This gave the folk high schools an important function in creating a wider understanding of the principles and general significance of science.

La Cour came to Askov at the invitation of the school's principal, Ludvig Schrøder when the school expanded its programme by establishing Askov Extended Folk High School and it was decided that the school would also teach advanced mathematics and physics. This would call for someone who was not only professionally competent and capable of teaching, but who also had a broad, popular appeal. Poul la Cour accepted the position without a moment's hesitation, which also meant abandoning any hopes of a prominent position at the Danish Meteorological Institute. He authored a good part of his teaching material himself, including *Historisk Matematik* (*Historical*

Mathematics), the first edition appearing in 1882. Generally the folk high schools and the agricultural high schools, such as Ladelund in southern Jutland and Dalum on Funen, played a vital role in spreading scientific and technical learning to the Danes. They may have lowered the bar as far as the popularity and accessibility of science was concerned, but their ambitions always remained high.

The position of science was also strengthened in Denmark's ordinary educational system. In 1850 natural philosophy and natural history entered the curriculum in secondary schools, and eventually also in the country's *realskoler* – equivalent to the German *Realschulen* – and technical schools. The Danish University Extension programme was founded in 1898, offering a wide variety of highly popular science-related special courses and lectures. Around the turn of the century, science, its impact and its findings were on the agenda in so many places and so many connections that most Danes were familiar with the concept of "general scientific edification". The larger range and number of educational institutions also made people more progress-minded than they had been in the original Grundtvigian tradition that founded the folk-high-school movement. This tradition had focused mainly on concepts like "personal development" and "lifelong learning" through history and literature, storytelling and song. In 1902, Georg Brandes gave a speech based on the previous year's introduction of parliamentarism into the Danish political system. Formative education was apolitical, Brandes claimed, and hence the academics were the ones who could keep the country together and unite its different population groups, liaising between the rural populations and the working classes. "Our task is to bring enlightenment to them both, scientific culture to them both." Grundtvig had done an admirable job, but time had shown that it was not enough. "The difference between the formative education that emanates from the folk high schools and that which emanates from us is twofold. It can be compared with the difference between the astronomy Tycho Brahe learned and that which Copernicus believed in." To Tycho, the Earth was the centre. To Copernicus, the Earth was nothing more than "a mote in the cosmos". This had a crucial impact on a person's intellectually formative processes. "For us, all is nature," Brandes declared, and "all intellectual science is natural science, including history. There is, for us, no natural history and supernatural history contradicting one another. All history is natural."[388]

Having a basic knowledge of science could also help one to put life into perspective. In 1917, for instance, the weekly *Familiejournalen* recommended that its readers, young and old, try their hand at astronomy. This idea sprang from the condemnation of the way young people tended to simply let their

18.5 *The scientific discussion about prehistoric man was quickly integrated into popular Danish culture. As the text accompanying this colourful depiction of mankind's early ancestors drily remarked: "Life for the earliest humans was no laughing matter".*

own minds be the leading force in their lives, ignoring all the things they could not understand. There was nothing more therapeutic for such "impetuous youths" than being introduced to astronomical learning, readers were informed. The sources were legion, so there was no excuse for not following the magazine's recommendation, whatever one's initial level. There was a veritable wellspring of knowledge in the field, *Familiejournalen* exclaimed: books, images, journals, courses and lectures that could help one along. Admittedly, the infinite universe could seem a bit overwhelming, but such studies served a valuable function:

> Astronomy teaches humility, and humility we need. Set against this world of stars, what previously seemed to us great and immensely important – the admiration and praise of others, for example – loses its significance. We learn to rearrange our values; our gaze has expanded, our measuring stick, which has thus far conformed to the trifles of our daily lives, must be replaced with a better one. Little by little, but ever so surely, an understanding of eternity's magnitudes begins to dawn on us.[389]

18.6 *Spectres, prehistoric monsters and creatures from outer space were favourite topics in popular-science literature. One article from 1907 introduced Danish readers to Martians.*

The guidance from *Familiejournalen* and its "Adviser for city and country house and home" certainly ran more in the vein of correctional education than personal edification, with an unmistakably religious flavour. However, it was just as clear to the ideologists of the new enlightenment as it was to the physics professor and the popular weekly magazine that science and a well-rounded personality belonged together. At this point the education and formation of a whole person had come to mean the development of a person who in the name of free thought was prepared to deal with the modern, scientific and technological world of the twentieth century.

Towards the end of the period there was no doubt about the position of science or its great significance to Danish society, culture and thinking, even though perhaps things had not gone quite so well as some had hoped during the most optimistic years. The facets of scientific education had been cut into shape and were firmly established alongside the historical, linguistic and literary facets that were also part of "a good education". The two sides now constituted the two axes in a complete classical education. At least this was the opinion of the bacteriologist Carl Julius Salomonsen. But although those who had championed science's inclusion in the accepted formative curriculum had won important victories within certain limits, that is precisely where these victories remained: within, and only within, a certain limited scope.

> Despite all that has been achieved in the area of broadening scientific learning here in this country since the beginning of the battle, the natural sciences have still not become what the best among the pioneers had hoped for. They have not become a new harmonizing and liberating factor in the intellectual environment thought to be far too unilaterally typified by Greek and Latin language studies. Rather, they won their powerful position as the handmaidens of technology.

Although science was everywhere, it was technology that had gradually come to dominate the self-image of the Danes and their perception of the world. Having been virtually synonymous throughout the period, "science" became more obscure compared with "technology", which was immediate, visible and directly applicable. According to Salomonsen, this also had an impact on the nature of general education in Denmark. "A new formative ideal is currently gaining ground, characterized by three contemporary circumstances: The zealous enthusiasm for sports, the rapid development and rising importance of all things technical, and the ever-more inclusive division of labour. One could call this new formative ideal one of acrobatism, technicism, specialism."[390]

The sciences were very much a part of the specialization Salomonsen spoke of, and furthermore they were inherently associated with technological achievements and even, perhaps more surprisingly, with the "zealous enthusiasm for sports". In this respect Salomonsen was in tune with the novelist and later Nobel laureate Johannes V. Jensen, who just a few years earlier had suggested that Leonardo da Vinci, Charles Darwin and Niels Finsen were among the chief pioneers of sports and outdoor living. It could well be that scientific edification in Denmark around World War I had not become the harmonizing and liberating element in Danish intellectual life that some had so ardently desired. On the other hand, science and its icons were present everywhere in the daily life of the Danes – in work and play, in society and culture.

Expanding the bounds of science

Occasionally science provoked strong reactions. It offered new discoveries and new insights that had wide-ranging consequences for the way the Danes viewed themselves and the world around them, for nature, for the beginning and the end of the world as we know it. This abundance of new knowledge created a fertile environment for sombre scepticism and great expectations. This in turn raised two parallel questions: Could all the things that science ascertained really be true? And was that really all there was? Even while science was expanding knowledge of nature, the Earth, and the universe, many people felt that science was too conservative, too one-sided and too restrictive. And so, paradoxically, while science constantly expanded the boundaries of knowledge it constantly ran into them.[391]

In the wake of this debate – philosophically expressed as a dichotomy of belief versus knowledge – came the questions of whether it was not better to steer clear of certain topics; whether there was no limits to scientific knowledge; or whether scientists should unabashedly continue their campaign and slowly but surely wrest from other fields the authority over questions and answers that had previously fallen outside their domain.[392] In the decades between 1850 and 1920 many discussions concerning the boundaries and limitations of science were sparked. Moreover, by virtue of its new and greater authority, science was also called upon to ultimately sanction or condemn, passing judgment in matters that were otherwise far removed from daily life in the laboratory.

In 1917, for instance, the weekly magazine *Familiejournalen* informed its readers of a scientific experiment concerning divining rods. "The arms of science are far-reaching," the journalist wrote. "It knows no respect for old, time-honoured rules and customs whose justification lies in something inexplicable, something mystical."[393] It was therefore only natural that scientists should embark upon a scheme to find out whether it was all just nonsense, and if not, what the explanation might be. The magazine's Danish readers were subsequently informed of a large-scale experiment led by the French biologist and speleologist Armand Viré in collaboration with a number of people purported to be the most famous water dowsers in France. According to the article, the dowsers were very successful in finding water in rock caves, hidden wedges and pieces of buried iron. Only copper was difficult for them. Generally, however, the findings were so convincing that the French scientists concluded that the divining rod did, in fact, work. The only question that remained was how? In their search for a natural explanation for this phenomenon, all they could say was that some connection existed between the movements of the rod and the breathing of the water dowser. Determining the nature of this connection called for further investigation.

This story of divining rods is much more than a colourful example from a popular magazine printed during World War I. First of all, many respectable scientists regarded water dowsing as a serious business and firmly believed it was a reliable method. One of them was J. J. Thomson, the British Nobel laureate in physics who in the late nineteenth century made his name by discovering the electron.[394] Secondly, dowsing was just one of many "fringe sciences" that were regarded either as sciences in their own right or as areas subject to scientific investigation. From 1850 until the end of World War I, the public exhibited a growing fascination with phenomena that lay either on or beyond the borders of the natural world, and within this realm, psychic phenomena enjoyed a special standing.

"Spiritualism before the tribunal of science"

The most remarkable critical investigations in Denmark were carried out by the psychologist Alfred Lehmann, who set up his laboratory in 1886 in the basement of the upper-secondary school Metropolitanskolen, located in central Copenhagen. His was only the second psychology laboratory in the world, supported financially by the Danish ministry of culture and the Carlsberg Foundation. Lehmann modelled his new facility on the experimental psychology laboratory founded in 1879 by Wilhelm Wundt in Leipzig, where Lehmann had worked in 1885–1886. Lehmann graduated from upper-secondary school as a mathematics major in 1876, in one of the first forms graduating after the Danish grammar-school reform of 1871.[395] Having completed his studies of advanced technology at the Polytechnical College in 1882, Lehmann had acquired the necessary scientific and practical approach to the new science of psychology, or "psychophysics". Lehmann's dissertation from 1884, *Farvernes elementære Æsthetik* (*The Elementary Aesthetics of the Colours*) allowed him to combine his background in advanced technology with his interest in philosophical topics. In 1888 his expertise in chromatology and chemistry made him a valuable consultant on a project concerning colour-determination techniques for margarine, under the Danish home office. Nevertheless, it was experimental psychology that would define his career.

Psychology was counted among the natural sciences. This was considered a matter of course for the new discipline, which was defined as "the natural science of human nature". Lehmann's establishment of his psychology laboratory – which moved to rooms in the old polytechnical college building in 1891 and was taken over by the University of Copenhagen in 1893 – marked the dawn of empirically based psychology in Denmark. It was included quite naturally in many contexts alongside the other sciences. One example is Lehmann's and R. H. Pedersen's investigations into psychology and meteorology in *Vejret og vort Arbejde: Eksperimentale Undersøgelser over de meteorologiske Faktorers Indflydelse paa den legemlige og sjælelige Arbejdsevne (The Weather and Our Work: Experimental Investigations into the Influence of Meteorological Factors on the Bodily and Mental Capacity for Work)* from 1907. Besides this, however, the scientific status of psychology also meant that many of the numerous reports dealing with particular psychic phenomena and featured in newspapers, magazines and journals could now be subjected to scientific scrutiny.

This task was readily taken up by Lehmann in *Overtro og Trolddom fra de Ældste Tider til vore Dage (Superstition and Sorcery from the Oldest Times to the Present Day),* which earned him a reputation as Denmark's most well-founded scientific critic of spiritualism. In the first edition appearing in 1893–1894 he argued that it was necessary to actively counteract spiritualism. Certainly, most daily newspapers still voiced critical reservations, but Lehmann feared that in line with what he had observed abroad, these critical voices would soon fall silent. "The waves of superstition will engulf us, and the same phenomena that have, thus far, kept a suitable distance, will begin to be seen amongst us." Lehmann defined spiritualism as a theory of the nature of spirits and their relations to human beings:

> The human soul is immortal and, following corporeal death, it is capable of connecting with the living and instigating a series of physical and psychic phenomena that human beings are not able to bring about, at least not according to our current knowledge of the forces of nature and mental processes. In order for the spirits, the souls of the deceased, to connect with the world of human beings, there is usually need of a specially gifted person, a so-called medium. Mediumism – the potential to become a medium – is, however, present to a greater or lesser degree in every person, although even the best natural predispositions must be developed through practice.[396]

According to Lehmann, this summed up the spiritualists' beliefs. His motive was to expose spiritualism as humbug, and as the consequence of "magic now raging like an epidemic in the major centres of culture." The open attack caused strong reactions in spiritualist circles, not least after the book was translated into German, Swedish and Hungarian during the 1890s. The German edition in particular, *Aberglaube und Zauberei* appearing in 1898, established Lehman internationally as one of the most fervent critics of spiritualism at the turn of the century; a critic both celebrated and notorious. In the preface to the second edition, a buoyant, self-assured Lehmann explicitly thanked "my esteemed opponents, the spiritualists, for the excellent service they have done me". Their meticulous reading and subsequent criticism of his work had been of invaluable help, enabling him to correct all the small errors they had so pedantically pointed out.

The spiritualist movement began in the United States in 1848, when the daughters of a Methodist farmer were allegedly found to possess special psychic abilities. In December 1847, the Fox family had moved into a house vacated by its previous owner because of inexplicable noises. No such sounds bothered the new inhabitants until the early spring. Then, on 31 March 1848, the date modern spiritualism was born, the girls "discovered" that the rapping noises seemed to convey meaningful messages and were able to "answer" their calls. If they knocked five times, then five knocks were returned in reply. When questioned about the girls' age, the rapping noises would give the correct response. This was soon interpreted as the girls being uniquely linked with the spirit world. News of the Fox sisters spread like a wildfire throughout the United States and Europe, not least because of the sensational claim that the spirits told them a murder had been committed in their house. In the wake of this news there appeared a host of psychic mediums with similar or even more amazing powers. Contact with the deceased had been established, the gateway between their world and ours had been opened, and there was no end to the answers they could give. At least many people were convinced this was true – among them a number of scientists. One of the most famous mediums was a Scotsman named Daniel Dunglas Home, who could make a harmonica play without anyone visibly touching it, make tables levitate, and make spirits materialize before an astounded – and paying – audience.

Home was also one of the most important mediums used by the British physicist and chemist William Crookes in his investigations of these phenomena. Crookes was so deeply convinced by Home's performances that he felt moved to announce the discovery of a "psychic force", supposedly a fundamental force of nature, like gravity. However, unable to convince the majority of his peers that his findings were credible, Crookes was obliged to retract his claims, leaving his integrity and his scientific career severely damaged.[397]

Crookes was not alone in his fascination with the abilities of these mediums, nor was he alone in expending time and laboratory resources to study them. The Society of Psychical Research, founded in 1882, was based in London and had many prominent contemporary scientists among its members. In 1905 Alfred Lehmann and Julius Schiøtt, the director of the zoological garden in Copenhagen, initiated a Danish equivalent, Selskabet for Psykisk Forskning, which counted about 200 active members in the early decades. In 1921 the society organized the first international conference on psychical research. Both the British and the Danish society are still in existence today.

As in most North American and European countries, spiritualism gave rise to public debate and attention. Stories of "the psychic force" were regularly reported following the Fox sisters' sensational claim and the family's clever

18.7 *The bacteriologist Carl Julius Salomonsen's scientific attack on modern art caused incredulity, indignation and amusement. The cartoonist Storm P. illustrated how "Professor Salomonsen conducts studies for his pamphlet on Dysmorphism".*

marketing of their skills, providing Danish readers with everything from percussive noises to automatic writing from the mid-nineteenth century. Looking back in 1912 on these events, the historian Christian Villads Christensen described the arrival of spiritualism in Copenhagen as a topic that generated fully as much attention, and quite as much conversation, as the "ladies' emancipation":

> It was in 1850 that the *table dance*, which had begun three years earlier in America, reached us here. Professor Holten and Senior Master Petersen of Metropolitanskolen assumed the task of investigating this matter and published an account of their findings: 'A newspaper was laid on the table, and it turned round while the table stood motionless; the same hap-

> pened to a hat ... We soon arrived at the conviction that the movements we had seen found their cause neither in electricity nor in any other similar force of nature, but solely in bumps or pushes that the operators, unbeknownst to themselves, communicated to the table'.[398]

The conclusion was not as exciting as many had hoped. Still, it did not put an end to the debate even though, as Lehmann noted, the newspapers were, by and large, reserved and sceptical in the decades that followed. This moved the spiritualist protagonist H. L. Hansen to raise a voice of complaint. Hansen was a self-proclaimed communicator of the borderland between the physical and the spiritual realms, translating the spiritualist writings of Alfred Russell Wallace as *Spiritualismen for Videnskabens Domstol (Spiritualism before the Tribunal of Science).* Besides being the co-discoverer with Charles Darwin of the theory of evolution by natural selection, Wallace was an ardent spiritualist.[399]

Wallace was aware that he was fighting against the common scientific opinion of alleged spiritualist phenomena. The gist of much of the criticism was that adherents of spiritualism based their support on belief alone. As interested readers could learn from Wallace's arguments, however, this was also true of gravity, and of the wave theory of light. Only a tiny minority of people had themselves conducted experiments, and therefore "for the most part they followed a blind belief resting upon authority". Most people regarded spiritualism as an altogether different matter. Such phenomena – rapping and knocking, extinguished candles, floating tables, writing pencils, levitating mediums – were

> so peculiar, so unbelievable, so contradictory to their usual manner of thinking, so apparently against the scientific thought that permeates our age, that they were neither able nor willing to accept them after hearing testimony second-hand, as they do with almost every other sort of knowledge.[400]

With his translation, Hansen fought against "an almost unisonous chorus of squealers, from the tone-setting Head Squealers in the Copenhagen newspapers right down to their echoes around the country".[401] Danish newspapers often used words like thimbleriggers, madmen, swindlers, charlatans and con artists to describe the spiritualists, generally taking a critical stance. But in spite of this, people were still attracted to spiritualism, even though it was an easy target for condemnation and ridicule. Lehmann joined in the fray, championing the cause of science, but for many Danes the fascination remained intact.

The limits of science

Between 1850 and 1920 a variety of fashions, schools and trends formed many favourite topics of conversation. Mysticism, fear and indignation were some of the most appealing objects of debate, leading people to discuss subjects like hypnosis, sleep-walking, telepathy, extraterrestrial life, and the end of the world. Perhaps more surprisingly, science also came to play a role in connection with the new artistic styles of the early twentieth century.

Like Alfred Lehmann, the bacteriologist and art enthusiast Carl Julius Salomonsen was on a mission of his own in the name of science. He was not targeting superstition, but battling a very different and, in his view, positively pathological feature of his time. Giving a lecture at the Society for the History of Medicine in January 1919 entitled "Contagious Mental Illnesses Past and Present, with a Focus on New Trends in Art", Salomonsen went straight for the jugular of the artistic avant-garde as represented in styles like cubism and futurism. A great and well-connected admirer of naturalist art, Salomonsen was also singularly well versed in the various artistic genres. However, this did not enable him to understand or accept the new clash with naturalism as a genuine innovation of art, or as a contribution to it.

The bacteriologist's diagnosis was clear: Modern artists lived in a warped world of the imagination, where they cultivated all things hideous and deformed, eschewing what was beautiful and natural. The problem, he said, was that

> based on certain artistic doctrines, which especially distance themselves from the lifelike reproduction of the world around them pursued by certain earlier schools, they consciously and systematically offer distorted and unnatural, and most often also unlovely or hideous, representations of the products of nature and art.[402]

The trend could be likened to a mental illness and, on top of everything else, Salomonsen claimed that it was contagious. He labelled the disorder "dysmorphism" and continued to track it down in modern poetry. The concept of dysmorphism gave rise to a lively debate about art, medicine and science in which very few backed Salomonsen's position. Predictably, his arguments fell on stony ground among artists, but even among contemporary physicians they found little support. Oluf Thomsen, who in 1920 followed Salomonsen as professor of general pathology at the University of Copenhagen, rejected the idea of artistic insanity, contagious or not, and the chief physician Hjalmar Helweg wondered how Salomonsen could arrive at a diagnosis without having exam-

18.8 *Depicting the end of mankind, the French astronomer and popular-science writer Camille Flammarion made the Danes shiver. The caption under this illustration in the popular weekly magazine* Familiejournalen *simply read: "And thus will be the end."*

ined the patients in question. Most people held that by regarding modern art as pathological, Salomonsen had overstepped not only the boundaries of science, but also the boundaries for what scientists could attempt to deal with scientifically. Salomonsen himself believed that he was merely doing his scientific duty by utilizing and disseminating his immense and indisputable knowledge of modern art and modern science.

Other scientists continued to explore the planet, filling in the dwindling number of white spots that remained on the map. They brought home curious new species of plants and animals. In the course of the nineteenth century people had grown accustomed to being surprised and astonished. Our own planet was teeming with life, and the rapid development of astronomy, thanks not least to the introduction of new technologies, brought other planets closer to Earth. This in turn fuelled the idea that other life forms might exist out there, somewhere. Mars was the favourite seat of speculation and people enthusiastically discussed whether it might be home to forests, fields, channels, and even entire cities and civilizations. In scientific circles and among the general population many found life on Mars to be a reasonable proposition. Many even believed that the Martians were attempting to communicate with the neighbouring Earthlings, but that we were simply unable to respond. The topologist Poul Heegaard, for instance, in his *Populær Astronomi (Popular Astronomy)* from 1902, told his readers that a Frenchwoman had left her entire fortune to the first person who could make contact with another planet, other than Mars. Apparently, establishing contact with our red neighbour was considered all too probable, and all too easy. As Heegaard nevertheless sceptically concluded, though without categorically rejecting the idea:

> And so the prospects are poor as concerns our receiving news of genuine signals from Mars any time soon. However, absolutely denying the possibility that such signals should come about is not something we dare to do. One cannot mention any compelling evidence against the belief that intelligent beings are to be found on Mars. Quite the contrary! And what is more, over the last century natural science has often surprised humanity with its discoveries.[403]

In other words, there was really no telling what might be out there, especially with the innumerable scientific contributions that had expanded the limits of human cognition and the realm of human imagination. Throughout the nineteenth century the debate about extraterrestrial life had gone on in popular and scientific circles. Not only enthusiasts and amateurs, but many professional scientists as well took such discussions very seriously and argued in support of one view or the other, for example in the proceedings of the Royal Danish Academy of Sciences and Letters.[404]

Between 1850 and 1920, science brought many new thoughts and ideas that only gradually gained acceptance. One thing was the demonstrable age of the Earth, which was substantially extended with every passing decade. A growing body of geological, biological and physical evidence made it increasingly difficult to continue to believe in an Earth, and a life, created by God within the past 6,000 years. Some people were infuriated, others horrified, but although humankind's vanity was tarnished by being scientifically degraded to just one species among many, the situation was still tolerable to most Danes. Their faith in God might be slightly shaken, but humankind and the world would go on.

It was quite another thing when science also claimed it was capable of predicting the end of the world. The idea of the Earth facing its ultimate demise seemed well suited to the biblical cosmology, but when science entered that debate, things suddenly became much more tangible and concrete. The field of physics presented a number of ominous apocalyptic scenarios. In one, physicists demonstrated that the Sun was slowly cooling down, which would eventually render it incapable of heating the Earth, thereby destroying the conditions for life. In another physics-based scenario, according to the second law of thermodynamics the universe would continue to seek equilibrium, eventually resulting in a uniform temperature throughout the entire universe. In short, things would get very, very cold, and as in the scenario of the failing Sun, life would cease to exist. A third popular scientific apocalypse envisioned the Earth being struck by a gigantic meteor that would instantaneously eradicate all life on the planet.

All three of these disturbing scenarios were depicted with unnerving clarity in the French astronomer and popular-science writer Camille Flammarion's dark futuristic fantasy *La Fin du Monde (Omega: The Last Days of the World)* from 1893. The book appeared in Danish in 1894 as *Verdens Undergang*, with a second Danish edition in 1910. Flammarion evoked the image of humanity languishing, trapped on an Earth that was slowly losing its water through evaporation, a planet that was freezing to death, that was struck by a devastating comet, scorched by the Sun's merciless flames; an Earth suspended in a solar system of planetary corpses that hung, dark and lifeless, slowly revolving around an extinguished star; a universe reaching its logical conclusion in thermodynamic heat death and snuffing out all and any possibility of life. Here was all the death and destruction anyone could imagine, and its substance was supplied not by the Bible, but by science. Admittedly, Flammarion seemed equally willing to include picturesque and romantic portrayals, but his narrative was rooted in the scientific ideas and theories of his time, with heat death as a good example:

> If we assume that we can trust the validity of our science, mechanics, physics and mathematics, and if we regard the currently valid laws of nature and of our own thinking as incontrovertible, then this must be the fate that is reserved for the cosmos.[405]

The end of the world was popular material, and Flammarion's visions made for interesting conversation. His books sold well. In Denmark, *Familiejournalen* made use of their unsettling illustrations, and the Danes were equally entertained and educated by the scientific thrills.

Both the world and science had their limits. They were not always easy to define, or to abide by. In Poul Heegaard's opinion life itself presented a difficult question – for the individual human being and for science. In his words, there was not "any definite boundary for what is to be called life, and what is not; a difficulty founded on the circumstance that there is absolutely no person who knows what life actually is."[406] Julius Thomsen, well-known scientist, industrialist public figure, and director of the Polytechnical College (see chapter 13 and elsewhere), harboured similar sentiments. At the inauguration in 1890 of the college's new buildings in central Copenhagen, Thomsen, speaking before a large audience that included members of the royal family, talked about the growth of science and the development of technology. But underneath his enthusiastic praise for science and its practitioners, who had made such a huge impact on Danish society during the nineteenth century – and on Thomsen himself – a more cautious tone was detectable.

Thanks to science, the capabilities of humankind were now so great that it seemed almost overwhelming to a late nineteenth-century mind. And yet, argued Thomsen, there were still things that no person could do. Certainly people could get an egg to hatch or a seed to sprout, but no one had "the capacity to form a cell that was able to live, or a seed that was able to sprout".

> And I think this ought to be kept in mind; for in our day and age, science is presenting so many wonderful findings, in the field of physics and that of chemistry, that it easily kindles the greatest aspirations and leads to an over-estimation of the reach of science; to an idolatry against which one must caution fully as much as H.C. Ørsted once zealously denounced the superstition of his times at the founding of the Polytechnical College.[407]

Science had made enormous advances from 1850 to 1920. It had discovered new worlds deep under the oceans and in the far reaches of outer space. It had altered preconceived ideas and profoundly changed the lives of every Dane, man or woman, adult or child. People lived in a new world; a different world that made the rural country of the 1850s seem distant and strange. The vast and varied fields of science had promised much and delivered more. The success stories far outnumbered the failures. Even though some voices called for concern, for most Danes the great expectations of the age remained unspoiled. A great many things could be done, and many of them were done. But not all. During the second half of the nineteenth century and early decades of the twentieth, the overall tone dominating Danish academic, intellectual and cultural debates about science and technology was buoyant and triumphant. But beneath the surface a certain dissonance alerted the attentive listener that indeed, science had its limits and its limitations.

PART IV

Boundless knowledge

1920–1970

Henry Nielsen and Kristian Hvidtfelt Nielsen

Profiling the period 1920–1970

Around 1920, science still only accounted for a very modest share of Denmark's national economy. By 1970, massive investments during the 1960s in scientific education and research had profoundly altered that situation. One major factor behind the drastic changes during this period was the shifting political and economic trends that affected Denmark, and consequently also Danish science. Generally speaking, it makes good sense to divide the period into three phases.

The first runs from 1920 to 1940. These two decades were influenced by several important factors: a Danish policy of neutrality, a political shift towards centre/left with the Social Democrats as the dominant political force, the global economic crisis of the 1930s and, in reaction to the ensuing depression, the earliest beginnings of what would come to be widely known, three decades later, as "the welfare state". In 1920 the scientific institutions in Denmark were small and inexpensive for the state to run – but the level of prestige that accompanied science was rapidly rising, owing in large part to the brilliance of Niels Bohr.

The second phase begins in 1940 and goes on until 1955 or so. It was heavily marked by the German occupation of Denmark, which lasted from April 1940 to May 1945, and by the judicial purge that followed the country's liberation. There was also the irrevocable breakdown of Danish neutrality when the country signed the North Atlantic Treaty in 1949, plus the troubles of an ailing Danish economy plagued by chronic foreign-exchange problems that left no room for renovating or replacing the country's worn-down production facilities. The European Recovery Program of the early 1950s, better known as Marshall Aid, gave the Danish economy a sorely needed shot in the arm, but there was still no adequate surplus for the sort of radical upgrade that Danish science would need to keep up with developments in a number of other countries; an upgrade that Danish scientists had long been urging.

The third phase stretches from the mid-1950s to 1970. Politically, these were stable years dominated almost entirely by the Social Democrats. Economically, this phase was characterized by an exceptionally long period of sustained growth, which fuelled Denmark's rapid transformation from a decidedly rural economy to a modern industrialized economy, and its development of the welfare state. The economic revival enabled the state to finance a virtual explosion in education and research, the likes of which Denmark had never seen, and which particularly benefited the fields of science and technology.

Expansion and internationalization

Because this outline of the period must be brief, it necessarily leaves out many details from the complex and multi-faceted reality of Danish science. In the bigger picture, one of the most noteworthy features is the massive scientific expansion that took place. For example, during the five-year interval 1920–1924, only 68 people took a master's degree in science from the University of Copenhagen. The corresponding figure for 1970–1974 was 952, or roughly 15 times as many.[408] Another example is the number of full-time employees teaching science in 1921 at the University of Copenhagen: 20. In 1973 the same institution had a total of 471 full-time and associate professors teaching science subjects.[409] And a third example: In 1938 the total appropriation in the Danish Finance Act for all of the country's institutions of higher education was less than 20 million Danish kroner (DKK), whereas the corresponding figure for 1970 had risen to more than DKK 500 million (both stated in 1955-DKK).[410]

It is no less interesting to chart the striking development of these figures year by year. As for the national allocations for education and research, their growth was very modest in 1938–1955, whereas in 1955–1970 it was massive. As will be shown repeatedly in the following, this indicates that the years from about 1955 to 1960 in many ways constitute a unique period in Denmark's history. It was unique because it was precisely at this junction that Denmark entered a period of 15–20 years of high economic growth. Unique because during these five years or so, the country's chiefly Social Democratic governments took the initiative to drastically increase public spending on scientific education and research, citing as its justification that this was a necessary step, indeed a crucial prerequisite for Denmark's future as a modern industrial nation.

Another striking feature was the increasing degree of internationalization. Acknowledging science as an international activity was not a new thing. For several hundred years the practitioners of science had studied, tested and built upon the results of their professional peers, paying no heed to geographic origin or location. Exalted scientific academies like the Royal Society in London and the Académie des Sciences in Paris accepted both their own nationals and foreign scientists as fellows. Promising young scientists from Denmark, like the physicist Hans Christian Ørsted (in the early 1800s) and the chemist Niels Bjerrum (in the late 1800s), basically regarded prolonged stays with the great contemporary minds in their respective fields as a matter of course. The last two decades of the nineteenth century gave birth to a tradition of regularly recurring international scientific congresses, first in mathematics and later in astronomy, physics, chemistry and biology. Crowning this development came the initial awarding in 1901 of the Nobel Prize in certain fields, bound by the covenants in Alfred Nobel's will that "in awarding the prizes no consideration whatever shall be given to the nationality of the candidates, so that the most worthy shall receive the prize, whether he be a Scandinavian or not."[411] Many intellectuals saw the natural sciences, with their tradition of openness and the free exchange of ideas across borders, as models for the various efforts being made to promote peace and harmony between nations.

With the sudden outbreak of World War I in 1914, all ideas of peaceful co-existence fell into ruin. From one day to the next, the majority of scientists in the warring nations abandoned their lofty ideals, uncritically adopting the chauvinistic policies of their own governments instead. One of the most infamous examples was the "Aufruf an die Kulturwelt!" from August 1914, in which 93 well-reputed German scientists rejected all claims of German troops demonstrating brutal conduct in Belgium, denouncing them as malicious slander.[412] British and French scientists responded in kind, and as the hostilities dragged on the animosity grew between the former intellectual allies in step with the appalling losses of life in the trenches. When the Great War, "the war to end all wars", finally came to an end in November 1918, this hatred had grown so strong that British and French scientific societies zealously sought to exclude German and Austrian scientists from participating in future international congresses.[413]

A different feeling prevailed among scientists in the neutral Scandinavian countries and in the Netherlands.[414] These nations had not experienced the war first-hand, and even before it erupted many Dutch and Scandinavian scientists had established close contacts with colleagues abroad, not least in Germany. After World War I many scientists in Scandinavia and the Netherlands felt isolated in their small, domestic professional communities and longed to participate in international conferences organized by German, British or French colleagues without taking the risk of being boycotted by the opposite party. One feature that prominent scientists in Denmark and in the rest of Scandinavia and the Netherlands shared in the 1920s was their dedication to actively bridging the gap, or rather the abyss, that had opened between their colleagues in the nations involved in World War I. The Danish physicist

Niels Bohr is one outstanding example. Throughout the 1920s and 1930s, Bohr's new Institute for Theoretical Physics served as a meeting place for nuclear physicists from around the world; as a sort of sanctuary where the international ideal of open science was held in high esteem, and put into practice. Many other Danish scientists followed Bohr's example.[415]

Later in the twentieth century the concept of internationalization took on a new and very different meaning. It sprang from Germany's defeat in both world wars, and from America's huge investments in developing science and technology after entering the twentieth century. The fact was that at the end of World War II, the United States stood alone as the undisputed world leader in natural and medical science. In a certain sense the internationalization of science became synonymous with its "Americanization". During the 1920s and 1930s young American scientists had travelled to Europe to study with world-renowned scientists there at the best institutes, but after 1945 the tables had turned. Young European scientists, fed up with working under miserable conditions in their own small, overcrowded laboratories, were drawn to the rich, well-equipped American universities, where the most successful might even hope for a permanent position. The English phrase "brain drain" was incorporated into the Danish language – an appropriate metaphor for a very real problem. More than one field saw scientific talent siphoned off by more attractive opportunities elsewhere before Danish science began to seriously expand in the late 1950.

Selected themes

The enormous growth and increasing internationalization of science that took place from 1920 to 1970 is the overarching theme in this fourth part of our history, which revolves, chapterwise, around five selected themes.

Theme 1: From private philanthropy to public funding. The first area we have chosen to focus on is the economic foundation of scientific research. During the period under review, especially before 1955, Danish science was generously supported by a number of private foundations whose importance can hardly be overestimated. It is true that the Danish state usually paid the expenses associated with running the university institutes, including teachers' wages, but the amount of "free" research funding available to the institutes was extremely small. The Danish zoologist and physiologist August Krogh is a good example. Krogh's annual research grant from the Danish state in 1920 –

the year he won the Nobel Prize in physiology or medicine – amounted to approximately DKK 2,000, and that was including equipment replacements and repairs.[416] Because the Danish economy was weak for most of the inter-war period, it was rarely possible to persuade the national authorities to assume full responsibility for building new research institutions or financing specific research projects, no matter how relevant and well documented a request might be.

That is what made the International Education Board (IEB) a vital source of funds for Danish scientists in the 1920s. Financing from the IEB, which was a branch of the Rockefeller Foundation, also supported many other Nordic and European research institutions.[417] Thanks to these grants from across the Atlantic, Denmark was able to realize important new building projects under the auspices of the university: the Institute for Theoretical Physics (today the Niels Bohr Institute), the Institute for Physical Chemistry, and the Rockefeller Institute, which was actually a complex of five scientific and medical research institutes in the fields of physiology, biochemistry and biophysics, as described in chapter 21. Along with the Division of Medical Education, which was another branch of the Rockefeller philanthropy, the IEB funded the construction of the Rockefeller Institute with a total of more than USD 400.000, which at the time corresponded to approximately DKK 2.4 million.[418]

Denmark also had its own philanthropic foundations that played a crucial role in the development of Danish science, as was also the case in Swedish physics, for instance.[419] Had it not been for the Carlsberg Foundation, Danish scientists would have been unable to carry out of series of ambitious research projects in the inter-war years, including very expensive scientific undertakings like the marine research voyages on the *Dana* in the 1920s and the Arctic explorer Lauge Koch's three-year expedition to Greenland in the 1930s.

As the Danish economy gained momentum from the mid-1950s onwards, and as the political recognition of science's enormous importance for Denmark grew, within a few years the country's public grants for scientific research and development had doubled and redoubled several times. This made the Danish state the largest contributor by far to the efforts of the Danish research community, even though the national authorities had no actual research policy in the modern sense of the word. Naturally, over the centuries science as a social institution has always been embedded within the political system, in Denmark and elsewhere. In reality, however, science and politics long remained two relatively autonomous spheres of activity.[420]

There was no sign of their merging until the mid-1960s, by which time the Danish state had become so deeply involved in science that public expenditure on research threatened to swell out of all proportion. That was when influential politicians and bureaucrats realized it was necessary to control development more effectively. But although the first small steps towards an actual research policy were taken in the mid-1960s, in practice it would be several decades before Denmark had implemented any sort of reasonably coherent policy for the area. Throughout the period, until about 1970, the Danish state allocated virtually all public research grants directly to the relevant scientific institutions, thereby leaving it up to the representatives of the sciences to use their separate grants as advantageously as possible. The large Danish foundations made use of the same scientific representatives, or of their colleagues, which meant that the actual distribution of research funds – by far the most crucial implement of control – was placed in the hands of a network of prominent scientists. Very few people perceived this as a significant problem. Quite the contrary. In fact, to put it mildly, most scientists were only too pleased that politicians and other groups refrained from interfering with their research decisions. This was a task best left to the foremost representatives of science, operating, of course, within the given financial framework.

Theme 2: Peaks in Danish university research. The examples given above of research initiatives that were wholly or partially backed by private funding played an important role in the golden age that a number of Danish scientific disciplines experienced in the 1920s and 1930s. This was particularly true of theoretical physics (Niels Bohr), physical chemistry (Niels Bjerrum, Johannes Brønsted and Søren Peter Lauritz Sørensen) and experimental physiology (August Krogh). Four Danish scientists within these fields received Nobel Prizes between 1920 and 1943, which is a large number for a country Denmark's size.[421] Its scientific activities were concentrated in and around the capital, and the research communities there enjoyed an excellent international reputation during the inter-war years. So much so that when the Rockefeller Foundation's IEB thoroughly assessed the state of European science around 1926, its findings showed the Copenhagen research groups in the above-mentioned disciplines to be among the peak performers in the topography of international science – which is why they received such generous IEB support.[422]
After World War II there were a number of shifts in the Danish scientific landscape. The glory days that many disciplines had enjoyed in the preceding decades began to fade. Other disciplines like theoretical nuclear physics flourished, resulting in such achievements as a Nobel Prize in physics for Aage Bohr, received jointly with Ben Mottelson and James Rainwater. Physiology research in Denmark maintained its strong position, but the focus shifted to

cell physiology in work carried out by prominent scientists such as Hans Henriksen Ussing and Jens Christian Skou (with the latter receiving the Nobel Prize in chemistry for 1997).

Theme 3: Scientific exploration and big science. Like most other countries in the Western world, the Danes sent out a variety of scientific expeditions between 1920 and 1970. Most of them, notably those focusing on botany, zoology, geology and geophysics, were destined for Greenland, which, like the Faeroe Islands, went from being a Danish colony to being an autonomous province within the Kingdom of Denmark during the period under review. Other notable Danish expeditions had ambitious programmes for marine research in the North Atlantic and elsewhere.

It may seem rather odd, particularly in light of Denmark's difficult economic situation prior to 1955, that the country chose to spend as much money as it did on scientific expeditions, particularly when those very years heard complaints voiced by almost all scientific institutes at university level, bemoaning the catastrophic lack of tenure positions and up-to-date scientific equipment. Part of the explanation has to do with national pride: a small nation wanting to show the world that even when times were hard, it was capable of launching spectacular scientific ventures. However, a closer look at the background of some of the most important expeditions of those years reveals national territorial interests as well. It is reasonable to surmise that these interests were probably what motivated the Danish authorities to be so unusually proactive in realizing expeditions of this nature.

During the five decades of this period, Danish scientists also participated in other types of scientific research outside their own country, in big-science projects carried out at large international laboratories.[423] The phenomenon that was termed "big science" did not reach Denmark until 1950 or so, after a number of European countries had recognized, and admitted, that they had fallen hopelessly behind their American counterparts in the fields of experimental and high-energy physics. These countries consequently began an effort to establish a large, joint research laboratory on European soil. Danish scientists were also a part of this effort, which ultimately became the Conseil Européen pour la Recherche Nucléaire – better known as CERN, located outside Geneva. Niels Bohr, the grand old man of Danish physics, was a key player in the founding of CERN, and he would have preferred to see the laboratory located in Copenhagen.

By the late 1960s, Denmark had signed on for two more European big-science projects: the European Space Research Organization (ESRO) and the

European Southern Observatory (ESO), a large, jointly operated European astronomical observatory in Chile. Both of these projects were to some extent modelled on CERN, perhaps because, quite apart from its scientific successes, CERN was an eminent example of a European organization that was capable of competing internationally – which essentially meant competing on a par with the United States of America.

Theme 4: Pure and applied science outside the universities. In Denmark, as in other countries, a good deal of scientific research takes place outside the universities, in public and private bodies. Traditionally, most of the research at the Danish universities has been basic, or "pure" science, while most non-university research has been applied. But during the period described here, this distinction became increasingly meaningless, with much basic research going on in non-academic environments and much applied research going on at universities and other places of higher learning.

This fourth thematic chapter describes key elements of the research going on from 1920 to 1970 in various, and highly varied, research institutions operating outside Denmark's universities: Risø Nuclear Research Centre (today Risø National Laboratory), the Nordic Insulin Laboratory and the Carlsberg Laboratory. Their stories illustrate how difficult it can be to distinguish between pure and applied science. In addition, they show how some of these institutions had to go through long, exhausting struggles, internal and external, to finally arrive at a suitable research profile that was capable of delivering results.

Theme 5: Science for the people. Between the two world wars Danish books, newspapers and literary journals almost exclusively presented science in a positive light.[424] The public came to know scientists as intelligent and deeply committed individuals whose sole concern was to cajole nature into divulging its deepest secrets. The commercial exploitation of research results was left to engineers and businesspeople, as science was to be kept pure and unsullied. Furthermore, Danish scientists were often favourably highlighted as dynamic champions of peace and reconciliation between German, French and British physicists during the 1920s, or as idealistic protectors of German scientists fleeing the Nazis in the 1930s.

After World War II the picture remained largely the same, even after the public learned of the physicists' vital contributions to developing the atomic bomb during the war. In the words of J. Robert Oppenheimer, the father of the A-bomb, that was when "the physicists have known sin" – though evidently Danish scientists and journalists did not believe this was true of Danish physicists. It was certainly true that Niels Bohr had been in Los Alamos, but had he not worked diligently to avoid having the terrible power of this new weapon unleashed in the war? Had he not fought a lonely and, unfortunately, vain battle to prevent the nuclear arms race between East and West? And did he not make sure to dissociate his famous physics institute on Blegdamsvej in Copenhagen from any and every link to commercial and military research?

In Denmark, confidence in the practitioners of science was virtually unlimited during the 1950s and 1960s. Apart from the occasional dissident, it was not until the late 1960s and early 1970s that a group of concerned citizens, including many biologists, began questioning the increasing pollution of nature and the ongoing development towards a "plutonium society" with all the risks that entailed. In the eyes of the Danish people, nuclear power was so intimately linked with physics and chemistry that representatives of these sciences suddenly found themselves in the harsh and hostile spotlight. It was at this juncture that the image of "hard science" started to slide, commencing a decline that would last for several decades.

Denmark's research and education policy throughout the inter-war period can best be characterized as utility-oriented. There was a tendency to associate university research with culture, which was considered to be in the same category as consumption and, by extension, luxury. University research was allowed to live its own life, and seen through modern eyes, the commercial and technological opportunities that such research represented was apparently of little interest to politicians.[425] This type of thinking was clearly present, although naturally the various areas of scientific research were perceived in different ways. The general trend between the wars was that utility-oriented fields were given a higher priority than fields that were purely scientific. The subjects used to educate students as certain types of professionals – engineers, pharmacists, doctors, veterinarians – did well during this period. At the opposite end of the scale was the University of Copenhagen, which from the early 1920s to the mid-1950s had to fight tooth and nail for every grant increase it received. We can best illustrate the priority that utility-oriented subjects enjoyed by looking at the construction projects carried out during the inter-war years. The largest publicly funded projects were the expansions of the Royal Veterinary and Agricultural College, and of the Danish Technical College (known as the Polytechnical College until 1933). The Agricultural College was hugely expanded in the early 1920s, and research efforts there intensified. The 1942 estimate for total annual allocations for agricultural testing and research activities was 2–3 million Danish kroner (DKK), much of which was coordinated through the Agricultural College.[426] An ambitious expansion at the Technical College was initiated in 1929, the first stage involving the construction of laboratories for technical chemistry for what was then a colossal grant of DKK 1.25 million. When the impressive complex on Øster Voldgade in central Copenhagen was finished in 1954, expenses added up to DKK 13.6 million – a staggering amount at the time.[427] The number of laboratories had grown from 15 to 32, and overall, during the 1930s the interest that the Danish politicians exhibited in the Technical College continued to grow.

The Danish historian Jørgen Holmgaard has described the country's political climate during the inter-war years as "research-sceptical", and he has pointed out that the politicians succumbed to the allure of the efficient, business-oriented, applicable research and education going on in certain types of professional colleges.[428] This description may sound surprising, given that during this period the Danish natural and medical sciences experienced a large and lasting resurgence. The early part of the period saw Nobel Prizes awarded to the physiologist August Krogh (1920), the physicist Niels Bohr (1922) and the cancer researcher Johannes Fibiger (1926), who were later joined by the vitamin researcher Henrik Dam (1943). Danish science could count itself among the global elite in several fields, and a number of scientific institutions expanded with large new building complexes. During the inter-war period, Denmark's combination of up-to-date research facilities and famous scientists ensured an influx of promising junior researchers from abroad, as described in greater detail in chapter 21.

This paradox can be explained, in large part, by the Danish scientific community's ability to attract capital from domestic and foreign foundations, resulting in a steady and substantial flow of supplementary financing to bolster the meagre public funds.

Doing research for private money

There is no comprehensive record of the many different private grants Danish science received between the two world wars. The following list is therefore not exhaustive, but it does give a reasonable picture and shows certain consistent features. With respect to new construction projects, the single most important factor was a series of American donations from the private Rockefeller Foundation, given through a branch it founded in 1923 called the International Education Board (IEB). Generally the Rockefeller Foundation primarily supported medical research, whereas the IEB's activities revolved around an ambitious plan to promote scientific education and research at an international level. The Rockefeller Foundation and the IEB backed a large number of research environments throughout Europe, but there is no doubt that Copenhagen was one of the locations that received the most generous support.[429]

Niels Bohr's Institute for Theoretical Physics was built in 1920–1921. At the time, the actual construction costs were mainly financed by the Danish state, although private fund-raising efforts had resulted in DKK 80,000, which covered about one sixth of the building project's total costs.[430] When the institute expanded in 1924, the main source of financing was an IEB grant of 40,000 US dollars (USD), but once again the Danish state contributed significant extraordinary grants for operations, furnishings and equipment.[431] On the other hand, the construction of Copenhagen's large new physiology institute – the Rockefeller Institute, as it was named in gratitude to its principal benefactor – was funded almost exclusively by grants from the Rockefeller Foundation (USD 300,000) and the IEB (USD 107,000), topped up with smaller donations from domestic sources like the Carlsberg Foundation and the Rask–

20.1 *Powerful men in the world of Danish science: the board of directors, meeting at the Carlsberg Foundation on 7 October 1942. The five board members – from left to right, with field and period of service in parentheses – are Poul Tuxen (Oriental philology, 1933–1952), Johannes Hjelmslev (mathematics, 1918–1950), the chairman of the board Johannes Pedersen (Semitic philology, 1926–1955, chairmanship 1933–1955), Niels Bjerrum (chemistry, 1931–1956) and Knud Jessen (botany, 1937–1960). Upholding tradition, the chairman presides at the head of the table, with the other four members seated in order of seniority.*

Ørsted Foundation.[432] Just a few years later, in 1927, the Danish chemist Johannes N. Brønsted was able to move into his new Institute for Physical Chemistry, built with an IEB grant of USD 100,000. The medical science community in Denmark also received substantial support from the Rockefeller Foundation, which in 1924 donated USD 200,000 for an expansion of the State Serum Institute, as well as USD 90,000 to the University of Copenhagen for the construction of the Institute for Human Genetics and Eugenics in 1938.[433]

In addition to these major construction projects, the Rockefeller Foundation also supported a number of regular research projects in Denmark. One was a collaborative effort in the mid-1930s by Niels Bohr, August Krogh and the Hungarian-born chemist George Hevesy to apply isotope indicators in biology, as described in chapter 21. Another example was the biochemist Kaj Ulrik Linderstrøm-Lang, who worked at the private Carlsberg Laboratory and received a fixed annual research allowance from 1937 onwards.[434] Also, from 1917 to 1970 the Rockefeller organization gave about 50 Danish men and women of science personal travel and research grants,[435] as well as enabling of large number of foreign scientists, including about 150 Americans, to visit Danish research institutions for varying lengths of time.[436]

Private Danish foundations also helped to expand scientific research during the inter-war years. The Carlsberg Foundation, established by the brewer and philanthropist J. C. Jacobsen in 1876, was of unparalleled significance to this development, serving as a veritable gold mine for Danish science. The foundation supported natural sciences and humanities alike, and its activities were strictly confined to national pursuits, although foreign scientists could receive support for their work, provided they were affiliated with Danish institutions.[437] The Carlsberg Foundation, whose influence on Danish science probably peaked between the two world wars, employed its resources in many different ways. It supported building projects, research work and publications, and it granted sizeable annual pay supplements to a number of permanent recipients to make up for their relatively small public salaries.[438] Buildings made possible by donations from the Carlsberg Foundation include its own Biological Institute (1932) and the university's Institute of Mathematics (1932). In this way the foundation's patronage ensured that despite political lethargy and zealous cost cutting, Danish science not only kept afloat, but also made some headway through the stagnant waters of the inter-war period.[439]

As a matter of fact, the Carlsberg Foundation's creation of the Biological Institute illustrates how foreign and national philanthropy worked together, not only aiming to make existing peaks higher but also, more ambitiously, to construct new mountains in the national academic landscape. This institute was set up in 1932 for Albert Fischer, an eminent experimental cell physiologist of international standing. During a stay at the Rockefeller Institute for Medical Research (RIMR) in New York from 1920 to 1922, Fischer had studied with the French-American surgeon Alexis Carrel and learned and refined the new technique of in-vitro tissue culturing – the fascinating yet somewhat Faustian technology of growing tissue cells in test tubes. In 1926 a special guest department was set up for Fischer at the Kaiser-Wilhelm-Institut für Biologie in Berlin, where he applied his tissue-culturing techniques on cancer cells. Around that time Fischer approached the University of Copenhagen, backed by RIMR director Simon Flexner and the Rockefeller Foundation, with plans

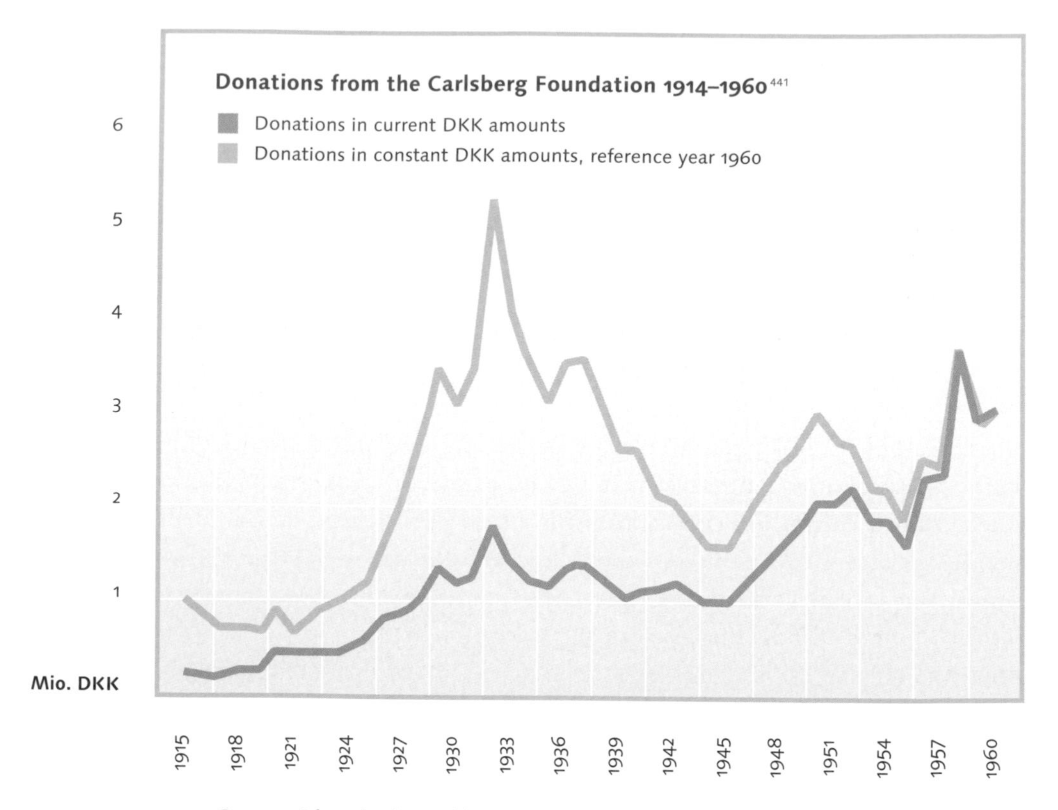

20.2 *Grants paid out by the Carlsberg Foundation 1914–1960.*

to continue his cell and cancer research at the Danish university. Negotiations were unproductive until the Carlsberg Foundation stepped in, offering to pay for buildings and equipment (DKK 586,000, or about USD 146,000) and securing governmental support in the form of a construction site. The Rockefeller Foundation supplied the new institute with DKK 1 million (USD 250,000) in working capital; a huge sum that secured Fischer working conditions and a degree of autonomy at which other scientists could only marvel. Modelled after the Kaiser Wilhelm Institutes as a freestanding, independent research institute, the Biological Institute of the Carlsberg Foundation constituted a remarkable organizational innovation in the Danish research system, all the more so for its creation in the midst of the depression.[440]

The Danish state's tight-fistedness was most often explained and justified by dispassionate financial concerns. In effect, the existence of the Carlsberg Foundation also facilitated this stance, enabling the national authorities to invoke their empty coffers time and again, passing responsibility on to private initiatives. As long as the foundation deemed its financial basis and allocable funds sufficient, this was not a major problem – not until the rampant inflation of World War II had drastically eroded its annual disposable sums. As Figure 20.2 shows, donations reached a new record low in 1945. The impressive peak in the early 1930s resulted from extraordinarily large grants, combined with the effective deflationary policy implemented by the Danish government in the mid-1920s.

Although the Carlsberg Foundation was not the only private body to support Danish science, it was clearly the largest, and consequently its donation policy was crucial to the types of science that could be cultivated in Denmark. The foundation's board of directors is appointed by, and chosen among, the members of the Royal Danish Academy of Sciences and Letters, which has therefore had an exceedingly strong influence on the development of science in Denmark throughout the years. In addition, the activities of the Carlsberg Foundation are not targeted to support commercially oriented research. That is why successive directors of the Technical College long hoped for the establishment of a similar body to support technical research. In the early 1930s, this dream was at long last fulfilled when a number of sizable Danish foundations were created with charters that also covered this goal: the Tuborg Foundation (1931, assets of DKK 2.5m), the Laurits Andersen Foundation (1931, assets of DKK 6m), the Otto Mønsted Foundation (1934, assets of DKK 23m) and the Thomas B. Thrige Foundation (1934, assets of about DKK 10m). Although these amounts hardly seem impressive today, the personal fortunes of electrical industrialist Thomas B. Thrige and margarine magnate Otto Mønsted placed them among the very wealthiest people in Denmark. Taken together these foundations were almost as large as the Carlsberg Foundation, which was extremely important given that technical research, like scientific research, was strongly dependent upon private patronage.[442]

The conclusion is clear: The inter-war years were the halcyon days of private philanthropy. The pattern observable in Denmark is also evident in countries like the US, the UK and Sweden. When the national authorities held back, private capital lent a helping hand – and vice versa.

Neutrality and internationalism

One body that already began to reinforce Danish science's international profile after the end of World War I was the Rask–Ørsted Foundation, set up by the country's national government in 1919. This foundation backed collaborative international efforts, scientific and organizational, in which Danish scientists participated. It also provided funding for international conferences held in Denmark and for foreign scientists invited to Denmark to teach and lecture. Besides all this, it supported the publication of Danish works translated into major foreign languages. In other words, the objectives of the Rask–Ørsted Foundation were simultaneously national and international. The main idea was to help Danish science to cooperate internationally, and in this respect nationalism and internationalism went hand in hand, reinforcing one another. The foundation's most important contribution was probably the many grants and scholarships that enabled Danish scientists to travel, study and work abroad, and helped their foreign peers to do the same here. From its establishment in 1919 and up to 1941, the Rask–Ørsted Foundation enabled about 100 scientists from abroad to work in Danish institutions. It is worth noting that another 150 guests visited and worked in Denmark in those years thanks to grants from the Rockefeller Foundation.[443]

Right from the start the Danish politicians were very mindful of the Rask–Ørsted Foundation, whose early directorship included such prominent figures as the country's prime minister, Niels Neergaard, and its later foreign minister, Peter Munch. The foundation's initial endowment – totalling DKK 5 million, generating an annual yield of about DKK 250,000 – was taken from the proceeds of Denmark's 1916 sale of its possessions in the West Indies, which today are the three largest US Virgin Islands. The creation of the Rask–Ørsted Foundation, named after the philologist Rasmus Kristian Rask and the physicist H. C. Ørsted, was a direct consequence of the total collapse of international scientific cooperation during and after World War I, which had killed virtually all such efforts in one fell swoop, causing chauvinism and patriotism to flare up among scientists on both sides. When the Great War ended there was no sign that cordial relations would develop again any time soon. In the Nordic countries, which had remained neutral during the war, scientists and politicians began discussing in the autumn of 1917 how the region could best help to quench the smouldering hostility and re-establish international cooperation in the scientific community. Ever since the mid-1800s, Germany's large and amply funded scientific institutions had been the heavyweights in the world of science, but in late 1917 it looked like the central powers would come out of the war greatly weakened. Therefore the Nordic plans, formulated during the final year of World War I, were in a sense opportunistic, tinged with the hope that at least in certain selected fields the Nordic region might be able to take over the key position previously held by Germany, becoming a "central place for international scientific work".[444]

The establishment and active involvement of the new Rask–Ørsted Foundation in 1919 gave Denmark an international angle at a time when national chauvinism was the order of the day. It was not until the mid-to-late 1920s, and at a variable pace, that international cooperation in the different fields of science slowly rose from the ashes once again. The financial support of the Rask–Ørsted Foundation ensured that Danish scientists came to play an important role in this rebirth. Among other things, as described in chapter 21, the foundation was of vital importance for Niels Bohr's ability to attract talented foreign scientists to his new institute in Copenhagen. Quite apart from that, it was also extremely valuable as a symbol. Frederik Graae, the department head at the Danish ministry of education and long-time member of the Rask–Ørsted Foundation's board of directors, declared in 1941 that the foundation's existence and activities were essential to the American donations that so greatly benefited Danish science during the inter-war years.[445] Viewed as a single, active and practically applicable expression of science policy, the creation of the Rask–Ørsted foundation was in tune with the country's foreign policy after the Great War. And this was hardly a coincidence, since the actual political force behind the foundation was Peter Munch, who served as Denmark's minister of defence during the Great War and went on to serve as foreign minister, becoming the single most influential player in Danish foreign policy during the inter-war years. Seen in this light, the Rask–Ørsted Foundation was one aspect in the Danish policy of neutrality that had Munch as its principal engineer.

The first Danish research council

Denmark's first national public research council, the Technical Research Council, was set up in 1946 and must be seen against the backdrop of World War II and the years preceding it.

In the inter-war years Denmark was still a distinctly agricultural nation, and so the agricultural crisis of the 1930s was an eye-opener for the Danish government, revealing with brutal clarity how vulnerable the country's economy was to shifts in global trade. This motivated the prime minister Thorvald Stauning – Denmark's popular, long-standing and first-ever Social Democratic leader – to set up a think tank in 1937. This new body, called the Production and Natural Resources Commission and headed by the governor of the national bank, C. V. Bramsnæs, was tasked with making proposals as to how the country's trade and industry could adapt to a world whose largest economies were preparing for war. The most important questions were how Danish agriculture could reduce imports of foreign feedstuffs, and whether the agricultural sector could expand its role as a supplier of raw materials to Denmark's domestic industrial sector. One avenue explored was milk protein, which could be processed into casein wool to yield a new synthetic fibre called "lanital", which some textile manufacturers considered promising at the time. Another was cattle cakes, an important item on Denmark's foreign-trade budget, which could be partially replaced by protein-rich feed grown and refined in Denmark. The most pressing issues revolved around agriculture, as rising prices for raw materials, particularly feedstuffs, began to spell trouble for Denmark's foreign-exchange balance in the late 1930s.

After the war broke out in 1939 and Germany occupied Denmark in April 1940, the country's connections with the British market and most of its external supply channels for raw materials were blocked. Suddenly the problems that the Production and Natural Resources Commission had to deal with took on a new urgency. The commission therefore proposed in 1940 that Denmark create a "special research institute" that could investigate the country's pressing technical and economic difficulties and seek out completely new resources and methods of production. In the event of German victory, Denmark would be forced to fundamentally change the structure of its trade and industry, which for many decades had been organized to reflect the country's huge agricultural exports to the UK. Working with the Technical College, the Agricultural College and the Danish Academy of Technical Sciences (founded in 1937), the commission published a *Recommendation concerning technical scientific research* in 1942, inventively proposing how such a research institute could be organized with due consideration to the country's existing business structure and research bodies.

The recommendation sparked a protracted debate among scientists, businesspeople and the national authorities. Curbing their aspirations, the parties agreed to set up a research council with a less ambitious profile. As it turned out, the plans for the research council were not implemented until after the war. The bill simply ended up in parliamentary limbo in the Folketing when the Danish government stepped down in August 1943 to protest a German ultimatum demanding, among other things, the death penalty for apprehended Danish saboteurs. But once the Technical Research Council finally became a reality in 1946, it commanded the attention of the political establishment. This is evidenced by its membership, which included the governor of the Danish national bank and the prime minister's permanent secretary. The council's annual appropriation was DKK 250,000, with another DKK

100,000 available for scholarships. These funds were mainly used to expand research activities at the Technical College and at the Academy of Technical Sciences, but the council also supported agricultural research and basic scientific research.

The Technical Research Council made the greatest impact through its role in distributing the funds Denmark received as Marshall Aid. Under this great American effort to alleviate the aftermath of World War II, officially enacted as the European Recovery Program, the Danish research council was able to distribute approximately DKK 46 million, or 2.7% of the total Marshall Aid, for research purposes from 1948 to 1956. Amounting to a good DKK 5 million per year in extra grants, this aid was immensely important to Denmark's universities and colleges, whose annual operating costs in 1950 totalled about DKK 30 million. Reflecting the aims of the European Recovery Program the aid money was mainly channelled into agricultural and technical research, but it was indisputably the largest infusion of funding that Danish science had ever received.[446]

Out of the shadow of war

The two atomic bombs dropped over Japan in August 1945 brought an abrupt end to World War II. Along with other new military technologies like radars and computers their use had palpably, and painfully, proved what an impact the large-scale mobilization of science could have. "Dropping the bomb" had demonstrated that even highly esoteric scientific disciplines could unexpectedly gain enormous practical importance. It was precisely this point that contemporary scientist emphasized again and again. Science had delivered the world from barbarism and seemed, in 1945, to be one of democracy's strongest shields. As a consequence, after World War II the status of science in general, and of physics in particular, fell just short of hero-worship – with scientists as the supermen of the new atomic age.[447] One Danish scientist, remarking on this shift in the way science was perceived, pointed out in 1958 that "nature's noble tiller" had now become "the nation's hero, and was cast by the press and radio as a sort of *Übermensch*".[448] In the post-war years people became immensely interested in all things technical and scientific, and many Western countries made massive investments to expand their scientific and technical research, especially in areas relating to nuclear technology, as discussed in chapter 23.

From a political point of view, science in Denmark made it through the German occupation virtually unscathed. Upon his return from the United

Frihedsraadet, da det var forsamlet til sidste Møde. Fra venstre til højre: Professor Husfeldt, Trafikminister Alfred Jensen, Minister Frode Jacobsen, Ingeniør Foss, Professor Fog, Redaktør Schoch, Professor Chievitz, Professor Bodelsen, Minister Arne Sørensen — fem Videnskabsmænd af 9 Medlemmer

Danske Videnskabsmænds Indsats i Danmarks Frihedskamp

Skønt der ikke eksisterer en Liste over alle de Videnskabsmænd, der deltog i Danmarks Frihedskamp, er det næppe for meget sagt, at der ikke i Danmark var een eneste større videnskabelig Institution, som ikke havde Medlemmer i det illegale Arbejde....

20.3 *A number of Danish scientists were active in the Danish resistance movement. This, combined with Niels Bohr's contribution to the American nuclear weapons programme, had Danish scientists riding on an unprecedented wave of success after World War II. Nevertheless, on the whole it would still prove difficult in the early post-war years to translate this goodwill into larger grants for Denmark's scientific institutions.*

States after the war, Niels Bohr was celebrated as a national hero for his contribution to the Manhattan Project. When Denmark's occupation ended in May 1945 the country's population was divided by an invisible line, with noble members of the resistance on one side and contemptible collaborators on the other. Danish scientists ending up on the wrong side of that line were few and far between. In fact, many scientists had played a prominent role in the Danish resistance, and five of the Danish Freedom Council's nine members were university-educated. And at the Rockefeller Institute the physiologists Richard Ege, Poul Brandt Rehberg and several of their staff had helped Jews living in Copenhagen to escape by sailing across the strait of Øresund to Sweden, and they also assembled care packages and sent them to the Danish Jews held in the concentration camp of Theresienstadt.

There were many other examples, and they all pointed in the same direction. During the war, the nation's scientists, who had previously been considered absent-minded and aloof, showed their true colours as active champions of democracy and freedom, and so they were represented as men with exceptionally high moral standards; men whose courage, decency and capacity for action were larger than life. This was just the sort of thing Denmark needed, with the war over and its sitting authorities tainted, its politicians widely regarded as morally flaccid after their policy of acquiescence and cooperation with the occupying forces. Compared with Denmark's political establishment, its scientific community was seen after the liberation as a sort of alternative elite.[449] Based on these moral considerations alone, it was obviously reasonable to reach the same sort of conclusion as Børge Michelsen, a journalist at the popular weekly magazine *Illustreret Familie-Journalen*: that Denmark's previous "tight-fisted policy" towards science had to change, and that in the future science must offer "reasonable salaries and decent working conditions in well-equipped, well-organized laboratories".[450]

A research policy for a new era?

Even though the Danish authorities had held university research in an iron fist throughout the inter-war years, this had not really animated the scientists to take any offensive countermeasures. The university's board tacitly accepted these conditions, seeking instead to defensively safeguard the institution's autonomy.[451] Then came World War II, which fundamentally changed the situation. A new generation of younger scientists was on the rise, and they set high goals for their research. During and after the war this generation began a determined attack on the Danish state's short-sighted and utility-oriented education and research policy. They adamantly voiced demands for better working conditions in science and urged Denmark to follow a more proactive research policy encompassing the entire range of scientific activities, from basic science to applied research.

The decades just after World War II were the heyday of "the linear model of innovation". This model claims that new inventions arise out of technical development efforts, which in turn spring directly from basic research. While the linear model itself was not new, Denmark's pronounced focus on basic research was, as was the idea that the national authorities had a special obligation to broadly finance and support basic research. No one could say in advance which discipline would be the source of the next technological break-

Og regnen kom . . .

Paa Zoologisk Museum maa man nu træffe interimistiske beskyttelsesforanstaltninger, hver gang det bliver regnvejr.

20.4 *The scientists' persistent pleas for funding in the post-war years were often the brunt of humorous comments in the press. Here the well-known cartoonist Bo Bojesen finding inspiration in the Zoological Museum's overcrowded exhibits and dilapidated condition. As the caption explains: "And the rain came ... These days the Zoological Museum has to apply provisional protective measures every time it rains."* Politiken, *28 September 1950.*

through. Such was the lesson in research policy that the US believed it had derived from the work done to develop the atomic bomb.[452] In Denmark these new research-policy views were most forcefully promoted by a group of young scientists who called themselves the Study Group Science and Society.[453] This group, which included a number of the most prominent contemporary Danish scientists, found inspiration in works like the Marxist scientist J. D. Bernal's *The Social Function of Science.* Led by Rehberg, the study group began its activities in late 1943 and early 1944, although the initiators quickly became involved in the resistance.

When the occupation of Denmark ended in 1945 the study group swelled to 25–30 members and went public with its views, embarking upon a long, slow crusade to improve conditions for Danish science. Simultaneous efforts were made in several areas, one of which was meant to convince politicians that in the future, the national treasury would have to increase the resources made available to basic scientific research efforts. The conditions governing scientific work had to be made a matter of public concern, the study group declared, petitioning for an official commission that would be charged with assessing the situation and plotting out a course for the future. In the longer term it was also important to give the public a positive impression of science, and the vehicles most vital to that part of the crusade were popularization and information. The study group therefore sought to mobilize Danish scientists at large, asking for their help to improve science's image among politicians and the Danish populace. In the post-war years, the fingerprint of the Study Group was clearly visible on Danish newspapers, which were virtually bursting with commentaries and feature articles on science and the conditions under which it operated in Denmark.

The impetus behind such ideological thinking on science notably came from the UK and the US, where national funding for research had increased during the war. The unprecedented offensive by Danish scientists in the early post-war years must also be seen as a consequence of the severely eroded value of the Carlsberg Foundation's assets, caused by the inflation Denmark suffered as a result of the German occupation. What is more, when the war ended it was rumoured that in the coming years the Rockefeller Foundation would concentrate its donations on the European countries that had been hardest hit.[454] If such a situation arose and the Danish state did not step into the breach, Danish science would be facing a catastrophe. Several members of the Study Group – among them Rehberg and Kaj Linderstrøm-Lang (head of the Carlsberg Laboratory's chemical department) – worked in areas that relied upon the Rockefeller Foundation's financial support. The gravity of the situation was compounded by other factors as well. For one thing, the expenses associated with the ever-more sophisticated laboratory equipment continued to grow. For another, the collective working patterns associated with big science meant that even small countries like Denmark were obliged to continuously mobilize larger volumes of material and financial resources if they wished to remain anywhere near the cutting edge of international research.

The Danish State Research Foundation

In the autumn of 1945, just months after the occupation had ended, the Study Group began lobbying for the establishment of a publicly funded, state-run research council to promote basic scientific research. At the same time, a similar demand was made in the Folketing by two Danish politicians with a special interest in science: the conservative MP Flemming Hvidberg (a professor of theology, and spokesperson for the university elite) and the communist MP Mogens Fog (a professor of neurology).[455] It was therefore quite natural for the parliament's Science Commission, set up in 1946, to discuss this issue. Accordingly, in the winter of 1947, Rehberg was invited to go before the commission and present the Study Group's views on the matter. However, no further progress was made at the time, as the commission's work on this particular issue ground to a halt until 1950.

The Nobel laureate August Krogh, who deeply sympathized with the young scientists' demands for better wages and working conditions, attempted to get the wheels moving again in 1948–1949. He had already been trying to rally support for this cause in the ranks of the Royal Danish Academy of Sciences and Letters for several years. Krogh believed that the Royal Academy ought to play an active role in Danish research politics, and also that it ought to welcome more young members, thereby giving proponents of an active approach greater weight in Danish science, but his ideas were politely rejected by the academy's governing board. Finally, in January 1949, Krogh was so fed up that he renounced his membership of the Royal Academy, publicly and volubly, demanding that the academy "move out of the deep shadow of exclusivity, and publicly place its insight and professional capability at the service of society". The time had come, he proclaimed, for the Royal Academy to roll up its sleeves and make its voice heard in the debate about the position of research in Denmark; to work to promote the idea of a scientific research council and seek the position as the government's official adviser on questions concerning research. Incidentally, in the 200-plus years that the Royal Academy had existed, Krogh was the first to voluntarily resign. The academy's governing board, headed by its president, Niels Bohr, and its secretary, the mathematician Jacob Nielsen, was intractable and categorically rejected Krogh's demands.[456] Although it was undoubtedly well-intentioned and idealistically motivated, Krogh's protest hardly promoted the cause of the young scientists he had intended to help. A regrettable side-effect of the incident was that it revealed the differences that existed in Denmark's scientific community.

Finally, in the winter of 1950–1951, the Science Commission creaked into motion again, although it took something of a debacle to get the wheels rolling. It all began with a radio feature produced by Viggo Clausen under the title "Alma Mater is crying", which was broadcast on national Danish radio on 10 January 1951. Clausen followed up on Børge Michelsen's previous analysis in emphasizing the poor working conditions of students and researchers at academic institutions in Denmark. This effectively jump-started a joint initiative,

„Kortsynet Sparsommelighed vil blive kostbar“

Studenternes og Videnskabens Alvorsord til Regering og Rigsdag i Gaar

Seks til syv Tusinde deltog i Demonstrationstoget og frøs tappert

Seks til syv Tusinde Par blaafrosne Hænder klemte sammen om Stænger med brogede Plakater eller stak dybt i Frakkelommer, og lige saa mange Par kolde Fødder trampede i Gaar Eftermiddags af Sted fra Frue Plads via Raadhuspladsen til Christiansborg Slotsplads.

Demonstrationen for bedre Vilkaar for Videnskab og Studerende blev, hvad Tilslutning angaar, en stor Sukces, men hvor var det dog koldt.

Trods Næser, der efterhaanden antog alle Regnbuens Farver, holdt Indehaverne af de hvide Studenterhuer dog tappert ud i de næsten tre Timer, Demonstrationen varede. Det vilde dog ogsaa være skammeligt andet, thi forrest stred ingen ringere end Rector magnificus, Professor, *H. M. Hansen*, sig frem mod den bidende Sydost sammen med Kollegerne fra Landets øvrige højere Læreanstalter, og *Pallas Athene*, staaende strunk og stolt paa sin Vogn, ligesom hypnotiserede alle til et »Fremad, frem!« Af klimatiske Grunde havde den yndige Dame, der, som det fremgik af Plakaterne, nødigt »vilde brændes af«, dog iført sig lune Bælgvanter.

Demonstranter paa Motorcykle. Studenten i Sidevognen gaar med »sin Fars aflagte«.

Det tog Tid at faa Rækkerne formeret foran Universitetet. Mest smertefrit gik det for Kategorien: »De, som ikke fik Raad til at studere.« En stor Firkant, indhegnet af et Tov, der »skulde have været 50 Gange saa langt« og blev baaret af fire Studenter, symboliserende disse Hundreder eller Tusinder. Omsider lød Ordren til Start. Først to ridende Betjente, saa Honoratiores, og derefter de mange, mange Studerende, ordnet efter Læreanstalt og Fakultet. I Spidsen for de Veterinærstuderende en Kammerat paa en hvid Ganger og indimellem Biler med realistiske Hentydninger til Videnskabens og dens Dyrkeres trange Kaar. Studenterviddet boltrede sig. Her gik en »med sin Fars aflagte«, en Rustning fra Riddertiden, og der havde en arm Student opgiver Ævred og hængt sig til Skue for alt Folket. *Jensen-Broby* blev ikke glemt og heller ikke de lysere Sider af Tilværelsen. Mellem de store, farveprangende Plakater dukkede pludseligt en Miniature op. Stilfærdigt og beskedent krævede den »Mere Øl«.

Ikke falsk Alarm

Da Fortroppen naaede Raadhuspladsen, forlod Musikkonservatoriets Elever som de sidste Frue Plads, og da Rektorerne drejede ind paa Slotspladsen med den kilometerlange Hale af syngende Studenter efter sig, blev Sangen overdøvet af en Brandudrykninge øredøvende Sirenehylen. Udrykningen gjaldt Slottets Nabo, Finansministeriet. Det var falsk Alarm, men det er Studenterdemonstrationen ikke, forsikrede Professor *Brandt Rehberg* i Højttaleranlægget.

Paa det Tidspunkt havde Deputationen allerede længe været indendørs, hvor Tingenes Formænd, *Gustav Pedersen* og *Ingeborg Hansen*, i Spidsen for Rigsdagens Præsidium og Partiernes Formænd tog imod, og Rektor, Professor *H. M. Hansen*, tog Ordet.

Han pegede bl. a. paa, at et frodigt videnskabeligt Liv er nødvendigt for at opretholde og styrke baade den aandelige og den materielle Kultur. Den foreslaaede Optagelse af et Præmieobligationslaan paa 25 Mill. Kr. vilde være tilstrækkeligt dels til Udvidelser og Nybygninger af videnskabelige Institutioner, dels til en øjeblikkelig Hjælp til Studenterstipendierne. En saadan Hjælp er ikke alene nødvendiggjort af de seneste Aars Prisudvikling, men ogsaa af den stadige Forringelse af Legatmidlernes Betydning. Studenterne maa i Dag i Modsætning til alle andre ubemidlede Samfundsgrupper se tilbage paa Forholdene i Slutningen af det 19. Aarhundrede som en relativt gunstig Periode.

Rektor understregede endvidere Nødvendigheden af, at der tilvejebringes Forskningsfond i Lighed med det allerede bestaaende Fond for de tekniske Videnkaber. Carlsbergfondets og de andre eksisterende Fonds Midler strækker ikke længere til. En kortsynet Sparsommelighed i Dag vil blive kostbar for det danske Samfund i Fremtiden, sluttede Professor H. M. Hansen.

Paa Rigsdagens Vegne svarede Gustav Pedersen. Hans Stilling var ikke saadan, at han kunde give Løfter af nogen Art, men han kunde love, at Problemerne vilde blive behandlet. Rigsdagen glemmer ikke, hvad videnskabeligt Arbejde og Forskning betyder for det danske Samfund, og det maa være en Pligt at huske, at Resultaterne af videnskabeligt Arbejde ikke altid kan gøres op i Kroner og Ører med Krav om Balance. Rigsdagen vil forstaa, at Ungdommen maa have Kaar, der gør det muligt at yde den nødvendige Hjerneindsats. At man ikke kunde give konkrete Løfter, betød ikke, at man ikke havde Vilje til at behandle Problemerne med den dybe Alvor, de kræver, sluttede Formanden.

Dermed var Modtagelsen til Ende, og Kommunisten *Inger Merete Nordentoft* »brændte« tilsyneladende »inde« med en til Lejligheden forberedt Tale.

Professor Bohrs Forslag

Den næste, man søgte Foretræde hos, var Undervisningsminister *Flemming Hvidberg*, som for en Nemheds Skyld indfandt sig i Samtaleværelset, hvor Mødet med Rigsdagen havde fundet Sted. Ministeren pegede paa den fortvivlede økonomiske Situation og fremhævede, at han ikke blot mente at have undgaaet deraf følgende Nedskæringer, men oven i Købet havde Haab om nogle ekstra Bevillinger til videnskabelige Formaal. Fra Professor *Niels Bohr*, som ikke deltog i Demonstrationen, havde han modtaget et Forslag om et Statens Forskningsfond, som skulde skaffe Midlerne til det videnskabelige Arbejde i de Bygninger, som det foreslaaede Præmieobligationslotteri skal tilvejebringe Pengene til.

Ogsaa Statsministeren og Finansministeren tog venligt mod Deputationen, og da Klokken nærmede sig fem, kunde Rektor begive sig tilbage til Slotspladsen, hvor Temperaturen var lav, men Humøret højt. Samtlige Studenter havde hoppet i Takt til Musikken for at faa lidt Varme i Kroppen, og et enkelt Sted gik Plakaterne op i Luer, medens de omkringstaaende nød det varmende Baal.

Rektor meddeler Demonstranterne Resultatet af Besøget paa Christiansborg.

Hvad kunde Rektor saa sige, da Klapsalven, han blev hilst med, døde hen? At man var blevet venligt modtaget, samt at Statsministeren havde erklæret, at han vilde sætte Pris paa, at Videnskabskommissionen snart maatte afslutte sit Arbejde og udsende en Betænkning.

Bifaldet var behersket. Politiet kunde i Aftes meddele, at alt var forløbet stille og fredeligt.

Edm.

Fra Modtagelsen i Rigsdagen. Rektor, Professor H. M. Hansen, taler ved Mikrofonen. Yderst til venstre Folketingets Formand, Gustav Pedersen, fhv. Statsminister Hans Hedtoft og Rektor, Professor Anker Engelund. Bagved skimtes fhv. Undervisningsminister Jørgen Jørgensen og Direktør, Professor Thorkil-Jensen, Landbohøjskolen.

20.5 *Coverage of the demonstration on 2 February 1951, under the heading "Short-sighted Thriftiness Will Be Costly". Students and science professionals marched side by side to protest what both groups saw as the Danish government's ongoing policy of starving the universities and their sister institutions. The professors called for larger research grants, while the students asked for more state scholarships.* Politiken, *3 February 1951.*

coordinated by Rehberg, that featured lectures on the radio and barrages of letters to the editors of leading newspapers, drawing attention to the plight of Danish science.[457] By this time the Royal Academy had also changed its position and had begun to actively exert pressure to resolve the predicament. In a radio speech broadcast eight days after Viggo Clausen's feature, Niels Bohr implored Denmark's politicians to increase research grants if they wished their country to avoid being "quickly left behind".[458] Things came to a head on 2 February 1951 when a large demonstration of students and scientists – six or seven thousand strong – marched from the university buildings near the cathedral square at Frue Plads to the seat of parliament at Christiansborg Castle. There the government received a delegation that demanded more money for research and scholarships.[459]

No promises were made that day, but in the long run the government could not ignore such a massive show of opinion. Flemming Hvidberg, who at this point had become the new conservative minister of education, quickly seized the opportunity to propose setting up a public science foundation. The following year the Folketing passed legislation that made the Danish State Research Foundation a reality. This body, occasionally referred to as "the state's Carlsberg Foundation", was roughly the same size as its esteemed private counterpart, with an annual total of DKK 2 million at its disposal. The funds were allocated to give the natural sciences 42%, the humanities 21%, medical science 15%, agricultural science 15%, and the social sciences 7%. The distribution of grants in each of these five areas was handled by a specialist committee of five or six scientists appointed by the various university-level institutions. Allocations were made exclusively on the basis of applications, meaning that the Danish State Research Foundation had no authority to independently initiate research, or to put together large, interconnected research programmes, or to otherwise actively manage or control research activities.

Presenting the Danish State Research Foundation bill to the Folketing as minister of education, Hvidberg cited the motivating circumstance that a variety of factors, including price developments and "the ever-growing need in scientific research for costly equipment", meant that the Carlsberg Foundation and other sources of private patronage were no longer sufficient. The situation now called for a "forceful and sustained effort on the part of the state". This was absolutely necessary, as the official phrasing ran, "first and foremost ... to promote the Danish intellectual community's interests", and it further confirmed that "Arranging the external conditions that will enable Danish science to inwardly and outwardly claim a position befitting an old cultural nation must, broadly speaking, be the concern of the state."[460] This was effectively a recognition of the Danish state's obligation to support scientific research, even though the general view of science was still mainly coloured by cultural concerns, as is evident from the concluding remarks in the minister's motivation. Certainly, he essentially stated that the Danish State Research Foundation's grants would only target basic research ("pure science"), but he added that basic research also had other advantages that were less obvious:

> Also, public enlightenment activities, which in our day quite rightly are considered very important, will come to benefit from such an improvement of the conditions of science, since a precondition for such enlightenment, and for the activities of our schools, is a scientific research community that is alive. Concerning the basic sciences, it will often not be possible to demonstrate any material useful effect of such research in the shape of a direct, tangible result. And yet it is beyond all doubt that in the long run a flourishing of these sciences, as a consequence of their influence on applied science, will promote societal interests of an economic and social nature.[461]

This brief passage actually expresses two competing and very different views of science. On the one hand, there is the view of educational democracy (contained in the reference to "public enlightenment activities" and "the activities of our schools"). This is opposed, on the other hand, by the linear model's firmer, more utility-oriented view – even though the linear argument was presented with certain degree of reservation. In summing up, one can say that the Danish State Research Foundation was established at a time when the linear model was still waiting in the wings, unable as yet to outshine the traditional cultural, national and educationally democratic ideas used to justify the research being done at university level.

The great expansion

Although the funds made available through the new Danish State Research Foundation hardly lived up to the aspirations of the most ambitious scientists, its very creation marked the beginning of a time when the demands of science were more readily heard. Grants continued to increase. After a brief period of liberal–conservative government, the Social Democrats once again came to power in 1953 and subsequently ruled the country for a decade and half. Denmark's economy was looking up, and for the first time in many years the politicians had the economic leeway they needed to improve the conditions of science. As early as 1954, and despite considerable opposition from the University of Copenhagen, the new minister of education, Julius Bomholt, pushed through the establishment of a faculty of science at the University of Aarhus, founded in Denmark's second-largest city in 1928.[462] In Copenhagen, the biologists received a grant to build a large, comprehensive Central Institute of Biology.[463] But it was only after the mid-1950s that the funds really started flowing.

For decades the thriftiness of the Danish finance ministry and the Folketing's finance committee had caused a culture of cautious prudence to trickle down through the chain of allocation that stretched from central administration to institute or department. For fear of receiving nothing at all if their requests were seen as exorbitant, an entire generation of department heads had become used to shaving their applications down to an absolute minimum. It is against this background that we must consider a peculiar message issued by the finance minister Viggo Kampmann – the new Social Democratic government's strong man. Kampmann was the first to signal that a new day was dawning. He did so in a feature article laid out as a sympathetic dialogue with Rehberg and Linderstrøm-Lang.[464] According to Kampmann, the time had come to "catch up with the omissions that the occupation and post-war years have brought with them regarding the expansion of our scientific apparatus." Future investments in research and education would rise to a completely new level. Kampmann, who would also serve as Denmark's prime minister from 1959 to 1962, had a surprisingly solid grasp on matters of science and research. In fact, of Denmark's many prominent politicians, he probably exhibited the greatest benevolence towards science, and the most genuine understanding of the area's special needs. Besides writing insightful articles on the nature of research, Kampmann possessed a singular will to address its problems, often leading to solutions that were quick, effective and unorthodox.[465] All manner of activities relating to culture and science received generous support.

Actually the wave of change that Kampmann proclaimed in 1957 had already begun in 1955, when Denmark created its Atomic Energy Commission, whose mission was to design and begin to assemble a research organization that could smooth Denmark's transition into the atomic age. Since chapter 23 has more on the entire process, it will suffice here to say that by the end of 1955 the Atomic Energy Commission had already presented and gained acceptance for its plan and schedule to construct a large nuclear research facility on the secluded peninsula of Risø in Roskilde Fjord. The magnitude of the projected construction and operating costs reflected how important a united Folketing found nuclear research at the time. However, it was not merely an interest in promoting atomic energy that drove the government's decision to move ahead with the Risø project. Over and above this, the government – and Kampmann in particular – intended that this decision of principle was to break down the political barriers that remained, enabling Denmark to realize more and larger future investments in research and education.[466]

By 1955, Denmark was on the brink of being seriously outpaced in many fields of research. That is why the government followed up on the Risø decision in the summer of 1956 by setting up a body known as the Expert Commission; a body that was instructed to look into Denmark's future needs for technical and scientific labour and submit proposals for modernizing the entire range of technical educations, covering everything from laboratory technicians to civil engineers. Even before a new, comprehensive recommendation appeared in 1959, several of the Expert Commission's proposals had been fully implemented, including one for a new bachelor-level engineering degree whose graduates were known as "academy engineers". Based on the commission's work, a variety of outdated programmes were modernized and the total capacity for technical and engineering education substantially increased. At the same time a decision was made to move the Technical College to a new and more spacious location at Lundtoftesletten north of Copenhagen. The expansion of the college took place from 1960 to 1976, the campus ultimately covering about 300,000 square metres at a total cost of DKK 780 million (in 1976 prices) – and the project reaching a scale that still remains unsurpassed in the history of Danish campus construction.[467]

Denmark's two academic universities met with an equally consistent approach in line with the newly proclaimed policies. Taking their cue from the technical subjects, they built their requirements on the significance of the technical education programmes, which inversely relegated the humanities and the social sciences to second rank.[468] A committee was formed to review the universities' wish list of new buildings for the decade 1958 to 1968, the forward-looking Kampmann already having briefed himself on these wishes in order to expedite the process. This quickly paved the way for construction of the Hans Christian Ørsted Institute (cost DKK 33 m), in which most of the University of Copenhagen's exact sciences were brought together under one roof, and the new Zoological Museum (cost DKK 23 m), as well as major enlargements to the faculty of science at the University of Aarhus. The new attitude to science did not go unnoticed by the press. In the autumn of 1959, for instance, the newspaper *B.T.* remarked on how "Science at last finds itself with the wind at its back. This is most clearly seen on Nørre Fælled in Copenhagen", where a whole new neighbourhood of scientific buildings was taking shape.[469] Just a few years earlier, the rapid expansion taking place after

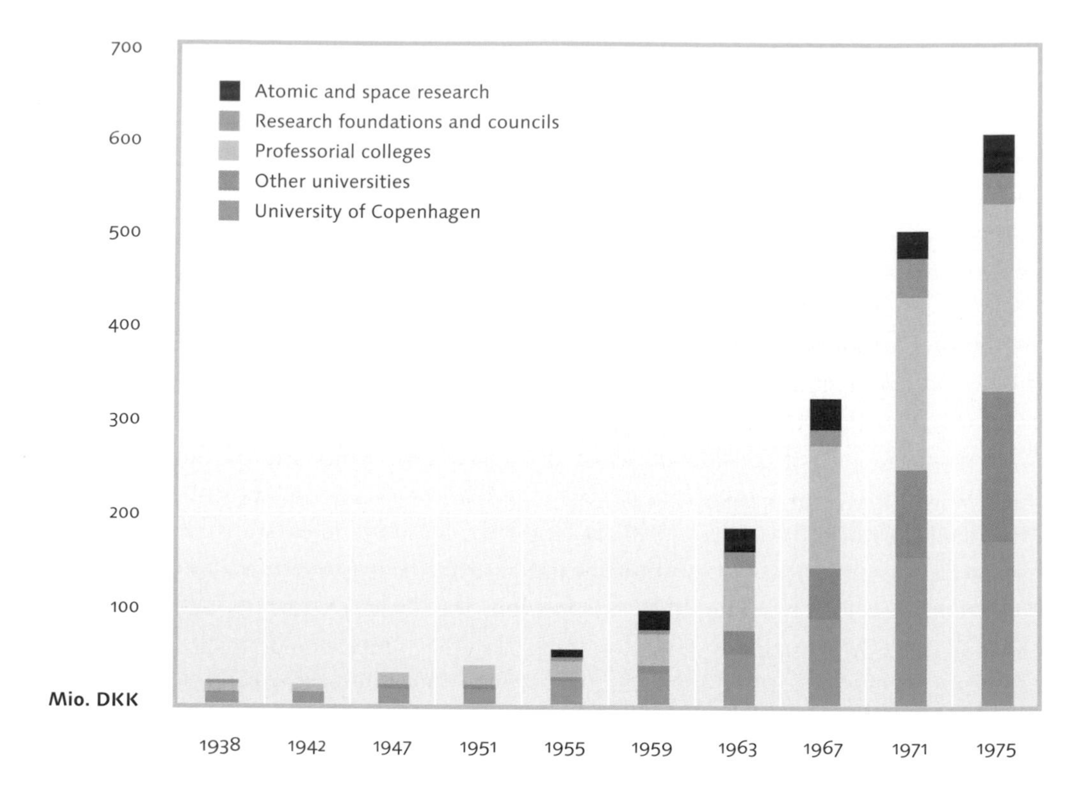

20.6 *Grants allocated under the Danish appropriations act to the various areas of research and education. Amounts in constant Danish kroner (DKK), reference year 1955. Even in constant prices the expansion from the mid-1950s onwards was impressive, with grant amounts doubling every five years between 1955 and 1971. Notice the large share of total public science grants going to nuclear and space research.*[470]

1957 had been absolutely inconceivable, which is why 1956–1957 was truly a watershed as far as Danish education and research policy is concerned. Figure 20.6 gives a good picture of the impressive growth.

It is not easy to define or explain all the different factors that combined to produce these changes. The Danish economy was booming, to be sure, and national projections on the lack of technicians and scientists were an important element.[471] In a global perspective the Cold War also played a significant role. Beginning in the late 1940s, through bodies like NATO and the European Recovery Program, the US was firmly persuading its allies in Western Europe to spend more money on science and technology.[472] Another factor that cannot be disregarded is the pervasive atomic euphoria that followed the

Geneva conference of 1955.[473] These factors were already exerting an influence well before the USSR claimed its first victory in the space race after successfully sending its famous Sputnik satellite into orbit around the Earth in October 1957. In the light of Russia's achievement, the Western world's technological edge seemed to be gravely jeopardized, obliging the West to further accelerate its own technical development efforts. In the years that followed, the OECD strongly advised the international community to regard research and education as productive investments.[474] Also around this time, trade and industry in Denmark overtook agriculture as the country's largest export sector. Industrial policy rose to the top of the political agenda, as indicated by measures like the creation of the Expert Commission.

Finally, the tactical considerations of the Social Democrats also had an impact. During the post-war years the party had at last abandoned its traditional views on distribution policy. According to its new welfare policy, social improvement would primarily be achieved by expanding production and growth. Rather than pursuing ideal distribution, the Social Democrats now considered it more important to generate a bigger pie so there would be more to go around. This in turn brought rationalization and technological progress sharply into focus among the majority of Denmark's politicians. In addition to being the party of the working class, the Social Democrats were increasingly becoming the political choice for white-collar workers and civil servants – groups for whom a good education was the key to success. The balance of power between city and countryside had shifted. The industrial society had become a political reality, and also a reality in terms of research policy, in that it unleashed explosive growth in the spheres of research and education. In fact, this was precisely the point that Kampmann emphasized on several occasions in 1957.[475]

Reinforcing the research-council system

In Denmark, as in many other countries, the 1960s were a decade of prolific growth. The explosive expansion in research and education intensified the political and administrative need for high-level advisory services and a coordinated research policy.[476] The 1960s brought a number of changes to the Danish research system, the most important of which deserve to be briefly outlined here.

The Technical Research Council was reformed in 1960, and the name prefixed by the word "Danish". The council's future role would combine political

Year	Research foundations: total budget in DKK m	Universities and similar institutions: total budget in DKK m	Share of funding (%)
1938	0.25	7.8	3.1
1942	0.25	11.9	2.1
1947	0.60	20	2.9
1951	0.75	31.8	2.3
1955	2.9	45.1	6.0
1959	6.0	82	6.8
1963	16	197	7.5
1967	28	476	5.6
1971	66	973	6.4
1975	94	1,788	5.0

Table 20.1 *Public grants to national research foundations, and to universities and other institutions of higher education (DKK current amounts).*

advisory services with overall responsibility for coordinating allocations to the entire area of technical research. New faces and new initiatives were introduced in a constellation that gave more power to the civil servants and to Danish trade and industry. The rejuvenated research council did not confine itself to supporting targeted technical research, but also worked to promote major research programs that cut across the traditional borders dividing basic and applied research. One noteworthy initiative from the Danish Technical Research Council helped to launch Danish space research in the 1960s.[477]

The ministry of education wished to see similar advisory bodies operating in other research areas. This led to the formation in 1965 of the Joint Research Committee, formed to advise the ministry on research issues of a general and cross-disciplinary nature. For the first time, here was a body composed of civil servants as well as privately and publicly employed scientists from all major research fields. In at least one respect the Joint Research Committee was extremely influential: In 1968 it succeeded in transforming the five science commissions of the Danish State Research Foundation into five actual research councils, which would be able to take independent initiatives on a par with the Danish Technical Research Council. The crucial thing was that each research council became responsible for its own scientific field and would serve as an advisory body to the relevant government ministries.[478]

Previously, the initiatives had issued from the scientists themselves, but now more than ever before, politicians and civil servants were beginning to set the agenda. No one had, as yet, begun to speak of the Danish state actually steering research in a certain direction – though this discussion did arise around 1970 – but it was clear that politicians and civil servants increasingly wished to justify and rationally prioritize among the various fields. This in turn meant that from the mid-1960s onwards, the relationship between science and politics in Denmark was rapidly metamorphosing into the more tightly knit research system that finally emerged in the 1980s.

As table 20.1 shows, Denmark's state-run system of research councils and foundations did not have an overwhelming impact on the period under review here – 1920 to 1970 – given that the funds allocated through the national system at no point exceeded 7–8% of total public expenses to the country's university-level institutions.[479]

That is not to say, however, that the actual influence of the research council and foundations was accordingly limited, since only about half of the institutions' funds were used for research. More importantly, the money from the research councils and foundations were dynamic. They gave the research system a certain degree of flexibility, which despite its limitations made it possible for new disciplines and research areas to gain a toehold before they were institutionalized at the colleges and universities – a process that is highly complicated and often takes a very long time, as new fields often lack outspoken advocates in the decision-making bodies of the educational institutions. When it comes to research, Denmark's universities were (and still are) conservative, whereas the research councils were intended to serve as dynamic players in the scientific arena.

Peaks in Danish university research

21

The International Education Board (IEB) was an essential part of the American philanthropy that reached out to Europe in the 1920s. Around 1926–1927 the IEB, a body within Rockefeller philanthropy, conducted an extensive study to assess the state of science in Europe. The object was to single out the best and most fruitful European scientific environments and offer them financial support, "to make the peaks higher".[480] The IEB believed that if selectively given a helping hand, such elite institutions would be able to attract and host promising young American scientists, who would almost inevitably wish to study in Europe at some point. The IEB also believed that such institutions could serve as a model for American science, which was growing at a dizzying pace. The authors of the IEB's study concluded that Copenhagen was a world-class player in the fields of theoretical physics (Niels Bohr), physical chemistry (Niels Bjerrum, Johannes Brønsted and Søren Peter Lauritz Sørensen) and experimental physiology (August Krogh), and that a number of other Danish scientists also stood out in the global scientific landscape.[481] Between 1920 and 1950, four Danes received Nobel Prizes in physics or in physiology or medicine. Seen in an international context this is quite a high ratio,[482] and it supports the assessments of the IEB and others who ranked Copenhagen among Europe's most important scientific centres.

Naturally, the pre-eminence of Denmark's great scientific minds should not be allowed to overshadow the fact that throughout the period under review here, most of Danish science was much less prominent on the global topographical map of science. While a fully nuanced picture of Danish science from 1920 to 1970 would need to include a description of the many humbler disciplines and their practitioners and achievements, this work entertains no such ambition.

What it does seek to do, however, is to answer questions such as these: What went on at the institutes specifically constructed to cultivate theoretical physics, physical chemistry and experimental physiology in the inter-war years – a period later referred to as "the golden age of Danish science"? Why did these three areas become scientific strongholds between the two world wars?

What happened after their "golden age"? Did these particular fields retain their high standing, or did different fields rise to prominence in the post-war years?

Niels Bohr and the spirit of Copenhagen*

Internationally speaking, throughout the 1800s "physics" had been almost synonymous with "experimental physics". Indeed, the new, impressive institutes of physics constructed in Germany and elsewhere during the latter half of the nineteenth century were largely reserved for experimental physics, with the study of theoretical physics regarded as secondary by comparison. Theoretical work that was not carried out with the object of "explaining" well-known experimental data, or that led to predictions of experimental outcomes that could not be readily verified in the laboratory, were regarded as speculative and largely inconsequential. There was a definite tendency to equate physics with precise measurements.[483]

The second half of the 1800s saw a nascent reaction against this approach, and a few universities in Germany and elsewhere did, in fact, create chairs in theoretical physics. A number of fortuitous professorial appointments – placing exceptionally gifted physicists like Max Planck in Berlin, Arnold Sommerfeld in Munich and Albert Einstein in Zurich (and later Prague and Berlin) – enabled theoretical physics to gain professional recognition before World War I, and allowed several highly esteemed institutes of theoretical physics to be established. These institutes gave top priority to developing theories of physics, although often the same institutes constructed laboratories and employed experimental physicists who would work closely with the theoreticians.[484]

Denmark followed this new physics trend in 1916, when an extraordinary chair in theoretical physics was created for Niels Bohr, then 30 years old, who had gained international renown with his treatises from 1913 on the nuclear atom, recently proposed on experimental grounds by Ernest Rutherford and his collaborators in Manchester. Bohr, backed by the University of Copenhagen and the Danish state, was able to realize his wish to found an institute of theoretical physics. The intention was not only to promote international cooperation, thereby reinforcing Denmark's role as a pioneer of international reconciliation after the war, but also to bring gifted theoretical and experimen-

---

* This section was written in collaboration with Finn Aaserud. For more details, see Aaserud 1991 and Aaserud and Nielsen 2006.

21.1 *Niels Bohr's lecture on 3 March 1921 at the inauguration of the new Institute for Theoretical Physics, located on Blegdamsvej in Copenhagen. The audience included university professors and politicians, most prominently the Danish minister of education Jacob Appel.*

tal physicists together.[485] Like Planck, Sommerfeld and Einstein, Bohr regarded theoretical physics as a branch of physics in which the questions that theoreticians posed to nature ought to be immediately subjected to experimental verification. He therefore insisted that at any given time his institute must also have experimental physicists and equipment that was directly linked to the theories being explored. In other words, from the outset Bohr's institute, inaugurated in 1921, was comparable to a number of other theoretical physics institutes set up in pre-war Germany.

Bohr had already earned a reputation as an excellent physicist by the early 1920s, and by the time he received the Nobel Prize in physics in the autumn of 1922 for his atomic theory, he had already become one of the international heavyweights in theoretical physics. Scientists from around the globe flocked to his institute in Copenhagen, which was already bursting at the seams in 1923 – and Bohr was able to expand the following year thanks to a generous grant from the IEB. Within a few short years his Institute for Theoretical Physics became a Mecca for physicists, comparable in stature with the established theoretical institutes led by Arnold Sommerfeld in Munich and Max Born in Göttingen.

21.2 *The original staff at Bohr's institute. Standing, from left to right: Jacob C. Jacobsen (Denmark), Svein Rosseland (Norway), Georg von Hevesy (Hungary), Hans Marius Hansen (Denmark) and Niels Bohr (Denmark). Seated, from left to right: James Franck (Germany), Hendrik Anton Kramers (the Netherlands) and Betty Schultz (Denmark, secretary).*

Incorporating and interpreting quantum mechanics. Bohr may have been awarded the Nobel Prize for his theory of the atom, but physicists still faced the task of resolving the inconsistencies that the quantum concept posed for classical physics. In the early 1920s Bohr's theory was still a lumpy amalgamation of old and new, creating the need for a completely new quantum physics that could totally replace the semi-classical theories. In 1924, the Institute for Theoretical Physics produced an extreme outcome in its quest to understand the old concepts in light of the new physics. The young American physicist John Slater, who was visiting the institute on an IEB scholarship, suggested an idea that caught the attention of Bohr and his Dutch assistant Hendrik Kramers. In the resulting publication, known as the "BKS paper", Bohr, Kramers and Slater proposed that the conservation of energy and momentum did not apply absolutely, but only statistically. This break with classical physics was so radical

21.3 *The German physicist Werner Heisenberg, mainly known for his key role in developing quantum mechanics, worked at Niels Bohr's institute from September 1924 to April 1925, then later as an associate professor at the University of Copenhagen from May 1926 to June 1927 before accepting a professorship in Leipzig. He returned to the institute in Copenhagen several times on visits of varying length. Here, Heisenberg and Bohr engaged in animated conversation in the institute's lunchroom in the mid-1930s.*

that even Bohr's closest colleagues found it unacceptable. The ill-fated BKS paper was soon eclipsed by further developments, however; developments in which Bohr and his institute were also heavily involved, and in which the German physicist Werner Heisenberg would be a pivotal figure.

Bohr and Heisenberg first met in Göttingen in the summer of 1922 at a lecture series by Bohr on his physics that later became known as the *Bohr Festspiele.* Heisenberg made such a formidable impression on Bohr that the young German physicist was immediately invited to visit the institute in Copenhagen, which he did in the spring of 1924. Heisenberg followed up his initial stay with several more, and he later served as Bohr's assistant. During and immediately after his first visit he developed the so-called quantum mechanics on an abstract mathematical basis, elaborating a theoretical foundation that made quantum physics a complete theory, independent of classical physics.

21.4 *Niels Bohr and Albert Einstein attending the Solvay Conference in Brussels in October 1930, where they discussed the interpretation of quantum mechanics.*

With quantum mechanics providing a secure platform on which to base quantum-physical thinking, speculative contributions like the BKS paper had become irrelevant.

The formal foundation was now in place, and yet interpreting the theory in terms of physics was problematic. Shortly after the publication of Heisenberg's groundbreaking paper on quantum mechanics from 1925, the Austrian physicist Erwin Schrödinger formulated an alternative theory based on a more traditional mathematical foundation that was easier for most physicists to understand. Schrödinger's alternative regarded continuous waves as fundamental to physics, whereas Heisenberg's work was based on particles and discontinuous processes. Nevertheless, Schrödinger and others quickly demonstrated that the two theories were mathematically equivalent – which actually exacerbated the problem of interpreting the new physics.

While developing his quantum mechanics Heisenberg had relatively little contact with Niels Bohr, but in his subsequent efforts to clarify the interpretation of his theory he and Bohr worked closely together. Based on the fundamental uncertainty relations between momentum and position, which Heisenberg deduced in Copenhagen, the two physicists worked to resolve significant discrepancies. They arrived at what was later called the "Copenhagen interpretation", which has since dominated the world view in much of the physics community.

The Copenhagen interpretation can be seen as a compromise between Schrödinger's wave theory and Heisenberg's particle theory, based on Bohr's concept of complementarity, which implies that a physical phenomenon can appear in two mutually exclusive forms depending on how one chooses to set up the experiment. Light, for instance, can appear as waves in one experimental set-up and as particles in another, and both forms are necessary to achieve a complete description of the phenomenon of "light".

The new developments not only led to an intense debate about the fundamental problems of physics. They also had an impact on practical physics research. Now that a feasible alternative to classical physics had been found, the work of the theoretical physicists mainly consisted in applying the new quantum physics to an impressive number of areas, such as calculating transition probabilities in sodium and solving the long-standing problem of the specific heat of the hydrogen molecule. This stage in the development of quantum mechanics led to hectic activity at the physics institutions it affected. Bohr's institute, for one, reached an all-time high of visiting scientists (24) and publications (47) in 1927, after which both indicators plummeted in 1928 (to 10 and 21, respectively). This lower level persisted into the 1930s, which may suggest that the institute ought to have sought new research directions.[486] Heisenberg's move to a permanent position in Leipzig in 1927 and the arrival of the Swedish physicist Oskar Klein as Bohr's new assistant also signalled the beginning of a new era at the Institute for Theoretical Physics. At the time there were no clear pointers showing which direction the institute ought to take, but momentous events in world politics would have a decisive influence on the choices Bohr made in the mid-1930s.

Changing course towards nuclear physics and experimental biology. When Adolf Hitler seized power in Germany in 1933, the consequences for Bohr's institute were rapid and concrete, as many of the German visitors were young Jewish physicists who could not find employment under Hitler's new regime. Bohr helped Jews who were fleeing Nazi Germany, using money he obtained through existing foundations and through organizations created specifically for that purpose. Thus, Bohr was a member of the Danish Committee for the

21.5 *The second in a long succession of informal physics conferences held at Bohr's institute took place in 1930. The influential gathering included some of the most brilliant minds in the world of physics. Front row, from left to right: Oskar Klein, Niels Bohr, Werner Heisenberg, Wolfgang Pauli, George Gamow, Lev Landau and Hendrik Kramers.*

Support of Refugee Intellectual Workers, which enabled several young physicists to stay temporarily at the institute in Copenhagen. Bohr's excellent contact with Rockefeller philanthropy in the US also made it natural for him to apply to its Special Research Aid Fund for grants to enable two scientists of his own generation to stay at the institute. Both of these experimental physicists, James Franck from Göttingen and Georg von Hevesy from Freiburg, had previously stayed at the institute after its establishment more than a decade earlier.

When Bohr applied for funding to assist Franck and Hevesy, he had not yet decided to change the course of the institute's theoretical and experimental activities to any great extent. During a visit to the US earlier in 1933, Bohr had resumed contact with the people behind the Rockefeller Foundation's general support program for science – which must not be confused with its temporary and much smaller Special Research Aid Fund. The IEB, which had so far been responsible for Rockefeller philanthropy's support of international science, was in the process of being dismantled, with the Rockefeller Foundation

21.6 *Niels Bohr and the experimental physicist Jacob C. Jacobsen (no relation to the famous brewer and philanthropist) captured in 1938 in front of the newly installed cyclotron in the basement of Bohr's institute.*

assuming responsibility for supporting science both inside and outside the United States. Under the leadership of Warren Weaver, the Rockefeller Foundation's donation principles differed greatly from the IEB's: While the IEB had based its support of basic science in Europe strictly on quality considerations, Weaver wanted to build up biology from his point of view that it was lagging far behind physics in its development. He therefore put together a comprehensive "experimental biology" programme, which would subsequently prove to facilitate the rapid evolution of molecular biology that took place during the 1950s.

Like the IEB, the restructured Rockefeller Foundation was always pleased to hear from leading scientists, and Niels Bohr was no exception. Even so, it was clear that any support given to Bohr and his institute would have to comply with the foundation's basic progamme. Following a series of meetings between Bohr, Hevesy and the physiologist August Krogh, the three prepared a joint application in the spring of 1934, proposing a precisely defined experimental project that was fully compatible with the Rockefeller Foundation's intentions, in that it would investigate the potential use of Hevesy's radioactive indicator method for medical and biological purposes. Their efforts were rewarded with a grant of just under a quarter of a million Danish kroner (DKK), to be spent over a period of five years. During that time Hevesy, with Bohr's and Krogh's help, was able to interest a number of medical and physiological institutions in Copenhagen in his method and establish a successful research programme.

Although at times Bohr gave the Rockefeller Foundation the impression that the new project would involve a complete transition of the institute's research priorities to experimental biology, and although he undoubtedly took the potential biological applications of Hevesy's method very seriously, his main motivation for the new project was no doubt to initiate the transition of his institute to nuclear physics. Firstly, the Rockefeller-funded project rested on the latest developments in nuclear physics, originating in the spring of 1934, when Enrico Fermi and his colleagues in Rome had mastered the art of producing radioactive isotopes of virtually any element by bombarding it with neutrons. It was this achievement that made it possible to produce radioactive isotopes of biologically relevant elements – a capability that Hevesy had sought for many years. Secondly, the equipment that the trio in Copenhagen had specified in the application for their experimental biology project happened to be identical to the instruments needed to study the atomic nucleus. Almost DKK 70,000 of the Rockefeller money was allocated to the purchase of a cyclotron, the sort of particle accelerator developed by the experimental physicist Ernest Lawrence just a few years earlier in California. In fact, Bohr was in close contact with Lawrence, who sent one of his junior physicists to Copenhagen to help install the cyclotron there. Even though the application stated that the Copenhagen cyclotron was to be used for producing biologically relevant radioactive isotopes, it was primarily an instrument designed for nuclear physics, and after it was put to work in 1938 it was alternately used for investigations into biology and nuclear physics.

So it was that in the mid-1930s, Bohr was able to bring to fruition a new, intimate relationship between theory and experiment on the basis of nuclear physics, which at this point was coming to dominate modern physics in general. He also appealed to other foundations to further this goal. One source of support was the Thomas B. Thrige Foundation, which donated the large magnet for the cyclotron and also contributed specialists to help install it. As the name indicates, this foundation was intimately linked with the Thomas B. Thrige electrotechnical company.[487] Prior to the outbreak of World War II,

Bohr also succeeded in obtaining financial support for other major installations – most notably a high-voltage laboratory and a van de Graaff accelerator – confirming the institute's status as one of the leading European centres in the field of nuclear physics.

The golden age of chemistry: Brønsted, Bjerrum and Sørensen*

Despite the very limited number of chemists whose contributions could reasonably be expected to promote chemistry as a science, chemists were a highly visible group within the Danish scientific community, and internationally active as well. Their distinctive profile was especially attributable to the quality of the research done by Johannes Nicolaus Brønsted, Niels Bjerrum and Søren Peter Lauritz Sørensen. Each of these three chemists was a professor at his own scientific institution, and each had earned an outstanding scientific reputation among the world's best and brightest. Together, they formed the solid core of Danish chemistry's "golden age", as the period around 1920–1940 is often, and aptly, called. This chapter concentrates on Brønsted and Bjerrum, while Sørensen, who was privately employed at the Carlsberg Laboratory and mainly worked in the field of biochemistry, is treated in chapter 22.

Johannes N. Brønsted: A new acid–base concept. In the 1920s, Brønsted focused on investigating the progression of chemical reactions, which allowed his experimental brilliance to truly shine. One example was his partial separation of a non-radioactive element (mercury) into its isotopes, an experiment he did in collaboration with the Hungarian chemist Georg von Hevesy, who worked at Niels Bohr's institute from 1921 to 1925.[488] Brønsted is doubtless best remembered today for introducing a new concept of acid and base in 1923.[489] The publication of his theory coincided more or less with that of the British chemist Thomas Martin Lowry – hence the designation Brønsted–Lowry acids and bases. Brønsted was nominated twice for the Nobel Prize in chemistry, in 1929 and 1933. On the second occasion he was nominated for his works concerning acid–base catalysis and made it all the way to the Nobel committee's final selection round, though not across the finish line.[490]

---

* This section was written in collaboration with Anita Kildebæk Nielsen and Torkild Andersen. For more detailed accounts (in Danish), see Kildebæk Nielsen 2000 and Kildebæk Nielsen and Andersen 2006.

As Brønsted's reputation was growing after World War I, an increasing number of foreign chemists made requests to visit and study at his laboratory, which was really quite small and ill-suited for that purpose. He had already repeatedly tried to persuade the University of Copenhagen to build a new institute for physical chemistry, which would provide an appropriate physical framework for his research and allow him to receive more guests than his old laboratory could accommodate. However, the university did not give his applications sufficient priority for him to receive a share of the modest amounts that the Danish state had set aside for new construction projects. That is why, in 1925, Brønsted decided to follow Bohr's example from 1923 and seek support for such a project from the IEB.

The IEB reacted quickly, sending Augustus Trowbridge, its European science representative, to visit Brønsted in September 1925. This visit was followed by several consultations, and by correspondence between Trowbridge and Wickliffe Rose, then president of the IEB. It is evident from their letters that they agreed that "Copenhagen is very outstanding even internationally, because of the presence of Niels Bohr" – but equally evident that they did not see eye to eye on their assessment of Brønsted. As Rose wrote, "I have made considerable inquiry here [in the United States] concerning Brønsted and his proposal. All the scientists conferred with concur in the judgement that Brønsted is one of the foremost men in physical chemistry today." Trowbridge, on the other hand, found that Brønsted "is not a great personality and not one who will arouse enthusiasm in his field, as Niels Bohr has done so notably in his own, but he certainly should be counted among the first half dozen men in his line in the world" Nevertheless, in his overall assessment Trowbridge did state that "there is no doubt that among the physical chemists Bronsted is rated very high. Chemistry generally is relatively strong in Copenhagen with BJERRUM, SORENSEN and BIILMANN, all of whom have more than local reputation."[491]

Despite certain reservations, in November 1927 the IEB decided to support the construction of a new Institute for Physical Chemistry, which Brønsted would run. However, letters kept in the IEB archives leave no doubt that one important reason why its board ultimately approved Brønsted's application was that the new institute would be located next door to Niels Bohr's internationally renowned Institute for Theoretical Physics, and so the IEB anticipated added benefits from synergies between the two.

Niels Bjerrum: The theory of strong electrolytes. Niels Bjerrum became a professor of chemistry at the Royal Veterinary and Agricultural College in 1912. His

21.7 *Professor Niels Bjerrum was once described by his colleague Kaj Ulrik Linderstøm-Lang as an "inspired and inspiring teacher, not only for the younger generation, but also for his contemporaries and for his older colleagues". Here, Bjerrum lecturing to "the younger generation" of organic chemists in the 1930s.*

appointment enhanced the status of chemistry at the Agricultural College, and also raised the college's standing in the Danish chemistry community. Drawing on his experience from the university, Bjerrum reformed the way chemistry was taught, and also wrote a textbook on inorganic chemistry that appeared in multiple impressions and was translated into a number of languages.[492] He organized study groups for his most eager students and weekly seminars where students and assistants could go over the newest scientific findings in chemistry and quantum theory. In short, Bjerrum successfully established an active research environment for chemistry at the Agricultural College, attracting international interest as well as foreign visitors, mainly from the Nordic region and Central Europe.

This attraction was intimately linked to Bjerrum's own scientific renown. Like Brønsted and Sørensen, Bjerrum made a name for himself doing research in the branch of chemistry that investigates electrolytes (meaning acids, bases and salts). That is what makes it relevant to talk about an actual Copenhagen research programme in this field during the 1920s; a programme that set the stage for Denmark's golden age of chemistry. One of Bjerrum's lasting achievements was his discovery that so-called strong electrolytes are completely dissociated into electrically charged particles (that is, split into ions) in aqueous solutions. When he first proposed this around 1910, it was considered contradictory to the generally accepted theory of electrolytes presented by the great Swedish chemist Svante Arrhenius in the 1880s, and according to which no electrolytes could be completely dissociated. In 1916, however, Bjerrum was able to materially substantiate his hypothesis, which became generally accepted around 1920.

Bjerrum was nominated twice for the Nobel Prize in chemistry for his contributions to developing the theory of strong electrolytes, first in 1927 and later in 1931. Both nominations resulted in highly positive assessments of Bjerrum's works, but like Brønsted, he too was passed over in the Nobel chemistry committee's final selection round.[493]

August Krogh and zoophysiology*

August Krogh was undoubtedly the most important Danish physiologist of the 1900s. His research in the 1910s was particularly concerned with the branch of his field known as "work physiology", as he sought to understand how the need of an organism (animal or human) for supplying energy to its internal metabolic processes is fulfilled. One of Krogh's central questions was how muscles can absorb adequate amounts of oxygen in connection with physical exertion. It was through these investigations that Krogh realized what is common knowledge today: that the blood's rate of circulation rises dramatically when the muscles are working, during cycling and running, for instance, and that this increase in rate is absolutely necessary to fulfil the organism's greater need for energy. He additionally demonstrated that the energy was consumed in the working muscles, and that the body could effectively use almost all of the oxygen in the blood if needed.[494]

Krogh mainly concentrated on the capillaries, the finest and most delicate part of the circulatory system, as he sought to understand their role and functioning in the organism's transport of oxygen and other blood constituents. These studies of muscle tissue at work and at rest, and the resulting discovery of how capillaries adjust to high and low demands for oxygen and nutrients, would eventually earn him the Nobel Prize. Briefly told, what Krogh could

---

* This section was written in collaboration with Anita Kildebæk Nielsen. For a more detailed account (in Danish), see Kildebæk Nielsen and Vorup-Jensen 2006. In English, see Kildebæk Nielsen 2001.

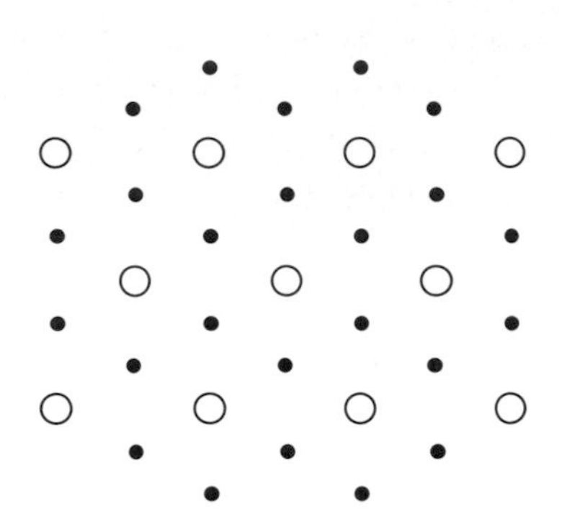

21.8 *Schematic representation of August Krogh's capillary model. The large open circles symbolize active capillaries that allow blood to flow through, while the small solid circles symbolize closed, inactive capillaries. All capillaries are alternately open and closed, and the number of open capillaries rises as the organism's energy requirement increases. One of Krogh's experiments demonstrated that the number of open capillaries in frog muscle could swell from 10 to 325 per square mm as requirements increased.*

prove was that the capillaries alternately opened and constricted the flow of blood, meaning that the system had an enormous built-in surplus capacity: The number of open capillaries could be increased as the need for oxygen and nutrients rose. He achieved his scientific results by combining theoretical insights with his talent for laboratory work, which included devising and designing new and more precise instruments.

Krogh began to publish his findings in 1918, and just two years later they earned him the 1920 Nobel Prize in physiology or medicine.[495] The money that accompanied the prize gave Krogh the financial freedom he had longed for, and much of it was used to fund further research. As we shall see, however, Krogh also had a knack for using the prestige that comes with every Nobel Prize.

Rockefeller-funded physiology. Once Krogh had gained international renown in the early 1920s, his zoophysiology laboratory quickly proved how sorely inadequate it was in light of the rising demand for his expertise. Many young physiologists from Europe and the US applied to stay with Krogh as visiting scientists, but a good deal had to be turned away simply because "there is never room for more than one at a time and even to have one causes some inconvenience and requires some resignation on the part both of the laboratory staff and the guest".[496]

In 1923 Krogh contacted the American director of the Rockefeller Foundation, George Vincent, to ask whether the foundation would provide funding for a new zoophysiology laboratory.[497] Krogh's application was assessed jointly by the Division of Medical Education (DME), a body under the Rockefeller Foundation, and the closely related IEB. During the assessment process in

21.9 *The heading and lead photo in one of the 100-plus articles and notices appearing in the Danish newspapers in the days after the announcement of Krogh's Nobel Prize (or "Nobel award", as it was widely called at the time). This picture of "The Nobel Prize winner in his study" accompanied an interview with the laureate.* Nationaltidende, *29 October 1920.*

1924, a suggestion was made for an institute that would house not only Krogh's activities, but also physiology at large at the University of Copenhagen. As the director of the DME wrote to his counterpart at the IEB:

> I am convinced Krogh deserves better quarters, but to help him alone might delay for years the development of physiology in the Medical School [...] The ideal plan would be for IEB and DME to cooperate with Government to bring the three efforts in physiology [those of Krogh, Henriques and Lindhard] together in one building at hospital site [...] each would have an autonomous department in the building [...] If this plan cannot be worked out I doubt value of helping Krogh alone – and delay might be advisable.[498]

The two other laboratories mentioned in the letter from the DME were headed by Valdemar Henriques, who had succeeded Christian Bohr (the father of Niels Bohr) as professor of physiology at the faculty of medicine in Copenhagen, and by the sports physiologist Johannes Lindhard. After the idea of a joint institute was well received by the relevant parties at the university, the professor of biophysics Hans Marius Hansen applied to have his department

21.10 *Aerial photo from 1928 of the new Rockefeller Institute for physiology, located on Juliane Maries Vej, at the time just outside central Copenhagen. In front of the institute are the two professorial residences, initially occupied by the head of the university's Zoophysiology Laboratory, August Krogh (the house at upper left) and the head of its Medical-Physiological Institute, Valdemar Henriques.*

included in the forthcoming project, which was later also joined by the university's new Institute of Biochemistry. Thus, the Rockefeller institute of physiology in Copenhagen ultimately became home to five departments from the faculty of mathematics and science and the faculty of medicine. The grant received final approval in November 1924, with the DME and the IEB providing a total of USD 400,000 (roughly DKK 2.4 million), and the University of Copenhagen (effectively, the finance committee of the Folketing) pledging to take care of the ongoing, and increasing, costs and making an appropriate construction site available, the latter amounting to about DKK 300,000. As it was not possible to allocate this amount from the Danish budget, it was procured instead as grants from Denmark's two principal science funds, the Carlsberg Foundation and the Rask–Ørsted Foundation.[499]

The new physiology institute – which soon became known as the Rockefeller Institute – enabled Danish physiologists, for the first time ever, to work in good physical surroundings, while also improving opportunities for cooperation in the field. In terms of research, Krogh and his Laboratory of Zoophysiology earned considerable attention from abroad and welcomed many guests. Most of the visiting scientists were educated in medicine, which was apparent from the laboratory's research profile: Despite its name, more than half of the scientific works published by the laboratory dealt with human physiology and comparative physiology.[500] In addition, it is noteworthy that Krogh himself did not confine his research activities to these fields, but also published a number of papers on general physiology and methodology. As the Danish physiologist Carl Barker Jørgensen concluded, "Krogh had a massive influence on all areas of research at the Laboratory of Zoophysiology."[501] Krogh was also a key player in developing insulin in Denmark, as described in chapter 23, and his rare gift for visualizing and constructing new instruments provided his laboratory with an extra source of income. For three decades, Krogh remained an important figure in the international physiology community, thanks in part to his sale of scientific equipment to foreign colleagues and his extensive professional network.

Factors contributing to Denmark's scientific golden age during the inter-war years

In connection with the 100th anniversary of the Nobel Prize in 2001, the Nobel Museum in Stockholm assembled an impressive exhibition called *Cultures of Creativity*, which has since toured the world. This exhibition's commendable effort to identify the stuff Nobel Prize winners are made of focuses on the particular brand of creativity that a number of Nobel laureates have exhibited, and on the features that characterize the research environments that have fostered many prize winners. Copenhagen is mentioned in the *Cultures of Creativity* exhibition and the catalogue.[502] Not surprisingly, the section on Copenhagen concentrates almost exclusively on Niels Bohr and his institute, explicitly emphasizing, as its particularly characteristic and creative element, the free and unfettered dialogue that he loved so well and always encouraged between physicists discussing the perplexing issues around which their professional lives revolved.

As historians of science, we certainly agree that this analysis pinpoints one of the essential qualities of Bohr's institute in the 1920s and 1930. It is nevertheless important to bear in mind, and to highlight here, a number of other factors that strongly contributed to creating this golden age in inter-war Denmark, not only in physics, but also in the fields of physical chemistry and zoophysiology.

21.11 *Niels Bohr (at right) with his younger brother Harald Bohr, a gifted mathematician who played an important role in the internationalization and institutionalization of mathematics in Denmark during the first half of the twentieth century. The Carlsberg Foundation had been already pledged in 1929 to found the construction of a new institute of mathematics, but for several years the university was unable to locate an appropriate site. In 1934 Niels Bohr suggested building the institute, which had long been in the pipeline, next to his own institute on Blegdamsvej. Harald Bohr became the leader of the new institute, and various features, such as an underground tunnel, were included to accommodate the close working relationship between the brothers and their respective institutes.*

Danmark – a neutral nation. Just after the end of World War I, the Allied Powers decided to replace the defunct International Association of Academies, founded in 1899, with a new organization called the International Research Council, which did not accept scientific societies and academies from the Central Powers as members. This decision did not sit well with Danish scientists, who had become well-integrated in the international scientific community during the preceding decades, and who perceived every constraint on international cooperation as a step backwards. They suddenly realized that Denmark's status as a neutral nation during World War I opened opportunities for the Danish scientific institutions, and those of other neutral countries, to become important meeting places for scientists from both sides of the bloody conflict that had now come to a close.

21.12 *Another strong family in Danish science: Elis Strömgren with his two sons, Erik and Bengt (seated) in Elis' office at the University of Copenhagen's astronomical observatory around the mid-1920s. Elis, who was appointed professor af astronomy in 1907, held that position until 1940, when Bengt took over the professorship. While Elis played an important role in keeping the international collaboration in astronomy alive during and after the World War I, Bengt became a leading figure in twentieth-century astrophysics. One of his earliest and most important contributions was to calculate the hydrogen and helium content of the Sun and similar stars using the newly developed quantum mechanics. In 1951, Bengt Strömgren was headhunted to become the director of the Yerkes and McDonald Observatories in the United States. In 1957 he took over the prestigious posistion as professor of astronomy at the Institute for Advanced Studies in Princeton, wich he held for ten years. He then returned to Denmark to become the director of Nordita, the Nordic Institute for Theoretical Physics.*

There was a will to exploit this opportunity among leading Danish scientists like Niels Bohr (physics), his brother Harald Bohr (mathematics), Einar Biilmann (chemistry), Elis Strömgren (astronomy), August Krogh (animal and human physiology) and many of their peers. What is more, these scientists had the great fortune that a number of Danish and foreign foundations (chiefly the Carlsberg Foundation, the Rask–Ørsted Foundation and the IEB) were highly receptive to the scientists' appeals for financial support to expand existing institutions in Denmark and build new ones, and to give scholarships to foreign scientists wishing to visit. The Danish government also espoused this cause, establishing the Rask–Ørsted Foundation in 1919 with the object of supporting "Danish science in connection with international research".[503] As the Australian historian of science Peter Robertson has described it, this state-run foundation was "the first established by any country aimed primarily at supporting scientists of other countries".[504] During the first decade of the Bohr institute's existence, the Rask–Ørsted Foundation financially enabled 13 foreign physicists – including the likes of George Gamow (Russia), Georg von Hevesy (Hungary), Oskar Klein (Sweden) and Svein Rosseland (Norway) – to work in Copenhagen for long periods of time, and also supported brief visits to the institute by a large number of internationally renowned scientists.[505]

Eminent scientists. Had it not been for the large number of contemporary scientists of international standing congregating in Copenhagen during the inter-war years, it is highly unlikely that anything resembling Denmark's scientific golden age would have occurred at all. In point of fact, the individual reputations of the most prominent scientists in the Danish capital, most notably Bohr and Krogh, were such that each alone was powerful enough to attract promising young scientists as well as esteemed senior colleagues from abroad for shorter or longer stays in Copenhagen. Some guests – such as Hendrik Kramers, Werner Heisenberg, Otto Frisch, James Franck, Georg von Hevesy and Oskar Klein – even remained at the Copenhagen institutes long enough to genuinely contribute to the city's excellent scientific environment.

Local networks of scientific institutions. It was certainly an important factor that all of the relevant scientific research institution in Copenhagen, including the Astronomical Observatory, Niels Bohr's Institute for Theoretical Physics, the Institute for Physical Chemistry, the Rockefeller Institute, the Carlsberg Laboratory and the scientific institutes at the Royal Veterinary and Agricultural College, were located in relatively close proximity to one another, allowing scientists to easily participate in each other's colloquiums and even to conduct cross-institutional research projects. Not all opportunities were utilized, however, and it is a well-known fact that although the IEB board's expectation to scientific synergies were partly what motivated them to support the construction of Johannes Brønsted's new Institute for Physical Chemistry near Niels Bohr's institute on Blegdamsvej, in reality the two famous professors were not close at all, and there was little or no cooperation between scientists from their two institutes.

On the other hand, the spirit of co-operation reigning among Bohr's institute, the Rockefeller Institute, the Carlsberg Laboratory and the Agricultural

College, as well as several institutions of medical science, was so pronounced that it is fair to say there existed a close-knit professional network of scientific institutions in Copenhagen during the inter-war years. The Hungarian chemist Georg de Hevesy, who worked in the Danish capital from 1920 to 1926 and from 1935 to 1943, was one of those who grasped the potential in exploiting initiatives like the experimental biology project supported by the Rockefeller Foundation in 1935. Having cooperated with physician Ole Chievitz – head medical doctor at the Finsen Institute and a close friend of Bohr and Bjerrum – to use radioactive phosphor to demonstrate that the phosphor in animal bones is continuously replaced, Hevesy started a research program in the last half of the 1930s that explored the use of many different radioactive isotopes as tracers in biological material. The size and the success of this project was due in large part to Hevesy's ability to collaborate with scientists at many different research institutions in Copenhagen, including those mentioned above, allowing him to have multiple experiments running simultaneously in the various laboratories around the city.[506] Had the scientific climate not been conducive to cooperation, and had the distances been much greater, such projects would not have been possible.

Scientific leadership and fund-raising. Brønsted's and Krogh's applications to the IEB, each seeking funding for a new scientific institute in his field, were born of the realization that the University of Copenhagen had no intention of allocating funds for such building projects, which the scientists themselves considered both reasonable and legitimate. These two applications, submitted in 1923 and 1925, respectively, were both crowned with success after extensive consideration by the IEB, bearing witness to the two men's willingness to pursue unconventional means to achieve their professional ambitions. A similar procedure was followed several times in the inter-war years when the need arose for new equipment or guest scholarships for foreign colleagues, though such applications were usually submitted to the Carlsberg Foundation or the Rask–Ørsted Foundation, not to potential benefactors abroad.

When it came to scientific fund-raising in Denmark, Niels Bohr was the undisputed champion. As the reputation of his work and his institute grew, at home and abroad, Bohr constantly found himself in need of funds for expansions, new instruments and visiting scientists. Fortunately, he had a knack for seeing new financing opportunities: the Danish state, the University of Copenhagen, the Carlsberg Foundation, the Rask–Ørsted Foundation, the Thomas B. Thrige Foundation, the IEB and the Rockefeller Foundation, and a host of others. More often than not, they looked favourably on his applications. Yet despite Bohr's outstanding reputation, he did not expect funding from potential donors to simply fall into his lap. On the contrary, in his applications Bohr always made a point of conscientiously and meticulously formulating the institute's financial and material needs. In fact, the great physicist would often work just as diligently on the wording of his applications as he did on his many scientific publications, which usually underwent numerous revisions before leaving his desk.

Even though Bohr found practical and administrative work extremely important, he did not involve the institute's young guests in the general planning process. Often his visitors would only be spending a few short months in Copenhagen, and Bohr's approach allowed them to dedicate a maximum of time and effort to the pursuit that had drawn them there in the first place: physics. Even the permanent staff at the institute were not expected to spend much time on administrative tasks. Naturally, Bohr would include them in discussions about ordering new equipment, building or expanding the institute or moving into new research areas – as he did when much of the institute's research shifted to the field of nuclear physics and experimental biology in the mid-1930s. No minutes were taken of such discussions, however, and the final decision rested solely with Bohr. This was a state of affairs that everyone accepted, and Bohr remained the uncontested leader of the institute from its founding until his own death in 1962. That is also why the Bohr institute's history from 1921 to 1962 is, above all, the story of one man and his life's work.

The Carlsberg Foundation's enormous importance to Danish science has already been emphasized numerous times in the preceding chapters. Its significance peaked between the two world wars; a period during which donations from the Carlsberg Foundation accounted for a large part of the independent, or "free", research funding in Denmark. Traditionally there have always been close ties between the Royal Danish Academy of Sciences and Letters and the Carlsberg Foundation. From 1931 to 1962, for instance, Niels Bohr was the president of the Royal Academy while a close friend of his, the chemist Niels Bjerrum, served on the Carlsberg Foundation's powerful board of directors from 1931 to 1956. If nothing else, at least these personal connections were not detrimental to applications coming from the informal network of scientific professors described above.[507]

The Copenhagen spirit. In light of Niels Bohr's extensive administrative efforts, the memoirs of those who visited his institute as young scientists in the interwar years show a conspicuous absence of remarks concerning that aspect of the institute's life. They describe only the purely scientific work and the unique spirit that reigned there.[508] "Der Kopenhagen Geist", a phrase coined by Werner Heisenberg, was first and foremost characterized by the informal, jovial and playful mood that allowed young physicists complete freedom to cultivate their scientific interests under the leadership of Bohr, who had a way of stimulating each individual's creative capacity.

The British physicist Nevil Mott, who worked at the Institute for Theoretical Physics in 1928, described physics in Copenhagen as "a social phenomenon" in which "half of one's time was spent discussing", in stark opposition to his experiences at Cambridge.[509] The Austrian physicist Otto Frisch, who spent several years at the institute in the 1930s, recalls in his memoirs how he went to attend a colloquium shortly after his arrival. Great was his surprise upon stepping into the designated room, where he found Bohr eagerly engaged in a discussion with his Russian colleague Lev Landau, who was comfortably reclined on a table. Frisch had never seen anything like this in Hamburg or London, and it took him a while to get used to "the informal habits at the Institute for Theoretical Physics in Copenhagen, where a man was judged purely by his ability to think clearly and straight".[510]

The following story from Heisenberg, who came to Copenhagen in the spring of 1924, illustrates another side of the atmosphere Bohr created. Newly arrived, the young German physicist waited for a couple of weeks for an opportunity to speak with the famous Dane about the physics-related problems he was grappling with at the time. When Bohr was finally ready, he invited Heisenberg on a three-day ramble across north Zealand, where the two spent much of their time conversing not about physics, but about history, literature and philosophy.[511] Physics was certainly important to Bohr, but it was part of a larger cultural framework, and he therefore expected the institute's employees and guests to take an interest in many other things as well. This facet of the Copenhagen spirit, which Heisenberg found fascinating, is one that Frisch also emphasized when, writing after Bohr's death, he described what it was like to visit Bohr in the Carlsberg Foundation's honorary residence, where he and his family lived from 1931 to 1962:

> Here, I felt, was Socrates come to life again lifting each argument onto a higher plane, drawing wisdom out of us which we didn't know was in us (and which, of course, wasn't). Our conversations ranged from religion to genetics, from politics to art, and when I cycled home through the streets of Copenhagen, fragrant with lilac or wet with rain, I felt intoxicated with the heady spirit of Platonic dialogue.[512]

Another facet of the spirit of Copenhagen was how, at any given time, Bohr had a "helper", at times jokingly referred to as his "slave", who would take notes as the master thought out loud, preserving his words for posterity as the great man wandered round and round the table persistently and unsuccessfully attempting to light his pipe. This was hard work, and the two would often toil into the night – with Bohr proclaiming the next morning that, having "slept on the matter" he had decided they would have to begin again from scratch. Serving as Bohr's helper could certainly take its toll on the assistant's own professional progress, but his slaves did not complain. Quite the reverse, as it was a great honour, a sign of recognition, to be chosen for this task. Bohr's practice of relying on others for assistance also helped to create the extraordinary atmosphere that his institute was known for around the world.

Danish science after World War II

Because no actual, major military engagements took place in Denmark during World War II, the country suffered far less destruction than most of the other countries affected. Indeed, in almost every respect the Danes were luckier than other European peoples. Nevertheless, the years immediately after the war were difficult, marked by political insecurity in the areas of foreign policy, by a shortage of goods and products and by permanent exchange-rate difficulties that prevented the successive governments from efficiently modernizing Denmark's manufacturing sector, which was in poor condition. And with inflation having greatly reduced the Carlsberg Foundation's pool of allocable resources as compared with the inter-war period, Danish science experienced several lean years – until eventually the Marshall Aid programme and the Danish State Research Foundation (described in chapter 20) created a budding optimism in the scientific community in the early 1950. This was followed by the donation of public funds on a scale never seen before, allowing Danish science to grow immensely during the latter half of the 1950s.

The post-war years gave Danish science a dramatically different political and economic framework. That alone makes it interesting to look at how the country's scientific summits from the inter-war years fared after the World War II. Did they still stand tall? Did completely new peaks emerge? Or did all of the high points erode and blend into a uniform and gently rolling scientific landscape reminiscent of the Danish countryside itself?

Bohr's institute and theoretical nuclear physics. Just after the end of World War II, Niels Bohr's institute was in jeopardy of losing the unique position its founder had built up in the 1920s and 1930s. The new generation of post-war physicists were not particularly interested in philosophical interpretations of the principle of complementarity, preferring more result-oriented research that could be pursued in the sort of modern, well-equipped laboratories found in many other locations, not least the US. With its ample funding, its excellent laboratories and its brilliant scientists – many of whom were European refugees – America had become the world's new scientific superpower. Nonetheless, it was in this climate that Niels Bohr, his son Aage Bohr (a gifted physicist in his own right) and the institute's core staff succeeded in turning things around and once again giving the Institute for Theoretical Physics a new profile. The fruitful cooperation that Aage Bohr and the American physicist Ben Mottelson began in 1951 to investigate the structure of the atomic nucleus was the beginning of an exceptionally fertile period for Danish physics, which lasted until about 1970.

In the 1950s and 1960s the institute was still a world-class centre of science, "the Mecca of all physicists", as the Austrian-American physicist Victor Weisskopf put it as late as 1966.[513] On 7 October 1965, the eightieth anniversary of Niels Bohr's birth, the institute officially changed its name to the Niels Bohr Institute. Niels Bohr himself had served as head of the institute until his death in 1962, and as his successor, Aage Bohr upheld the distinctive spirit and time-honoured traditions that his father had valued so highly. Nuclear physicists from all over the world sought out Copenhagen as a favourite place to work and exchange ideas, and it was the Niels Bohr Institute that ultimately wrote the book on nuclear physics – quite literally, in fact, when Bohr and Mottelson published their monumental two-volume work *Nuclear Structure* in 1969 (vol. 1) and 1975 (vol. 2).[514] It seemed only logical that the Nobel Prize in physics for 1975 should go to the three individuals who had played a crucial role in developing the collective model of the atomic nucleus: Aage Bohr, Ben Mottelson and James Rainwater.

Perhaps it goes without saying that the excellent research and the legendary atmosphere of free and open discussion at the Niels Bohr Institute were a decisive motivating factor for the many foreign physicists who chose to visit or work at this particular institute between 1950 and 1970. Nevertheless, there were other crucial factors as well. One of them, documented by the esteemed historian of science John Krige, is that the Ford Foundation's work and travel scholarships allowing scientists to work at the Copenhagen institute – which from 1956 to 1970 totalled roughly USD 650,000 – were not merely a result of the foundation's desire to promote high-quality scientific research. Those managing the American foundation were equally attentive to the opportunities such trips presented for promoting Western values in locations that were known to be meeting places for leading intellectuals from East and West.

The Ford Foundation regarded the Niels Bohr Institute as one of the Western world's most important hubs for disseminating free, democratic ideas among influential figures in the communist countries. The institute had a long-standing tradition of receiving physicists who lived behind the Iron Curtain. In the Ford Foundation's view, when physicists from the East were allowed to meet and mingle with their peers from the West and freely discuss science, social issues and democracy, those from the East would develop a more positive view of Western values.[515] Although the political position of his American benefactors could hardly have been in absolute harmony with Niels Bohr's ideas of an "open world", he still regarded the Ford Foundation's support as an opportunity to practise science while at the same time promoting his own ideas.

Following in Krogh's footsteps: Ussing, Skou and Danish physiology after World War II. * Like theoretical physics, physiology also remained a high plateau in Danish science during the second half of the twentieth century. It was an impressive ascent that led from the founding of August Krogh's chair in zoophysiology in the early 1900s to the emergent molecular understanding of physiology that was reached in the late 1960s; a long and precipitous path, reflecting the biological sciences' rapid development between 1920 and 1970. August Krogh was still a dominant feature in the contours of Danish physiology, not merely during his lifetime but for more than two decades after his death in 1949. It was Krogh who in the late 1930s recruited a promising young physiologist named Hans Henriksen Ussing for a project on the transport of water through amphibian skin.[516] This project, as well as a later one on protein synthesis, employed the hydrogen isotope deuterium in the form of heavy water, a technique that Ussing had learned from the Bohr institute's Georg von Hevesy. According to Ussing himself, it was also Krogh who got him on track in his investigations of inorganic ion transport through cell membranes. This clearly indicates that Krogh also took on the role of mentor.[517] Krogh had long supported the view that the cell membrane was permeable to ions such as sodi-

---

* This section was written in collaboration with Thomas Vorup-Jensen. For more detailed accounts, see Vorup-Jensen 2001, and (in Danish) Kildebæk Nielsen and Vorup-Jensen 2006.

21.13 *Danish press photo of the year 1997. According to the prize committee, photographer Jonna Keldsen succeeded in "relating Skou to a theory that is difficult to understand for ordinary people."*

um. In fact, the American physiologist Robert Dean had even proposed the existence of what he called a "sodium pump", which ostensibly maintained a low sodium concentration internally in the cell by pumping sodium out of the cell to counterbalance incoming sodium from the bodily fluids. In a remarkable talk he gave at the Royal Society in London in 1946, Krogh voiced his support for the existence of an active cellular ion transport mechanism.[518]

Other physiologists were more sceptical to the idea of active transport, believing that the cell's total energy consumption for this process would be greater than the energy from the available nutrients.[519] However, by means of detailed calculations Ussing concluded that the cell did have sufficient energy available to maintain active transport. Ussing realized, of course, that he would have to verify his calculations by measuring the transport of ions across the cell membrane. Working with the chemist Karl Zerahn, Ussing devised an apparatus for this particular purpose, which could measure the movement of

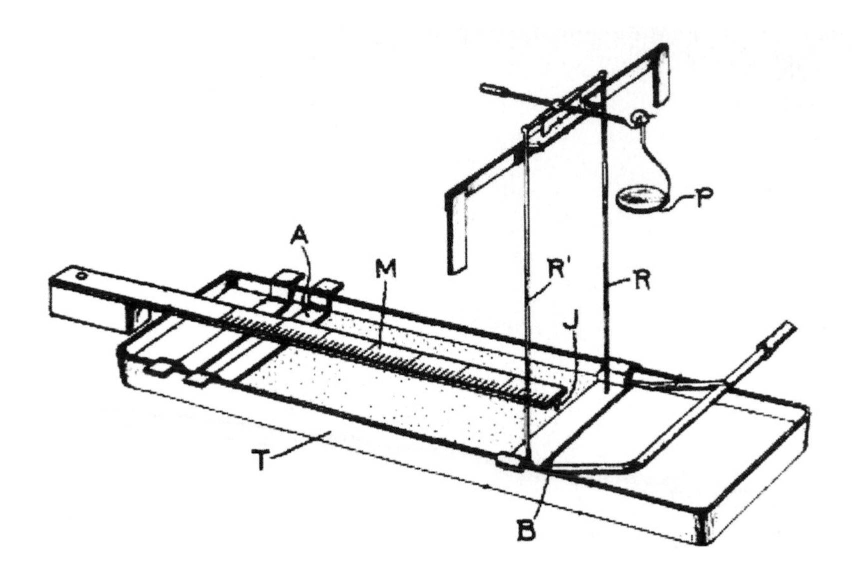

21.14 *Langmuir trough reproduced from the eminent American chemist Irving Langmuir's original article. Two floating barriers (A and B) delimit the surface (the dotted area) in a low-sided tray filled with liquid (T). Trough barrier B is linked to a pressure sensor. Two glass rods (R and R') are fitted through B. The system's flexibility means that adding weights to the pan (P) will press barrier B forwards, in the direction of A. By moving barrier A towards B, barrier B is pressed back to a point (J) where the pressure against B can be calculated as a function of the mass of the weight in P and the height of R. The surface area of the lipid layer is determined by looking at the scale M.*

electrical charge through an epithelium by simply short-circuiting the fluids on either side of the skin sample. Known as "the Ussing chamber", this apparatus may well have been the period's most important analytical instrument for achieving an understanding of active ion transport. In addition, having been quoted more than two thousand times, Zerahn and Ussing's publication went on to become a citation classic.[520]

Around 1950 at the University of Aarhus, another young Danish physiologist by the name of Jens Christian Skou had also taken an interest in the active transport of sodium and potassium ions. Skou's angle of approach was quite different from Ussing's, however, as he was investigating what it was that made local anaesthesia work. Skou wondered whether the mechanism causing the anaesthetic effect was related to the anaesthetic's penetration into the nerve-cell membrane, which, like membranes in other cell types, mainly consists of lipids. He attacked the problem using a Langmuir trough, an apparatus, shown in figure 21.14, in which a layer of lipids impermeable to water was spread on top of an aqueous layer. Given that the surface was delimited by the shallow trough's sides and the floating barriers, one of which was attached to a pressure-measuring device, the components in any non-aqueous substance (say, an anaesthetic) added to the aqueous solution beneath the lipid layer would penetrate this upper layer, causing the pressure against the barriers to rise.

Skou worked until 1957 to describe the enzyme activity in the cell membrane in order to understand the uptake of anaesthetics in the cell. He especially concentrated on an enzyme that split a chemical compound called adenosine triphosphate (ATP), which in turn influences cell metabolism. This was a difficult task, but Skou discovered that the presence of both sodium and potassium ions was a necessary parameter for high enzyme activity. Skou wrote an article on his discoveries, proposing almost in the last sentence that this activity might indicate a relation between the enzyme and the transport of sodium and potassium. As it turned out, this proposal expressed remarkably innovative thinking on the functioning of the cell membrane, as no one had previously had an eye for the transmembrane operation of the sodium–potassium pump. The Nobel Prize in chemistry awarded to Skou in 1997 confirmed the international significance of his work from the 1950s, in a sense rounding off the important contributions Danish scientists had made to the development of physiology in the twentieth century.

The lack of chemistry schools. While theoretical physics and physiology maintained their stature and upheld Denmark's scientific reputation at home and abroad, the same cannot be said of chemistry. Certainly, Danish chemistry did not collapse, but when Bjerrum, Brønsted and Sørensen retired during or just after the World War II, it became evident that the country's chemistry community had no obvious successors of the same international stature. Good chemistry work was still being done, but the golden age of Danish chemistry ended with the war. The question is: Why? One main reason might be that the eminent scientists who for decades had been at the heart of the thriving chemistry community had done little in the way of assuming leadership, or forming schools. Unlike the thinking of Niels Bohr and August Krogh, the ideas of Denmark's most eminent chemists had only a very limited influence on the generation that succeeded them. This trend has been evident in Danish chemistry, for better and for worse, throughout the nineteenth and most of the twentieth century.

*Ice-core research – a new Danish peak.** Whereas Danish chemistry was unable to maintain and expand its international profile during the latter half of the twentieth century, other fields of scientific research became more visible, and at least one new and quite prominent Danish peak arose: ice-core research.

The birth of Danish ice-core research can be dated very precisely to the summer of 1952, when a young biophysicist named Willi Dansgaard was working with a new mass spectrometer that the Laboratory of Biophysics (part of the Rockefeller Institute in Copenhagen) had received under the Marshall Aid program. A mass spectrometer can be used to distinguish molecules with different weights from one another in a given sample – different oxygen isotopes in a piece of ice, for instance. The new instrument was not intended to be used for examining ice samples, but for biological and medical research. Dansgaard advertised the method in several chemical and medical journals, but it soon became apparent that there was little interest in using light isotopes in these particular fields, which left Dansgaard with a highly sophisticated, state-of-the-art instrument and no research project to use it for.

Before beginning his job at the Laboratory of Biophysics, Dansgaard had worked with the weather service at the Danish Meteorological Institute, where he had developed an interest in meteorological phenomena. That is where he got the idea of using the mass spectrometer to analyse rainwater samples. And so, on a summer's day in 1952, while a strong warm front hung over Denmark, Dansgaard collected rainwater in his garden, later taking the samples back to the laboratory. His analysis of the isotope composition in the rainwater samples seemed to show a link between the temperature at which the raindrops had coalesced and the rainwater's content of the heavy oxygen isotope ^{18}O – meaning that perhaps ^{18}O would be a useful indicator of air temperature.[521]

Over the following months Dansgaard substantiated his theory by analysing rainwater samples from all over the world, and his measurements suggested that there was indeed a connection between air temperature and the content of ^{18}O in the sample. Dansgaard published his findings in 1954, further proposing that the demonstrated connection between air temperature and ^{18}O content in precipitation could be used to reconstruct temperatures that had occurred in the past. All it would take was a location where ancient precipitation was preserved, like the immense sheet of ice that covers central Greenland, and where each consecutive year's snowfall overlays that of the year before.

---

* This section was written in collaboration with Maiken Lolck. For more detailed accounts (in Danish), see Lolck 2007 and Kragh, Lolck and Nielsen 2006.

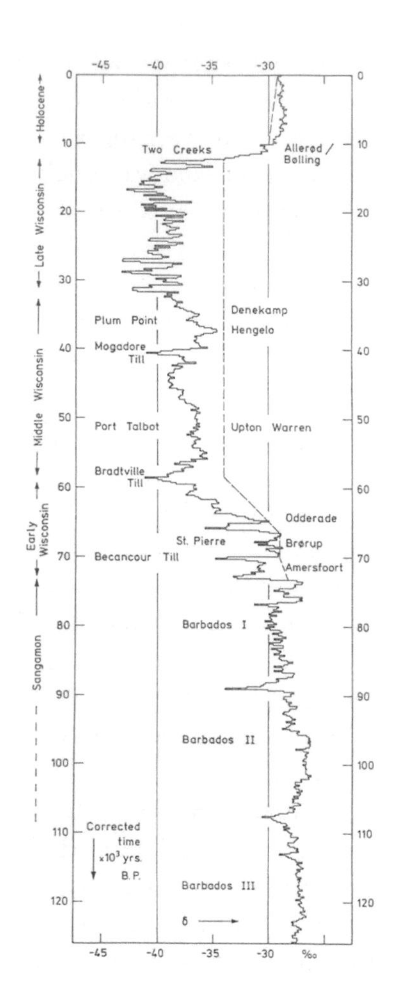

21.15 *Isotope curve from the Camp Century ice core. The curve shows how the isotopic composition of the ice varies in the annual layers of snow deposited over the past 120,000 years. Because isotopic composition is an indicator of the air temperature at the time when the precipitation fell, the curve reflects the temperature variations throughout the most recent ice age and the current interglacial period. The time axis is vertical, with the most recent readings at the top.*

Realizing Dansgaard's vision of drilling deep into the ice cap of Greenland or Antarctica to extract ice cores would be a costly venture. Certainly the research foundations in Denmark did not have the kind of funding such a project would require – but fortunate favours the bold. In the late 1950s, during the Cold War, the American military was strongly present in Greenland, because of its location on the shortest air route between the US and the Soviet Union. To operate militarily in Greenland required knowledge on the Arctic environment, and therefore comprehensive glaciological studies were initiated.At the time the US had plans for a colossal project code-named "Iceworm", which would create a network of tunnels deep in the ice stretching from Narsarsuaq in the south of the country to Thule in the far north. The projected tunnel system, totalling 4,000 km in length, would be used to conceal up to

600 missiles armed with nuclear warheads. The missiles would be hidden beneath the ice, even while the huge extent of the tunnel network would make it impossible for the enemy to strike them all in the event of a nuclear war. To construct such a tunnel network the US military had to know the properties of the ice. That is how in the early 1960s, backed by American cold-war funding, a team of American scientists and military specialists began to drill the world's first deep ice core.[522]

The ice core was extracted near Camp Century, situated near Thule and close to the edge of the Greenland ice sheet. The camp itself was entirely carved out beneath the ice and could accommodate up to 250 people. It contained all of the necessary facilities, including a library, a gym, workshops, a hospital clinic and a shop, and its energy was supplied by a mobile nuclear reactor. Camp Century was part of the US military's activities in Greenland, but much of the research carried out at the camp – including the studies of the ice cores that had been extracted – were more in the vein of basic research. And even though much of the glaciological research had no direct military application, the Americans did not refrain from emphasizing the patriotic significance of doing research under the inhospitable conditions in northern Greenland, which they described as "one of the last frontiers left on the earth."[523] The drilling of the deep ice core at Camp Century was concluded in 1966, when the drill bit finally reached bedrock at a depth of 1,390 metres.

Because the Americans were not particularly interested in carrying out an exhaustive scientific study of the entire ice core, and because Willi Dansgaard and his team in Copenhagen possessed the necessary expertise (not to mention an intense desire) to do so – and because the US wished to demonstrate their goodwill towards Denmark, which officially had supremacy over Greenland – the Danish researchers were allowed access to the Camp Century ice core and conducted a systematic isotope analysis.

Three years after the drilling operation at Camp Century was over, the isotope analysis results of the core material had been completed. The isotope curve reproduced here illustrates the temperature variations in northern Greenland over the past 120,000 years. Among the many recognizable climate events are the transition from the latest ice age about 10,000 years ago and the relatively warm period during the Middle Ages. Besides confirming the scientists' climatic knowledge of the past 100,000 years or so, the curve provided a wealth of information on climate change at a level of detail never seen before. The layers from the last few centuries could show temperature changes from one year to the next, and the scientists could even distinguish between winter snow and summer snow.[524]

But the most astonishing finding in the ice core was its evidence of climate changes that seemed to occur very suddenly during the most recent ice age. This evidence was corroborated by a new ice core extracted ten years later, which proved that violent climate change can take place – even without the influence of humankind. The ice cores have revealed that the temperature in the North Atlantic region can vary by about 10 degrees of celsius within a single century, and perhaps within a single decade. This knowledge, and the continuing study of ice cores from Greenland and Antarctica, still play a vital role in the scientific investigation of the Earth's climate, and in the ongoing debate about climate changes in the future.

Concluding remarks on peaks, plains and valleys

During the twentieth century, Danish science developed immensely. It was a century that witnessed not only dramatic expansion, but also the consolidation of scientific internationalization. These processes were not always problem-free. In this chapter we have sought to describe the background for some of the peaks in the Danish scientific landscape of the twentieth century, well aware that these are the exceptions that confirm the rule – the rule being that in an international perspective, Danish science has largely remained on the middle ground. It is also important to stress that the scientific summits Denmark has occasionally reached could only have been scaled with the aid of helpful national and international scientific networks and strong financial sponsorship.

Prior to World War II, Danish scientists successfully built up international centres for theoretical physics, physical chemistry and experimental chemistry. All three of these centres depended heavily upon the generosity of private philanthropists and foundations, at home and abroad, as the Danish state was not particularly interested in research, or at least not in providing financial support for scientific research. The IEB – which in the inter-war years had presented a favourable assessment of Danish science in the relevant disciplines – was the decisive player in developing the potential areas of strength that Denmark had to offer. Obviously, the gift that Danish scientists, especially Niels Bohr, had for fund-raising and good research leadership also played a significant role. In fact, Bohr's organizational talents may have been fully as important as his scientific brilliance when it came to establishing his "Copenhagen school of physics". What is more, Bohr had the strength to consistently carry through his own personal leadership style, creating and nurturing the special "Copenhagen spirit" that made his research community a living legend.

The most successful Danish scientists were able to found schools, which must certainly have had a beneficial impact on their own research. As the post-war changes in the Danish chemistry community show, creating a school may well be an important factor in maintaining and expanding a research field – bearing in mind that Danish chemists during the period did not cultivate this aspect of their research to the same degree as their contemporary colleagues in other fields had done. Such schools help to reinforce and sustain areas of scientific strength, which, over time, often achieve their own momentum, enabling them to guide the direction of research as it advances.

Another factor that has a certain preserving and consolidating effect on the dynamics of research is known, at least in the sciences, as "the Matthew effect".[525] As the name suggests, it refers to the famous Bible verse Matthew 13:12, where Jesus says: "For whosoever hath, to him shall be given, and he shall have more abundance: but whosoever hath not, from him shall be taken away even that he hath." In the world of Danish science, the Matthew effect has mainly been evident in issues of scientific priority, in determining whether a scientific result should be attributed to one scientist or another. A well-documented tendency generally runs through the history of science of giving first priority to scientists who are already famous, rather than giving credit where it might more appropriately be due, or where it ought to have gone to a completely different person who was less well known. On the positive side, the Matthew effect can also help famous scientists, such as Niels Bohr or August Krogh, to obtain financial support and sustain attention within their fields. This in turn makes it easier for such prominent scientists to strengthen and expand their own particular disciplines, though not infrequently to the detriment of others.

Scientific exploration and big science

22

On the threshold of the twentieth century a distinction had begun to emerge between field science and laboratory science.[526] Since many scientists did both kinds of work the boundary was blurred, and it was hardly made clearer by the fact that the field sciences had, to a certain extent, internalized the methods and conventions known from the laboratory.[527] Historically the two types of research are closely linked, but there are also obvious differences, which mainly stem from their physical locations. Laboratory sciences take place within the framework of a scientifically organized indoor space, to which there is limited and strictly controlled access for humans and objects, and in which the problems, methods and items dealt with are comparatively easy to regulate, adapt, arrange, and systematize. Ever since the laboratory revolution in the nineteenth century, the quality of placelessness has promoted the high epistemic standing of laboratory research among scientists and the general public.[528]

Once science moves outside the laboratory's four walls, things begin to look very different. Unlike their counterparts indoors, the field sciences depend far more upon factors that are beyond control. Not infrequently, geographical, social and political circumstances play a decisive role in defining the questions a scientist can hope to answer by doing research in the field. This has forced the field sciences to invent and refine what has been referred to as "practices of place",[529] which are based on the difference and diversity of the various fields, and which incorporate the complex natural and social conditions applying in the specific field being examined. Practices of place are often markedly heterogeneous, boldly blending elements from other fields sciences with practices borrowed from the world of laboratory science. Some practices of place crystallize into completely new scientific cultures that "transcended the old, simple distinction between the laboratory and the field".[530]

Moreover, throughout history national policies and private financial interests have shaped the field sciences to the extent that historians of science are beginning to speak of "the colonial turn".[531] This concept refers not only to the historical demonstration of the field sciences' intimate association with the colonial interests of different countries,[532] but also to a number of studies that clarify how the field sciences themselves have been shaped by the colonial powers' encounters with the foreign regions they colonized.[533]

The focus of this chapter is not field research as such, but rather the more specific segment of the field sciences that scientific expeditions represent. Narrowing the scope even further, we will concentrate on two important Danish scientific expeditions carried out in the twentieth century: the Danish Three-Year Expedition to King Christian X Land (1931–1934) and the Danish Galathea Deep-Sea Expedition, also known as the second Galathea Expedition, (1950–1952). Both undertakings plainly illustrate many of the issues outlined above.

The Three-Year Expedition to King Christian X Land, 1931–1934*

Between the two world wars Denmark's research policy measures, new institutions, new technology and increasing internationalisation set new standards for the research being done in Greenland.[534] The early dog-sled expeditions had relied heavily on ethnography and archaeology as ancillary sciences to support their geographic and cartographic efforts. In the 1920s and 1930s geology was the magnet that drew scientists to the far north. And whereas the "polar research" of years gone by had been the preserve of small groups of death-defying, fur-clad loners mushing their way through the wilderness, the Greenland expeditions of the modern age were large groups that exploited the new technologies to their fullest, enabling them to do their fieldwork in a safe, comfortable environment – relatively speaking.

In several respects, the Three-Year Expedition to King Christian X Land (that is, North-East Greenland) from 1931 to 1934 is the earliest point at which the new day and age had fully impacted Danish exploration in Greenland. And yet, although it was a prestigious effort and an impressive feat, as well as one of the greatest scientific successes Denmark had ever seen, the many changes (technical, political, social and cultural) to Danish Arctic science that the three-year expedition built upon and stood for also led to harsh disputes that would leave a bitter stain on Danish geology and Greenlandic exploration for decades to come.

---

* This section was written in collaboration with Christopher Jacob Ries. For more detailed accounts (in Danish), see Ries 2003 and Nielsen, Ries and Østergaard 2006.

22.1 *The departure or arrival of an expedition vessel always gathered a crowd at the quayside. Here the Arctic explorer and geologist Lauge Koch, returning to Copenhagen in 1933 with the Three-Year Expedition.*

Arctic research and geopolitics. Up until 1880 the Danish interests in Greenland had been concentrated around the colonies that had grown up on the country's west coast. The next development, following in the late nineteenth and early twentieth centuries, was a series of major Danish expeditions to the harsher east coast of the great island. From around the turn of the century, however, Danish exploration efforts moved further up into North-East Greenland, a region where the Americans had already been prominently represented for about two decades by Robert Peary's numerous expeditions.

Twenty years of Danish–American competition in exploring northern Greenland finally reached a conclusion in 1917 when Denmark ceded its overseas possessions in the West Indies to the US, which in return paid DKK 25 million and acknowledged Denmark's sovereignty over Greenland in its entirety. At last, with the Bicentenary Jubilee Expedition around the northern tip of Greenland in 1921–1923, led by Lauge Koch, the heroic days of the legendary dog-sled expeditions in North Greenland finally came to an end. The focus of Danish exploration subsequently shifted to North-East Greenland, where Norwegian marine-mammal hunters and fishermen had begun to spread their activities in the North Atlantic up to, and even onto, the Greenland coast.

In the spring of 1924, Denmark opened talks with Norway to negotiate a treaty on East Greenland. Norway now openly contested Denmark's control of the region, and although the Danish negotiators maintained their country's sovereignty over the whole of Greenland they were unable to force the Norwegians to relinquish their claim. In reality the issue remained unresolved, even though Norway successfully petitioned for a twenty-year right to hunt, fish and set up permanent camps, and to establish radio stations and do scientific work in the uninhabited areas of East Greenland. Shortly after the two nations had signed the East Greenland Treaty, which became effective from July 1924, the Danish presence in the region was underscored by a colonization expedition in 1924–1925 to the Scoresbysund area led by well-known writer and polar explorer Ejnar Mikkelsen. In 1925, 83 native Greenlanders were moved from Ammassalik to the newly established settlement of Scoresbysund (today Illoqqortoormiut), which lay father north.

A new era. Over the following years, both Denmark and Norway increased their hunting, fishing and exploration activities in the northern and eastern reaches of Greenland. It was the Danish geologist Lauge Koch who spearheaded Denmark's push into East Greenland with his expeditions in 1925–1926 and 1929. Koch was also the Danish prime minister Thorvald Stauning's choice in 1930 to head the large and prestigious Three-Year Expedition that ran from 1931 to 1934. The geopolitical implications of the Danish thrust into the region were impossible to overlook, and in the very year that the Three-Year Expedition was launched, Norway struck back by declaring a large part of East Greenland to be Norwegian territory. Denmark immediately retaliated, filing a suit against Norway with the Permanent Court of International Justice in The Hague. The extensive findings resulting from the Three-Year Expedition, which were presented as evidence during the ensuing legal proceedings, contributed to the judges deciding the case in favour of Denmark. Norway subsequently ceased all scientific activities in the region, apart from meteorological investigations.

Another thing that made the results of the Danish expeditions to Greenland so relatively important during this period, politically speaking, was their enormous volume and their superior scientific quality, even when measured by international standards. One crucial precondition for achieving such results was the tense situation in the North Atlantic, which made national investments in scientific exploration an important strategic move in the geopolitical contest for sovereignty over Greenland. For Denmark there was the additional factor that in light of the crisis during the inter-war years, the country had great expectations to the potential benefits of drawing on the island's natural resources, which had so far been largely unexploited, cryolite extraction being the most notable exception. As Jens Daugaard-Jensen, the director of the Danish agency known as the Administration of Greenland, often put it, Greenland was "Denmark's big lottery ticket". If the persistent rumours of valuable deposits of oil, coal, cryolite and gold did turn out to be true, Greenland might be transformed from a burden into a boon for the Danish national economy.

The great East Greenland expeditions of the inter-war years became national initiatives that had Danish business interests thronging round, all hoping to share in the prestige and the profit. This meant that the expeditions' generous public funds were bolstered by significant private investments. The Three-Year Expedition of 1931–1934 ended up costing a good DKK 1.3 million. Nearly one third of this amount was paid by private investors, and just over one fourth was paid by the Carlsberg Foundation, with the Danish treasury forking out the remainder. The Two-Year Expedition of 1936–1938 amounted to just under DKK 700,000, with the Danish state paying a good half million, while a single wealthy clothing wholesaler called "Bulldog" Bryde-Nielsen contributed the rest. Investments of this magnitude made it possible to really increase the pace and the productivity of Danish Arctic research. Aeroplanes and aerial photography were revolutionizing topographic and geological cartography, and for the first time, the construction of permanent research stations in the field – complete with running water, electricity, radio stations and well-equipped laboratories – made it possible for scientific specialists to live and work in the field year-round with a minimum of distraction from the harsh Arctic climate.

The establishment of a high-tech infrastructure in the field made it possible to rapidly amass enormous amounts of data, while at the same time endowing the expeditions with an aura of industrial efficiency that went well with the contemporary expectation that technological progress improved the quality of life. From Greenland, Koch sent "radio letters" home to the Danes, describing his spectacular reconnaissance flights over (virtually) unknown territory. By promoting geological sensations to the public – like the expedition's finds of gold, or the mysterious fossil of a four-legged fish – such reports contributed to the general mood of popular support for the expedition project.

Geologists in an uproar. Clearly, it was not for nothing that the press touted the Three-Year Expedition as heralding "a new era" in Danish Arctic research. The huge amounts of collected material required a corresponding need for storage space, labour and laboratory facilities to record and process the new collections. Given that geology was the primary discipline in the work being done in East Greenland, it is understandable that the Danish geology community hoped to see new glory days for their science – and for Danish geologists.

At that point in time, the scientific hub of Danish geology was the university's Mineralogical Museum on Øster Voldgade in central Copenhagen. In the early years of his career Lauge Koch had cultivated his professional network daily at the museum, where he had also recruited most of his expedition members. However, after the special Three-Year Expedition Committee was set up in 1930 under the auspices of the Commission for Scientific Research in Greenland, Koch left the museum for a new office in the Administration of Greenland's buildings in 1931. As the period's leading figure in Denmark's exploration of Greenland, and by virtue of his good connections to Thorvald Stauning, the prime minister and minister for Greenland, and to Jens Daugaard-Jensen, the director of the administration, Koch was in a position to exert a substantial influence on contemporary research policy – such as it existed at the time. Thus, in more than one sense, Koch came to challenge the status of the Mineralogical Museum as Danish Arctic geology's premier institution.

Koch had found new colleagues in Sweden, the UK, Germany and Switzerland, and the number of Danish participants in his expeditions dropped suddenly and dramatically. The establishment of permanent research stations in the Arctic meant that many of the preliminary investigations could now be carried out by experts in the field. Also, after Koch's move to the administration's buildings in 1931 the collections gathered on his expeditions were stored in a locked warehouse in the Copenhagen docks, and any samples requested by the foreign experts Koch had employed were sent from the warehouse to their various European universities for examination and processing. The Danish geologists – whose results, in Koch's opinion, failed to comply with the quality standards that were necessary, scientifically and politically – were no longer granted access to Koch's expedition samples, and repeated appeals to have his collections handed over to the Mineralogical Museum were rejected with reference to Koch's plans for a grand, new "Greenland museum".

In addition to this, Koch retained absolute control over the publication of the expeditions' findings. Customarily, from 1878 onwards all Danish expeditions to the far north had published their results in the periodical *Meddelelser om Grønland* (*Monographs on Greenland*), edited by the Commission for Scientific Research in Greenland. In December 1930, however, a decision was made to give Koch complete editorial control of several volumes of the series, which were reserved for publishing the results of the Three-Year Expedition. And since no contrary decision was taken in the interim, this arrangement also came to apply to his Two-Year Expedition, which lasted from 1936 to 1938. The circumstance that Koch, without official permission, termed his expeditions Grønlands Geologiske Undersøgelser (Geological Survey of Greenland) – which had a decidedly official ring to it, mirroring the venerable institution called the Geological Survey of Denmark – merely twisted the knife that his Danish colleagues felt he had already embedded in their back.

The tension in the geoscientific community reached a climax in December 1935, when Lauge Koch sued eleven of Denmark's most prominent contemporary geologists for defamation following the appearance of their jointly written, and extremely critical, review of his book entitled *Geologie von Grönland*. Over the next few years "the Lauge Koch Controversy" regularly made front-page headlines in the Danish newspapers as it made its way through the Danish legal system. Having lost his case at the high court in 1936, Koch lodged an appeal, and on 21 June 1938 the Danish supreme court returned its Solomonic verdict: Part of the criticism could rightly be termed as "slander" and part of it could be seen as falling within the limits of honest and fair scientific critique. However Solomonic the verdict may have been, one thing is clear: The Lauge Koch Controversy created a deep and lasting rift in Denmark's geological community, and Denmark's research in Greenland, its repercussions affecting the scientific field for decades to come.

When in 1946 Denmark actually established the precursor of a body that later was to become the Geological Survey of Greenland, the initiative came from the three of Koch's staunch opponents from the 1930s. This official national institution, whose main objective was to conduct investigations in West Greenland, was closely associated with the Mineralogical Museum and with the Geological Survey of Denmark, founded in 1888. Koch – who was not invited to participate in its work – continued as a geological consultant with the Administration of Greenland and completed a series of intensive summer expeditions beginning in the late 1940s and ending in the summer of 1958. That is why, for more than a decade, Denmark had two mutually independent, publicly funded geological survey organizations working in Greenland, side by side – or, strictly speaking, working on opposite sides, as Koch stayed on the east coast and the official geological-survey people on the west coast. Cooperation was out of the question, and so, for 25 years, Danish geology was unable to draw upon Koch's extensive findings and experience from Greenland. A full three decades would pass before Koch – for the first and last time since the trial – would once again lecture to the students at the institute of geology in Copenhagen. This was in 1964, the very year he died.

However bitter the battle between Koch and the other Danish geologists, there is no doubt that the modernization of the Arctic expedition that was Koch's hallmark during the inter-war years was a powerful inspiration for the newly established Geological Survey of Greenland after World War II. The scientific findings and results from Lauge Koch's numerous expeditions to East Greenland were presented as a series of treatises and maps, most of which were published in the *Monographs on Greenland* series. As it turned out, this extensive body of basic knowledge was fundamental when finally, in 1968, the national Geological Survey of Greenland began investigating the Caledonoid formations in East Greenland found between 70° and 81° N – ten years after Lauge Koch had stopped working in the region.

The Danish Galathea Deep-Sea Expedition, 1950–1952

On 15 October 1950 the Danish Galathea Deep-Sea Expedition – widely known as the second Galathea Expedition – set out on what, at the time, was Denmark's hitherto largest and most costly scientific undertaking.[535] The naming of the frigate *Galathea* was a historic tribute as its predecessor, the Danish corvette *Galathea*, had circumnavigated the globe in 1845–1847 on a journey to accomplish scientific and other missions. Besides the frigate's crew, the second Galathea Expedition consisted of a scientific unit and a press unit. This structure reflected the dual nature of the expedition, which was firstly to conduct scientific deep-sea research and secondly to increase the world's awareness of what Denmark had to offer, culturally and commercially. The idea of linking a national public relations campaign with scientific exploration was something entirely new, as was the presence of a press corps within the formalized framework of a scientific expedition.

Two of the people behind the expedition – the travel journalist Hakon Mielche and the editor-in-chief of the popular daily *Ekstra Bladet*, Leif B. Hendil – were seminal in shaping the new public-relations venture.[536] This part of the expedition was conceived by the ubiquitous Mielche, who in 1941 had read a summary in *Ekstra Bladet* of a lecture by Anton F. Bruun, who would later head the Galathea Expedition, but was employed at the time in the mollusc department at the Zoological Museum. Popularization was one of

22.2 *The frigate* Galathea, *off Campbell Island in the sub-Antarctic ocean south of New Zealand.*

Bruun's hobbyhorses, and the topics he had touched upon to entertain his audience included deep-sea research and sea monsters. Bruun assured his listeners that anyone taking the trouble to trawl the ocean depths not only would bring up animals never before beheld by human eyes; they would also capture creatures that may have given birth to the myth of the great sea serpent. Bruun's argument was based on his participation in an earlier exploration voyage at sea, the Dana Expedition, 1928–1930, which had retrieved a giant eel larva in the waters off Cape Town that might conceivably have the potential to grow to enormous proportions – hence the sea serpent conjecture. Mielche contacted Bruun to verify the passage he had read in *Ekstra Bladet*, which had indeed quoted Bruun accurately, and Mielche jumped at the chance for a sensational headline: "Danish scientists snag sea serpent". Mielche and Bruun jointly decided to initiate a scientific deep-sea expedition that would continue the efforts of the Dana Expedition, but which would moreover concentrate on the plants and creatures living on the seabed, making a special effort to investigate the unplumbed depths of the oceans.

Bruun set up the necessary network in the scientific community, while Mielche acted as fund-raiser. This brought Mielche into contact with a group of adventurous Danes – including Leif B. Hendil, who came up with the idea of a Danish Expedition Foundation. Hendil's intention was that this foundation would raise money for Danish expeditions from Danish expatriates, who could donate goods and produce for sale on the war-starved Danish market. His plan was put into effect right after the end of World War II, with the Danish state permitting the duty-free importation of goods donated to benefit the Expedition Foundation. Under this programme, cigarette sales alone at these "legal" black-market prices brought in almost DKK 1.2 million.

All of those engaged in planning the expedition agreed that because of Denmark's immediate surrender to Nazi Germany in 1940 and the government's policy of collaboration with the German occupation forces up until August 1943, it would be important after the war to rebuild Denmark's national self-esteem and its international reputation. Mielche was, moreover, convinced that a second Galathea Expedition would not only improve Denmark's image, but also bring scientists closer to "the man on the street" – whom Mielche said had no interest whatsoever in the natural sciences, finding Tarzan more amusing than Einstein, and seeing Superman as far more advanced than Niels Bohr. Mielche considered it a great shame that scientists and the general public lived in two separate worlds, for "behind the great, grey wall of science there are hundreds of adventures that are fully as exciting as Tarzan's crocodile hunts in a synthetic jungle, and more romantic than the blonde secretary getting engaged to the boss's blue-eyed son".[537] The Galathea Expedition was one such adventure.

The scientific results. Whether the expedition's scientific work and findings were just as exciting as so many Tarzan stories is not for us to say, but Mielche's three books about the expedition make it amply clear that occasionally even he found it difficult to see the adventure and romance in the perpetual trawling, the murky, muddy water and the endless mounds of material the expedition continued to dredge up.[538] The expedition's chief objective was to collect biological samples of life from the deep sea and the ocean floor. The expedition participants also carried out investigations involving hydrographics and Earth magnetism, as well as collecting samples in coastal areas and on land wherever this was possible and desirable. Finally, while sailing in the Pacific Ocean the expedition was associated with ethnographical, zoological and botanical studies conducted on Rennell Island, located in the Salomon Islands. The following are just a few of the expedition's highlights and most significant results.

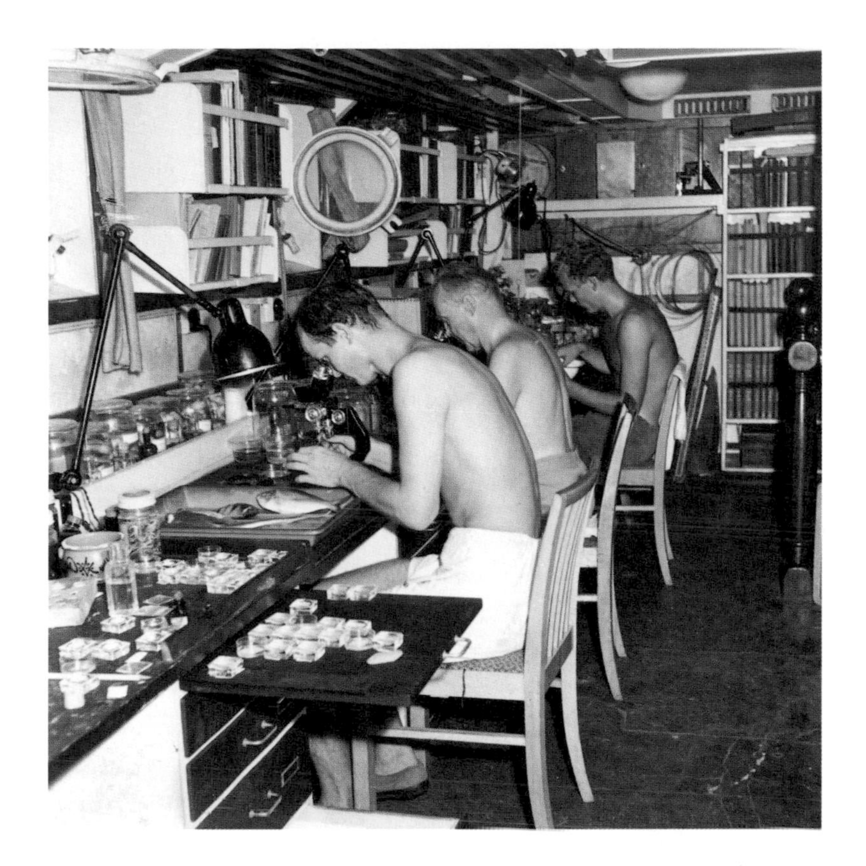

22.3 *At work in the laboratory of the* Galathea. *The ship's scientific crew eagerly inspected every trawl that returned from the deep undamaged. The contents were initially sorted on deck, then moved to the laboratory for processing. In the foreground the zoologist Torben Wolff, at centre the zoologist Paul Lassenius Kramp, and in the background the laboratory assistant Svend Aage Horsted. Following meticulous on-board examination, the material was preserved in glass containers and shipped back to Denmark for further study. This Galathea expedition (the second of three to go by that name) brought back a total of 12,000 samples.*

Several different types of trawls and scrapers were used to gather specimens. One was the so-called Petersen bottom-sampler, named after Carl G. J. Petersen, who had used it to determine the volume of available plaice food in the waters of the Limfjord in North Jutland. The Galathea Expedition used the Petersen bottom-sampler to determine the density of animal life on the ocean floor. All in all, the expedition carried out 28 successful quantitative seabed samplings at depths below 2,000 m, retrieving 7 of these from trenches more than 6,000 m below the surface. No one had expected to find such a relative wealth of life forms in the deep, which is why the expedition members were quite taken aback at the results from two bottom samples from the Banda Trench off South-East Asia: Retrieved from 7,280 and 6,580 m, the samples contained 60 and 55 individuals per square m, respectively, representing 35 and 40 different species. The average result for deep seabeds was 10 individuals per square metre. The result obtained by the Galathea scientists was a surprisingly high number compared with the population density just a few hundred metres down in the waters of northern Europe and the Mediterranean Sea, which is only a few times larger. In other words, the Galathea Expedition discovered that the deep oceans were teeming with life, despite extreme conditions: crushing pressure, frigid temperatures and profound darkness. Because of the extreme living conditions in these ultra-abyssal regions, expedition leader Anton Bruun in 1956 proposed calling the animal life found below 6,000 m "the hadal fauna", after Hades, the Greek god of the underworld.

The expedition's greatest achievement was, perhaps, that in the face of enormous technical difficulties, it proved that life – sea cucumbers, bristle-worms, mussels, snails, sea anemones and tiny crustaceans – existed even at the greatest depths then known to man, at about 10,500 m below sea level in the deepest part of the Philippine Trench. By pressurizing samples of seabed material at 1,000 atmospheres, the American microbiologist Claude E. ZoBell, who joined the expedition in Bangkok, demonstrated that the deepest ocean trenches are home to barophile bacteria. These microorganisms not only possess a remarkable ability to convert organic matter without light and under extreme pressure, but may also serve as food for one-celled organisms that are fed upon by higher life forms. The discovery of life at these profound depths earned the expedition considerable attention in the international community of marine biologists and in newspapers all over the world.

The expedition also found an unexpectedly high presence of phytoplankton in the oceans – news that the press found rather less interesting. These plant plankton, which mainly consist of microscopic algae, are found in the upper layers of seawater (from the surface to a depth of about 100 m), and carry a heavy responsibility: They alone manufacture the organic matter that feeds all other life in the ocean. Scientists already knew of the algae-production mechanism in coastal areas, but they were not aware of the mechanism at work in pelagic waters, simply because it operates down to far greater depths in the oceans and thus is much smaller and more widely dispersed, rendering standard chemical analytical methods useless. The expedition's determination of the biomass production of phytoplankton was a pioneering effort. The method employed was developed by the botany professor Ejnar Steemann

Nielsen of the Pharmaceutical College in Copenhagen, and it was based on the radioactive substance carbon-14. By measuring how much carbon-14 the algae absorbed during their growth, which took place under strictly controlled conditions aboard the ship, it was possible to quantitatively determine the quantity of biomass being produced in the world's oceans. The results were far higher than anticipated, showing that the production of pelagic plant matter rivalled that occurring on dry land – approximately 40,000 million tonnes of organic matter each year – meaning "there is not the slightest doubt that the most productive areas of the sea can produce as much as our best cornfields".[539]

Publicity. The Galathea Deep-Sea Expedition was a revolutionary effort in combining field science with press coverage and scientific communication. As mentioned earlier, besides the expedition's scientific tasks there was a parallel agenda of representing Denmark and giving the country positive publicity. Everywhere it went, the large expedition vessel *Galathea* and its mixed crew of 90–100 people were the centre of attention, usually inviting locals to "open-ship" events or receptions and making official visits ashore. Sometimes they would show films or participate in activities like garden parties, combined lecture-dinners, excursions and football, or hand out leaflets and other information about Denmark. Another sort of representative role was more directly linked to the expedition's scientific work on board and, as hinted at above, had far less to do with socializing. This was the coverage of the expedition in the Danish and foreign media.[540]

In the early phases of the expedition, the press service was harshly criticized in the Danish media, which accused Mielche of orchestrating a monopoly situation. Mielche and his colleagues also ran into other problems. The first leg of the voyage, hugging the western coast of Africa, did not involve any major scientific activities, and much of the work that had to be done was either representative or trivial or preparatory in nature. No sea serpents turned up, and Mielche and his "man on the street" longed for a taste of scientific adventure:

> During that period we caught not a single animal that was more than 5 centimetres long, but then again, we caught loads of those. Glass upon glass was filled, the mass of mud samples destined for geological and chemical testing in Copenhagen continued to grow. (…) In despair we, in the press service, invented one sensation after another: tartan-patterned flatfish with horse's manes, giant eels that stuck their heads over the rail, discovery of the lost Atlantis, an old copper weathercock brought up in the sled trawl – all of it was callously cast aside, for the press section's way to the telegraph station is paved with matter-of-fact scientists with no sense of drama and sensation.[541]

22.4 *The sea serpent J. C. Jensen-Broby and the good ship* Galathea. *The cartoon's original caption reads: "When the Galathea Expedition was almost sunk. Jensen-Broby, chairman of the Finance Committee, did his best to stop the Galathea Expedition."*

This passage shows that Mielche may have inflated his own expectations, and those of others, as far as the sensational nature of working aboard the *Galathea* was concerned, and it also clearly demonstrates the division of labour between the expedition's scientific staff and the press staff. It was the scientists, not the reporters, who ultimately decided what could go out to the public about the scientific work being done.

The Galathea Deep-Sea Expedition never did find the sea serpent that Bruun had hoped would cross their path, but although the monster remained in the realm of the unknown, it had served a purpose. As Mielche observed in his memoirs:

> And the sea serpent? So sorry, I had almost forgotten it. And yet it had done its duty the very instant we set out from home. It had been the catalyst that got the whole idea going. I take full responsibility for so shamelessly exploiting the poor creature. But Bruun did believe that one day it would be found – and he stood by that conviction until the day he died.[542]

The sea serpent's catalytic effect may have got the expedition going, but it was no safeguard against financial troubles along the way. In the summer of 1950, just months before the expedition's scheduled departure, the Korean War broke out, followed by a bout of high inflation and sky-rocketing prices. And even as the expedition was preparing to investigate the Philippine Trench, the Danish daily *Berlingske Aftenavis* was able to reveal, on its front page, that the expedition would come to cost a further DKK 1 million, which put the whole undertaking at risk of sudden and unexpected termination. The Danish ministry of defence proposed in its application to the parliamentary finance committee that the state cover DKK 825,000 of the amount (of which DKK 600,000 was directly attributable to price hikes), and that the remainder be covered by the Expedition Foundation.[543] This left the government with two options: Either recall the expedition, or recommend that the finance committee allocate the extra funding. The government chose the latter, in light of the fact that the expedition had already come so far that untimely termination would be meaningless. The Danish government later provided another extra allocation, but nevertheless the expedition had to be cut short by about three months because of insufficient funds. Despite its financial problems, the expedition was still a success when it came to scientific results and publicity – which more than 50 years later helped to pave the way for Denmark's launch of a third marine exploration venture in 2006: the Galathea III Expedition, which returned to Copenhagen on schedule in 2007.[544]

A little country in a world of big science*

The Danish state's heavy commitments to the second Galathea Expedition was one of the first signs of a new approach towards publicly financing scientific research projects, following a shift that took place in the post-war years. It was also during this period that Danish scientists and the Danish state became involved in the Conseil Européen pour la Recherche Nucléaire – CERN – one of the first big-science projects in Europe, the organization of which was characterized by its demand for close cooperation between the natural sciences and public and private investors. Historically, the term "big science" was probably

---

* This section was written in collaboration with Klaus Rasmussen and Simon O. Rebsdorf. For a more detailed account (in Danish), see Rasmussen, Rebsdorf and Nielsen 2006. See also Rasmussen 2002.

coined by the American physicist Alvin M. Weinberg, who was head of the large Oak Ridge National Laboratory in Tennessee.[545] Weinberg saw big science as being intimately linked to the enormous role that science itself played in the development of space rockets, high-energy accelerators, nuclear reactors, and similar large-scale installations. Others have since used the term "big science" with a slightly different meaning, to denote the new era's drastic increase in the number of scientific projects, scientific publications and team efforts that have hundreds of contributing participants.[546]

Big science is also a dynamic and variable concept, but it always describes a sort of science that is somehow extensive, voluminous or highly profiled, and historical studies underscore the multiplicity of ways in which science can be "big".[547] The historian of science Peter Galison, for instance, points out that "big" in big science "connotes expansion on many axes: geographic (in the occupation of science cities or regions), economic (in the sponsorship of major research endeavours now costing on the order of billions of dollars), multidisciplinary (in the necessary coordination of teams from previously distinct fields), multinational (in the coordination of groups with very different research styles and traditions)."[548] By virtue of its magnitude or its importance, big science helps to create a relationship of mutual interdependence between the natural sciences and the society that surrounds them.

One important category of big-science projects mainly has to do with coordinating efforts from many different institutions in order to achieve tangible, specific goals like developing an atomic bomb, or sending someone to the moon. Big science is also used in research projects that are more basic – modern cancer research, the human genome project, the publication of a new star atlas – where the result of the project is most often scientific publications or databases. Finally there is a kind of big science that aims to achieve greater basic knowledge while retaining a significant technological component, and this is where projects that revolve around advanced equipment or machinery like satellites, telescopes and particle accelerators belong.[549] This last category of big science is particularly prone to attract criticism, since the most easily and readily visible elements in such projects are the costly technological facilities, not the scientific results that the machinery can help scientists to discover.

Alvin Weinberg was critical of significant aspects of big science, which he believed was suffering from no less than three serious ailments: "journalitis", "moneyitis" and "administratitis". According to Weinberg, "journalitis" is characterized by the tendency for the tone of scientific articles to lean towards popular journalism when the underlying research is dependent upon massive public support. The two other diseases, "moneyitis" and "administratitis", are also associated with the tendency such projects have to involve excessive costs and a disproportionately large administrative bureaucracy.[550] The enormous technical installations linked with such projects would come to be regarded as monuments of Western civilization, Weinberg believed, even going so far as to compare them with Egypt's magnificent but useless pyramids. Ultimately, science itself would suffer when, at some point, society began to shy away from financially supporting a scientific community that seemed to be doing nothing but building enormous and inscrutable machines.

The history of Denmark's participation in CERN begins shortly after World War II, but as the following account shows, it was not until the late 1960s that Danish physicists began to feel any symptoms of the big-science malaise Weinberg had described.

Prelude to CERN. The scepticism Weinberg expressed in 1961 was virtually unknown among American scientists of the 1940s and 1950s, and during those decades physics in the US took an especially big step towards big science. This trend worried many European scientists, however – not because they feared big science, which was popular on both sides of the Atlantic, but because they felt they were lagging behind for lack of money. The predominant feeling in the US and Europe was that big science was good for physics, but that Europe would have to pull itself together in order not to be drained of talented physicists, who would inevitably seek greener pastures in America if nothing was done. Meanwhile, some of the influential Danish physicists took a less categorical view of the situation. Writing of theoretic nuclear physicists in 1946, Leon Rosenfeld, a Belgian-born professor of physics at Niels Bohr's institute, said that they "would require evidence from a domain of energies still much higher than those we have been considering, but not outside the range of the modern accelerators".[551] In other words, Rosenfeld's message was that there was no reason to panic.

Over the following three years, nuclear and other physicists in Denmark kept their composure, whereas other countries, most notably France, had a growing number of scientists advocating a radical strengthening of European physics. That is why in 1949 and early 1950 French physicists, working closely with officials in the French foreign ministry, sought out their peers in a number of other European countries to propose the construction of a large, joint research laboratory, equipped with experimental nuclear reactors and high-energy accelerators. Their idea met widespread scepticism. One of the coolest receptions came from the Dutch president of the International Union for Pure and Applied Physics (IUPAP), the physicist Hendrik Kramers, who had been

Niels Bohr's first assistant back in the 1920s. Kramers mainly feared that a joint laboratory would draw some of the already sparse financing away from the national research programmes. The French idea also got the cold shoulder in the UK. Being Europe's best-outfitted player in terms of the technical hardware needed to compete in the development towards big science in physics, the UK did not see it as strictly necessary to cooperate with continental Europe. What is more, the radar and nuclear weapons programmes of World War II had linked British scientists closely with the US, which is why they found it at least as natural for them cooperate with the Americans.

One thing seriously hampering the plans for a joint European effort in nuclear and particle physics was that in the early post-war years, the US generally discouraged scientific initiatives that might threaten America's desire to hold on to its nuclear secrets for as long as possible. Consequently, the Europeans would need an opening from the Americans before they could make their first move.

A project takes shape. Because most prominent American scientists were closely associated with colleagues in Europe, it is understandable that many US researchers wanted Europe to become part of the modern, cost-intensive development of physics that was under way. The resulting pressure on the US government certainly contributed to a statement made by the Americans at UNESCO's fifth general assembly in Florence in June 1950, which urged the Europeans to establish a large regional research centre. Responsibility for bringing the proposal to fruition was offered to the French physicist Pierre Auger, the head of UNESCO's science department. Auger eagerly took up the challenge, working resolutely throughout the 1950s to realize the huge project, whose ideological underpinnings partly came from the original French idea.[552] The great UNESCO-based project concentrated on setting up a joint (Western) European laboratory housing the most powerful particle accelerator in the world. No Danes were present when the project was discussed at several meetings in 1950 and 1951.

Nevertheless, the Danes – more specifically Niels Bohr and a few other from the circle around him – kept up with developments thanks to Kramers. Bohr and the other Danes supported the idea of a joint European effort, but they also agreed that the stated aims of the UNESCO project were not the right way to go about things: The massive scale of the project was simply unrealistic. They therefore suggested that the project begin by constructing a small accelerator before possibly, at a later stage, moving on to building the world's largest. The smaller-scale installation would enable scientists to gather valuable information while giving them time to thoroughly consider the required specifications for the next machine.

Auger accepted the idea of first building a smaller accelerator, but he maintained that the large machine initially planned was to be begun immediately afterwards.[553] At almost the same time, a predictable quarrel broke out over the new laboratory's location. The cities of Geneva in Switzerland and Mulhouse in France quickly put in bids to host the facility, whereas Kramers and Bohr argued that the best place for the laboratory would be close to an existing, recognized research institution – in which case Niels Bohr's institute in Copenhagen fit the bill very nicely. While the different campaigners parried, fell back and regrouped, the Danes increasingly began to have second thoughts about the ambitious undertaking, and scientists in Britain, the Netherlands, Norway and Sweden also began to voice doubts as to whether the UNESCO project was what European nuclear physics truly needed. Bohr and Kramers raised the ante by recommending that the entire project be mothballed, partly to bide their time and partly to obtain funds to intensify intra-European exchange programmes for scientists. Such an effort, they argued, focusing on people instead of machines, would provide European nuclear physics with new and much-needed momentum while the scientific community deliberated what installations would actually be necessary in the long term.[554]

In short, the Danes involved in the discussion were in the vanguard of a group of scientists who were attempting to slow down the UNESCO train, which was going full steam ahead – though they would soon learn that their efforts were too little, too late. Denmark was really the only country where the scientists internally agreed that the UNESCO project was overly ambitious and premature. Sweden, Norway, the Netherlands and the UK all had roughly equal numbers of supporters and opponents of the grand project. In this context Denmark was remarkably unified in its messages, which must particularly be attributed to Niels Bohr's unique position in the Danish science community.

Once it became clear that no joint opposition against the UNESCO project could be organized, the project pushed on, virtually unaffected, towards the realization of a great, joint European research laboratory. Facing this situation Denmark, unanimously urged by its scientists, chose to join up for the initial stages of the European project, the motivation being that cooperation was desirable in all events, and that the nation's reputation would suffer if the Danes remained on the sidelines.

Denmark's fight for a piece of the pie. During the first half of the 1950s the contents of the UNESCO project were determined at a series of meetings in the

CERN Council. Niels Bohr and his institute were Denmark's greatest asset in these negotiations, but as explained below, Bohr's eminent status could be an obstacle as well as an advantage.

While attempting to modify the project, Bohr and Kramers had primed the situation so that Bohr's institute could play an important role in the project's final formulation. Whereas the other participating countries promised financial support to the UNESCO project, Denmark initially promised to put the Bohr institute's facilities and expertise at the project's disposal – an offer that placed the project's supporters in something of a dilemma. On the one hand, no one wished to reject an offer from Bohr, who was, after all, the grand old man of European nuclear physics. On the other hand, however, the project's most ardent proponents were afraid that a close link between the joint project and Bohr's institute would enable Bohr to swing developments in the direction of more international cooperation among scientists at the expense of the large European accelerator. The result was a compromise, with the Copenhagen institute becoming the seat of a theoretical study group for a number of years, until the central laboratory had become a reality.[555]

Copenhagen fought out the final round of the location battle with Geneva, Paris and the Dutch city of Arnhem. The decisive negotiations were particularly concerned with which criteria would be used to make one city preferable over the others. These criteria fell into two groups that can reasonably be described as the hard, objective scientific criteria and the softer, more general criteria. The first group had to do with the given location's geology, infrastructure and proximity to a scientific centre, while the second group had to do with issues like language, culture and traditions. It was soon ascertained that all four candidate cities largely fulfilled the objective scientific criteria, which brought the softer, more general criteria to the forefront. It is interesting in this context that the Danish negotiators confined their arguments almost completely to the first group of criteria, simply arguing that it would serve physics best if the European laboratory were located near the Institute for Theoretical Physics in Copenhagen. What they did not do, however, was to emphasize the distinctive atmosphere, the "Copenhagen spirit", which many physicists considered a distinctive quality that was unique to Bohr's institute, and which would have been one of the "soft" advantages of a Copenhagen location. Seeing that in due course Geneva was chosen to host the facility – the city's benefits including its traditions for housing large international organizations and its bilingual Franco–German population – it is not unjust to ask whether the Danish negotiators made a mistake in focusing merely on the objective criteria. And yet in all fairness, the answer to this question is "no". Consider the fact that CERN was the first (international) big-science project being co-organized by the Danes, which meant they had no prior experience with the priorities that would be decisive in shaping the future of such a project. Then consider the fact that the "soft" qualities of the Copenhagen location were brought up not by the Danish delegation, but by negotiators from other countries, which if anything would surely have reinforced the Danish candidacy.[556]

In relation to these analyses of negotiation tactics, it is reasonable to adjust our optics to take in a wider perspective and ask whether the Danish negotiation strategy might have been completely inconsequential to the outcome of the process. The Polish-born physicist Stefan Rozental, another of Bohr's close associates at his institute in Copenhagen, has expressed that the laboratory's location was really decided beforehand, since France insisted that it must be located in a French-speaking area.[557] This single reason is hardly sufficient to explain the outcome entirely. However, when coupled with the misgivings that many European physicists, especially the younger ones, had about seeing Bohr as the organizer of a large laboratory, and with the security concerns so typical of the times about Copenhagen's vulnerability, due to its close proximity to the Iron Curtain, a fairly clear and consistent picture begins to materialize. In retrospect it is highly unlikely that CERN would ever have been located in Denmark, no matter what arguments the Danish negotiators had used.

Even so, the negotiations were by no means unimportant. The Danes, unable to have the laboratory come to Copenhagen, pragmatically sought to make the best of the situation. One concrete result of the Danish efforts was that the theoretical study group in Copenhagen was allowed to continue its work far beyond the 18 months originally commissioned. One can, of course, choose to see this as a modest consolation prize for a thwarted candidate. However, given that during its existence from 1953 to 1957 the Copenhagen study group formed the framework around most of CERN's scientific activities, the Danish physicists had good reason to interpret the negotiation results as positive for their science – although they would obviously have preferred to see the laboratory itself located on their own home turf.

The early years of a permanent CERN. The theoretical study group in Copenhagen was formally dismantled in 1957, by which time the construction of the laboratory in Geneva was so advanced that it could house a new and larger theoretical group. Back in Copenhagen, the work that had so far taken place under the auspices of CERN was carried on by a new body called the Nordic Institute for Theoretical Atomic Physics (NORDITA). Here the joint European cooperation was replaced by a joint Nordic cooperation, following an

22.5 *Throughout the 1960s, CERN's most important installation was a proton synchrotron (PS, for short) that could accelerate protons up to 28 GeV (billion electron Volts). Here a small section of the circular tunnel that houses the particle accelerator. The PS was inaugurated in 1960 and long remained the most powerful piece of equipment at CERN.*

idea that had arisen as early as 1953, but been shelved so as not to damage the Nordic influence on CERN during its early stages of development. The disbanding of the theoretical study group and founding of NORDITA largely put an end to the Danish players' involvement in CERN. The people around Bohr applied themselves to new NORDITA-related tasks, and were also spending time and energy on the Risø project, which is described at length in chapter 23. This came to overshadow the CERN activities, and although naturally Denmark was represented at the meetings in the various CERN bodies, the country was no longer the sort of proactive presence it had been.

We believe this is linked to the sort of work CERN was mainly focusing on around 1957–1960: handling the technological challenges that arose from the construction of the small and the large accelerator. Something the Danish physicists definitely were *not* was "machine physicists" on the scale that these accelerators demanded. The Danes simply had nothing to offer the project at this stage, when the interplay between physics and technology, between scientists and engineers, was of paramount importance. Another reason why very few Danish physicists were involved in the research going on at CERN prior

Country	Ratio
Denmark	0.3
Norway	0.2
Sweden	0.4
UK	3.3
France	4.0
Italy	2.8

Table 22.1 *Ratio between contribution to domestic CERN-related research and membership contribution to CERN. (Source:* Nordiske Betænkninger *1964:3, pp. 8-9).*

to 1960 was that the organization was chiefly investigating nature's smallest components – the elementary particles – rather than atomic nuclei, which at this point were the primary area of interest for the Danish physicists working at Bohr's institute in Copenhagen.

Little wonder that the Danes' sparing use of CERN caused doubts about whether the country's financial contribution to the project was justified. Because the other Nordic countries were discussing similar issues, the Nordic Council created a task force in 1963 to examine the matter. The chairman was Gunnar Källen, a professor of physics at Lund University, and the two Danish members were the physicists Bernard Peters and Knud H. Hansen, both employed at the Niels Bohr Institute. The task force submitted its report in 1964, finding that relative to their membership contributions, Norway, Sweden and Denmark each spent only one tenth of what the large countries were spending on CERN-related corollary research – that is, domestic research projects to ensure efficient utilization of the country's contribution.[558]

This disturbing circumstance had contributed greatly to the small countries' limited involvement in CERN's activities. The report recommended that each of the Nordic countries provide larger corollary research grants, and that the Nordic region build an accelerator with an energy level of about 10 GeV, which would better enable physicists in the Nordic countries to benefit from CERN's large accelerators. Though the former recommendation had some effect, the proposal for a Nordic accelerator was never realized.

Even while the task force was preparing its report, Denmark's efforts to get more out of its CERN membership were building up speed. The Danish government had just set up what became known as the Accelerator Committee in

22.6 *The 1960 inauguration ceremony for CERN's proton synchrotron. Niels Bohr, seated on the podium, on the speaker's right when seen from the audience, had the privilege of symbolically starting up the large new accelerator – despite his initial opposition to CERN in the form it ultimately took.*

November 1963. One important outcome of this committee's work was that from 1964 onwards, the Danish Finance Act contained an item called "CERN corollary research", designating an amount that the Accelerator Committee was free to use as it saw fit. Although the amount never surpassed ten per cent of Denmark's annual payment to CERN, it had a significant positive impact over the following years on the Danish physicists' ability to valuably participate in the various research projects going on in Geneva.

Likewise, 1964 also saw the commencement of ISOLDE – one of the great Danish–Nordic projects at CERN.[559] The Danish professor Peder Gregers Hansen was one of the driving forces behind ISOLDE, and his group wished to investigate the physical properties of extremely unstable nuclear isotopes. This could be done by coupling an isotope separator directly to the CERN synchro-cyclotron and employing a technique of Isotope Separation On Line – giving the project its name. But why did this mainly Nordic research project not get under way until the mid-1960s, instead of pre-1960? For one thing, it was only during the 1960s that new and more efficient solid-state detectors were developed, providing greatly improved possibilities for analysing the often extremely complicated decay patterns from the unstable isotopes ISOL-DE was studying. For another thing, until the early 1960s the CERN synchro-cyclotron had been occupied by groups of researchers who were using it to study particle physics. At that stage, when CERN's proton-synchrotron was ready to run actual experiments in 1961, most of the activities relating to particle physics were transferred, making room for experiments in nuclear physics at the synchro-cyclotron.[560]

CERN moving into the 1970s. The greater Danish involvement in CERN during the last half of the 1960s meant that physicist working in other areas than nuclear and particle physics began to take an interest in what was going on at the pan-European laboratory – and to demand greater influence on the ever-rising Danish contribution. One sign of this was that in 1969 Jørgen Bøggild, an official Danish delegate at the CERN Council meetings, was obliged to spend far more time than before explaining and defending CERN's ambitious expansion plans at home.[561] The most significant element in these plans was the construction of two immense – and immensely expensive – accelerators. One was an ISR (Intersecting Storage Rings) installation, which could be used to study collisions between protons and antiprotons moving towards one another at high energies. The other was an SPS (Super Proton Synchrotron), which, as the name indicates, largely resembled the facility's proton synchrotron, but produced particles with energies about ten times higher. The ability to work at such high energy levels was crucial, the CERN enthusiasts argued, if Europe wished to compete with the US, the USSR and Japan in investigating nature's smallest building blocks. As they explained, the smaller the details one wishes to study, the higher the energies applied to the accelerated particles must be, and the greater the investments needed. The opponents of CERN did not dispute these facts, nor did they deny the potential for valuable technological spin-offs that such advanced projects might hold. What the opponents typically did do, however, was to argue that the huge sums in question would be better spent on other scientific projects, or on projects that would benefit society at large in a far more immediate and obvious fashion. This debate, conducted by scientists, politicians and representatives of the general public, was by no means unaffected by the prevalent criticism during those years of large scientific institutions, be they Danish or international. In the late 1960s, the goals and financial exigencies of science had become a topic of public debate in Denmark; a development that had already been presaged in the US when Weinberg voiced his misgivings about big science in 1961. By the end of the five decades under review, Denmark's scientific community was forced to accept that the implications of big science now reached far beyond science itself – a fact that became increasingly apparent in the ongoing debate about Denmark's continued commitment to CERN.

23 Pure and applied science outside the universities

After World War II science acquired an exceptionally high status as the key driver of social progress in general and technological progress in particular. One factor contributing to this was a report prepared by the influential engineer and administrator Vannevar Bush, who had also mobilized scientists during the war for the Manhattan Project. Shortly before the war ended he had submitted his report, entitled *Science – The Endless Frontier*, to the American president Harry S. Truman.[562] The central point made in this report, which was prepared by a number of America's top scientific and medical researchers, was that the development of the radar and the atomic bomb had clearly demonstrated the scientific community's ability to solve seemingly insurmountable technological problems, provided it was given free rein and sufficient resources to take up the challenge. There was no doubt that similar results would be achievable in peacetime. The philosophy of the Bush report was that modern technological advances are, fundamentally, dependent upon new scientific results. "New products, new industries, and more jobs require continuous additions to knowledge of the laws of nature, and the applications of that knowledge to practical purposes," as the report's summary states. The best thing a modern, forward-looking society could do was, therefore, to generously support basic scientific research, and then to make sure that industry was given the best possible conditions for translating the scientific results into new, scientifically based products. On the other hand, Bush reasoned, the national authorities had to be wary of supporting both basic and applied science, since if the two types of research were present in the same institution, then "applied research [would] invariably drive out the pure".[563]

The Bush report's one-sided emphasis on the decisive importance of pure science in advancing society became trend setting in post-war America and in Europe. Within a few years, all of the countries in the West had multiplied public spending on scientific education and basic research several fold. Further evidence of the Bush philosophy's clout is that many of the atomic energy commissions and experimental nuclear facilities set up in Western countries ten to fifteen years after the war were managed by famous scientists. Examples

23.1 *The main building of the Carlsberg Laboratory, from 1897, made up its entire physical framework until an extension was built in 1976.*

include John Cockcroft in the UK, Frédéric Joliot-Curie in France, Werner Heisenberg in Germany, and Niels Bohr in Denmark.

Not until the mid-1960s were any serious doubts officially raised as to the credibility of the Bush report's claim that modern technological progress is primarily driven by free and independent basic research. When at last the claim was examined, however, it was done quite thoroughly, in a large-scale study called *Project Hindsight*, which analysed the most significant sources for numerous fairly recent technological innovations within the American defence industry. The study's conclusions were striking: Less than 1 percent of the innovations examined could be attributed to a scientific breakthrough stemming from pure science conducted relatively recently, meaning within the past 20 years, whereas the majority of innovations had resulted from applied science. *Project Hindsight* and similar research impact assessments triggered a development that led to a dramatic shift over the following decades in the allocation of public funding, which was moved from unsupervised basic research over to basic research that was utility-oriented and conducted within special research programmes overseen by the national authorities.

In retrospect, it seems somewhat curious that the Bush report carried so much weight. Ever since the earliest industrial research laboratories were created in the late 1800s, leading scientists in private and public facilities had been aware that the relationship between technology and science was considerably more complex than the Bush report implied. This is borne out in recent historical studies of the work done in the Research and Development (R&D)

departments of large companies – for instance the chemical corporation Bayer in Germany, and the Bell Telephone Company and General Electric in America – prior to a series of technological innovations that would have a major impact on modern society as we know it today.[564]

Even though the scientific research departments in Danish companies, private and public, had usually been significantly smaller than corresponding departments in the foreign companies mentioned above, it is easy to find Danish examples of non-academic undertakings whose R&D efforts shed an interesting light on the relationship between basic and applied research. For our purposes we have chosen three example – two financed privately and one publicly – which together draw a varied picture of non-academic research in Denmark during the period under review. The private undertakings are the Carlsberg Laboratory and the Nordic Insulin Laboratory, and the public organization is Risø Nuclear Research Centre (now Risø National Laboratory).

As the review of these examples will show, the research going on from 1920 to 1970 at these three Danish institutions was very different in nature, but in all three locations it was quite significant and can best be characterized as "utility-oriented basic research", having been conceived in the conviction that the relevant research held an enormous practical and financial potential.[565] The three examples also demonstrate that the research actually going on at a given institution is often influenced by that institution's own history, which may include provisions laid down by its founder and social circumstances determining its establishment.

The Carlsberg Laboratory: envisioning utility-oriented basic research*

Located on the extensive brewery grounds in Valby, then on the outskirts of Copenhagen, the Carlsberg Laboratory was founded in 1875 by the renowned brewer and philanthropist J. C. Jacobsen. According to his visionary ideas, the laboratory would pursue basic research that was linked to the brewery operation.[566] The laboratory consisted of two departments, one for chemistry and one for physiology, each with their own department head. The guidelines for the laboratory's work were set out in its statutes, which the foresighted brewer

---

* This section was written in collaboration with Henrik Knudsen. For further details (in Danish), see Knudsen and Nielsen 2006.

himself had formulated. As the Danish historian Kristof Glamann has pointed out, Jacobsen was inspired in a great many things by the French chemist Louis Pasteur and his thoughts on the interaction between science and practice.[567] There was one serious problem, however, in that the statutes did not clearly define the degree of freedom the laboratory heads would have to act and to decide the direction of their research. Studying the processes of malting, brewing and fermentation was, incontestably, their main concern. However, to effectively attract the brightest minds, Jacobsen had also explicitly written into the statutes that in addition to the brewery-related work, the heads of both laboratories would also have enough autonomy to develop their scientific talents in other directions, leaving the question of how the two contrasting provisions were to be weighted.

On the one hand, the working methods and the atmosphere in the laboratory reflected the traditional academic style. Pursuant to Jacobsen's instructions – and contrary to conventional industrial practices – all of the laboratory's findings were to be made public. There was additionally a strong emphasis on the staff's participation in study trips and on their open sharing of knowledge with colleagues. Moreover, routine operational analysis work lay outside the laboratory's scope of activity, since Jacobsen had already set up a special laboratory for this purpose in 1886. On the other hand, as noted earlier, Jacobsen did intend for the laboratory to give first priority to brewery-related research. In order to underscore this he required that laboratory assistants must spend three months working in the brewery itself, which taught them the fundamentals of brewing before they went to work in the research laboratory.

In fact, in its early years the laboratory's research concentrated almost exclusively on issues related to beer brewing. Its first results counted discoveries that were very valuable to the brewery, technically and economically. The chemist Johan Kjeldahl, for instance, soon succeeded in developing an improved analytical method for estimating the nitrogen content in organic material. Widely known as "the Kjeldahl method", this important technique soon became universally adopted. And the mycologist Emil Christian Hansen was the first to successfully culture a pure strain of brewers' yeast, distributed under the name of "Carlsberg yeast". Using pure strains made for much better products and more stable fermentation processes. Hansen suggested to Jacobsen that the brewery immediately establish a factory to produce and distribute pure strains to brewers all over the world. Jacobsen did not agree, however, and maintained his position that whatever was discovered at the Carlsberg Laboratory should be shared with the world. Adhering to this principle was

Chemistry Department	Physiology Department
Johan Kjeldahl (1875–1900)	Rasmus Pedersen (1876–1877)
Søren P. L. Sørensen (1900–1939)	Emil Christian Hansen (1877–1910)
Kaj U. Linderstrøm-Lang (1939–1959)	Johannes Schmidt (1910–1933)
Martin Ottesen (1959–1976)	Øjvind Winge (1933–1956)
	Heinz Holter (1956–1971)

Table 23.1. *Heads of the Carlsberg Laboratory's two departments.*

tantamount to giving away large sums of money, but Jacobsen did not mind. On the contrary, well pleased with the work of his scientists, he had a new, expensive and impressive building constructed in 1897 to house the Carlsberg Laboratory.

From the 1920s and into the 1970s the Carlsberg Laboratory enjoyed considerable international recognition, not only for its industrial contributions but also, and perhaps most of all, for its activities in basic science. Throughout this period the laboratory remained able to attract the very best researchers in the scientific fields it studied. At the same time, the laboratory's intimate link with the brewery was weakened in several decisive respects. One of the first indications of this disassociation was that the laboratory assistants' mandatory training period in the brewery was discontinued, which happened during World War I.

The two scientists heading the departments at the beginning of the period, the chemist S. P. L. Sørensen and the biologist Johannes Schmidt, were both excellent researchers. Sørensen specialized in protein and enzyme chemistry, and his investigations into the synthesis of amino acids, the building blocks of proteins, earned him international renown. Sørensen was particularly interested in the effects of changes in hydrogen-ion concentration on enzymatic processes, and in 1909 he introduced the concept of pH as an easy and convenient method of expressing hydrogen-ion concentrations. He also made significant improvements in the colorimetrical methods for pH determination, introducing a new system of standard indicators covering the pH spectrum from zero to fourteen. Sørensen was a Danish pioneer in the field of biochemistry, and after the end of World War I, foreign scientists flocked to the laboratory to study Sørensen's techniques. Even though the nature of Sørensen's research unmistakably leaned towards pure science, his efforts also touched upon areas

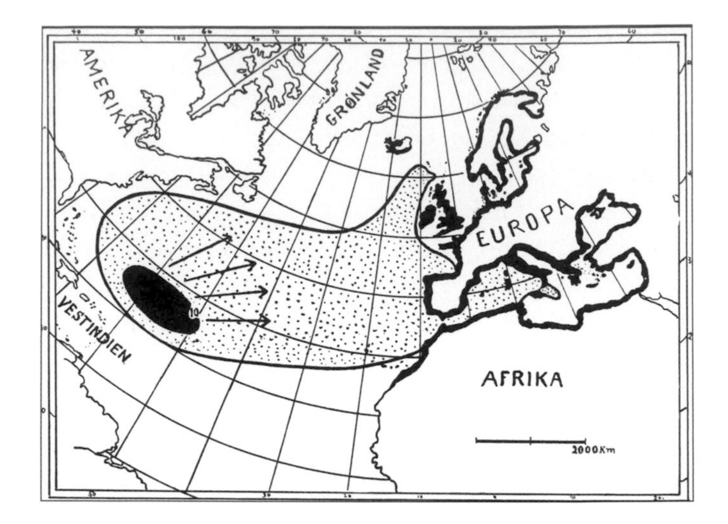

23.2 *The distribution of the European eel. The spawning grounds in the Sargasso Sea are indicated in black, as are the countries in which the eel is found. As the arrows indicate, the tiny eel larva journey from the waters where they were spawned towards the coasts of Europe and Africa.*

that had great practical significance for the brewery. In this context, the ability to measure pH throughout the various stages of beer brewing was highly significant, and Sørensen's methods for measuring pH values were soon incorporated into the production process.[568]

Things were very different during those years at the physiology department, where Johannes Schmidt, a biologist to the core, strained the laboratory's statutory provisions to their limit. Schmidt, like Sørensen, was an internationally recognized scientist. Early on he carried out cultivation tests on hops, but his actual scientific interests lay in marine science and oceanography. He later began to cultivate this area with great enthusiasm, concentrating all of his energy there after World War I. Each year, almost without exception, Schmidt spent several months on scientific expeditions. He was the first to substantiate the claim that the European eel breeds in the Sargasso Sea, as shown in figure 23.2. His achievements also include the discovery of an oceanic ridge on the floor of the Indian Ocean, now known as the Carlsberg Ridge.

Schmidt's marine-biology research was completely unrelated to the practical problems of the brewery and so went against the spirit of the laboratory's status. This predicament was resolved in 1931 with the founding of a special sub-department dedicated to yeast physiology, which was meant to balance out the lack of pertinent research resulting from Schmidt's forays into marine biology. During the 1920s, the brewery's ordinary R&D laboratory had also been substantially expanded. In other words, what the Carlsberg Laboratory did not deliver, the brewery was prepared to obtain elsewhere.

The account of Johannes Schmidt's marine biology research illustrates how difficult it was to preserve the delicate balance between pure science and applied research that was stipulated in the Carlsberg Laboratory's statutes. In practice, pure and applied science were segregated, and each was given its own institutional framework. History repeated itself in the laboratory's next generation, though the roles were reversed. Schmidt's successor in the Physiology Department, Øjvind Winge, who was a botanist and geneticist, was committed to practical cultivation work, and he was deeply involved in producing new varieties of hops, aspiring to create one that would be better suited to the Danish climate than traditional varieties – though his efforts were never successful. Denmark was simply not a good place to grow hops. Within genetics, however, Winge achieved an abundance of seminal results, working with his assistants to become the first to prove that yeast can reproduce sexually and not just multiply vegetatively, as scientists had previously thought. The discovery of sexual reproduction among such organisms had suddenly made it possible to systematically cultivate completely new strains of yeast.

While Winge's main goals were utility-oriented, the activities of Kaj Ulrik Linderstrøm-Lang, who was appointed head of the Chemistry Department in 1939, were more tangential to the brewery's activities. Linderstrøm-Lang was a highly gifted scientist who made many original contributions to protein chemistry, most significantly the introduction, in 1951, of the distinction between primary, secondary and tertiary structures of proteins – a distinction that is still in use today. After World War II, Linderstrøm-Lang mainly focussed on the factors that stabilize protein structures on the secondary and tertiary levels. He and his colleague and close associate, the Austrian born chemist Heinz Holter, share the credit for making the Carlsberg Laboratory an international centre for protein chemists in the post-war years, with researchers from Denmark and abroad making their way to Valby to discuss their projects and problems. Linderstrøm-Lang's numerous talents also included fund-raising. While he was department head, many of the laboratory assistants' wages were being paid with external funding contributed by sources like the Rockefeller Foundation and, later, the National Institutes of Health in the US. The department's research at the time was of the highest scientific quality, but its actual value to the brewery itself was negligible. And, in the phrasing of the sitting chairman of the laboratory's board of directors, because his findings shed more light on the scientific discipline of human and animal physiology than they did on the practical work with yeast, barley and hops, perhaps the "greatness and scope of Linderstrøm-Lang's later work" was not always "fully appreciated" by the board. It should also be noted that Linderstrøm-Lang's successor, Martin Ottesen, pursued a research programme much more in tune with the brewery's core business.[569]

23.3 *The Carlsberg Laboratory had many visitors from Denmark and abroad under the leadership of Søren P. L. Sørensen and Kaj U. Linderstrøm-Lang. This aspect of laboratory life must have suited Linderstrøm-Lang's gregarious nature, and he is seen here (at far right) in the late 1950s with a mixed group of staff and guests in the assistants' laboratory. Fifth from left is the Austrian-born chemist Heinz Holter, who became affiliated with Carlsberg Laboratory in 1930. Holter and Lang also co-authored a number of articles on microchemistry published between 1934 and 1943. From 1956 to 1971, Holter headed the Carlsberg Laboratory's physiology department.*

As an industrial research institution the Carlsberg Laboratory had certain advantages: It had ample funds, and working conditions there were unconstrained. This combination gave the institution greater flexibility than corresponding institutions that were purely academic, where changes are often slower in coming. That is why in several instances, for particular fields, the Carlsberg Laboratory played an active role in the national process of institutionalization and the formations of disciplines. As a case in point, S. P. L. Sørensen can be considered Denmark's first professional biochemist, while Øjvind Winge's position and work in genetics was a mainstay during the emergence and consolidation of genetics as a university discipline in Denmark.

The Carlsberg Laboratory was enormously successful. Far more so, in fact, than most other academic institutions during the same period. However, when considered as a model for organizing an industry-related, utility-oriented undertaking focused on basic research, Jacobsen's vision was not always easy to put into practice. The interests of science and business often pull in opposite directions – although at times they can go hand in hand, as exemplified in the history of the Danish Nobel laureate August Krogh and what came to be known as "the Danish insulin adventure".

The Danish insulin adventure: two academics as entrepreneurs*

On 12 December 1922, August Krogh and his wife Marie – a practising physician, and a prolific medical researcher with nearly 70 scientific publications to her name – returned to Denmark after a lengthy lecture tour of the United States and Canada. Among the things they brought back was the exclusive right to manufacture insulin in Scandinavia. The very next day, August Krogh met with a number of people to discuss the specific details of how to get an insulin production line up and running. Thus began one of the most amazing industrial export ventures in twentieth-century Denmark.[570]

Insulin is a crucial metabolic regulator hormone in the body, and it is produced in the pancreas of humans and animals. The most common type of diabetes is caused by a decline in the body's production of insulin, which is why people with diabetes must be supplied externally with insulin to ensure their long-term survival. In the early twentieth century, scientists were homing in on the discovery of insulin in several locations. It was known that the pancreas produced a substance that was capable of lowering the blood-sugar concentration in diabetics. The first ones to demonstrate and isolate insulin in 1921–1922 were a little-known group of scientists based in Toronto, Canada.[571]

August and Marie Krogh heard of the Toronto group's discoveries during their North American tour in the autumn of 1922, at which time preparations were already under way for the industrial production of insulin in Canada, the US and the UK. Krogh was interested in insulin, not least because of his wife Marie, who had herself been diagnosed with diabetes and urged him to write to the head of the Toronto group. While visiting the group in Canada, he successfully negotiated exclusive rights as the sole manufacturer of insulin in

* This section was written in collaboration with Henrik Knudsen. For further details (in Danish), see Knudsen and Nielsen 2006.

Scandinavia. After the couple's return to Denmark, August Krogh teamed up with a young physician named Hans Christian Hagedorn, who had earned his doctorate the previous year defending a dissertation on diabetes, and because Krogh soon resumed his regular scientific work it was Hagedorn who was in charge of organizing the actual production of insulin.

On paper, the process of manufacturing insulin seems straightforward: The raw material, pancreatic glandular tissue, is minced, and the insulin extracted using alcohol acidified with hydrochloric acid. The insulin can then be obtained by evaporation and crystallization and subsequently purified, though the details of the process can vary. This is where a problem arises, namely that of rendering the insulin in a form that is practical and appropriate for those using the preparation. Finally, that problem overcome, the product must be standardized to ensure uniform dosage.

In the winter of 1923 the manufacture of insulin was still a new activity. There were many unsolved problems, and there was a huge need for further research and testing. It was one thing to have a process that worked in the laboratory, but quite another to set up a smoothly running, full-scale production line. Manufacturers in the US and Canada, for instance, had seen their extraction of insulin inexplicably fail for months at a time. The pancreatic material used for extraction in the US and the UK came from slaughtered cattle, but the Danes quickly changed over to pancreases from pigs, which were easily obtainable as a result of the country's large pork industry. The Danish team also ran into problems in its developmental work, but they were solved one by one, thanks to the team's solid insights into biochemistry, biology, pharmacy and medicine, and to their abundant supply of raw materials, enhanced with a fair measure of technical and organizational talent.

The many experiments were expensive. August Kongsted, a co-founder of the leading Danish pharmaceutical company Løvens Chemical Factory, agreed to supply the project with additional capital in return for having the Løvens brand – a lion – stamped on the finished products. The Nordic Insulin Laboratory was now a reality, and soon Hagedorn and Krogh were able to achieve a higher yield and a purer product than their foreign counterparts. In late 1923, diabetes patients in the US had to pay about 4 Danish kroner (DKK) for 100 units of insulin, and patients in the UK about DKK 3.50, whereas the same dose in Denmark cost only DKK 2.50.

Early insulin treatment had the patients inject themselves with a fluid that contained tiny insulin crystals. The effect was relatively brief, however, as the microscopic insulin crystals were rapidly dissolved, making it necessary for the patient to have another injection after just 8 hours. Obviously, any manufac-

23.4 *August and Marie Krogh (at left, with case) disembarking in Copenhagen on 12 December 1922 following a lecture tour in North America. One of the things they brought back was the exclusive right to produce insulin in Scandinavia, which would set in motion a unique Danish industrial adventure. This photo appeared the next day in* Politiken *("America-Day in the Free Port") and* Berlingske Tidende *("Danish Science Celebrated in America").*

turer who could prolong the effect of their insulin would have the upper hand, which had teams around the world racing to solve the problem. The Nordic Insulin Laboratory passed the finish line first, in the mid-1930s. By binding the insulin to protamine – a cationic peptide molecule usually found and isolated from fish sperm – the laboratory succeeded in producing a sparingly soluble crystalline compound that remained effective for up to 12 hours after injection, limiting treatment frequency to two injections per 24-hour period instead of three.

The Danish team also suffered serious setbacks, however. A dispute arose in 1924 between Hagedorn and one of the laboratory's key employees, the pharmacist Thorvald Pedersen, who was dismissed without further ado. In his capacity as pharmacist he had signed a confidentiality agreement stating that he would not use what he had seen in the laboratory to produce insulin on his own. Harald Pedersen, who was Thorvald's brother, also worked at the laboratory, but because he was a technician he had not signed such an agreement.

Following the break with Hagedorn, the Pedersen brothers therefore set up as independent insulin manufacturers in a newly built facility they called the Novo Therapeutic Laboratory. The Pedersen brothers offered to sell the entire operation to Hagedorn, who declined. This sparked an intense rivalry between the two companies that lasted until 1989, when they merged under the name of Novo Nordisk.

But what was it that made a well-consolidated university professor and Nobel laureate like Krogh set aside his own pure scientific research and launch himself into a business-oriented scientific venture that must initially have seemed quite unmanageable? Krogh himself stated a series of technical arguments that favoured starting an independent production of insulin in Denmark: the existing lack of reliable shelf-life information, poor product transportability, uncertain supplies of imported insulin. Krogh was also concerned with keeping amateurs and frauds out of insulin manufacturing. Another weighty argument may have been the medical needs of his wife Marie Krogh, whose diabetes gave August Krogh a very personal interest in making the new wonder drug available in Denmark.

Finally, there was Krogh's life-long preoccupation with the practical opportunities that his research opened to him – and, indeed, to his country, which he sought to serve through relevant research during both world wars.[572] Money was not the primary motivation for Hagedorn or Krogh, and for many years their production of insulin was a non-profit undertaking. In fact, the Toronto group had stipulated as a requirement that no one was to reap any personal profit from their discovery. Accordingly, the Danish domestic market was supplied with insulin at cost price, and the income from the company's rapidly rising exports was used to set up the charitable Nordic Insulin Foundation. This foundation's legacy, carried on today by the Novo Nordisk Foundation, has led to major contributions to biomedical research in Denmark and the other Nordic countries.

The ground-breaking research leading to the discovery of insulin took place in Canada, and yet a Danish company succeeded in setting up an industrial-scale production of insulin at such an early stage that it became one of the leaders in an ever-growing global market for the new product. Beyond illustrating how important it is for a small country to have talented scientists who are receptive to new impulses from the outside world, this story also shows that even world-class scientists can be interested in transferring their scientific expertise, working methods and prestige to an industrial undertaking. That even they may be prepared to set aside their "pure" science – at least for at time – in order to promote a particular effort in applied science.

Risø Nuclear Research Centre: the quest for a suitable strategy

During the politically unstable period just after World War II, three of the smaller European countries – the Netherlands, Sweden and Norway – embarked upon projects aimed at utilizing nuclear energy for civilian and, to a certain extent, military purposes.[573] Initially all three had hoped for technical support from the US, but when these hopes were dashed by the American policy of nuclear secrecy, the projects were converted into national efforts instead. Sweden had significant amounts of uranium ore on its own soil, and Norway was one of the few countries in the world that produced heavy water, so both of these Scandinavian countries had been dealt a good hand when it came to realizing their nuclear ambitions. Still, it was only in 1950 when the Netherlands chose to play their trump card, making roughly 6 tonnes of natural uranium available to the Norwegians, that the two countries were able to collaborate on building the Halden experimental reactor, which was inaugurated in November 1951. Just under three years later, Sweden followed suit with a small research reactor built into the bedrock under the Royal Institute of Technology in Stockholm.

No similar initiatives were taken in Denmark in those year, which is fairly surprising given Denmark's total dependence on imported coal and oil to cover its energy needs. The lack of naturally occurring uranium and heavy water must certainly have been a serious drawback for the country, but the passive Danish attitude towards investigating the possibilities of atomic energy can hardly be explained without considering Niels Bohr's position in Danish society as a whole. Bohr was in a class all his own; a national treasure. No responsible Danish minister would ever think of taking any action concerning the topic of "atoms" without consulting the nation's own legendary physicist. But this was precisely where Bohr's weak spot lay: in the exploitation of nuclear energy. As a member of the team of physicist that had largely been responsible for developing the atom bomb in Los Alamos, Bohr had come under scrutiny from the American security authorities. Because of activities that included communicating with the Soviet nuclear physicist Peter Kapitza, Bohr had been obliged to promise not to divulge any atomic secrets, much less take part in any work to help develop atomic energy. Denmark, in turn, was obliged to remain on hold until the US gave the go-ahead in 1953. As a permanent secretary in the administration, Hans Henrik Koch, a close associate and confidant of Niels Bohr and later an active member of in the Danish Atomic Energy Commission, diplomatically put it:

23.5 *The Risø complex as it appeared in October 1957. Many of the office and laboratory buildings were already in use. In the distance, at the far left end of the Risø peninsula, the round building nearing completion was to house the reactor DR2, and work on DR3 had not yet begun. The DR1 reactor, installed in the small, square building visible in front of DR2, was already up and running.*

> Indeed, it was not until others had taken up the idea that we [Denmark], too, ought to begin working with the practical, peaceful use of atomic energy, and particularly after Bohr had been brought to understand that leading circles in Britain and later the United States would look with sympathy and support upon such Danish efforts, that Bohr wholeheartedly espoused the idea of a Danish atomic energy programme.[574]

A few months before the American president Dwight D. Eisenhower addressed the United Nations General Assembly and delivered his famous *Atoms for Peace* speech on 8 December 1953, there were signs coming from Washington that hinted at an imminent relaxing of the rigid American nuclear policy. These signals motivated the Danish Academy of Technical Sciences (a private organization founded in 1937 with strong ties to the industrial sector) to propose, at a meeting in early November 1953, that Denmark appoint an atomic committee "to follow the international work developing the industrial use of atomic energy and, based on this, to make such proposals as might be appropriate for a potential Danish working programme".[575] The academy's council agreed to urge Niels Bohr and J. C. Jacobsen, both professors from the Insti-

23.6 *The inauguration of Risø on 6 June 1958 was attended by a number of prominent guests, headed by their royal highnesses King Frederik IX and Queen Ingrid (at left), Finance Minister Viggo Kampmann (at centre, wearing glasses) and Professor Niels Bohr (at right).*

tute for Theoretical Physics, and Torkild Bjerge, a professor at the Danish Technical College, to sit on the committee. All three accepted the challenge. The resulting body, called the Atomic Energy Board, broadened its base when it immediately chose to invite civil engineer Haldor Topsøe on board. Topsøe, a young industrialist who also ran his own consulting firm, was well acquainted with both Bohr and Bjerge.

Denmark was clearly a late starter in the nuclear-energy race, but when the country finally started moving, it quickly got up to speed. The time had come to catch up with Norway and Sweden, which had a good head start. Denmark's new Atomic Energy Board made use of Bohr's good relations with Lewis Strauss, the chairman of its American counterpart, and with leading figures in the corresponding British organization to achieve favourable arrangements for supplies of enriched uranium and heavy water in sufficient amounts to allow Denmark to begin constructing its own experimental atomic reactors. These arrangements were, however, contingent upon the formation of an official Danish atomic energy organization mandated by the Danish government to enter into binding contracts with the national atomic energy organizations in the US and the UK.

In early 1955, the Atomic Energy Board therefore approached the Danish government stating that it had now fulfilled its mission, and that further progress depended on the government's will to provide moral and financial support. The board received a warm welcome from the social-democratic government, and the finance minister Viggo Kampmann, whose ministry would be responsible for atomic issues, was among the most enthusiastic. Kampmann was an ardent supporter of the cause, belonging as he did to the group of modern social-democratic economists who were convinced that the path to greater wealth and increased welfare for Danish society as a whole followed a path of increased education and research, particularly in science and technology. Kampmann readily promised that he would do what he could to obtain the necessary funding of about DKK 100 million – the Atomic Energy Board's estimated cost of building an atomic research facility that would bring Denmark neck-and-neck with the other Scandinavian countries.

During the spring of 1955 Denmark appointed a Preparatory Atomic Energy Commission, chaired by Niels Bohr, that was empowered to contract with the US and the UK for the delivery of two (soon to become three) nuclear reactors for a Danish atomic research site. The contracts were signed in June 1955, and in October of that year the commission decided that the new facility would be built on Risø, a peninsula in Roskilde Fjord 40 kilometres west of Copenhagen. Meanwhile, the political parties were involved in intense negotiations to settle on a more permanent solution; intense, though not overly problematic, as they took place in the atmosphere of global atomic optimism that followed in the wake of the Geneva Conference on the peaceful exploitation of atomic energy held in August 1955. On 21 December 1955, the Folketing enacted legislation to establish Denmark's permanent Atomic Energy Commission, whose object would be to act "for the promotion of the peaceful exploitation of atomic energy for the benefit of society".[576]

The permanent Atomic Energy Commission (AEC) consisted of 24 members selected to represent all of the stakeholders in Danish society. Initially the AEC consisted of 10 scientists from Danish universities and colleges, 7 industrialists, 3 representatives of the power industry (the Danish electricity plants), 3 union representatives and 1 civil servant (the permanent secretary H. H. Koch). This composition clearly reflects the dominant position the scientists held, thereby also reflecting the conviction of Denmark's social-democratic government as to the decisive role science played in modernizing Danish society. The AEC convened for the first time on 13 February 1956, choosing Bohr as its chairman and appointing an Executive Committee under the leadership of Koch to handle the AEC's day-to-day administration.

Denmark's new Atomic Energy Commission had no problem staying busy, with its agenda including: (1) hammering out the details of Denmark's cooperation agreements with the US and the UK, (2) constructing the extensive research complex on the Risø peninsula, (3) hiring hundreds of employees, and (4) last but not least, devising a strategy to ensure that the work carried out at Risø would, in fact, be "for the benefit of society". As for the first three items on the agenda, there can be no doubt that the AEC did a fine job, and even worked so quickly that for a while the Danes took to using the new phrase "Risø tempo" of things that were done with exemplary speed and efficiency. On 6 June 1958, roughly one and a half years after the establishment of the AEC, Risø was officially inaugurated. At the time, two of the site's experimental reactors (DR1 and DR2) had already been installed, to be followed by the third and largest reactor (DR3), which went critical on 16 January 1960. The completion in 1963 of the Risø hot-cell plant was regarded as the final phase of construction, bringing the research centre up to the size planned by the AEC. Total investments at the time amounted to about DKK 150 million, and annual operating costs were roughly DKK 40 million in current amounts – making Risø a definite peak in Denmark's scientific research landscape. By 1963 the staff at Risø had swelled to about 750, a third or so of whom were scientists and engineers. Most of the people in this group were young and had graduated from the University of Copenhagen or the Danish Technical College only a few years earlier.

Enthusiasm and trust. On 6 March 1956 the AEC appointed Torkild Bjerge as the technical-administrative director of the Risø Nuclear Research Centre, and J. C. Jacobsen as the director of research. Working with the state's representative H. H. Koch, the two physicists began to interview and employ staff and to build up the scientific environments that would be endeavouring to fulfil Denmark's expectations.

Although it may seem peculiar, it is nevertheless a fact that neither Bohr, Bjerge, Jacobsen or Koch ever prepared a detailed plan of how Risø was to be organized and which tasks the new institution was to embark upon.[577] Presumably the four key figures were in such agreement on the general outline and goals that they never felt the need to go into greater detail. Minutes of AEC meetings leave no doubt that especially Bohr and Koch were passionately opposed to any planning that might involve long-term constraints. As noted, Bohr, Bjerge and Jacobsen were all physics professors at institutions of higher learning, and they therefore found it natural, perhaps so much so that it went without saying, that Risø would be organized in the same way as a large university institute. The result was a number of departments (six), each led by a head with wide-ranging powers, who would answer to a joint governing body that mainly had a coordinating administrative function.

Once the governing body had decided on the future departments at Risø and appointed the six department heads, it largely left it up to each of them to put together their own department, which included making plans for research projects and employing the necessary scientists and technicians. There was a rule, however, that all expense-related decisions, whether they involved construction, plant or employees, had to be presented to the governing body at one of the "Friday sessions" held weekly at Risø. Koch would meet with the two directors, Bjerge and Jacobsen, and all of the department heads. There are no minutes of these Friday sessions, but accounts from people working at Risø at the time all agree on one significant point: Only very rarely were requests from the department heads refused. The usual procedure was that a list of all requests – sometimes diplomatically reworded following suggestions by Koch – was passed on for approval by the Executive Committee of the AEC, which normally convened in Copenhagen on the following Tuesday. By virtue of his inclusion in the "Friday club" and the "Tuesday club", Koch played a crucial role in smoothly and painlessly resolving the multiplicity of problems that arose.

Experimental work commenced in the years following 1956, as the departmental scaffold was gradually fleshed out with employees. And Denmark and the Danes had enormous expectations to the new nuclear research facility. From the instant the ceremonial shovel first pierced the topsoil in March 1956, the newspapers made sure that their readers were kept up to date on everything happening on the Risø peninsula. Topping-out ceremonies, appointments of directors and heads of department, reactors going critical, experiments that would excite the imagination – all links in a chain of events that was meant to quickly and safely guide Denmark into the atomic age.

Tentative efforts to submit critical reports or indicate potential dangers connected with the peaceful exploitation of nuclear energy were ruthlessly rebutted, either by the AEC or by Risø's directors, with all of the authority these two powerful authorities could muster. This period was typified by positive attitudes toward science and technology, not least towards the peaceful utilization of nuclear energy, which was regarded as the only possible means to save industrialized civilization from an impending global energy crisis. The nuclear physicists and reactor engineers who were to realize this miracle were the new heroes of society. The enormous appropriations made to create and operate Risø – without any corresponding requirement in the form of clear-

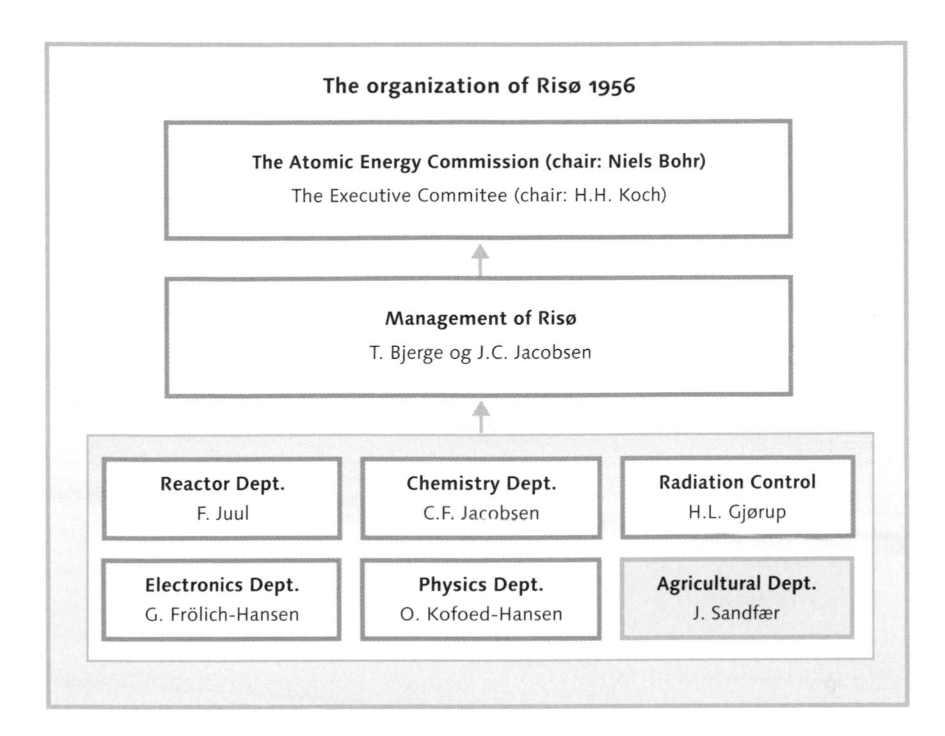

23.7 *The administrative organization of Risø after all department heads had been appointed in November 1956. In addition to the five departments originally planned, Risø also set up a department for agricultural testing at the behest of key players Kampmann, Bohr and Koch, who wished to ensure support for the Risø project from Venstre – the Danish liberal party, which is historically rooted in Denmark's agricultural community. The Agricultural Department's activities were coordinated by the Royal Veterinary and Agricultural College, not by the Atomic Energy Commission.*

cut goals for the new institution – demonstrates the extent to which Danish society and government were willing to place their trust in Risø's scientific and technical elite.

Basic and applied research: the Bush philosophy with a Danish twist. The vast majority of research going on at Risø during the institution's first 10–15 years was rooted in the Bush report's philosophy that modern technology is a direct outcome of applied science. It was a natural consequence of this philosophy, and of the predominance of scientific researchers (particularly physicists) in the AEC and the upper echelons of Risø, that there was a strong emphasis on basic, or "pure" research from the very beginning. Not surprisingly, pure research was mainly cultivated in the Physics Department and the Chemistry

Department. A review of the research projects in both departments around 1960 has shown that more than half of them must be categorized as pure science. In other words, their primary goal was to achieve greater insights into the workings of nature, not to contribute to solving specific technological problems.

The basic research done at Risø from 1956 to 1963 was mainly aimed at utilizing the facility's three experimental reactors, which produced neutrons as well as many different kinds of radioactive isotopes. The neutrons could be removed from the reactors in the form of intensive neutron rays, a technique used to measure their lifetime, to examine their interaction with various atomic nuclei, and to study physical and chemical transformations in irradiated materials. The radioactive isotopes were chiefly used for basic research in nuclear physics, but some of them were also well suited as radioactive tracers for the plant and animal experiments going on in the Agricultural Department at Risø. Even though the basic research projects at Risø sometimes ran into problems, most of them were fairly successful. This was mainly because the senior scientists had talent, experience and helpful contacts at scientific centres in Denmark and abroad, but it was also because Risø had high-quality laboratories and equipment, including its own powerful, state-of-the-art neutron ray facility.

According to the Bush report, scientific breakthroughs coming from pure scientists are the root source of progress. This is followed by the involvement of other type of researchers, typically engineers, who are able to point out potential uses of discoveries and develop models that work in the laboratory. Finally, practically oriented employees in relevant commercial undertakings further develop the laboratory models into robust appliances that may prove to be competitive in the marketplace. This line of thinking is, of course, a simplified representation of the way things were perceived at Risø, but it largely reflects the approach that most of the facility's six departments took during the first decade of the national laboratory's existence.

We know today that this one-dimensional perception of the relationship between science and technology is greatly over-simplified. That is why, looking back, it is not surprising at all that most of the Risø departments found it difficult to present genuine successes when it came to applied science; "successes" meaning examples of research done at Risø that led to saleable products manufactured by Danish companies. The Electronics Department at Risø is a good example. Led by the engineer G. K. Frölich-Hansen, this department's employees developed a series of instruments designed to measure different types of radiation. Frölich-Hansen himself saw this work as lending a helping

23.8 *Risø's gamma-radiation facility during installation in April 1957. At right agronomist Jens Sandfær, the head of the Agricultural Department, and behind the gamma-ray source Vagn Haahr, a scientist working in the same department. At centre civil engineer Henry L. Gjørup, the head of the Radiation Control Department. This installation was mainly used to study the effects of various radiation doses on selected plants during the growth season. The relative dose is easily determined, as it is inversely proportionate to the square of the distance to the gamma-ray source.*

hand to the Danish electronics industry, which he encouraged to develop and market Risø's products. The only problem was that the Danish electronics companies were not interested. Not feeling that they had a problem, they saw Risø's concern as condescending, something they would much rather do without. The aggravation pervading the electronics industry is evident in the entire sector's remarkable intransigence towards Risø in 1963, which was also the *annus horribilis* of another Risø department, as described later in the chapter.

Fortunately there were high points, too, many of which were accomplished by the Agricultural Department. Its research was supervised by the Royal Veterinary and Agricultural College, which had a long-standing tradition for doing applied research and closely collaborating with the agricultural sector's own organizations. Building on this tradition, it was natural for Danish agriculture to regard the Agricultural Department at Risø as yet another location – one that enjoyed the use of hitherto inaccessible technical facilities – that could do agronomic R&D work. A gamma radiation source was set up on one of the experimental fields at Risø in 1957, enabling the scientists to subject selected crops to controlled doses of radiation and thus investigate the potential of gamma radiation in the service of selective plant breeding and disease control. In addition, Risø carried out a series of experiments using radioactive tracers to study biological processes in plants, animals, and the topsoil. These and numerous other experiments received good coverage in the press. It was easy to explain their purpose to the layperson, and because the Agricultural Department's down-to-earth experiments were closely linked to agricultural practices, it was often cited as an example the other departments at Risø would do well to follow.

Another high point was the Accelerator Group, which gained independent status as a department in 1962 under the leadership of the Icelandic physicist Ari Brynjolfsson. As early as 1958, Risø had ordered a 10 MeV accelerator with a view to irradiating and preserving foods and sterilizing medical equipment. The day after the finance committee had approved the DKK 2.5 million purchase, the newspaper *Børsen*, citing the Danish Meat Research Institute as its source, was able to inform its readers that "Danish slaughterhouses will soon be entering the atomic age. This means that within a few years, foods will come to market that have been preserved by means of atomic irradiation and can remain fresh longer, without being boiled or frozen or otherwise altered".[578] The article in *Børsen* is typical of the widespread euphoria of the 1950s and 1960s, and the lack of critical distance to the experts' visions of the future – especially visions that had anything to do with atoms. Within a few years it turned out that irradiation was not going to be the future of longer shelf-life for foodstuffs, which could be better and more cheaply preserved by a simple method like freezing. On the other hand, the Accelerator Department was highly successful in promoting its commercial irradiation service to sterilize many of the disposable hospital products (plastic syringes, catheters, blood transfusion kits, and so on) that were conquering the markets in those years.

Reactor-engineering research and development. Beginning at the planning stage, the people behind Risø agreed that the new experimental complex had to keep up with nuclear-reactor developments. In the summer of 1957, a year before Risø's inauguration, the newly hired staff in the Reactor Department found themselves facing a critical choice: Should they put all their efforts into developing an independent Danish reactor type? Should they try to join a large, foreign-dominated reactor development project? Or should they employ their limited resources to conduct a series of less ambitious research projects linked to existing reactor types?

Based on their deliberations, the young reactor engineers chose to initiate an investigative project to study which materials could be used for building a

Deuterium-moderated Organically-cooled Reactor, and they named their project DOR. They would also study other factors, such as the reactions of fuel rods, moderators and coolants subjected to conditions of intense radiation. Moreover – and perhaps most importantly – the DOR project was intended to give Risø's untested reactor engineers the practical experience they would need if the were to attempt to create any sort of serious cooperation with the Danish electricity plants and the industrial sector to design and build a nuclear-powered reactor.

There were three reasons for choosing to go with the DOR, rather than any other reactor type: Firstly, given the choice, the enthusiastic reactor engineers at Risø preferred to work with a type of reactor that had not yet been studied elsewhere, at least not in any great depth. Secondly, they deemed that this type of reactor would not present any great constructional problems, which ought to give Danish industry a fairly good chance to participate in developing the potential DOR power plants of the future. Thirdly, the DOR was capable of running on natural uranium. Bearing in mind the uranium deposits identified on Greenland, this implied that Denmark's energy supply might have the potential to become self-sufficient.

The directors of Risø found these arguments so compelling that the investigative project was allowed to progress smoothly for the first 3–4 years. Staff working on the DOR project carried out a series of detailed technical studies and model-based calculations aimed at solving the problems that would inevitably crop up in this type of reactor. Nothing about the project was very dramatic – but then again, none of its findings were ever close to becoming immediately useful either.

Then, in 1960, something happened that would decisively alter the course the Reactor Department had followed so far. The head of the Physics Department, Otto Kofoed-Hansen, had spent several months in the autumn of 1959 travelling around and studying various atomic facilities, most notably General Atomics in San Diego, California, where he had had the opportunity to compare the development of reactors in the US with the R&D efforts going on at Risø. The comparison did not reflect favourably on Risø. Upon his return to Denmark, Kofoed-Hansen penned a confidential memo to the directors of Risø, harshly criticizing the facility's DOR project, or at least the way in which the project was being managed. According to the memo, the Reactor Department at Risø lacked all of the things that General Atomics had, namely: (1) experienced staff, (2) contacts in the industry, and (3) a well-defined project. Kofoed-Hansen rounded off his memo by imploring "our topmost administration" to help chart the way back onto to the right course, since, as he put it,

23.9 *Discussing reactor physics in a working group for Risø's DOR (Deuterium-moderated Organically-cooled Reactor). In the foreground at right civil engineer Povl Ølgaard, the leader of the Reactor Department's section for reactor physics.*

"researching out into the blue is a game of chance, where we, with our small potential, have a first-class opportunity to draw a big blank".[579]

Although there is no trace of the memo in the form of any decisions taken to the minutes by the directors or board of Risø or the AEC, there are indications that it made an impression. At any rate, in August 1961 the AEC wrote an official letter to Danatom, an organization formed by the electricity plants and a number of industrial enterprises as early as 1957 in order to stay abreast of reactor developments around the world. Right from the start Risø had been annoyed with Danatom, never really coming to terms with the fact that the electric utilities and the industry were not content with getting their advice and information from Risø alone. Now here was Risø, swallowing its pride and asking Danatom, the country's small, well-connected organization, if they might like to participate in developing an actual experimental DOR.

After mulling over the offer for several months, Danatom gave its reply. It was a resounding "no". Danatom had no confidence in the DOR concept, but if Risø insisted on pursuing the work with a DOR-type installation, Danatom recommended that Risø seek out foreign expertise in the area, as "the possibility that we, here in Denmark, can independently lead the development of the DOR-type reactor to a level ahead of, or even equal to, what is happening in Euratom and Canada, seems slight, even if we were to direct all efforts into DOR."[580]

The response infuriated the Risø reactor physicists, who repudiated all of Danatom's arguments, and Risø's next move was to try to go it alone, without involving Danatom at all. In the autumn of 1962 – shortly before the death of Niels Bohr, the chairman of the AEC – Risø attempted, through the AEC, to establish a binding cooperation between Risø, the Danish industries and the electric utilities to build an experimental DOR plant that the electricity plants were supposed to order. Sufficient progress was made for the AEC to form a committee, the Atomic Power Committee, consisting of representatives of for the three involved parties, who were to investigate the matter further. But in February 1963, even before the committee had begun its work, a frustrating and protracted conflict broke out between Risø and the Association of Danish Electric Utilities. Lasting five years, this conflict ultimately put and end to Risø's dreams of playing a leading role in a project to develop a unique Danish reactor.

It was not until 1967 – after yet another vain attempt on Risø's part to develop a new reactor type, this time based on a Swedish model – that a lasting peace was negotiated between Risø and the electric utilities. As a result, the distribution of the two institutions' responsibilities changed completely. Up to this point, Risø had sought to play a leading role in the prospective introduction of nuclear energy in Denmark, though it had nothing to show for its attempts except an ongoing conflict with the Danish electricity plants. Having learned from the experiences in other countries that had already built nuclear power plants, the utilities now offered Risø that it could handle the approving and supervisory function for any nuclear power plants built on Danish soil. On the other hand, the electricity plants stipulated that after such an agreement, clearly they, and they alone, would be the ones to decide when Denmark was going to introduce nuclear power, and which type of commercial reactor the country would opt for. From this point on, Risø's reactor people could only sit back and wait for the day to come when the electrical-utility people decided that the time was ripe. When that day eventually came, in August 1972, the time would have become over-ripe: By the time the utilities and Risø were finally ready to present a joint proposal for the introduction of nuclear power in Denmark, the global community was beginning to question the blessings of nuclear power. Objections were so widespread and so forceful that the proposal was blown away in the mid-1970s by the storm of protest rising from a majority of the Danish population.

Utility-oriented basic research. Despite the conflicts with the electric utilities, on the face of it the work going on at Risø between 1963 and 1970 did not change much. The truth is that the discord actually gave rise to soul-searching in several departments at Risø, and that the weighting of research aimed at reactor engineering, applied research and basic research was definitively changed in the course of the 1960s.

During the harsh confrontations between Risø and its opponents after the facility's break with the electricity plants in 1963, the critics of the experimental facility would unfailingly emphasize the reactor research and (portions of) the applied research at Risø as being irrelevant to Danish society. On such occasions, conversely, the basic research done at Risø usually received acknowledgement – whether it was because the heads of the Physics Department and the Chemistry Department (both heavy on the basic research) were scientists with incontestable international reputations, or because the critics did not possess the expert knowledge needed to assess the quality of the basic research. By being positive towards significant parts of the work going on at Risø, the critics were more at liberty to strike all the harder at those portions of the facility's programme that they mainly wished to do away with. At no time during the first 10–15 years of Risø's history was there any question of cutting the grants given to basic research projects. This gave both the department heads and the large staff in the departments for physics and chemistry an ever-stronger incentive to cultivate pure research. But how were things going in Risø's other, more utility-oriented departments?

As previously described, in 1961 the Accelerator Department had already offered Danish hospitals and pharmaceutical companies to sterilize their disposable equipment by means of irradiation using the electron accelerator at Risø. Not many customers made use of the service in the first few years, but when Risø came under the critical media spotlight in 1963, Ari Brynjolfsson (then head of the Accelerator Department) and Niels W. Holm (his successor in 1965) recognized the opportunity that this activity gave Risø to earn goodwill from the Danish hospitals and industries. Brynjolfsson and Holm decided that the accelerator facility must be made available to external customers, whatever the cost. As the demand for commercial irradiation steadily grew, from about 20 tonnes in 1961 to 270 tonnes in 1965, so did the inconvenience it caused for scientists from the other departments at Risø who wished to use the accelerator in their own research. The directors nevertheless remained

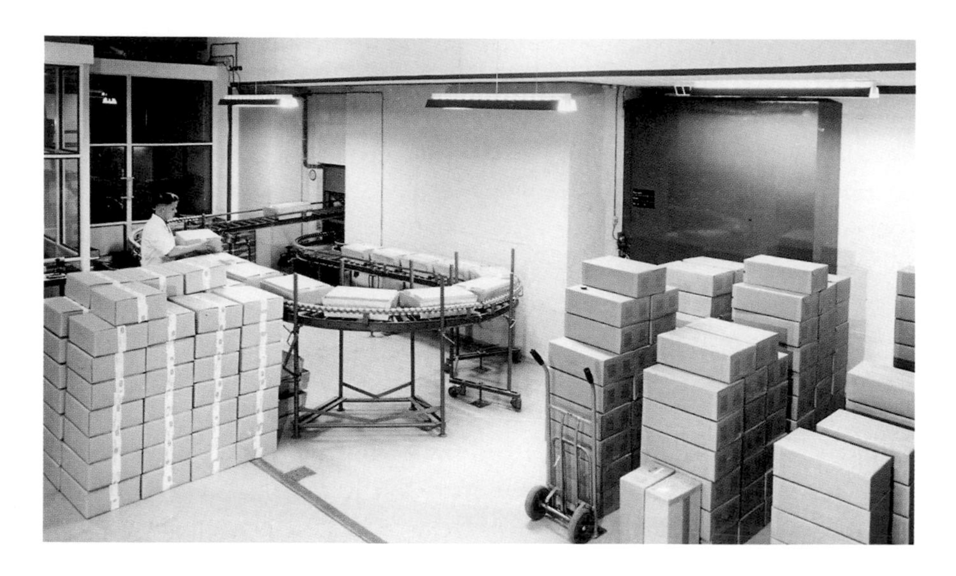

23.10 *Customers delivered material for irradiation in Risø's electron accelerator, packaged in standard cardboard boxes. The boxes were placed, unopened, on a conveyor belt that moved them through the accelerator's radiation field. The operator could regulate the speed of the belt to easily adjust the radiation dose. Risø's irradiation service was later transferred to a private company called Radest, founded in 1968.*

adamant about not rejecting commercial customers as long as there was no realistic alternative, particularly in light of the fact that Risø itself had created and spurred on the demand for irradiation. This distinctly service-minded approach fit in with Risø's serious efforts, begun as early as 1963, to set up a private irradiation enterprise. Quite a significant part of the work involving the accelerator was targeted at helping Danish businesses to bring new commercial products to market. When at last Radest A/S began to offer commercial irradiation in 1968, this new limited-liability company was given access to Risø's accumulated expertise on radiation dosage, dosage control and other relevant parameters.

As part of the DOR project the Reactor Department's metallurgy section, headed by the civil engineer Niels Hansen, had been working for several years with casing materials for the fuel rods intended for use in a DOR. When the project was terminated in 1964, Hansen succeeded in tearing his group loose from the Reactor Department and having it recognized as an independent entity: the Metallurgy Department. Hansen used the department's new-found independence to break with the reactor line known as DK-400, which the

Risø directors strongly supported. Hansen believed that for far too long Risø had been doing "applied research with no applications", and that the DK-400 project was in imminent danger of repeating the DOR scenario, the DK-400 being a heavy-water reactor, whereas the electric utilities and Danish industry were more interested in light-water reactors.

Hansen's view was backed by Helsingør Skibsværft og Maskinbyggeri (HSM), a Danish shipyard and machine company that wanted to enter the growing market for reactor equipment. This led to the establishment in 1963 of a long-standing cooperation between HSM in Elsinore and the unit that would later become the Metallurgy Department at Risø. Their goal was to develop fuel rods that could be used in Risø's own experimental reactors and in commercial light-water reactors. In technical terms the project was a success, as the partners were able to develop fuel rods that did well when tested in various foreign reactors. In commercial terms, however, the project barely got off the ground. One significant reason is that because Denmark itself never introduced nuclear power, there was never an actual domestic market for the product, and without a domestic market, the likelihood of doing well on the international market was virtually nonexistent. No one could have predicted this in 1963, however. And whether one chooses to call the development of fuel rods a success or a failure, the story serves to demonstrate that opinions diverged at Risø when it came to defining the meaning of research "for the benefit of society". It also demonstrates that a strong department head could sometimes set his own agenda, even at odds with the directors.

Niels Hansen was not the only example of this at Risø. The Electronics Department's course also took an interesting turn after the first department head, Frölich-Hansen, had retired and been succeeded by Jens Rasmussen. Rasmussen soon realized that the strategy followed so far in applied science had been patronizing, and therefore futile. Quite apart from the fact that the Danish electronics industry was not prepared to see itself as inferior, Danish companies would almost always lag behind the large foreign corporations when it came to marketing advanced standard metering equipment. Rasmussen therefore decided that in the future, the employees of the Electronics Department would be doing utility-oriented basic research in areas where they had special talents and had the potential to make a difference. One of these areas was improving safety in the workplace. Rasmussen was convinced that the safety conditions for people working in (nuclear) power plants and other large industrial installations was a field of research that had so far been neglected, and which would soon be drawn into the public spotlight.

The Electronics Department at Risø initiated the first research project of its kind in the mid-1960s. This was just around the time when Risø's relationship with the utilities had become extremely tense, and yet interestingly, Rasmussen and his people were able to obtain permission to conduct research in the control rooms of Danish electricity plants. Their studies looked at how operators interacted with the entire array of keyboards and panels that registered the condition of the electric system at any given moment, and with the instruments that enabled the operator to work the system. What made this possible, according to Rasmussen, was that his department never approached the electricity plants saying: "We're going to come and help you set up your control rooms." Instead, their attitude would be more like: "We are doing some research on the set-up of control rooms, and we need your experience."[581]

The findings of this study got so much attention that they further confirmed Rasmussen in his conviction that going forward, the Electronics Department must concentrate on utility-oriented basic research into new areas that held promise for the future. His philosophy would determine the sort of research done in the Electronics Department for many years to come.

Applied science – or basic science? Science and research are not institutionally limited to the universities and other places of higher education. Non-academic institutions also contribute, and basic-science activities can very well take place in research environments that fundamentally aim to address practical problems through applied science. Inversely, the application of scientific results sometimes springs from discoveries made in pure research. At other times, applied research is done in areas where no basic research has yet established the scientific underpinnings. The interrelation between pure research and applied research is always specific and depends on the historical context. The Carlsberg Laboratory, which in many ways holds a unique position in the history of Danish science, is also quite exceptional from the perspective of the pure vs applied science discussion. Here is a private research laboratory that has always expended considerable resources on research activities that had no prospective utility at all relative to the brewery corporation's core activity – but which has meant a great deal to Danish science as a whole. Although it arose from non-academic activities, the Carlsberg Laboratory still in many ways resembled an academic institution and offered its researchers a relatively high degree of freedom. Indeed, much of the work that went on at the laboratory was really more like basic research than applied research.

The history of insulin in Denmark is the story of how two prominent, university-educated scientists at the apex of their research careers heard about the revolutionary discovery of insulin in Toronto, obtained exclusive rights from the Canadian discoverers for producing the substance in the three Scandinavian countries, and set in motion a major research project to address the treatment of patients with diabetes. To achieve this goal they founded the Nordic Insulin Laboratory, which, although initially a non-profit enterprise, would eventually develop into one of Denmark's largest pharmaceutical companies. So clearly, applied research can begin with basic researchers who have no experience in the applied field, although, as this story also shows, the basic researchers were obliged to seek help and support from an existing pharmaceutical company before their efforts were crowned with success. Small wonder, given that basic science and applied science have different goals, use different methods and have very different criteria for success.

The history of Risø, the national laboratory complex constructed in the 1950s with the clear objective of guiding Denmark into the atomic age, illustrates how difficult it can be for a new research institution to work out a strategy that delivers convincing results "for the benefit of society". From the outset, Risø was dominated by people with a scientific university background, but without having any overall plan for the institution. Instead, the various heads were allowed to set up their individual departments more or less as they saw fit. That is why the history of the research that took place at Risø during the first fifteen years of its existence is not so much one history, but rather a series of diverging histories that, taken together, can provide interesting insights into the interplay – or lack of interplay – between basic and applied science.

Of the various departments, the Reactor Department suffered the hardest defeats. Basing its decisions on the Bush report's presumption that science played a pivotal role in developing new technology, but without any significant prior experience in reactor technology, the department optimistically pushed ahead in its efforts to develop a new type of reactor, placing its faith in the support it believed would come from the Danish electric utilities and the industrial sector. This support was not forthcoming. As early as 1963, the Reactor Department was forced into a defensive position, and consequently during the 1960s it was obliged to acknowledge that in its own case, the Bush philosophy had led to nothing but a dead-end street.

The other departments at Risø also had to adapt. Those least affected were probably the Physics Department and the Chemistry Department, both of which had aimed from the outset to become university-type institutes that mainly concentrated on pure science. Their development over Risø's first decade confirmed that this had been a wise course to follow. The Agricultural Department, with its close links to the Agricultural College, had no reason to make any radical modifications or innovations either. Meanwhile, several of the other departments had undergone a veritable metamorphosis, which although hardly visible from the outside nonetheless paved the way for the dramatic changes that struck Risø after 1976 – the year the Danish government dismantled the AEC and indefinitely postponed the decision to introduce nuclear power in Denmark. As recounted in this chapter, the metamorphosis had already begun around 1963 with several independent attempts to rethink the relationship between pure and applied research at Risø.

There were at least three departments – Electronics, Metallurgy and Accelerator – where the department heads had recognized that even though research was important to Risø's survival, the prospect of failure is painfully real if such research does not take place in cooperation with users who are known in advance to be interested in the product. Using the motto "No applied research without applications" the Metallurgy Department, under the leadership of Niels Hansen, therefore joined forces with the company HSM to develop fuel rods for existing reactor types. Likewise, the Electronics Department, headed by Jens Rasmussen, took the initiative to cooperate with the electricity plants in a new area they called "control-room research", while Ari Brynjolfsson and Niels Holm from the Accelerator Department intensified their cooperation with the Danish pharmaceutical industry and the hospital sector to research and optimize irradiation techniques for sterilizing hospital equipment.

The greater priority given to fundamental utility-oriented research in the three departments last mentioned, combined with the sustained focus on pure research in the traditionally research-intensive departments for chemistry and physics, probably did influence the tendency of Risø's critics to regard the facility's basic research in a positive light. However, this tendency was also a result of the conviction held by the heads of these departments that whereas basic research will always make you smarter, applied research done with no specific application in mind often turns out to be a waste of time and energy. Few would dispute the claim that accepting this insight – a process that took a long time to permeate the entire organization – was a crucial precondition for the survival of the complex known today as Risø National Laboratory, enabling it to weather the shifting economic, political and scientific climates it has seen during its 50 years in existence.

Science for the people

24

A wide range of new initiatives were taken between 1920 and 1970 to communicate science to a wider audience. Most of the ideas came from the scientists themselves, with the primary objective of informing the general public about splendid new scientific results. Throughout the period 1920–1970, science was largely presented in a positive light, but the with the atomic bomb, the arms race between East and West and the increasing problem of pollution came a nascent, nagging fear among the population that science might be used for destructive purposes. The almost one-sidedly positive promotion of science was therefore followed, during the 1960s, by a more multifaceted communication about science. Science journalists and an increasing number of scientists – biologists, in particular – began to focus on the side effects of technological development, and to pose critical questions to the sciences. At the same time, the phenomenon that in recent years has become known as "infotainment" began to gain ground as one of the standard modes of scientific communication.

Conveying science to the people

World War I – sometimes called "the war of the chemists" – seriously challenged people's faith in the future; a faith that had been pervasive before the war broke out. However, the war also underscored the enormous potential of science. "If the world could afford to subsidise research on the same scale, during a decade under peace conditions, our rate of progress would be beyond all belief", as the British chemist Frankland Armstrong wrote in 1924.[582] But even with less generous support, science was still a wellspring of new and astonishing results, not only in seemingly remote academic fields of research like nuclear physics and astrophysics, but also in more tangible classical disciplines such as geology, chemistry and physiology. But was it possible to keep the general population informed about it all? And if so, how best to approach this task? These were questions that concerned many scientists in the decades after World War I, and one tangible result of their deliberations was a variety of new publications, significantly including two series called *Kultur og Videnskab* (*Culture and Science*, 1923–1969) and *Frem* (*Forwards*, 1924–1928).

24.1 *In 1924 the social-democratic politician Harald Jensen – then a leading figure in the Danish Association of Typographers and newly elected to the Danish parliament – came to play an important role in founding the Workers' Education Association. He resolutely took the helm, serving as the organization's dynamic leader until his death in 1929.*

Kultur og Videnskab. This book series ran to a total of 72 books, 17 of which dealt with the natural and technical sciences. The 26 volumes on the social sciences dominated the series, followed by the humanities with 24 volumes. The health sciences and theology were most poorly represented, with 3 and 2 volumes, respectively. Throughout its lifetime, the *Kultur og Videnskab* books were published by the Education Association of the Students' Society, which was rooted in the social-democratic Workers' Education Association (WEA). This organization had grown out of the need for education among working men and women after World War I, by which time the Social Democrats had become the second-largest political party in Denmark and were regarded as ready and able to win an election and form a government in the near future. At any rate, having attended an international meeting in Brussels in 1922 about workers' education, Carl Valdemar Bramsnæs, the head of the Social-Democratic Workers' School, realized with embarrassing clarity that in this particular area Denmark still had much work to do: "Although there was no reason for enthusiasm over everything that was passed on [...], I did become aware that Denmark, with its strong and powerful workers' movement, had fallen behind. We had much to learn from other countries on this matter. And I travelled home, resolved to make our educational efforts a little more systematic."[583]

Bramsnæs was one of the founders of the WEA in Denmark, and he also became its first chairman. The WEA was actually a cultural superstructure for the entire workers' movement, and as such it received financial contributions from the Social-Democratic Association, the Danish Federation of Trade Unions and the Social-Democratic Youth of Denmark. It therefore had a secure foundation able to support a large-scale effort to inform and educate, "free of doctrines and propaganda" and in accordance with its broadly worded objects clause: "The object of the Workers' Education Association is to disseminate knowledge among the working population of Denmark."[584] The new association set up education committees across the country. Ten years later the number of WEA committees in Denmark had swelled to 126, and every year they arranged series of lectures attended by tens of thousands of people, and helped to assemble hundreds of study groups and evening classes for those who were particularly active and inquisitive – and harboured social-democratic sympathies.

Even if it had tried, the WEA could hardly have avoided finding inspiration in the successful efforts of the Danish folk high school movement, which was based on the ideas of the pastor and poet N. S. F. Grundtvig on enlightening and educating the country's rural population. Nevertheless, the WEA's day-to-day head, Harald Jensen, found it appropriate to emphasize that his organization was not directing the Grundtvigian ideas of personality formation at the working classes: "The saga is not able to awaken fervent enthusiasm. The knowledge of history that the worker seeks is not defined nationally or geographically. It spans as far and wide as the foot of humankind has trodden, in all places and in all times [...] For the twentieth century, it is not the poet who is the disseminator of history; it is the scientist, the researcher."[585]

In keeping with this manifesto, a large part of the WEA's educational work was focused on explaining scientific advances. An appropriate literature was needed for this purpose, and in the decades that followed the WEA published, or supported the publication of, a large number of books, including the *Kultur og Videnskab* series, intended for people who wished to study independently at home or in groups. With a few rare exceptions, the authors were well-known, respected scientists from the University of Copenhagen. Although most of these authors doubtless leaned towards the left side of the political spectrum, this is not discernible in the books they published through the WEA – at least not in their scientific works. The contents are sanitized, free of even the vaguest hint of political commentary. The editor-in-chief of the series must have decided that the opponents of socialism would have no gratuitous opportunities to accuse the Education Association of the Students' Society of smuggling political messages into the activities it was describing as informative and educational.

Although naturally the volumes in the series differ greatly, depending on the author and subject matter, a modern reader can only conclude that nowhere is the tone by any means patronizing. Overall the style is crisp, professional and academic. Each book introduces the reader to a huge amount of pertinent material, including a wealth of professional terminology. Illustrations are all but absent, as are examples and potential applications. What the series does is to convey pure scientific knowledge. The format is one-way communication from the omniscient scientist to the largely uninformed reader, who is, however, assumed to be capable of grasping the material by making a determined effort. This shines through in the following instruction, written by the chemistry professor Knud Estrup in his preface to the volume on atoms and molecules: "This little book is intended for use in such a way that one reads it from the beginning, slowly, and preferably not too far. At the next sitting, one begins again from the beginning, but reading a bit more than the first time, and so on and so forth. Only by comprehending the preceding content can one come to understand that which follows."[586]

Having given his readers fair warning, Estrup moves on at a terrific pace. In just 58 pages he succeeds in explaining why the hypothesis of the existence of atoms and molecules is necessary; introducing concepts like atomic and molecular weight, gram atom, gram molecule, molar volume and valence; clarifying the utility of chemical notation; using several different methods to determine Avogadro's constant; explaining the structure of atoms (including the basics of Niels Bohr's theory from 1913); and, finally, telling the reader in a few short lines about the enormous amounts of energy which, according to Einstein, lie hidden in every atom. Here – in the last four lines of his book – the author at last steps out of his role as the strictly objective scientist, with a comment that reveals a certain anxiety for the future: "When one sees what calamities the human race can bring about with the energy sources now at its disposal, it is perhaps best to hope that developments in this particular field might not come about too rapidly."[587]

Far from being unique, this example is quite typical of a recurring trend in all of the cultural and scientific books in the series. We know of no studies that could tell us how many people read or purchased the individual volumes, but it is hard to imagine that readers without some prior knowledge to build on would have gained much by studying the books on their own. On the other hand, these publications from the Students' Society may well have been a source of valuable information for students and educated readers searching for

24.2 *Cover of the first issue of the new* Frem, *from 1924. Showing a drive-through redwood next to the spire on Copenhagen's city hall, the cover asks: "Would you like such a tree?" This journal, published 1924–1928, was edited by Hans A. Svane, who was a deputy education officer, and Victor Madsen, director of the Geological Survey of Denmark.*

an unbiased, professional, brief introduction to a certain topic that lay outside their own field of specialization.

The new Frem. As a countermove to the WEA's book series, the apolitical University Extension organization took the initiative to revive the popular educational series *Frem*, previously published from 1897 to 1917. It consisted of slender subscription booklets that could be assembled into composite volumes in special binders. The new *Frem* was a solid effort in the spirit of progress, as is clear from the editors' introduction in the first booklet. "Everyone would like to move forwards in the world," it explains, "but there are so many things that stop us, like the lack of knowledge and experience". And it is little help that "Time passes so quickly and is so overwhelmingly full, with each day bringing new discoveries and inventions; new impression pressing upon us from all regions of the Earth, from past and present". Admittedly, schooling has given us a certain foundation on which we ourselves can build and develop further, but we still miss having "a guide and helper" who can lead us through the jungle of information. And yet there is hope, the editors say, just as there was hope for the Danes living in the late 1800s, who had access to just such a guide while "the old *Frem*" was published, enjoying a circulation of more than 125,000 around the turn of the century. Unmistakable proof of this, they tell us, can be found in the way they constantly bump into people "who, with the utmost gratitude, recall the two mottoes of the old *Frem*: 'Knowledge is power' and 'Light o'er the land'. People who know how great a part the knowledge they absorbed there has had in shaping their outlook on life, the lives and careers, so that in short, *Frem* was a milestone in their development."

At this point, however, in the mid-1920s, the times have changed. The sheer number of scientific results has exploded, the technical possibilities for creating spectacular publications have improved dramatically, the potential readership is larger and expects far more. This calls for a new approach to the task if the new *Frem* is to carry the same weight as its predecessor:

> ... the times we live in call for a different content, a different form, a different procedure than the times around the turn of the century. The *Frem* that is to do the same deed today as the old *Frem* did in its time must span a much larger area of knowledge and experience, and must allow the impressions to flow quickly and change often. It must follow the variegated life of our day and age and open the gateway to the wide world that surrounds us ...[588]

Both *Kultur og Videnskab* and *Frem* were predominantly written by professional scientists. However, unlike *Kultur og Videnskab*, whose editors spent little or no time on pedagogical considerations that would help their non-academic readers, *Frem* clearly bears the mark of a certain editorial influence from experienced publishers and educators. The articles in *Frem* were organized into various groups, the texts were easily understandable, and most of the articles were well illustrated, at least by contemporary standards. The author would usually approach the material from a journalistic angle, beginning with an interesting story and then gradually widening the perspective. A comprehensive index also enabled readers to use *Frem* as a work of reference. And so, according to the editors, there was nothing to hold people back from subscribing. Next to a picture of a steam locomotive going at full speed, and the heading "Won't you join us?", the text read:

> The crucial thing is that here a work is begun which, in the Nordic region, is something completely new, something quite unique with the special combination of instruction and entertainment that it offers. It is as easily read as a newspaper or a novel, and makes one wiser than either of those. It is like a great adventure train, carrying us through the vast kingdoms of the world, through distant lands, along foreign roads, a train of adventure that takes us through life in a completely new way, enabling us to see the enigmas and the wonders of nature and human life, and to see them in such a way that some of the light of the adventure shines through into our own daily lives.[589]

The editors included brief statements in the introductory booklet from a number of famous and lesser-known cultural figures who praised the initiative as an exceptional event in the history of Danish popular science. But how did the project fare? Was the new *Frem* able to live up to the great expectations the editors themselves had helped to generate? Judging from the number of readers, there is no doubt that *Frem* was a considerable success. About the year after its launch, the editors estimated their weekly readership at three to four hundred thousand – or about ten per cent of the Danish population.[590] However, because there was no weekly column where readers could voice their criticism and their praise, it is difficult to determine how great a role *Frem* actually played on the contemporary cultural scene. Nevertheless, one thing is certain: Not everyone was equally pleased with the initiative.

In September 1926, the social-democratic cultural historian and later minister of education Hartvig Frisch delivered a devastating critique of *Frem* and all that it stood for. He did admit that, naturally, it contained some good articles, but from an educational point of view the project was "cocaine for the people – without a plan, without a system, without a genuine will to instruct". Most of the material in *Frem* was "random bits and pieces", whereas true education requires "a construction of knowledge, in which one stone is laid upon another, so that finally one reaches a result, a platform higher up, whence the view opens out onto a new land. The only knowledge that is power is that which becomes one's own personal possession, and which one can use because one masters it."[591]

The diatribe that Frisch delivered was surely meant in earnest, but at the same time it signalled the beginning of a series of competing publications edited by Frisch himself in collaboration with the economist Magnus K. Nørgaard, who was also a well-known Social Democrat. Centred around a journal on culture and nature called *Gry* (*Dawn*), the next few years saw the publication of four works of popular science: *Kunsthistorie* (*Art History*, third edition) by the Swedish art historian Carl G. Laurin, *Teknikkens Vidunderland* (*The*

Technical Wonderland) by civil engineer Vilhelm Marstrand, *Nordens Dyreverden* (*Animals of the North*) by the zoology professor Ragnar Spärck, and finally *Europas Kulturhistorie* (*The Cultural History of Europe*) by Hartvig Frisch himself. Whereas *Frem* was a kaleidoscopic collection of articles on topics large and small from the world of science and culture, Frisch and his associates focused on just four major topics – which, however, were treated in depth. It was especially Frisch's own work on the cultural history of Europe that was enthusiastically, or at least respectfully, received by readers of all political colours when it appeared in 1928. Given that, unlike *Frem*, Frisch's work was republished several times, it is difficult not to draw the conclusion that history favours his approach to the popularization of science.

The rise of physics in the 1950s. After decades of stagnation, the mid-1950s witnessed a resurgent interest in technical and scientific education, most noticeably in the fields of mathematics, physics and chemistry. A remarkable reinforcement of science took place at all levels following the economic upswing and the growing optimism that pervaded Danish society, as described in chapters 20 through 23. The fortification of science included the introduction of new, scientifically oriented curriculums in the primary, middle and upper-secondary schools. Pupils were now supposed to learn science in the same way researchers did in their laboratories, and those with a special scientific aptitude or interest were to be encouraged to begin studying science after graduation. The publishers reasoned that, accordingly, there must be a considerable need for popular-science books that could be used independently or to supplement the school curriculum, to spur curious students on to investigate the fascinating world of science and technology. In the decade from 1955 to 1965 there came a steady stream of popular and semi-popular books on nuclear energy, space travel, research into the natural sciences broadly speaking, and, most significantly, physics – the quintessential natural science. Some of these books were written by Danish scientists. Most, however, were translated from original works in English, because that language offered such a wide selection, especially after "the Sputnik shock" in 1957 energized the Americans to invest massively in scientific education and research. Titles like *One Two Three … Infinity* (in Danish, *Videnskab er sjov*, 1960) by George Gamow, *Wonders of Nature* (in Danish, *Videnskabens vidunderlige verden*, 1959) by Jane W. Watson, and *The Evolution of Physics* (in Danish, *Det moderne verdensbillede*, 1961) by Albert Einstein and Leopold Infeld are just a few of the many titles in the rich science literature that was now becoming accessible to Danish children and young people in their own language.

I. Bernard Cohen, *Fysikkens gennembrud* (1962) [*The Birth of a New Physics*, 1960]
Donald J. Hughes, *Den fantastiske neutron* (1962) [*The Neutron Story*, 1959]
Arthur H. Benade, *Musik og lyd* (1962) [*Horns, Strings, and Harmony*, 1960]
Donald G. Fink and David M. Lutyens, *Fjernsynets fysik* (1962) [*The Physics of Television*, 1960]
C. V. Boys, *Sæbebobler og de kræfter, der danner dem* (1962) [*Soap Bubbles: Their Colors and Forces Which Mould Them*, 1958]
Robert Wilson and Raphael Littauer, *Cyklotronen og andre acceleratorer* (1963) [*Accelerators: Machines of Nuclear Physics*, 1960]
Alan Holden and Phylis Singer, *Krystallernes verden* (1963) [*Crystals and Crystal Growing*, 1960]
Patrick M. Hurley, *Hvor gammel er Jorden?* (1963) [*How Old is the Earth?*, 1960]
D. K. C. MacDonald, *Det absolutte nulpunkt* (1963) [*Near Zero; The Physics of Low Temperature*, 1961]
Alfred Romer, *Det aktive atom* (1963) [*The Restless Atom*, 1960]
René Dubois, *Pasteur og vor tids videnskab* (1964) [*Pasteur and Modern Science*, 1960]
Victor F. Weisskopf, *Viden og undren* (1964) [*Knowledge and Wonder: The Natural World as Man Knows It*, 1963]

Table 24.1 *Kvantebøgerne* – The series of 12 "quantum books" that appeared in Danish during the first half of the 1960s.

However, during the period under review the greatest single effort in the genre was known as *kvantebøgerne* – the "quantum books" – and it issued from Denmark's largest and most respected publishing house Gyldendal in the years around 1960. The Danish editorial committee was comprised of four recognized physicists: Niels Bohr, Henning Højgaard-Jensen, Aage Petersen and Søren Sikjær. All of the books in the series were translated into Danish from American-English originals that sprang from an initiative by a body called the Physical Science Study Committee. This committee consisted of prominent scientists and educators in the US who, from the mid-1950s onwards, exhibited "energetic efforts to reform the teaching of physics and stimulate an interest in the science of physics", as Niels Bohr wrote in his foreword to the "quantum books" series, explaining that "As a step in its efforts, the committee is publishing a series of smaller volumes meant for self-study and as a supplement to the physics taught in our schools. Each book treats its own topic, and the intention is that the series will gradually cover all of the areas of physics, from the world of the atoms to the depths of outer space, from the physics of the ancients to those of modern times."

With a cautionary word to his readers that they must obviously be prepared to make an effort in order to comprehend the contents of the "quantum books", Bohr concluded by expressing his wish that the books "in Denmark, too, would stimulate the educational work in this and other areas of science, the development of which holds rich promise for promoting welfare and culture".

Despite the high quality of the entire series, despite the generally excellent reviews the books received, and despite Gyldendal ensuring that they were all selected, edited and launched by the most eminent professionals in Denmark, the series never became the resounding success that its publishers and editorial committee had hoped for. Sales figures for most of the books in the series were dismally small, and predictably the books went on sale at cut-rate prices just a few years later. Perhaps after the Danish educational reform of 1958, the new physics instruction at the upper secondary schools demand so much work from the students that only very few had the extra energy to look for supplementary literature? Perhaps the market for popular-science books of this type, which predominantly give scientific explanations of various phenomena, had already been saturated in the 1950s? Perhaps the "quantum books" offered too much scientific knowledge and not enough social and cultural information? Whatever the explanation, by the time the last volumes appeared in the mid-1960s, the disappointing sales clearly signalled that the age of unbridled scientific enthusiasm was largely a thing of the past.

The rise of ecology in the 1960s. While the so-called hard sciences received most of the attention focused on science in the 1950s and early 1960s, the interest in the biological sciences mushroomed during the 1960s. This is reflected in the surprisingly large proportion of young people majoring in mathematics and science at the Danish upper-secondary schools, in the increasing number of biology students at the universities, and in the remarkable rise in the number of popular-science books that dealt with biology topics. Surely a variety of reasons combined to create this development, one of which is, undoubtedly, that citizens in the developed countries were gradually becoming more and more worried about whether nature would be able to tolerate the heavy environmental pressures that accelerating technological progress was exerting. More than anyone else, it was probably the American biologist Rachel Carson who helped to change the direction things were taking.

Carson's book *Silent Spring*, which appeared in the US in 1962 and was

24.3 *The American biologist Rachel Carson had already produced a considerable body of work before publishing her* Silent Spring *in 1962 – a work that catapulted her to international fame.*

published in Danish by Gyldendal in 1963 as *Det tavse forår*, is one of the most influential books of the twentieth century. By linking a large number of biological studies and collating their data on animal, bird and fish populations across the US over a prolonged period of time, Carson reached the extraordinary and unsettling conclusion that a number of species had seriously declined or disappeared altogether in areas with intensive use of chemical pesticides such as DDT. A number of these problems had previously been documented by other biologists, but Carson was the first to attempt to provide the public with a general overview of the state of affairs. And she predicted a future – perhaps not so distant – when spring would no longer be alive with the well-known sounds of birds and other wildlife, but filled with a great silence. The silence would reign not just in isolated locations, but across wide swathes of the American countryside, and by that time it would be too late to turn back. The damage would be beyond repair. But how could humanity have come so far down the path to destruction? Because, said Carson, "the practitioners of chemical control" have sought to achieve complete control over nature, but without having achieved an understanding of the extremely complex interaction that exists between the multiplicity of life on our planet, and without having any humility in the face of the power of nature. As she concluded:

> The 'control of nature' is a phrase conceived in arrogance, born of the Neanderthal age of biology and philosophy, when it was supposed that nature exists for the convenience of man. The concepts and practices of applied entomology for the most part date from the Stone Age of science. It is our alarming misfortune that so primitive a science has armed itself with the most modern and terrible weapons, and that in turning them against the insects it has also turned them against the earth.[592]

Carson's book commanded enormous attention wherever it was published. Most reviewers agreed that her documentation of the serious events she described in the book was sound. Not surprisingly, however, "the practitioners of chemical control" held that her conclusions were wildly exaggerated. She was speaking of isolated incidents, they claimed, the likes of which could easily be avoided in the future, as long as the insecticides were used correctly. In addition, the financial benefits of using modern chemical substances, and not traditional pesticides, were considerable and could not simply be ignored. Carson's professional peers were more inclined to agree with her views, and Danish biologists and physicians sided with her as well. One staunch supporter was the Danish professor and doctor of medicine Knud O. Møller, who wrote a six-page foreword to the Danish edition of *Silent Spring*. Here he presented a series of Danish "accidents" involving insecticides and pesticides, which had caused massive deaths among local bird and fish populations within the preceding 10–15 years, and he also cited one example of a 17-month-old baby who had died of parathion poisoning after putting soil from his mother's garden into his mouth. According to Møller, one had to be careful to not simply regard the problem as an American one:

> In our little, and in many aspects admirable, country, there is a sentence that has been uttered too often: 'Nothing like that is going to happen here.' Because of the thinking that lies behind that expression, many [Danes] awoke to a harsh reality on 9 April 1940 [the day Germany invaded Denmark]. Fortunately, this country has not yet witnessed such monstrous interference in nature's fixed cycles as the United States has; but, on the other hand, our reserves of wild plants and of wild birds, fishes and other animals are far, far smaller than in the United States, and hence more vulnerable. And yet, here in Denmark so much is already going on that anyone who loves the natural areas in our country must entertain the most serious fears that irreparable damage might occur if the utilization of chemicals is not changed.

Even though Møller takes great pains to repeatedly emphasize that Carson's book is about the US, and that Denmark has not yet progressed so far along "the chemical path" as the Americans, one can still detect a certain irony when he quotes an "influential person" from Denmark as having said that "Rachel Carson is hitting wide of the mark". Møller therefore concludes his foreword with the hope that the Danish edition of the book might serve as a warning that "*it is high time to begin curbing the use of herbicides and pesticides*".[593]

Seen in the bigger picture, Carson's book was a noteworthy precursor to a number of popular-science books that discussed ecological issues hitherto unknown to the Danish public, and which, bit by bit, revealed a yawning gap between the representatives for the hard and soft natural sciences. This gap became fully visible in connection with the *Limits-to-growth* debate and the nuclear power debate of the 1970s.

Popularizing science through journalism

Parallel to the considerable efforts Danish scientists themselves were making to popularize science, a large number of non-scientists were interested in science and research, and in communicating what was going on in these areas to various target groups. An important part of this "external" communication on science issues has gone through news channels like the press, radio and television. Because such coverage typically operates on the terms of journalism, the scientific knowledge is processed using another set of professional criteria, which are very different from strictly scientific criteria. The primary concern of journalists will be to their readers, listeners and viewers, and therefore such communication of scientific material must chiefly be adapted to reflect the interests and tastes of these groups, rather than to respect scientific precision and validity.

The spreading of science through journalism had helped to make it an integrated part of society. Some of the earliest professional science journalists made the social dimension of science clearer to a wider audience, emphasizing, among other things, the responsibility society and the politicians had to provide reasonable financial conditions for scientific research. Moreover, the journalistic popularization of science also highlighted its considerable potential as entertainment. The classical and, from a scientific viewpoint, principal aim of conveying scientific material is to inform and enlighten. As the sciences gradually began to make their way into radio and television broadcasting, people also began to think about science's entertainment value. In the late 1960s, a popular Danish television show called *Spørg Århus* (*Ask Århus*) introduced Denmark to a new phenomenon, putting scientists in the spotlight, where they cordially contributed to the nation's Saturday-night entertainment while sharing their scientific expertise, as described later in this chapter.

Danish science journalists. The transition to the twentieth century was a watershed for the Danish newspaper industry. It underwent a sweeping process of democratization, as printing techniques and reporting techniques alike were modernized, which also involved a remarkable expansion of the journalists' areas of interest.[594] The entire process was accelerated by World War I and its demands for even more news reporting on domestic and international events. On the whole, during the interwar years the Danish press experienced an enormous growth, not least the weekly magazines, which tripled their circulation during the 1920s and 1930s. Magazine sales were mainly generated by four major magazine publishing houses – Gutenberghus, Aller, C. W. Bærentzen and Berlingske – which were all developing the editorial, technical and marketing capabilities of their publications. Likewise, the daily newspapers also substantially increased their circulation with new initiatives in editing, technical solutions and public relations.

This upsurge of the printed media, combined with the ever-growing importance of the press to public life in Denmark, led to an increased interest in the professional qualifications of Danish journalists. Were they skilled enough and knowledgeable enough to handle the important function they performed? The Provincial Journalists' Association was particularly active in addressing the issue of a formal (higher) education for journalists. The association's initiatives included several courses for journalists, held in the 1920s and 1930s. It was only after World War II, when the University of Aarhus set up a formalized journalism course in 1946 that the education of journalists in Denmark acquired an actual institutional foundation, which eventually grew into the Danish School of Journalism. Danish journalists had previously been self-taught and were trained by senior colleagues in working methods and fields of specialization. Most journalists could write, and *would* write, about more or less anything, as long as it made for interesting news copy, although some did specialize in areas like politics, culture or science.

A few journalists were reporting on science even before World War I and in the early inter-war period. After World War II, science increasingly came into focus among journalists as a specific area of specialization, resulting in 1949 in a new association for science journalists. The founding members of the Danish Science Journalists' Association were: the chairman E. Schelde Møller (*Social-Demokraten*), the deputy chairman Niels Blædel (*Politiken*), Erik Olaf-Hansen (also *Politiken*), Børge Michelsen (UNESCO's department for the popularization of science, previously *Illustreret Familie-Journal* and *Ekstra-Bladet*), Otto Walsted (*Nationaltidende*), E. Schwencke (*Berlingske Tidende*), Sven Alkærsig (State Broadcasting Service), Eric Danielsen (*Land og Folk*) and Svend Aage Andersen (*Aftenbladet*).[595]

Børge Michelsen, alias Mikkel Borgen. The science journalists wrote and spoke about scientific findings and methods, but many also dealt with the social and political aspects of science in their professional reporting and in other walks of life. Børge Michelsen, whose work as a journalist is treated in this section, was probably the most deeply involved contemporary Danish science journalist when it came to the social and political implications of science. Like many scientists and others interested in the state of science around World War II, Michelsen found inspiration in the Irish physicist, Marxist and historian of science John Desmond Bernal.

In 1939, Bernal published his book *The Social Function of Science*, which changed the way many people regarded the relationship between science and society.[596] Bernal was one of the first to clearly see the sciences as small, social units embedded within a larger social system: society. He therefore saw science not as something that could or should function in social isolation, but rather as a crucial activity that society as a whole ought to promote with ample funding, and attempt to control in directions that were societally desirable.[597] Bernal's book encouraged considerations along these lines in many countries, including Denmark, and his urging for more public support for the sciences bore fruit, in Denmark and elsewhere, in the post-war years.

Michelsen became acquainted with Bernal's thinking while serving as the scientific editor at *Illustreret Familie-Journal* (later *Familiejournalen*, the name used hereinafter) from 1939 to 1945, and during this period he played an important role in promulgating Bernal's ideas in Denmark. Prior to this, Michelsen had studied comparative literature at the University of Copenhagen while working as a freelance reporter, until giving up his academic studies in 1936 and opting for a permanent position as a journalist at *Familiejournalen*. On his own initiative, Michelsen began to regularly write pieces on science in *Familiejournalen* in 1940 under the pseudonym of Mikkel Borgen. His alias had cropped up in the weekly magazine earlier that year in the byline of articles dealing with patents, opera and other topics. Over the next five years Michelsen dedicated his efforts under this pseudonym almost exclusively to science journalism appearing in *Familiejournalen*, while continuing to write on other topics under his own name.

Not surprisingly, Mikkel Borgen's articles closely resemble those of Børge

Michelsen in style as well as content. At this point the readers of *Familie-journalen* had come to know Michelsen as a sympathetic reporting journalist who travelled far and wide to visit companies in many different lines of business. Børge Michelsen had a penchant for seeking out little-known professions that his readers would find intriguing, like his piece on "the poignant workday of a Danish claims assessor" from January 1940. In April of that year he visited "my friend, the policeman on the corner", and in August he covered "Windows – look out!", another dangerous day in the life of a window cleaner.[598] Michelsen always put a distinctive slant on the workaday lives of the protagonists he chose, and he used an anecdotal style that made their lives seem appealing, and entertaining. His articles often had a Danish slant as well, underscoring contributions made by Danish individuals or industries at a Danish or, preferably, international level.

Similarly, science pieces by Mikkel Borgen were often based on visits he paid to scientists, conveying the sense of wonder that their peculiar preoccupations were presumed to arouse in the readers of *Familiejournalen*. Also, Borgen also had a habit of emphasizing international contributions made by Danish science. Many of Borgen's visits were later included in Børge Michelsen's trilogy from 1941 with the overarching title *Vore unge Videnskabsmænd arbejder* (*Our Young Scientists at Work*). The preface explains Michelsen's original motivation for writing these books:

> One day, a taxpayer was struck by the idea that it really would be quite amusing to find out what Danish science was spending its money on, and because that taxpayer just happened to be a journalist and was used to translating his ideas into action, he immediately set about satisfying his inquisitive nature.[599]

This point of departure is soon forgotten, however, as the reader delves into the contents of the books. Michelsen himself describes how he is captivated by the atmosphere in the laboratories and by the scientists' exciting stories. The reader is led through various accounts from the workshops of science, with special focus on young Danish researchers, and on their projects and their potential significance to science and society.

The trilogy contains very few references to the financial conditions of science, but Michelsen nevertheless concludes that "it is only a very small fraction of the taxpayers' money that goes to Danish science. The Danish society pays a cheap price for its scientific results. Whether this is worth it in the long run is not something I will take a position on here."[600]

But take a stand was precisely what Bernal had done with *The Social*

Function of Science in 1939, and later Michelsen took a stand, too. In a feature article printed in *Politiken* in 1943, Michelsen drew the readers' attention what he termed the "proletarization" of Danish science.[601] He demonstrated through a number of concrete examples how, since the turn of the century, the salaries of junior researchers in Denmark had fallen far behind compared with their foreign colleagues, and even compared with Danish craftsmen like carpenters and painters. Bernal had incidentally made the same point in his discussion of British science versus American and Russian science. As a result, Michelsen asserted, Danish scientists were forced to scratch out a living as best they could, making do with an absolute minimum until perhaps, at a ripe age, they could achieve tenure in one of the country's all-too-few professorships. And if they were finally so lucky as to get a chair, they could expect a life burdened with far too much teaching and far too many administrative responsibilities – all of this ensuring only one thing: that they would hardly have time to continue the research that was so vital to the future reputation of Danish science.

Michelsen's opinions did not fall on deaf ears, but resulted in deluge of letters to the editor, particularly from younger scientists. Now that someone outside their own circle had prised the lid off of the pressure cooker, it boiled over with personal accounts of miserably paid jobs, an absence of job security, and a general lack of recognition, all of which they had so far kept to themselves. For weeks and months after Michelsen's article, the heated debate simmered on between the rebellious scientists and the politicians and officials, who defended themselves by referring to the difficulties caused by the German occupation. Finally the parties agreed to form a commission that would scrutinize the conditions under which science had to work. Meanwhile, the break between the Danish government and the German forces in August 1943 meant that the commission was not actually appointed until World War II had ended. And because the commission did not exactly throw itself into the task, nothing was really done to resolve the grave financial problems facing Danish science until the early 1950s, as described in chapter 20.

In the protracted interim Michelsen regularly wrote articles and commentary in various newspapers chiding the responsible politicians and entrenched institutions like the Royal Academy of Sciences and Letters for passing too lightly over the matter of the young Danish scientists' wretched conditions. Michelsen took matters into his own hands when, after becoming the science editor of the daily *Ekstrabladet* in 1945, he founded the *Ekstrabladet* honorary science award of 10,000 Danish kroner, which was given to nine Danish scientists "of international standing", as the founding statutes prescribed. But the award was also a pat on the back for the young scientists who, in Michelsen's

1946	Gunnar Nørgaard	Geodetic Institute, for gravimetry and inventing a special gravimeter
1947	Gunnar Thorson	Zoological Museum, for marine larval ecology
1948	Heinz Holter	Carlsberg Laboratory, for cell research
1949	Peder Schultz	National Museum of Denmark, for discovering the Viking fortress of Aggersborg
1950	Hans Ussing	Zoophysiology Laboratory, for isotope research
1951	Svend Johansen	University of Copenhagen, for literary research
1952	Johannes Iversen	Geological Survey of Denmark, for pollen research
1953	Franz From	Psychology Laboratory, for psychology research
1955	Aage Bohr	Institute for Theoretical Physics, for atomic-nucleus research.

Table 24.2 *Recipients of Ekstrabladet's honorary of science award.*

opinion, were underpaid and whose interests in science were being grossly exploited. The motivation for the first science award, which went to the geologist Gunnar Nørgaard described how "Doctor Nørgaard, who with unflagging diligence has expended all of his time on scientific work and has therefore always lived a modest life, fulfils, to an exceptional degree, all of the conditions one might reasonably wish to see fulfilled by a scientist who is to be rewarded with the prize in question."[602] Michelsen did not fail to remark on the fact that giving the award to Nørgaard was part of the effort to keep him in Denmark, despite the poor conditions here, and despite his receiving far more attractive offers from abroad.

The *Ekstrabladet* honorary science award was presented once a year on 12 February, the newspaper's anniversary date, from 1946 to 1953 and in 1955 (the 1954 science award giving precedence to an honorary prize for the "golden wedding anniversary couple of the year" to celebrate the newspaper's own fiftieth anniversary).

Around 1946–1947, Michelsen became involved in UNESCO's efforts to increase "the popular understanding both of science *per se* and of science's social and international implications".[603] He became acquainted with the likes of the British biologist, humanist and internationalist Julian Huxley and the British biochemist Joseph Needham, best known for his works on the history of Chinese science. According to Michelsen, they were the ones who offered him the position as head of the newly created UNESCO body known as the Division of Science & Its Popularization.[604] Michelsen took up his new position in February 1948, with great expectations on behalf of science and, not least, its popularization. He wrote the book *Rejsende i Populærvidenskab* (*A Traveller in Popular Science*, 1954) on his work for UNESCO, initially outlining the importance of science to practical agriculture and disease prevention, and strongly underscoring the role of science in the battle against general ignorance, and against demagogues like Adolf Hitler. However, it was not the scientists themselves who were to convert science into such positive results directed at the wider population. According to Michelsen, this was to be done by popularizers of science, as exemplified by science journalists.

In reality, the popular-science journalist has more influence on how the scientific results are to be exploited than the scientist, who is comparable to an engine driver that does not always believe the train's departure times are as appropriate for the travellers as they should be. It is, however, beyond his power to change them. He must stick to the schedule if he wishes to keep his job.

> Can anything else be done about it? Yes, if the journalist takes on the role of engine driver and makes the travellers realize that there is something wrong with the schedule. Then they can have it adjusted if they band together, and go to the railway management and demand it.[605]

In other words, Michelsen was not merely confident that using and disseminating scientific knowledge, communicated by science journalists, would make the world a better place to be. He also hoped that by conveying science to the people, they would generally become conscious of its usefulness and development, thereby becoming motivated to seek influence on the direction and aims of progress. However, after three years in UNESCO's Parisian bureaucracy Michelsen had had his fill and voluntarily gave up his role as head of the division, opting for a position as a regular employee with UNESCO's science department in India. He spent a year there getting various popular-science initiatives going in several Asian countries, Australia and New Zealand before returning to Denmark in 1952 and resuming his career as a science journalist at the newspaper *Information* and elsewhere. For a time he edited a regular spread called "Videnskab og Samfund" ("Science and society") – and so Michelsen was still taking an interest in increasing the visibility and perhaps even strengthening the social function of the sciences.

Niels Blædel. Besides Børge Michelsen there was another prominent Danish science journalist who worked in the early 1950s to improve the financial situation for Danish science: Niels Blædel. He had been employed as a science

24.4 *The science journalist Niels Blædel was presented with the Danish Union of Journalists' prestigious Cavling Prize in 1952 for "his journalistic initiative and his talented contribution in fighting the battle to improve the conditions for Danish science" (anonymous quote, 1952). Writing in the Danish daily* Politiken, *Hans Marius Hansen the rector of the University of Copenhagen, congratulated Blædel, whose "tireless work for the issues which, at this very point in time, are crucial matters for Danish science – the construction of sorely needed laboratories, support for science through the establishment of a scientific foundation and scholarships for young scientists, support for our students so they can find the time to study – has been instrumental in creating the sentiment among the [Danish] population that paves the way for concrete political action. In this work, the journalistic quality of which is outstanding, he has always made sure to acquaint himself with the substance of the matter at hand, and everyone who has been associated with him agrees upon emphasizing his unfailing loyalty" (Hansen 1952).*

journalist at *Politiken* in 1947 after a brief stint as a science journalist at *Berlingske Tidende*, and like Michelsen he wrote about Danish science as a sector that was in dire financial straits. Blædel did not subscribe to Bernal's ideas on the social function of science, as Michelsen did, but like many others he was upset to see Danish scientists leaving the country, lured away by more attractive positions abroad. Examples were the physiologist Hohwy Christensen, who moved to Sweden in 1949 to join the Karolinska Institute in Stockholm, and the astronomer Bengt Strömgren, who left Denmark in 1950 for the US. That is why Blædel decided to delve further into the financial aspects of science in Denmark.[606]

Blædel successfully applied in 1950 for a travel grant from the Smithsonian Institution to cover a three-month study tour of the US, intending to see with his own eyes the conditions in the nation that was able to attract so many Europeans scientists. He visited several large American universities on his journey from coast to coast and was overwhelmed by the spirit of optimism, enthusiasm and initiative among the scientists he met. Upon his return to Denmark, Blædel put his experience to use by initiating a newspaper campaign in *Politiken* to improve the conditions that reigned in the world of Danish science. Blædel himself wrote a series of articles about the trials and tribulations of Danish scientists, also encouraging Danish and foreign scientists to write to the newspapers about conditions in Denmark compared with those abroad. He even launched the idea in early 1951 of starting a lottery to benefit the sciences. Blædel's campaign coincided with – and was partially coordinated with – the internal mobilization among dissatisfied Danish scientists; a mobilization that resulted in a large demonstration of students and scientists who marched together through the streets of Copenhagen on 2 February 1951 to rally outside Christiansborg Castle, the seat of the Danish parliament, as also described in chapter 20.

Niels Blædel continued to work as a science journalist at *Politiken* until 1970. For most of his time there he also edited the popular-science journal *Naturens Verden* (*The World of Nature*), which he took over in 1970 to publish independently through his own publishing company, Rhodos. Unlike Børge Michelsen, who often emphasized the responsibility of the science journalist to become thoroughly acquainted with the subject in advance, and to keep abreast of scientific developments, Blædel believed that the best science journalism built upon a total, albeit often feigned, ignorance of the scientific topic at hand.[607] Only this would enable the science journalist to ask the "stupid" questions that were needed to translate the language of science into something generally comprehensible. And although Blædel is often described as very loyal to the practitioners of science, in conveying science he was concerned above all with his readers.

The impulses that led Blædel into the field of science journalism originally came from the two atomic bombs dropped on Japan in 1945, and from the science-related questions they put on the public agenda. In Blædel's philosophy, the science journalist's role was to be the *general public's* spokesperson in the dialogue with the sciences, not vice versa, which is precisely why he believed that as communicators, journalists had to identify with their readers rather than with the scientists. Concern for the reader was crucial, and the best way to show it was to assume a stance that was just as uninformed as the reader's. Therefore, wrote Blædel in 1954, the ideal science journalist would not have any scientific background. That would merely prevent the communicator from asking the representatives of science the right questions. On the contrary, Blædel concluded:

> The journalist who has no scientific education is identical with the reader, with the broad readership, and he is the one the reader needs. That is why it was the journalists who pushed forward when the first atomic bomb fell, throwing up new boundaries between past and future. When twentieth-century humanity felt the ground shake, it had to be the journalists who conveyed the questions to science.[608]

Blædel proceeded to sketch out two paths for Danish science journalism: one that was both informative and entertaining, and one that was purely informative. The latter, which was the path Blædel took, had difficulty finding acceptance among the editors. One reason for these difficulties was the teething troubles that science journalism in general was going through, but more important was the growing demand that science should be entertaining, which made the more serious, informative approach hard to promote. Blædel feared that in the future, the entertainment approach would completely take over, not least in light of the new visual "methods of communicating": films and television.[609]

Spørg Århus. Niels Blædel was not the only sceptic when it came to entertainment as a means of successfully communicating about science. Learning demanded a measure of self-sacrifice, whereas entertainment was escapism. This was the basis of many science popularization initiatives in the 1950s and 1960s, when it was understood that science and technology were subjects that could only be properly conveyed in the form of factual information.

The provincial programming department of the Danish Broadcasting Corporation took a rather less rigorous approach: One Saturday evening, on 8 October 1966, a brief television presentation introduced a new Saturday-evening show called *Spørg Århus* (*Ask Århus*), adapted by Tage Elmholdt and Mogens Kjelstrup, who both worked in provincial programming. They had found inspiration in a Swedish show called *Fråga Lund* (*Ask Lund*), which had been a great television success in Sweden since 1962.[610] The idea behind *Fråga*

24.5 *On-screen image from the science show* Spørg Århus – Ask Århus – *which was launched in 1966 and quickly became a great success. The expert panel, recruited from the Århus area, consisted of "seven learned doctores" who would answer questions sent in by viewers. Seated left to right: Søren Sørensen (music science), Robert Bech (law), Jens Kruuse (literature and culture), Flemming Kiss-meyer-Nielsen (medicine), Kjeld Ladefoged (biology) and Olaf Pedersen (exact sciences). Standing: Knud Hannestad (history), the panel moderator, and Tina Wilhelmsen, the show's "charming hostess".*

Lund, and later *Spørg Århus*, was to have a panel consisting of people with an academic background respond to questions sent in by viewers. *Fråga Lund* had successfully married academic content with popular entertainment. This was also what the Danish adapters aspired to do, and they did not shy away from the idea of giving the entertainment aspect even greater weight in their show than it had in the Swedish original.[611] This was emphasized in the conclusion of each of the first nine *Spørg Århus* programmes, all broadcast live from Stak-laden, the spacious student canteen at the University of Århus. Once the trans-mission of the programme was over, the audience – which largely consisted of young people from the university's student body – was invited to participate in a prearranged dance held on location.

The Swedish programme was an indisputable success, at least in terms of viewer ratings. *Fråga Lund*, aired in one of the best prime-time slots, attracted about two million viewers, and questions to the panel literally poured in. One reason was the humorous way in which the six panel members handled the questions, which can hardly have suffered under the effects of the wining and dining that preceded each programme. The same model was used in Århus, where the "seven learned doctores", as the panel was always called, would meet on the Saturday of each session and round off their meal with coffee and brandy before going onstage. In other words, alcohol proved to be an element that while not insignificant was perhaps also not surprising in the communication and popularization of science as represented in television programmes like *Fråga Lund* and *Spørg Århus*.

Fundamentally speaking, good chemistry was the key to the considerable success of these programmes. The people initially chosen to participate in the Danish programme knew one another and the adapters already from campus life in Århus, so their tone was jovial and relaxed. The first six were Robert Bech (professor of law at the University of Aarhus), Knud Hannestad (professor of history, likewise in Århus), Flemming Kissmeyer-Nielsen (chief surgeon at Århus Municipal Hospital), Jens Kruse (cultural editor at the daily *Jyllands-Posten*), Søren Sørensen (professor of music at the University of Aarhus), and Keld Lagefoged (forest superintendent). The producers also got Olaf Pedersen (senior associate professor, and later professor, of the history of exact sciences at the University of Aarhus) to join in the fun, and to guarantee the factual correctness of matters relating to the natural sciences. These were the "seven learned doctores" who declared themselves ready to answer "all manner of questions from inquisitive viewers" when *Spørg Århus* first went on the air on Saturday, 22 October 1966.[612]

Besides serving on the answering panel, Hannestad also presided over the discussions that went on, assisted by the *Spørg Århus* hostess Tina Wilhelmsen. There was also a "question team", selected in advance for each programme, which was responsible for following up on viewer questions and posing unprepared questions to the panel members. Aided by a number of student assistants, the two producers and Hannestad would choose a number of viewer questions from the pile of letters that had come in. These were then presented to the panel well in advance of the programme, giving panel members time to prepare their answers – which must be neither too long nor too erudite, but might very well be illustrative or even humorous. The question team was supposed to make sure that the answers were understandable, and their follow-up questions gave the show an element of spontaneity and surprise.

The intention behind building the show around viewer questions was to "accommodate a healthy and natural curiosity among the public".[613] "Keep their sense of wonder awake" was the slogan Knud Hannestad used in the presentation programme, which also illustrated the sort of questions they were looking for. As part of the presentation, Hannestad and Mogens Kjelstrup gave four examples from *Fråga Lund*:

> What can be the reason for the unprecedentedly rapid development in all scientific and technical fields within the past 100 years? Why did nothing, or very little, happen during the preceding millennia? Are modern humans more intelligent than their forefathers?
>
> What happens, in strictly physiological terms, when a person washes? What does a person perceive as coldest – having clean feet, or dirty?
>
> What happens when a Nordic beech tree is transplanted to a warm African climate? Will it still act like a deciduous tree and lose its leaves?
>
> What will happen if a Swedish ant is suddenly moved to another country, say, Denmark? Would such a tiny ant, which has been exclusively 'raised' in Swedish, also be able to make itself understood in Danish?

The presentation on television also emphasized that *Spørg Århus* was neither meant to be a quiz show nor a test of intelligence between the participants or the panel, nor for that matter an advice-column programme that would address the "infirmities" of its viewers. The show was meant to be an outlet for the kind of questions that aroused speculation, which most people experience regularly when running into problems beyond their own sphere of knowledge. The issues could be small or large, serious or amusing. As Hannestad concluded on behalf of the panel:

> It is the panel's mission to answer the questions in the same spirit as they were posed. We shall try to respond to them, well, first and foremost as correctly and as exhaustively as we are able to, but we sincerely hope that we will also be able to do so in a genial and informal fashion ... We shall endeavour to make the programmes entertaining.[614]

The viewers' unanticipated eagerness to send in questions meant that the show was carried beyond the three programmes first planned for 1966. In December of that year, the producers decided to continue *Spørg Århus* throughout the winter, as "the nature of the viewers' involvement in the new question show has been so intense".[615] The Danish Broadcasting Corporation received about 4,000 letters from curious viewers for each programme, which made the channel decide to run the show during the 1967–1968 season.[616] Years later, in 1974, the idea was revived again in a series of programmes broadcast up to and including the Christmas season of 1976. This time the show was not transmitted directly from Stakladen as before, but at the Kasino studios in Århus. According to the adapter Tage Elmholdt, the move resulted from the television people's worries about how the university students would act in the audience.[617] The student rebellion had occurred between the first and second rounds of *Spørg Århus*, and apparently the risk of tomatoes and rotten eggs being thrown at "the learned doctores" was something the producers took very seriously.

Criticism, sensationalism and coexistence

When the first "real" science journalists began to make their presence felt in the major Danish newspapers and magazines, and later on radio and television as well, it was typical that they all sought to draw a positive picture of science and its representatives. As the examples in this chapter show, their efforts mainly went into communicating scientific methods and findings to a wider audience in a way that laymen could understand, and which at the same time was loyal to the practitioners of science. The examples also show how some of the most influential science journalists energetically sought to convince the public of science's vital importance to society, and to argue that Danish science had laid many golden eggs, despite the miserable conditions it had to cope with. It was rare indeed to see the nature or the utility of any sort of scientific research drawn into question. Researchers and journalists grazed peacefully side by side in the fields of scientific knowledge.

As the 1970s drew near, this once so idyllic and peaceful coexistence began to develop into a more complex relationship. Within a matter of years, factors like the student rebellion and its focus on what it called "critical science", as well as the burgeoning environmental debate and, not least, the debate over nuclear energy, created a new brand of science journalism that considered it fair and right to pose highly critical questions to scientists about their research. Such critical science journalists typically accused the scientists of concentrating on the minutiae of their profession in order to veil the potentially unfortunate consequences of their results. On the other hand, the scientists and technicians who found themselves under attack were inclined to think that the critical journalists were simply interested in sensationalist reporting and, also that these journalists were simply not competent to write about the complicated world of modern science. The trenches in this war were dug deep when the 1970s came around.

Gradually, as the 1970s wore on, the highly polarized disputes on environmental issues and nuclear power subsided. They were superseded by more moderate debates in which science journalism was once again able to play a constructive role as a critical mediator between science and the public sphere, as the last of the periods reviewed in this thousand-year history of Danish science drew to a close and a new era began to unfold.

Epilogue

This work is our best effort to present a broadly contextualized description of how science has developed in Denmark, beginning in the Middle Ages and ending in the 1960s. Obviously, development did not cease at that point, but the pace began to accelerate dramatically compared with the preceding decades. Over the last 40 years science in Denmark and around the world has been typified by an ever-higher degree of specialization, by increased focus on inter-disciplinary research projects, by continued internationalization, by the expansion of big science, and by many other trends as well. When standing amidst the rapid eddies and currents of these turbulent times it can be difficult to identify the features that most clearly typify scientific development over the last four decades, and all but impossible to determine which scientific research fields and breakthroughs offer the greatest future potential in a Danish context. Any conjecture along these lines would have to build upon findings arising from a major project to map out contemporary history; a sort of project that we certainly have not been able to carry out. The following is therefore nothing more than the personal observations of the authors when contemplating what most notably distinguishes science prior to 1970 in Denmark (and probably in many other countries as well) from science in the decades approaching and entering the new millennium. Generally speaking, the most important trend is a radically altered view of the utility value of science.

The old adage that there is nothing new under the Sun holds more than a grain of truth. As Parts I and II show, during the Middle Ages and the Renaissance the benefactors of science definitely expected tangible results – repayment, if you will – from the scientists they supported. The patrons, whether royal, aristocratic or ecclesiastical, expected the work of their scientists to have utility in the sense that it would lead to increased insight into God's Creation. One's life on Earth was considered to be a test, a sort of examination on the path leading to eternal life after death, and in order to pass this examination one had to believe the Word of God as written in His book, the Bible. Fortunately, faith could be bolstered by thoroughly studying the book of Nature – that is, studying the realm of science – because such studies could bring humankind greater insight into how God in His infinite wisdom had chosen to organize the world.

During the Enlightenment of the 1700s another view began to gain ground, namely that life on Earth is significant in and of itself, and that the living conditions of humanity could be improved through scientific research, by subsequently using scientific findings to benefit society as a whole. For many years the interaction between pure and applied science was quite modest, but the ties grew in strength and number, and from the late 1800s onwards a new trend emerged. A number of modern industrial companies, most notably in the fields of electrical engineering and chemistry, set up their own independent research laboratories. One very illustrative – and very early – Danish example of this is the Carlsberg Laboratory, which is described in several sections of Parts III and IV. Beginning in the 1870s, scientific disciplines like chemistry and physiology were systematically used to advance the technological development of Carlsberg's commercial activities, nudging the art of brewing closer to the realm of science. And for the scientists involved in these activities, the primary criterion for success was not to achieve excellent research results and convey them to their peers in the international scientific community, but rather to make a positive impact on their company's bottom line.

At the same time many university scientists were busy promoting "pure science", even though public grants were small and research funding still had to be sought from private patrons. Danish philanthropy was particularly fortunate to have the Carlsberg Foundation, which was of inestimable importance to Danish science from the late 1800s and into the 1950s. Proponents of the two different types of research, pure and applied, emphasized the usefulness of their brand of science, though naturally they defined its "utility" in different ways. Applied science was deemed to be useful from a narrow financial point of view, whereas pure science could claim usefulness because of its value to sustaining Danish culture, its contribution to greater basic knowledge about nature, and, of course, its usefulness in the service of formative ideals and actual education.

After World War II, and particularly from the mid-1950s onwards, Denmark saw a marked increase in public investments in scientific research. It is notable, however, that within a very broad and loose framework the politicians left it to a small, exclusive group of prominent scientists, headed by Niels Bohr, to decide how the large amounts of new funding were to be spent. This group decided to give top priority to pure, basic science, which it considered to be the most beneficial of all types of research, as did similar groups of scientific advisers in other countries. But why? For two reasons: firstly because basic science was expected to result in the greatest contribution to humankind's pool of knowledge and understanding, and secondly because the assumption was that pure scientific discoveries were guaranteed to lead, sooner or later, to practical applications that would contribute to society's rapid technological and economic development.

History has since shown that the linear innovation model, as this assumption is often called, did not adequately reflect the complexity of the real world. Excellent Danish basic research in the field of nuclear physics, for instance, failed to engender the successful development of Danish nuclear reactors, even though the reactor physicists and engineers at the Risø Nuclear Research Centre energetically worked towards that specific goal in the 1960s. On the other hand, in the post-war years a number of other Danes presented basic-research findings of an international calibre. One example is Willi Dansgaard and his associates, who spent several decades drilling kilometre-long ice cores out of the Greenland ice cap, measuring the content of ^{18}O in the succession of deep-core samples, and using the values for this oxygen isotope as a precise indicator of climatic variations in the past. Another example is Jens Christian Skou, who after years of studying cell membranes in the late 1950s discovered the mechanism known as the sodium–potassium pump, which maintains the salt balance in the cells of living organisms.

In 1997, Skou received the Nobel Prize in chemistry. Most people probably imagined that Skou, then a seasoned scientist of 79, would simply except the homage and enjoy the fruits of his labour – but that was not what happened. A week after the Nobel Prize was announced, Skou gave a celebratory lecture at the University of Aarhus. The audience, which included the Danish minister of research, was not a little surprised when Skou launched a frontal attack targeting several key elements in the new Danish research policy. Having first told the story of his groundbreaking discovery four decades earlier, Skou made quite a remarkable statement that echoed across the length and breadth of Danish society:

> The funding for research was meagre when I began researching in the 1950s, but I was fortunate enough to have lived under a grant system that allowed me to work undisturbed. Seven years passed before I was ready to write my first article. That would never have been acceptable under the Americanized grant system. I therefore make an appeal in favour of returning to the old system, which will give young scientists the opportunity to put their ideas to the test.[618]

Over the weeks and months that followed, the debate raged in the Danish newspapers and journals, and it was continuously fuelled by contributions from the seemingly indefatigable Nobel Prize winner and the many people who shared his views.

Be that as it may, developments since then do not seem to indicate that Skou or any other like-minded scientists will succeed in provoking significant changes in Denmark's research policy. Much has changed since Skou carried out his epoch-making research at the University of Aarhus in the 1950s, and up through the 1990s the new national research policy that gave priority to goal-oriented, strategic research efforts rather than "blue-skies" research began to gain considerable momentum.

What Skou reacted to so strongly was the reform of the Danish research system that took place in the 1990s.[619] An OECD assessment of Danish research policy carried out in 1987–1988 had expressed considerable amazement – and a certain degree of scepticism – at the scientist-based, highly pluralistic and non-coordinated decision-making structure underlying Danish research. One point of criticism was that seen as a whole, Denmark's research policy was a motley assortment of ministries, committees, councils, agencies, and groups. This jumbled picture was what prompted the OECD to recommend that Denmark implement a comprehensive (re)organization of the decision-making processes and advisory functions governing the country's public research funding and activities.[620]

In 1993, for the first time ever, Danish research got a ministry of its own, albeit initially a small one with limited resources, and therefore with a limited ability to influence the country's research policy. Nevertheless, during the 1990s this ministry played a key role in establishing what has since become known in Denmark as "programme policy", which brought a series of strategic research efforts into focus. The aim of these efforts was to break down the barriers that had become entrenched in the Danish research system since the 1960s, and all of the efforts rested upon the fundamental premise that scientific research was intended to promote industrial growth in a number of core areas such as information technology, biotechnology and materials research. The Danish programme policy was actually a measure aimed to counter the perception that research policy acted to support the goals and means of research itself. Precisely because research was now seen as a key techno-economic growth factor, it was vital to have a policy that would increasingly control what went on in the world of scientific research.[621]

The new utility value of science – to society, to businesses and to the marketplace – as attributed to scientific research in the late twentieth century has been underscored with renewed vigour in the early twenty-first century.

Research is one of the most significant instruments for increasing innovation and socio-economic growth, and it is the object of research policy to ensure that research does indeed make itself useful in this way. The boundaries between the scientific disciplines, which in many cases have solidified along lines defined by historical processes, are not always conducive to realizing the full potential of scientific utility. That is why cross-disciplinary research has been receiving increased attention as a productive way to create result-oriented, societally relevant, strategically valuable knowledge.[622] The most recent restructuring of the Danish advisory and funding system for scientific research took place in 2003. This reform was mainly driven by the wish to promote major inter-disciplinary research initiatives that not only would ensure Danish science a prominent position in the international arena, but would also serve a useful function for Denmark and the Danish people in the years to come. Whether the new research policy can fulfil these aspirations is a question best left to the historians of the future.

Notes for Chapters

NOTES FOR THE PROLOGUE

1 *Københavns Universitet, 1479–1979*, vols 1–14. Volume 7 deals with the faculty of medicine and volumes 12–13 with the faculty of science.

2 Meisen 1932. Shackelford 1994. Jamison 1982. See also Pedersen 1992.

3 Van Berkel, van Helden and Palm 1999. See also the essay review in Kragh 2000. The Belgian history is Halleux et al. 2001. On the general problem of writing national histories of science, see Crosland 1977.

4 Jamison 1982. On the concept of national styles of science, see Vicedo 1995.

5 Petersen 1893, p. 73.

6 Morsing 1937, p. 23.

7 Van Berkel, van Helden and Palm 1999, pp. 7–9.

NOTES FOR CHAPTER 1

8 Hartner 1969. Drachmann 1971.

9 Demographic data from Ladewig Petersen 1980. See also Johansen 2002.

NOTES FOR CHAPTER 2

10 McGuire 1982.

11 Sunesen 1985 is a Danish translation of the work, which is kept at the Royal Library in Copenhagen. For an analysis of Sunesen and his work, see Ebbesen 1985.

12 Along with the printed text from 1510, Kroon 1993 also contains a photographic reproduction of the manuscript.

13 See Bradley 1976.

14 On medicine in medieval Denmark, see Møller-Christensen 1944 and Gotfredsen 1973, pp. 107–122.

15 *Liber herbarum* is reproduced in Hauberg 1936.

16 Schæffer 1963.

17 Bagge 1984.

18 Pinborg 1981.

19 For a translation into modern Danish, see Boethius 2001.

20 Grant 1974, p. 48.

21 On Petrus de Dacia, see Pedersen 1968 and Meisen 1932, pp. 12–15.

22 Meisen 1932, p. 12.

23 Pedersen 1975.

24 The most complete work on Clavus is Bjørnbo and Petersen 1904.

25 The authoritative work on the history of the University of Copenhagen is *Københavns Universitet 1479–1979*, vols 1–14. Photographic reproductions of the university's charter and statutes for 1475–1482 are found in Pinborg 1979. See also Norvin 1929.

NOTES FOR CHAPTER 3

26 Hørby 1962.

27 For more on the Round Tower as an astronomical observatory, see Nissen 1937 and Thykier 1990, vol. 1.

28 Oliver 1703, pp. 1406–1408.

29 Schepelern 1971 and 1985. See also

Shackelford 1999. On early collections and museums, see Hooper-Greenhill 1992.

30 On Worm and the notion of "sports of Nature" (*lusis naturae*), see Kragh 2006.

31 On the Gottorp collection, see Schlee 1965, and on the early history of the Royal Kunstkammer in Demark, see Liisberg 1897.

32 Scherz 1971, pp. 9–137.

33 Lausten 1987, p. 143.

34 Helk 1987 includes detailed information about the educational travels of Danish-Norwegian students.

35 Ilsøe 1963 presents a complete account of foreigners' journeys in Denmark.

36 Fussing and Pihl 1955. Birch 1968, vol. 3, pp. 182–183.

37 Boyle 2001, vol. 3, p. 168. The French correspondent was Adrien Auzout, who had recently been elected a foreign member of the Royal Society.

38 Leibniz 1976–1995, vol. 4, p. 66.

39 Dahl 1957. Appel 2001.

40 Christianson 2000, p. 85. See also Mosley 2000.

41 Hall 1962, p. 191.

42 Ilsøe 1980. Ilsøe 1982.

43 Bartholin 1914 is a Danish translation of the Latin text.

44 Steensgaard 1995. Lassen 2002.

45 Arrebo 1965.

46 Worm 1965–1968.

47 On early scientific journals, see Kronick 1962 and Manten 1980.

48 Hall and Hall 1965–1986, vol. 9, p. 403.

49 *Acta medica* has never been subjected to the scholarly investigation it deserves. For an introductory survey, see Kragh 2003a.

50 Fussing and Pihl 1955, p. 92.

NOTES FOR CHAPTER 4

51 For a comprehensive analysis of this relationship 1536–1636, see Fink-Jensen 2004. See also Danneskiold-Samsøe 2004.

52 Norvin 1937–1940, vol. 2, pp. 9–40. On aspects of the reformed university in Copenhagen, see also Grane 1981.

53 Wear 1981, pp. 229–230.

54 Kusukawa 1999, p. 160.

55 Norvin 1937–1940, vol. 2, p. 27.

56 Rørdam 1889–1992, p. 24.

57 Schepelern 1971, p. 101.

58 Schönbeck 2004.

59 On Caspar Bartholin, see Grell 1993 and Ebbesen and Koch 2003, pp. 139–168.

60 Sinning 1991, pp. 39–40.

61 Steno 1997, pp. 159–160.

62 On Aslaksen and Mosaic physics, see Blair 2000.

NOTES FOR CHAPTER 5

63 Kragh 2002.

64 Modern biographies of Tycho include Thoren 1990, Wittendorff 1994 and Christianson 2000.

65 Swedish translation in Nordlind 1951. On Tycho biographies, see Kragh 2007.

66 Gingerich and Westman 1988.

67 On Tycho's cosmology and different versions of it, see Schofield 1981.

68 Sandblad 1943.

69 Quoted in Christianson 1979.

70 Moesgaard 1977.

71 Chapman 1984. Hashimoto 1987.

72 Kragh and Sørensen 2007.

73 Moesgaard 1975.

74 Pedersen 1987.

75 Pihl 1944, p. 78.

76 Rømer 1910. Cf. Middleton 1966.

77 Rømer 2001, pp. 583–584. Van der Star 1983.
78 Rømer 2001, pp. 371–375. Pedersen 1976.
79 Grell 1998a.
80 Ebbesen and Koch 2003, p. 274.
81 Brahe 1946, p. 132.
82 Figala 1972.
83 Shackelford 2003
84 Severinus 1979, p. 122.
85 Worm 1965–1968, vol. 1, p. 49.
86 Severinus 1979, p. 46.
87 Maar 1926. Borch 1983.
88 Borch 1983, vol. 3, pp. 65–66. Principe 1998, pp. 259–260.
89 *Philosophical Transactions* 1674, p. 299.
90 Maar 1926, p. 24.
91 Kragh 2003b, p. 59

NOTES FOR CHAPTER 6

92 Westfall 1994, p. 7.
93 Ladewig Petersen 1980, p. 135 and p. 379.
94 Bartholin 1991, p. 32. See also Lohne 1977.
95 For a meticulous analysis, see Danneskiold Samsøe 2004.
96 Hougaard 1983, p. 482.
97 Danneskiold-Samsøe 2004.
98 See also Grell 1993.
99 Bruun and Loldrup 1982, pp. 185–242.
100 Moe 1988, p. 137.
101 Baldwin 1998.
102 Godtfredsen 1957b. Garboe 1949–1950. On Bartholin and Harvey, see also French 1994, pp. 153–168.
103 Hovesen 1987, pp. 170–172
104 Steno 1987, vol. 1, p. 69.
105 Bartholin 1940. Garboe 1949–1950. See also Schioldan-Nielsen and Sørensen 1994.
106 The literature on Steno is extensive. See www.stenoarkiv.dk. See also Moe 1988 and Poulsen and Snorrason 1986.
107 Harvey 1957, pp. 107–116. Kragh 2003b.
108 Boyle 1772, vol. 2, p. 652.
109 Steno 1997.
110 Scherz 1968.
111 Garboe 1959, p. 36.
112 Steno 1969, pp. 161–163
113 On Steno and geological time, see Frängsmyr 1971.
114 Drake 1996, p. 117.
115 Huxley 1881, p. 453.
116 Steno 1987, p. 186 and p. 283.

NOTES FOR CHAPTER 7

117 Johansen 1979, p. 259.
118 Rømer 2001, p. 340.
119 Forchhammer 1869, p. 132.

NOTES FOR CHAPTER 8

120 The various tenure arrangements applied to professorships throughout the centuries are reviewed in Slottved 1978.
121 Quoted in Norvin 1937–1940, vol. 2, pp. 114–115.
122 Petersen 1893, p. 258.
123 On Oeder and the *Flora Danica*, see Anker 1951 and Skytte Christiansen 1973.
124 Snorrason 1974. Splinter 2007.
125 See Bostrup 1996, pp. 55–60.
126 On Pfaff and his work on the Volta pile, see Kragh 2003.
127 See Midbøe 1960 for a detailed account of the Norwegian society.
128 Quoted in Gilje and Rasmussen 2002, p. 383.
129 Amundsen 1957.
130 Quoted in Ebert 1988, p. 98.
131 For details on Wessel's mathematics and cartography, see Lützen 2001 and Branner and Lützen 1999.
132 Andersen 1968.

133 Pedersen 2001. For an English translation of the travel diary, see Pedersen 1997.

134 The comprehensive account, which spanned more than 600 pages, was translated into both English and German in 1801. A new annotated edition of the English translation is found in Crosland 1969.

135 Wagner 1999.

136 Quoted in Lundbye 1929, p. 39.

137 Winter 1965 and Winter 1990.

138 Smith 1950.

139 Nielsen 1950.

140 Schumacher's correspondence with Humboldt and Gauss is published in Peters 1860–1865 and Biermann 1979.

NOTES FOR CHAPTER 9

141 Gouk 1983.

142 Koch 1992, pp. 24–30. See also Andersen 1934, pp. 610–612; Kristiansen 1997, pp. 77–90. Quotes from these sources.

143 Cf. Spang-Hanssen 1965, p. 68.

144 Bostrup 1974; Albeck 1984, pp. 13–16.

145 Bostrup 1996, p. 109.

146 Quoted after Koch 2003, p. 268, which contains a review of Rothe as a natural philosopher.

147 See Larsen 1991. For the development after 1775, see also Haue 2003.

148 Larsen 1991, p. 107.

149 The note in *Dansk Litteratur-Tidende* is reproduced in Teuber 2002, pp. 105–106.

150 Harding 1924. Wagner 1999, pp. 137–142.

NOTES FOR CHAPTER 10

151 Thuborg 1951.

152 H. C. Ørsted, *Videnskaben om Naturens Almindelige Love* (Copenhagen, 1809), vol. 1, pp. 6–7.

153 On Steffens, see Hultberg 1973 and Koch 2004, pp. 31–56

154 *Forhandlinger ved de skandinaviske Naturforskeres andet Möde* (Copenhagen, 1841), pp. 25–42.

155 Christensen 1995.

156 Jacobsen 2001.

157 Ørsted 2003.

158 Ørsted 1966, p. 286 (first published in 1852).

159 Meyer 1920, vol. 3, p. 314.

160 Sylvest 1972, appendix 4, p. iv.

161 On Ørsted's discovery of electromagnetism, see Dibner 1961.

162 Jacobsen and Larsen 2002.

163 Ørsted 1820, p. 276.

164 See Dibner 1962, especially pp. 42–51.

165 Ørsted's article on "Thermo-Electricity" is reproduced in Meyer 1920, vol. 2, pp. 351–398.

166 Pihl 1972, pp. 46–47.

167 Dahl 1972. Caneva 1997.

168 Quoted from Dahl 1972, p. 48.

169 Dahl 1972, pp. 125–126.

170 Caneva 1997.

171 In an unpublished note from 1854, Colding suggested that an atom was composed like "an infinitesimally small planetary system".

NOTES FOR CHAPTER 11

172 Dahl 1941, p. 58.

173 Gosch 1873–1878, vol. II. 1, p. 400–402.

174 Ibid., p. 463.

175 Translated from excerpts in Christensen 1924–1926, vol. 1, p. 192.

176 Christensen 1924–1926, vol. 1, pp. 190–193; Meisen 1932, p. 80. For a different, more prosaic and critical assessment of Fabricius's thoughts on evolution, see Helveg Jespersen 1946.

177 The "Hermstaedt" to which Tychsen refers is

the German chemist Sigismund Friedrich Hermbstädt, who in 1792 published a German translation of Lavoisier's textbook and became a supporter of the new chemistry.

178 On Hauch, see Andersen 1996 and Jacobsen 2000.

179 See the detailed description in Jacobsen 2000.

180 Bostrup 1996.

181 Hultberg 1973. H. Steffens, "Et Bidrag til Hypothesen om den almindelige Organismus", *Bibliothek for Physik, Medicin og Oekonomie* 14 (1799), 215–241, published in German in *Zeitschrift für spekulative Physik* 1 (1800), 143–168.

182 Wilson 1972, pp. 392–396.

183 Clément 1926.

184 Clément 1920, pp. 21–23.

185 Forchhammer 1869, p. 344.

186 Wentrup 2001.

187 Garboe 1966.

188 On Schouw, see the brief articles in Meisen 1932, pp. 100–103, and in *Dictionary of Scientific Biography.*

189 On post-Humboldtian plant geography, see Nicolson 1996.

190 Quoted in Olwig 1986, p. 83.

191 Olwig 1980.

NOTES FOR CHAPTER 12

192 Woolf 1959, pp. 176–179.

193 Nielsen 1957.

194 Kragemo 1960, p. 111. The reference is made to the recognized French natural philosopher Jean Jacques Dortous de Mairan, who in 1733 presented a Cartesian theory of the northern lights in his *Traité physique de l'aurore boréale.*

195 Rostrup 1985, p. 189.

196 Barton 1998, p. 171.

197 Malthus had included demographical data from Denmark and Sweden in the empirical material on which he based his views on population growth. On Malthus and Denmark, see Christensen 1996, pp. 81–87.

198 Barton 1998, p. 29.

199 Johnston 1830 (special issue).

200 Kjølsen 1965.

201 Rasmussen 1997, p. 59.

202 Ibid., 1997, p. 70.

203 On the journey to Arabia, see Mortensen 1996 and, in particular, the magnificent work Rasmussen 1997, which contains a number of original sources and plentiful illustrations.

204 Niebuhr 2003–2004.

205 Fisher 1977. Lind and Møller 2003.

206 Steller 1993.

207 The expeditions to Iceland are described in Lomholt 1942–1973, vol. 3, pp. 101–118.

208 Bjerg 1984, p. 75.

209 Gad 1976, p. 454.

210 Kornerup et al. 1923.

211 Here after Thalbitzer 1932, p. 104. A part of the *Fauna Groenlandica* concerning mammals and birds appeared in a Danish translation in 1929.

212 See Sweet 1972.

213 On the history of cryolite, see Kragh and Styhr Petersen 1995, pp. 97–128, and Kragh 1996.

214 On Giesecke and his relations to Denmark, see K. J. V. Steenstrup's introduction to Giesecke 1910 (pp. i–xxxvii) and Garboe 1963.

215 Giesecke 1910.

216 Garboe 1959–1961, vol. 2, pp. 142–149. A detailed account of Hallgrímsson's life and work is given in Ringler 2002.

217 On foreigners travelling to Iceland, see Barton 1998, pp. 127–142.

218 Accounts of the expedition can be found in

Bille 1849 and Lomholt 1942–1973, vol. 3, pp. 361–382.

219 On P. W. Lund, see Holten and Sterll 2000.

NOTES FOR CHAPTER 13

220 Thomsen 1884, 1–2. This section is based on Kjærgaard 2006b.

221 Martin Knudsen's speech is quoted in Dorph-Petersen 1961, pp. 15–16.

222 See, for instance, Wengenroth 2003.

223 Hvidt 1990, p. 255.

224 Carl T. Dreyer, "Om Flyveindustrien og dens Betingelser i Danmark", *Illustreret Tidende* (29 May 1910, p. 432).

225 Mygdal-Meyer, pp. 38–39 and 179–180.

NOTES FOR CHAPTER 14

226 The following sections are based on Kjærgaard 2006c.

227 Thomsen 1856, pp. 3–4.

228 Thomsen is quoted in Lundbye 1929, p. 224.

229 See Kjærgaard 2002 for a more detailed description of the hierarchical structure of science in the 1800s.

230 See Cahan 2003.

231 Sand-Jensen 2006.

232 Pedersen 1942, pp. 12–16.

233 Kjærgaard 2006h.

234 Sand-Jensen 2003, pp. 21–29.

235 See, for instance, Kildebæk Nielsen 2000, chapters 4–5.

236 Kjærgaard 2006h; Sand-Jensen 2006, p. 163.

237 Skydsgaard 2006, p. 217.

238 Sørensen 2006, pp. 202–204.

239 Kjærgaard 2002; Kjærgaard 2006g.

240 Knudsen and Kjærgaard 2006.

241 Quoted in Dorph-Petersen 1961, p. 16.

242 Brandes 1903, p. 403.

243 Knudsen 2005, chapter 2.

244 The establishment of the Faculty of Science at the University of Copenhagen is dealt with in Broberg Nielsen and Slottved 1983, particularly pp. 61–73. See also Slottved 2000.

245 See Kjærgaard and Kristensen 2003, pp. 90–102.

246 Holter and Møller 1976.

247 Hessenbruch and Petersen 2001.

248 M. C. Harding, "Den Polytekniske Læreanstalt – under G. A. Hagemanns Styrelse 1902–1912", *Illustreret Tidende* (12 May 1912, p. 400).

249 Drachmann 1987, p.189, 200–201.

250 Rubin 1899, p. 24.

251 Quoted in Rubin 1899, p. 72.

252 Quoted in Sestoft 1972, p. 22.

253 Quoted in "Fra Rigsdagen", *Kvinden og Samfundet* (5 June 1915, p. 169).

254 Haue, Olsen and Aarup-Kristensen 1985, pp. 58–62.

255 Rosenbeck 1994, p. 17.

256 Knudsen 1984.

257 Larsen and Tønsberg 1984.

258 Larsen 1899.

259 Quoted in Jacobsen 1925, p. 118.

260 Petersen 1993, p. 385–386.

261 Quoted in Petersen 1993, p. 388. See also Jacobsen 1925, pp. 117–135.

262 *Anordning angaaende Kvinders Adgang til at erhverve akademisk Borgerret ved Kjøbenhavns Universitet.* Reprinted in Jacobsen 1925.

263 Jacobsen 1925, pp. 137–138.

264 Broberg Nielsen and Slottved 1983, pp. 85–86; Jacobsen 1925, pp. 166–178.

NOTES FOR CHAPTER 15

265 This section is based on Kjærgaard 2006c.

266 The issue of national styles in the sciences has been discussed by a number of scholars. See,

for instance, Kohlstedt 1987, Reingold 1991 and Vicedo 1995.

267 Bravo and Sörlin 2002.

268 Pedersen 1998.

269 Ries 2006.

270 Kildebæk Nielsen 2006.

271 Fischer and Kristiansen 2002.

272 Chambers 1864.

273 See Burton 1866; Flower 1869; Humboldt 1877; Frey 1878. Flower even used the term "kjökken-mödding people" to describe the prehistoric inhabitants on the island of Herm (Flower 1869, cxvii).

274 See Atkinson 1872; Laws 1878; Deans 1876; Gooch 1882; Tregear 1888; Milne 1881.

275 Spärck 1932, p. 118.

276 A notable exception is the publication of Steenstrup's kitchen-midden findings in a distinctly Danified French – *Les* kjøkkenmøddings *de l'âge de la pierre et sur la faune et la flore préhistorique de Danmark* – following his presentation at the International Congress of Archeology in Copenhagen 1869 (Steenstrup 1872). This should, however, be compared with his first publication on the subject, which appeared in Danish in 1848 (Steenstrup 1848). Darwin had a friend translate some of Steenstrup's work into English for him.

277 This was a paper on rotifers published in the German journal *Zoologischer Anzeiger*; Wesenberg-Lund 1898.

278 Sand-Jensen 2003, pp. 231–236.

279 Kildebæk Nielsen 2001.

280 Schmidt-Nielsen 1997, pp. 311–323.

281 See Richards 2003 and Oldroyd 2003 for a generel historiographic review of natural-history subjects in the 1800s.

282 See Sand-Jensen 2006, pp. 152–153.

283 J. C. Jacobsen held Louis Pasteur and, by extension, French research in high esteem. See Holter and Møller (eds) 1976; Kjærgaard 2006c, p. 51.

284 Quoted in Sand-Jensen 2006, pp. 160–161. For more on Wesenberg-Lund, see Sand-Jensen 2003.

285 See Sand-Jensen 2006, pp. 161–167; Sand-Jensen 2003, pp. 41–59.

286 Quoted in Sand-Jensen 2006, p. 161. See also Glamann and Glamann 2004, pp. 76–97.

287 Ibid., p. 165. See also Sand-Jensen 2003, pp. 48–49; Glamann and Glamann 2004, pp. 98–106.

288 Johannsen 1903a; Johannsen 1903b. See also Hjermitslev, Andersen and Kjærgaard 2006, p. 362.

289 For more on the reception of Darwinism in Denmark, see Kjærgaard and Gregersen 2006, Kjærgaard et al. 2008, Funder 2003 and Møller 2000.

290 The letter was reprinted as a facsimile in Steenstrup 1909.

291 Darwin 1851, p. 347.

292 For loan statistics, see Hansen and Olsen 1980, p. 65; Robson 1982, pp. 224–236.

293 Lütken 1863.

294 Kragh 2006, pp. 108–111.

295 See Buchwald and Hong 2003 for a historiographic discussion of the physical sciences in the 1800s.

296 Ørsted quoted in Kragh 2006, p. 96.

297 Ibid.

298 For more on Adolph Steen and Danish mathematics during the period under review, see Henrik Kragh Sørensen's chapter on mathematics and statistics in Peter C. Kjærgaard (ed.) *Lys over landet – Dansk Naturvidenskabs Historie*, vol. 3 (Sørensen 2006).

299 "Anomalous dispersion" denotes certain

refractive properties in some substances, where the refractive index grows with the wavelength near an absorption line and subsequently decreases. See Kragh 2006, p. 98

300 See Buhl 2005; see also Clark and Nielsen 1995.

301 For the background and the path to Bohr's discovery, see Heilbron and Kuhn 1969.

302 Kragh 2006, pp. 108–111.

303 See Nielsen 2006 and Skydsgaard 2006. The practical applications of chemistry are discussed in Kragh and Petersen 1995. For a historiographic review of the chemical sciences in the nineteenth century, see Bensaude-Vincent; and for a similar review of medical science, see Hagner 2003.

304 For more on Johan Kjeldahl, see Johannsen 1976 [1900]; on Emil Christian Hansen, see Klöcker 1976 [1910] and Glamann and Glamann 2004; on S. P. L. Sørensen, see Linderstrøm-Lang 1976 [1939].

305 Nielsen 2006, p. 130.

306 See Nielsen 2000, pp. 30–43; Kragh and Petersen 1995, pp. 91–93.

307 Skydsgaard 2006, p. 235.

308 Quoted in Skydsgaard 2006, p. 227.

309 See Kjærgaard 2006d, p. 266.

310 Quoted in Schmidt Kjærgaard 2006, p. 384.

311 Hessenbruch and Petersen 2001.

NOTES FOR CHAPTER 16

312 Quoted from Hvidt 1990, p. 176. This section is based on Kjærgaard 2006b.

313 Georg Brandes 1910, p. 271.

314 See Mørch 1982, pp. 180–188.

315 See Wistoft, Petersen and Hansen 1991, pp. 43–50.

316 For a more detailed description of how the Danes experienced the advent of electricity, see Kjærgaard 2009 (forthcoming).

317 Quoted from Hvidt 1990, p. 176.

318 *Illustreret Tidende* (12 March 1882, p. 295).

319 *Illustreret Tidende* (30 July 1882, p. 540).

320 *Illustreret Tidende* (10 July 1902, p. 718).

321 This section is based on Kjærgaard 2006c.

322 Poul Videbech, "Finsens Lysinstitut", *Illustreret Tidende* (18 August 1901, p. 720).

323 A. Fraenkel, "Gamle Carlsberg", *Illustreret Tidende* (7 November 1897, p. 97).

324 "Det botaniske Laboratorium", *Illustreret Tidende* (2 February 1890, p. 211).

325 "Statens Serum-Institut", *Illustreret Tidende* (14 September 1902, p. 786), H. O. G. Ellinger, "Den kgl. Veterinær- og Landbohøjskole", *Illustreret Tidende* (12 May 1919, p. 280) and Poul Videbech, "Finsens Lysinstitut", *Illustreret Tidende* (18 August 1901, p. 717).

326 For more information about Danish plans to build new places for science during the period, see Oxenløwe 2006. On the city's transformation from an "ailing patient" to a modern, technically streamlined metropolis, see Møller 2006.

327 See Kjærgaard 2006b, pp. 13–16.

328 Quoted in Hansen 1985, p. 44.

329 Ibid. p. 58.

330 Quoted in Dorph-Petersen 1961, p. 20.

331 See Hyldtoft 2003.

332 Letter to the Royal Danish Academy of Sciences and Letters, 25 September 1876, quoted in Pedersen 1942, p. 15.

333 Holter and Max Møller 1976.

334 Quoted in Pedersen 1942, p. 16.

335 See Hyldtoft 2003.

336 Glamann 1990, p. 210.

337 Quoted in Nielsen and Wistoft 1996, p. 145.

338 See Kjærgaard 2009 (forthcoming).

339 See *Politiken. Udstillingsbladet* (18 June 1909 and 21 June 1909); quoted from Schmidt Kjærgaard, p. 389.

NOTES FOR CHAPTER 17

340 This section is based on Kjærgaard 2006d.

341 "Ugen og Dagen", *Illustreret Tidende* (7 November 1909, p. 72).

342 See Browne 2005 for a discussion of the scientist's new role as a hero.

343 This section is based on Kjærgaard 2006c.

344 M. C. Harding, "Den Polytekniske Læreanstalt – under G.A. Hagemanns Styrelse 1902–1912", *Illustreret Tidende* (12 May 1912, p. 401).

345 "Statens Serum-Institut", *Illustreret Tidende* (14 September 1902, p. 786).

346 "Statens Serum-Institut", *Illustreret Tidende* (14 September 1902, p. 786).

347 Valdemar Greibe, "Steins Laboratorium 1857–1907", *Illustreret Tidende* (23 June 1907, p. 500).

348 Hauch 1929, p. 250.

349 "Aandens Arbejdere", *Blæksprutten* (1919).

350 "Niels R. Finsen", *Illustreret Tidende* (2/10/1904), "Julius Thomsen", *Illustreret Tidende* (21 February 1909), "Direktør Adam Poulsen", *Illustreret Tidende* (10 January 1907) and "Professor Segelcke", *Illustreret Tidende* (4 February 1900).

351 Stuckenberg 1910, p. 82.

352 Brandes 1910, p. 467.

353 "Meteorologisk Institut og den vaade Sommer", *Illustreret Tidende* (10 August 1902, p. 718).

354 Kjærgaard 2006b, p. 22.

355 Claussen 1983, p. 15.

356 Brandes quoted from Wagner 2006, p. 269

357 For details on Lütken and *Opfindelsernes Bog*, see Wagner 2006, pp. 277–280 and Hjermitslev, Andersen and Kjærgaard 2006, p. 354. For details on Trier and *Vor Ungdom*, see Lynning 2006, pp. 405–406.

358 Smiles 1877. *Egen Kraft* (*Self-Help*) was published in 1869, *Sparsommelighed* (*Thrift*) in 1877 and *Pligt* (*Duty*) in 1881.

359 Cantor 1996, p. 183.

360 Eugen Ibsen, "Professor E. A. Scharling", *Illustreret Tidende* (23 September 1866, pp. 427–428); see Wagner 2006, pp. 282–289.

361 See Wagner, pp. 289–294.

362 Ries 2006, p. 314.

363 Quoted in Ries 2006, p. 315.

364 Quoted in Ries 2006, p. 317.

365 Koch also damaged his public popularity by making critical comments in an obituary he wrote over Rasmussen; see Ries 2006, p. 317. He additionally put himself at odds with most of Denmark's public geological science community during a case concerning charges of plagiarism and scientific dishonesty, which ended with a ruling from the supreme court; see Ries 2003.

366 Kildebæk Nielsen 2006, pp. 319–320.

367 Ibid., p. 326.

368 Ibid., p. 328.

369 Johannsen 1925, pp. 75–76.

370 Jacobsen 1925, pp. 23–27; Rosenbeck 1994, p. 16.

371 Meyer 1925, p. 83.

372 Johanne Krebs, "Om Arbejdsfordelingen mellem Mænd og Kvinder", *Kvinden og Samfundet* (September 1885, p. 114).

NOTES FOR CHAPTER 18

373 Thomsen 1998, pp. 32–38.

374 This section is based on Kjærgaard 2006f.

375 Andersen 1861.

376 C. Been, "Den frie Udstilling", *Illustreret Tidende* (15 May 1910, p. 408).
377 See Kjærgaard 2006b, p. 33.
378 Quoted after Mørch 1982, p. 273.
379 Robert Storm Petersen, "Kjøbenhavns ForskønnelsesForening", *Svikmøllen* 1919.
380 Salomonsen 1917, p. 190.
381 "To the readers", *Illustreret Tidende* (2 October 1859, p. 1).
382 Ibid.
383 Ellinger 1897/1898, p. 1. This section is based on Kjærgaard 2006b, pp. 34–40.
384 Christensen 1904/1905, p. 107.
385 "Frem's Program", sample issue, 1897.
386 Brandes 1903, p. 416.
387 Jacobsen 1918, p. 366.
388 Quoted after Korsgaard 1997, pp. 274–275.
389 *Illustreret Familie-Journal* 1917, 41(1), p. 36.
390 Salomonsen 1917, pp. 190–191.
391 The following sections are based on Kjærgaard 2006g.
392 For more on the discussion between knowledge and belief, see Koch 2006.
393 "Manden med Ønskekvisten", *Illustreret Familie-Journal* (1917).
394 Thomson 1936, pp. 158–163.
395 For more on the adacemic school reforms, see Lynning 2006, pp. 399–402.
396 Lehmann 1920, vol. 1, p. V ("Forord til første udgave") and p. 352.
397 See Kjærgaard 2000, chapter 4.
398 Christensen 1912, p. 405.
399 See Kjærgaard 2004.
400 Wallace 1887, pp. 19–20.
401 Hansen 1887, p. 165.
402 Quoted after Abildgaard 1994, p. 142.
403 Heegaard 1902, p. 158.
404 Petersen 2001, p. 130.
405 Flammarion 1910, p. 282.
406 Heegaard 1902, p. 155.
407 Julius Thomsen quoted after Lundbye 1929, pp. 223–224.

NOTES FOR CHAPTER 19

408 The numbers are from *Københavns Universitet 1479–1979*, vol. 2, p. 573.
409 Ibid., p. 508.
410 Knudsen 2006, fig. 17.7.
411 Nielsen and Nielsen 2001, p. 9. See also the official Nobel Web site at http://nobelprize.org/alfred_nobel/will/index.html
412 von Ungern-Sternberg 1996, pp. 156–160.
413 Kragh 1999, pp. 143–148.
414 van Berkel 1999, pp. 170–172.
415 Pais 1991, pp. 375–406; Kildebæk Nielsen 2000, pp. 249–284.
416 *Københavns Universitet 1479–1979*, vol. 12, part 2, p. 452.
417 Kohler 1991, pp. 162–199.
418 Kildebæk Nielsen and Vorup-Jensen 2006, p. 169.
419 Lindqvist 1993, pp. 347–377, 398–417.
420 Pestre 1997, p. 63.
421 Nielsen and Nielsen 2001, pp. 379–384.
422 Söderqvist et al. 1998.
423 In 1961, Alvin Weinberg, the director of the American research centre Oak Ridge, wrote an article in the journal *Science* in which he declared, with regret, that large research teams cooperating on year-long experiments and working with extremely expensive equipment seemed to be a necessary stage in the development of science. He dubbed this phenomenon "big science" – an expression that has since become common usage in many languages.
424 Forman (1971) argues how, after World War I,

Germany's intellectual elite, particularly influenced by Oswald Spengler's *Untergang des Abendlandes* (1918 and 1922), developed an anti-rationalist and anti-scientific rhetoric that struck a chord among large groups in Germany.

NOTES FOR CHAPTER 20

425 This description is from Lundager Jensen 1996, pp. 31–32.

426 Hermansen and Lobedanz 1958, pp. 31–32. Produktions- og Raastofkommissionen 1942, p. 17.

427 Andreasen 1954, pp. 51–58.

428 Holmgaard 2000, pp. 676 and 678.

429 Kohler 1991, pp. 157–160.

430 Robertson 1979, pp. 24–33.

431 Aaserud 1990, pp. 19–25. See also chapter 21.

432 Schmidt Nielsen 1995, pp. 139–142; Aaserud 1998, pp. 207–211. See also chapter 21.

433 Figures from Kildebæk Nielsen 1998, p. 196; Madsen 1940, p. 109; Graae 1941, p. 19; Koch 1996, p. 150.

434 Holter and Møller 1976, p. 38.

435 The Rockefeller Foundation 1972, pp. 13–14.

436 Graae 1941, p. 14.

437 For details about the Carlsberg Foundation, see Glamann 2003.

438 Andersen 2004, p. 5.

439 Ibid., p. 6.

440 Max Møller 1982; Ballhausen 1982; *Det Kongelige Danske Videnskabernes Selskab – Oversigt over Selskabets virksomhed juni 1932–maj 1933*, pp. 143–147; The Rockefeller Foundation 1932.

441 Figures from the Royal Danish Academy of Sciences and Letters, *Det Kongelige Danske Videnskabernes Selskab – Oversigt over Selskabets virksomhed* (reference KDVS, various years).

442 Knudsen 2005, pp. 136–143.

443 Graae 1941, pp. 12–17.

444 Heiberg 1919; Paulsen 1919. Quoted after Stang 1917.

445 Graae 1941, pp. 14.

446 Knudsen 2005, pp. 191–323 and 362–387.

447 Kevles 1995, p. 334.

448 Hermansen and Lobedanz 1958, p. 40.

449 Thomsen 1986, pp. 195–202 and 205.

450 Borgen 1945.

451 Holmgaard 2000, p. 681.

452 See, for instance, Kevles 1995, pp. 302–310 and 324–348.

453 This section is based on Holter 1965; Knudsen 2000; Interview with P. Brandt Rehberg by Morten Ruge, Niels Blædel and Bent Wiberg, 1975, The Collection of audiotapes, Centre of Manuscripts and Rare Books, The Royal Library in Copenhagen.

454 Evidence that there were rumours going around Copenhagen of an imminent "radical change" in the Rockefeller Foundation's grant policy for Europe is found in an account of a journey to the UK (dated 14 December 1945) from the physiologist Einar Lundsgaard to the Royal Danish Academy of Sciences and Letters. Archives of the Royal Danish Academy (reference KDVS prot. nr. 1171/1946).

455 Knudsen 2005, pp. 336–339.

456 Krogh's protest is described in Blegvad 1992, pp. 45–48 and Schmidt-Nielsen 1995, pp. 233–236.

457 Interview with Niels Blædel by Henrik Knudsen, 25 October 2001. Gade Rasmussen 1975.

458 Radio speech by Niels Bohr, *Niels Bohrs tale i radioen 18. januar 1951*. Archives of the Royal

Danish Academy of Sciences and Letters (reference KDVS prot. nr. 660/1951).

459 Gade Rasmussen 1975.

460 Rigsdagstidende 1951–1952, supplement A, col. 3708.

461 Ibid., col. 3709.

462 Nielsen 2006, pp. 219–238.

463 Thomsen 1986, p. 245.

464 Kampmann 1957b.

465 Interview with P. Brandt Rehberg by Morten Ruge, Niels Blædel and Bent Wiberg, 1975, The Collection of audiotapes, Centre of Manuscripts and Rare Books, The Royal Library in Copenhagen.

466 Nielsen et al. 1998, p. 60.

467 Berg Hansen 1979, pp. 243–252.

468 Thomsen 1986, p. 254.

469 Quoted here after Thomsen 1986, p. 256.

470 Figures from Thomsen 1986, table 16, p. 80.

471 Berg Hansen 1979, p. 244.

472 Krige 2006.

473 Nielsen et al. 1998, p. 28.

474 See, for instance, OECD 1963.

475 Kampmann 1957a and Kampmann 1957b.

476 Grønbæk 2001, p. 86.

477 Grønborg 1996, pp. 32–33.

478 Grønbæk 2001, p. 88, and Fog 1977, pp. 134–136. Nørregaard Rasmussen 1968.

479 Figures from Thomsen 1986, p. 80. The discrepancies between this source and the figures stated here for national Danish research foundations and councils are the result of adjustments based on our own figures.

NOTES FOR CHAPTER 21

480 Kohler 1991, p. 149.

481 Kildebæk Nielsen 1998, pp. 180–200; Aaserud 1998, pp. 201–221.

482 Nielsen and Nielsen 2001.

483 Crawford 1987, pp. 30–59.

484 Jungnickel and McCormmach 1986.

485 Bohr 1921.

486 Robertson 1979, pp. 156–159.

487 A booklet published privately after World War II by the Thrige factories, ... *fra Thrige 6* (No 1, February 1953). It consists entirely of articles about the experimental equipment and the work done at Bohr's institute, all written by people working there, which provides further proof of the close links.

488 Kildebæk Nielsen and Kragh 1997.

489 Kildebæk Nielsen 1997a.

490 Nielsen and Nielsen 2001, pp. 265–266.

491 Quoted after Aaserud 1998, pp. 211–216 and Kildebæk Nielsen 1998, pp. 191–197. See also Kildebæk Nielsen and Kragh 1997.

492 Guggenheim 1960, p. 111.

493 Nielsen and Nielsen 2001, pp. 262–265.

494 Kildebæk Nielsen 2001, p. 435.

495 For a more in-depth review, see Kildebæk Nielsen 2001, p. 438 ff.

496 Kildebæk Nielsen 2001, p. 449.

497 One factor that may have prompted Krogh to contact the Rockefeller Foundation in 1923, and not at some other point in time, is the grant Niels Bohr had received that very year from the Rockefeller-related IEB to expand his Institute for Theoretical Physics.

498 Letter from R. M. Pearce to W. Rose, 2 March 1924. Folder 404, box 28, series 1, Appropriations, subseries 2, International Education Board, Rockefeller Archive Center, Sleepy Hollow, New York (RAC).

499 For a detailed account of the arrangement and outfitting of the new physiology institute, see Krogh 1930.

500 Barker Jørgensen 1979, p. 457 ff.

501 Barker Jørgensen 1979, p. 460.

502 Larsson 2001.

503 Quoted after the foundation's statutes (§1), printed in the annual report *Rask-Ørsted Fondet – Beretning for 1919–20.*

504 Robertson 1979, p. 50.

505 Ibid., pp. 156–159.

506 Aaserud 1991, pp. 220–228.

507 See the appendix "Bilag nr. 4a" in Glamann 1976. See also Andersen 2004.

508 The expression "the spirit of Copenhagen", or "Kopenhagener Geist" first appeared in print in a series of published lectures that Werner Heisenberg gave in 1929 at the University of Chicago. It should be noted that the original meaning of the phrase actually referred to the *physics* in Copenhagen, rather than city's scientific environment. Heisenberg 1930a, p. VI. The English version of this work, Heisenberg 1930b, retains the German expression.

509 Marner 1998, p. 171.

510 Frisch 1964, p. 138; Frisch 1979, p. 101. The episode is recounted in Aaserud 1991, p. 7.

511 Heisenberg 1971, pp. 51–64.

512 Frisch 1964, p. 138.

513 This quote is reproduced in Krige 1999, p. 357.

514 Bohr and Mottelson 1969; 1975.

515 Krige 1999, pp. 333–361.

516 Ussing 1980.

517 Ibid. This is also substantiated in Larsen 2002.

518 Robinson 1997 offers an excellent historic review of this topic.

519 Ibid.

520 Ussing and Zerahn 1951.

521 Dansgaard 2004, pp. 12–17; Dansgaard 2000, pp. 63–68. This story is also recounted in Kragh, Lolck and Nielsen 2006.

522 Lolck 2004, pp. 91–93; see also Lolck 2007.

523 Lolck 2004, p. 93.

524 The principal findings from the analysis of the Camp Century deep ice core appeared in two articles published in *Science* and *Nature*, respectively: Dansgaard et al. (1969); Johnsen et al. (1970). See also Lolck 2004, pp. 102–106.

525 Merton 1968.

NOTES FOR CHAPTER 22

526 Kuklick and Kohler 1996a.

527 Kohler 2002.

528 Ibid., p. 8; see also Shapin 1988; 1995. Kuklick and Kohler (1996 b:1) even argue that the reason historians of science have largely neglected the field sciences is that the epistemic and academic status of these disciplines is relatively low. This tendency is also evident in the Danish history of science, the history of field research mainly being treated in accounts written by the field scientists themselves, for instance Wolff 1967; 1979; and Noe-Nygaard 1979; with Ries 2003 as a recent exception.

529 Kohler 2002, p. 22.

530 Ibid.

531 Schiebinger 2005, p. 52.

532 For instance Worboys 1976; Pyenson 1985; Reingold and Rothenberg 1987; Gasgoigne 1996; Schiebinger 2004; Swan and Schiebinger 2005.

533 For instance MacLeod 2000.

534 The following account of Lauge Koch and modern Arctic research is mainly based on Ries 2003.

535 Bruun et al. 1956; Wolff 1967.

536 Mielche 1951.

537 Ibid., p. 28.

538 The following is mainly based on the articles in Bruun et al. 1956.

539 Steemann Nielsen 1956, p. 61.

540 Mielche 1952.

541 Mielche 1951, p. 134.

542 Mielche 1979, s. 232.

543 Bruun et al. 1956, p. 10.

544 See the homepage of the Galathea III Expedition at www.galathea3.dk.

545 Weinberg 1961.

546 Price 1963.

547 Galison and Hevly 1999.

548 Galison 1999, p. 2.

549 Kevles 1997, p. 290.

550 Weinberg 1961, p. 161.

551 Hermann et al. 1987, p. 10.

552 Ibid., pp. 82–90.

553 Ibid., pp. 138–142

554 Ibid., pp. 148–174.

555 Ibid., pp. 179–202

556 Ibid., pp. 237–246.

557 Rozental 1994, p. 137.

558 Källen 1964, p. 7.

559 See Hansen 1996 for a more detailed account.

560 Iliopoulos 1996.

561 Bøggild 1969.

NOTES FOR CHAPTER 23

562 Bush 1945. The full text is also accessible at http://www.nsf.gov/about/history/vbush1945.htm.

563 Even today, employing the concepts "basic research" and "applied research" is a widely used way of grouping different types of technical and scientific research activities. The Frascati Manual, first published by the OECD in 1976, defines basic research as "experimental or theoretical work undertaken primarily to acquire new knowledge of the underlying foundation of phenomena and observable facts, without any particular application or use in view", and applied research as "original investigation undertaken in order to acquire new knowledge. It is, however, directed primarily towards a specific practical aim or objective". See OECD 2002.

564 Noble 1977; Meyer-Thurow 1982; Wise 1985; Reich 1985; Debre 1998, Latour 1988, 1996; and others.

565 The concept of application-oriented or "use-inspired" basic research is treated in greater depth in Stokes 1997.

566 This section is chiefly based on Holter and Møller 1976.

567 Glamann 1988; Glamann 1991, p. 193.

568 Kragh and Styhr Petersen 1995, pp. 256–257; Kildebæk Nielsen 2000, pp. 341–372.

569 Holter and Møller 1976, p. 273; Linderström-Lang 1952; for a scientific biography of Linderström–Lang, see Schellman and Schellman 1997.

570 The beginning and development of Danish insulin production is treated in the following sources: Deckert 1998; Krogh 1924; Richter-Friis 1991; Schmidt-Nielsen 1995. Most of the information given here is from Deckert 1998.

571 Bliss 2000.

572 Krogh 1924; Knudsen 2005, pp. 60–61; Schmidt-Nielsen 1995, p. 181.

573 This section is based on Nielsen 2006a and Nielsen et al. 1998. The history of the nuclear programme in the Netherlands is recounted in Lagaaij and Verbong 1998. Sweden's entry into the atomic age is treated in several sources, including Lindström 1991, Kaijser 1992 and Agrell 2002. Norway's nuclear history is described in Forland 1987, Forland 1988 and Njølstad 1998.

574 Koch 1963, p. 302.

575 Quoted after Nielsen et al. 1998, pp. 34–35.

576 Quoted after ibid., p. 63.
577 Ibid. 1998, p. 98.
578 *Børsen*, 13 March 1957.
579 Quoted after Nielsen et al. 1998, p. 119.
580 Ibid., p. 121.
581 Ibid., p. 162.

NOTES FOR CHAPTER 24

582 Quoted after Gregory and Miller 1998, p. 27.
583 Quoted after Bomholt 1942, p. 260.
584 Ibid., p. 262.
585 Quoted after Skovmand 1949, p. 186.
586 Estrup 1923, p. 3.
587 Ibid., p. 58.
588 *Frem* 1924, No.1, pp. III–IV.
589 Ibid., p. XV.
590 Information given in *Frem*, B1, pp. 736–737. Number of subscribers not stated.
591 Quoted after Christiansen 1993, p. 97.
592 Carson 1963, p. 249.
593 Ibid., pp. 10 and 14.
594 Thorsen 1951 and Thomsen 1972.
595 Schelde Møller 1949. Niels Blædel has drawn our attention to the fact that he never acted as the deputy chairman of the assocation, let alone became a member. Interview with Niels Blædel by Kristian Hvidtfelt Nielsen, 28 October 2004.
596 Bernal 1939.
597 Freeman 1999.
598 Michelsen 1940a, b, c.
599 Michelsen 1941, p. 5.
600 Ibid., p. 7.
601 Michelsen 1943.
602 Michelsen 1946.
603 Michelsen 1947, p. 7.
604 Michelsen 1951
605 Michelsen 1947, p. 8.
606 Interview with Niels Blædel by Kristian Hvidtfelt Nielsen, 28 October 2004.
607 Blædel 1954.
608 Ibid., p. 122.
609 Ibid., p. 126.
610 The programme *Fråga Lund* was later revived several times. For more information, see http://www.svt.se/malmo/fragalund/historia.html.
611 Interview with Tage Elmholdt by Kristian Hvidtfelt Nielsen and Henry Nielsen, 21 July 2004.
612 Danish Broadcasting Corporation, Press Service, 1966.
613 Ibid.
614 Ibid.
615 Ibid.
616 Danish Broadcasting Corporation, Press Service 1967.
617 Interview with Tage Elmholdt by Kristian Hvidtfelt Nielsen and Henry Nielsen, 21 July 2004.

NOTES FOR EPILOGUE

618 *Politiken*, 24 October 1997, first section, p. 7.
619 Grønbæk 2001.
620 Aagaard, Kaare (2000). Dansk Forskningspolitik – Organisation, virkemidler og indsatsområder. Rapport 2000/9. Analyseinstitut for Forskning. Accessible on the Internet at www.afsk.au.dk/ftp/Forskningspolitik/forskpol.pdf
621 Grønbæk 2001.
622 Aagaard, Kaare (2003). Forskningspolitik og tværdisciplinaritet. Rapport 2003/7. Analyseinstitut for Forskning. Accessible on the Internet at www.afsk.au.dk/ftp/Forskningspolitik/2003_7.pdf

References

Part I

Appel, Charlotte (2001). *Læsning og Bogmarked i 1600-Tallets Danmark* (Copenhagen: Museum Tusculanum).

Arrebo, Anders (1965). *Anders Arrebo. Samlede Skrifter.* Vagn L. Simonsen, ed., vol. 1: *Hexaëmeron* (Copenhagen: Munksgaard).

Bagge, Sverre (1984). "Nordic students at foreign universities until 1660", *Scandinavian Journal of History 9*, 1–29.

Baldwin, Martha (1998). "Danish medicines for the Danes and the defense of indigenous medicines", 163–180 in Allen G. Debus and Michael T. Walton, eds, *Reading the Book of Nature. The Other Side of the Scientific Revolution* (Ann Arbor, MI: Thomas Jefferson University Press).

Bartholin, Erasmus (1914). *Bartholins Tale om det Danske Sprog.* C. Behrend, ed. (Copenhagen: Det Danske Sprog- og Litteraturselskab).

– (1991). *Experiments on Birefringent Icelandic Crystal.* J. Z. Buchwald and K. M. Pedersen, eds (Copenhagen: Copenhagen University Library).

Bartholin, Thomas (1940). *Skrifter om Opdagelsen af Lymfekarsystemet.* G. Tryde, ed. (Copenhagen: Løvens Kemiske Fabrik).

Birch, Thomas (1968). *The History of the Royal Society*, 4 vols (New York, NY: Johnson Reprint).

Bjørnbo, Axel A. and Carl S. Petersen (1904). "Fynboen Claudius Claussøn Swart: Nordens ældste kartograf", *Royal Danish Academy of Sciences and Letters, Historisk-Filosofiske Skrifter VI.2.*

Blair, Ann (2000). "Mosaic physics and the search for a pious natural philosophy in the late renaissance", *Isis 91*, 32–58.

Boethius de Dacia (2001). *Verdens Evighed. Det Højeste Gode. Drømme* (Copenhagen: Det Lille Forlag).

Borch, Ole (1983). *Olai Borrichii Itinerarium 1660–1665*, 4 vols, H. D. Schepelern, ed. (Copenhagen: Reitzel).

Boyle, Robert (1772). *The Works of the Honourable Robert Boyle*, vol. 2 (London: Johnston et al.).

– (2001). *The Correspondence of Robert Boyle*, 6 vols, Michael Hunter et al., eds (London: Pickering & Chatto).

Bradley, S. (1976). "Mandevilles Rejse: Some aspects of its changing role in the later Danish middle ages", *Medieval Scandinavia 9*, 146–163.

Brahe, Tycho (1946). *Tycho Brahe's Description of his Instruments and Scientific Work.* H. Ræder, E. Strömgren and B. Strömgren, eds (Copenhagen: Munksgaard).

Bruun, Niels W. and Hans-Otto Loldrup, eds (1982). *Thomas Bartholin. Cista medica Hafniensis*

(Copenhagen: Dansk Farmaceutforening).

Chapman, Allan (1984). "Tycho Brahe in China: The Jesuit mission to Peking and the iconography of European instrument-making processes", *Annals of Science 41*, 417–443.

Christianson, John R. (1979). "Tycho Brahe's German treatise on the comet of 1577", *Isis 70*, 110–140.

– (2000). *On Tycho's Island: Tycho Brahe and his Assistants, 1570–1601* (Cambridge: Cambridge University Press).

Dahl, Svend (1957). *Bogens Historie* (Copenhagen: Haase & Søns Forlag).

Danneskiold-Samsøe, Jakob (2004). *Muses and Patrons: Cultures of Natural Philosophy in Seventeenth-century Scandinavia* (Lund: Lund University).

Drachmann, Aage G. (1971). *Lægæst og hans Guldhorn. Betragtninger i Anledning af Willy Hartners Tolkningsforsøg* (Copenhagen: Gad).

Drake, Ellen (1996). *Restless Genius: Robert Hooke and his Earthly Thoughts* (New York, NY: Oxford University Press).

Ebbesen, Sten, ed. (1985). *Anders Sunesen. Stormand, Teolog, Administrator, Digter* (Copenhagen: Gad).

Ebbesen, Sten and Carl Henrik Koch (2003). *Den Danske Filosofis Historie*, vol. 2 (Copenhagen: Gyldendal).

Figala, Karin (1972). "Tycho Brahes Elixier", *Annals of Science 28*, 139–176.

Fink-Jensen, Morten (2004). *Fornuften under Troens Lydighed: Naturfilosofi, Medicin og Teologi i Danmark 1536–1636* (Copenhagen: Museum Tusculanum).

French, Roger (1994). *William Harvey's Natural Philosophy* (Cambridge: Cambridge University Press).

Frängsmyr, Tore (1971). "Steno and geological time", 204–212 in Gustav Scherz, ed., *Dissertations on Steno as Geologist* (Odense: Odense University Press).

Fussing, Hans H. and Mogens Pihl (1955). "Breve fra Thomas Henshaw og Rasmus Bartholin til Royal Society i London 1672–1675", *Danske Magazin 6*, Series 7, 82–106.

Garboe, Axel (1949–1950). *Thomas Bartholin* (Copenhagen: Munksgaard).

– (1959). *Geologiens Historie i Danmark*, vol. 1 (Copenhagen: Reitzel).

Gingerich, Owen and Robert S. Westman (1988). *The Wittich Connection: Conflict and Priority in Late Sixteenth-Century Cosmology* (Philadelphia, PA: American Philosophical Society).

Godtfredsen, Edvard (1957). "The reception of Harvey's doctrine in Denmark", *Journal of the History of Medicine and Allied Sciences 12*, 202–208.

– (1973). *Medicinens Historie* (Copenhagen: Nyt Nordisk Forlag).

Grane, Leif, ed. (1981). *University and Reformation* (Leiden: Brill).

Grant, Edward, ed. (1974). *A Source Book in Medieval Science* (Cambridge, MA: Harvard University Press).

Grell, Ole Peter (1993). "Caspar Bartholin and the education of the pious physician", 78–100 in O. P. Grell and Andrew Cunningham, eds, *Medicine and the Reformation* (London: Routledge).

– , ed. (1998). *Paracelsus: The Man and his Reputation, his Ideas and their Transformation* (Leiden: Brill).

Hall, A. Rupert (1962). *The Scientific Revolution 1500–1800* (London: Longmans).

Hall, A. Rupert and Marie Boas Hall, eds (1965–1986). *The Correspondence of Henry Oldenburg* (Madison, WI: University of Wisconsin Press).

Hartner, Willy (1969). *Die Goldhörner von Gallehus* (Wiesbaden: Franz Steiner Verlag).

Harvey, E. Newton (1957). *A History of Luminiscence from the Earliest Times until 1900* (Philadelphia, PA: American Philosophical Society).

Hashimoto, Keizo (1987). "Longomontanus' Astronomia Danica in China", *Journal for the History of Astronomy 18*, 95–110.

Hauberg, Poul, ed. (1936). *Henrik Harpestræng. Liber Herbarum* (Copenhagen: Hafnia).

Helk, Vello (1987). *Dansk-Norske Studierejser 1536–1660 fra Reformationen til Enevælden* (Odense: Odense Universitetsforlag).

Hooper-Greenhill, E. (1992). *Museums and the Shaping of Knowledge* (London: Routledge).

Hougaard, Jens et al. (1983). *Dansk Litteraturhistorie*, vol. 3 (Copenhagen: Gyldendal).

Hovesen, Ejnar (1987). *Lægen Ole Worm, 1588–1654. En Medicinhistorisk Undersøgelse og Vurdering* (Århus: Aarhus University Press).

Huxley, Thomas (1881). "The rise and progress of palaeontology", *Nature 24*, 452–455.

Hørby, Kai et al. (1962). *Academia Sorana. Kloster, Akademi, Skole* (Copenhagen: Gyldendal).

Ilsøe, Harald (1963). *Udlændinges Rejser i Danmark indtil År 1700* (Copenhagen: Rosenkilde og Bagger).

– (1980). "Universitetets biblioteker til 1728", 289–364 in Svend Ellehøj and Leif Grane, eds, *Københavns Universitet 1479–1979*, vol. 4 (Copenhagen: Gad).

– (1982). "Fra Rundetårn til Massachusetts: Nogle kilder til universitetsbibliotekets historie omkring 1660", *Bibliotek for Læger 174*, 69–86.

Johansen, Hans Christian (2002). *Danish Population History 1600–1939* (Odense: University Press of Southern Denmark).

Kejlbo, Ib (1972). *Historisk Kartografi* (Copenhagen: Dansk Historisk Fællesforening).

Kragh, Helge (2002). "Fra Harpestreng til Bohr: Dansk videnskab af verdensklasse", 11–24 in Jan Teuber, ed., *Højdepunkter i Dansk Naturvidenskab* (Copenhagen: Gad).

– (2003a). "Acta medica, Nordens første videnskabelige tidsskrift", *1066 – Tidsskrift for Historie 33*: 3, 13–23.

– (2003b). *Phosphors and Phosphorus in Early Danish Natural Philosophy* (Copenhagen: Royal Academy of Sciences and Letters).

– (2006). "The inventive nature: Fossils, stones and marvels in seventeenth-century Danish natural philosophy", *Cardanus 6*, 15–25.

– (2007). "Received wisdom in biography: Tycho-biographies from Gassendi to Christianson", 121–134 in Thomas Söderqvist, ed., *The History and Poetics of Scientific Biography* (London: Ashgate).

Kragh, Helge and Henrik Kragh Sørensen (2007). "An odd couple: Descartes and Longomontanus. A contribution to Cartesianism in seventeenth-century Denmark", *Ideas in History* 2, 9–36.

Kronick, David A. (1962). *A History of Scientific and Technical Periodicals* (New York, NY: Scarecrow Press).

Kroon, Sigurd, ed. (1993). *A Danish Teacher's Manual of the Mid-Fifteenth Century* (Lund: Lund University Press).

Kusukawa, Sachiko, ed. (1999). *Philip Melanchthon: Orations on Philosophy and Education* (Cambridge: Cambridge University Press).

Ladewig Petersen, E. (1980). *Dansk Socialhistorie*, vol. 3 (Copenhagen: Gyldendal).

Lassen, Trine (2002). "En 1600-tals borgers naturerkendelse: Univers og virkelighedssyn i Hans Nansens Kosmografi 1633", 119–142 in Charlotte Appel et al., eds, *Mentalitet og Historie: Om Fortidige Forestillingsverdener* (Ebeltoft: Skippershoved).

Lausten, Martin S. (1987). *Danmarks Kirkehistorie* (Copenhagen: Gyldendal).

Leibniz, Gottfried W. (1976–1995). *Leibniz. Mathematischer, naturwissenschaftlicher und technischer Briefwechsel*, 4 vols (Berlin: Akademie-Verlag).

Liisberg, H. C. Bering (1897). *Kunstkammeret. Dets Stiftelse og Ældste Historie* (Copenhagen: Det Nordiske Forlag).

Lohne, J. A. (1977). "Nova experimenta crystalli Islandici disdiaclastici", *Centaurus* 22, 106–148.

Maar, Vilhelm, ed. (1926). *Mindeskrift for Oluf Borch paa 300-Aarsdagen for hans Fødsel* (Copenhagen: Nyt Nordisk Forlag).

Manten, A. A. (1980). "Development of European scientific journal publishing before 1830", 1–10 in A. J. Meadows, ed., *Development of Science Publishing in Europe* (Amsterdam: Elsevier).

McGuire, Brian P. (1982). *The Cistercians in Denmark: Their Attitudes, Roles and Functions in Medieval Society* (Kalamazoo, MI: Cistercian Publications).

Meisen, Valdemar (1932). *Prominent Danish Scientists Through the Ages* (Copenhagen: Levin & Munksgaard).

Middleton, W. Knowles (1966). *A History of the Thermometer and its Use in Meteorology* (Baltimore, MD: Johns Hopkins University Press).

Moe, Harald (1988). *Nicolaus Steno. An Illustrated Biography* (Copenhagen: Rhodos).

Moesgaard, Kristian P. (1975). "Tychonian observations, perfect numbers, and the date of creation: Longomontanus' solar and precessional theories", *Journal for the History of Astronomy* 6, 84–99.

– (1977). "Cosmology in the wake of Tycho Brahe's astronomy", 293–305 in Wolfgang Yourgrau and Allen D. Breck, eds, *Cosmology, History, and Theology* (New York, NY: Plenum Press).

Mosley, Adam (2000). "Astronomical books and courtly communication", 114–131 in Marina Frasca-Spada and Nick Jardine, eds, *Books and the Sciences in History* (Cambridge: Cambridge University Press).

Møller-Christensen, Vilhelm (1944). *Middelalderens Lægekunst i Danmark* (Copenhagen: Munksgaard).

Nissen, Andreas et al. (1937). *Rundetaarn 1637–1937. Et Mindesmærke* (Copenhagen: Levin & Munksgaard).

Norlind, Wilhelm (1951). *Tycho Brahe: Mannen och Verket* (Lund: Gleerup).

Norvin, William (1929). *Københavns Universitet i Middelalderen* (Copenhagen: Gad).

– (1937–1940). *Københavns Universitet i Reformationens og Orthodoxiens Tidsalder*, 2 vols (Copenhagen: Gyldendal).

Oliver, William (1703). "Remarks in a late journey into Denmark and Holland" *Philosophical Transactions 23*, 1400–1412.

Pedersen, Kurt Møller (1976). "Ole Rømers opdagelse af lysets tøven", *Astronomisk Tidsskrift 9*, 160–166.

– (1987). "Une mission astronomique de Jean Picard: Le voyage d'Uraniborg", 157–203 in Guy Picolet, ed., *Jean Picard et les débuts de l'astronomie de précision au XVIIe siècle* (Paris: Éditions du CNRS).

Pedersen, Olaf (1968). "The life and work of Peter Nightingale", *Vistas in Astronomy 9*, 3–10.

– (1975). "John Simonis of Selandia", 142–143 in *Dictionary of Scientific Biography.*

Pihl, Mogens (1944). *Ole Rømers Videnskabelige Liv* (Copenhagen: Royal Academy of Sciences and Letters).

Pinborg, Jan, ed. (1979). *Universitas Studii Haffniensis* (Copenhagen: University of Copenhagen).

– (1981). "Danish students 1450–1535 and the University of Copenhagen", *Cahiers de l'institut du moyen-age grec et latin 37*, 70–122.

Poulsen, Jacob E. and Egill Snorrason, eds (1986). *Nicolaus Steno, 1638–1686. A Re-Consideration by Danish Scientists* (Gentofte: Nordisk Insulinlaboratorium).

Principe, Lawrence (1998). *The Aspiring Adept: Robert Boyle and his Alchemical Quest* (Princeton, MA: Princeton University Press).

Rømer, Ole (1910). *Adversaria.* Thyra Eibe and Kirstine Meyer, eds (Copenhagen: Royal Academy of Sciences and Letters).

– (2001). *Ole Rømer. Korrespondance og Afhandlinger samt et Udvalg af Dokumenter.* P. Friedrichsen and Chr. G. Tortzen, eds (Copenhagen: Reitzel).

Rørdam, Holger, ed. (1889–1892). "Aktstykker til universitetets historie 1621–60, 1637–60", *Danske Magazin*, Series 5, 1–28, 126–152, 217–242, 320–349.

Sandblad, Henrik (1943). "Det copernikanska världssystemet i Sverige", *Lychnos*, 148–188.

Schepelern, Henrik D. (1971). *Museum Wormianum. Dets Forudsætninger og Tilblivelse* (Copenhagen: Wormianum).

– (1985). "Natural philosophers and princely collectors: Worm, Paludanus, and the Gottorp and Copenhagen collections", pp. 167–176 in Oliver Impey and Arthur MacGregor, eds, *The Origins of Museums: The Cabinet of Curiosities in Sixteenth- and Seventeenth-Century Europe* (New York, NY: House of Stratus).

Scherz, Gustav, ed. (1968). *Steno and Brain Research in the Seventeenth Century* (Oxford: Pergamon Press).

– (1971). *Dissertations on Steno as Geologist* (Odense: Odense University Press).

Schioldann-Nielsen, Johan and Kurt Sørensen, eds (1994). *Thomas Bartholin: On Diseases in the Bible. A Medical Miscellany 1672* (Copenhagen: Munksgaard).

Schlee, Ernst, ed. (1965). *Gottorfer Kultur im Jahrhundert der Universitätsgründung* (Flensburg: Christian Wolff).

Schofield, Christine (1981). *Tychonic and Semi-Tychonic World Systems* (New York, NY: Arno Press).

Schæffer, Aage (1963). *Hofapotekere og Hofkemikere i Danmark ca. 1540–1660* (Copenhagen: Dansk Farmacihistorisk Selskab).

Schönbeck, Jürgen (2004). "Thomas Fincke und die *Geometria rotundi*", *NTM 12*, 80–99.

Severinus, Petrus (1979). *Petrus Severinus og hans Idea Medicinæ Philosophicæ: En Dansk Paracelsist.* E. Bastholm, ed. (Odense: Odense Universitetsforlag).

Shackelford, Jole (1999). "Documenting the factual and the artifactual: Ole Worm and public knowledge", *Endeavour 23*: 2, 65–71.

– (2003). *A Philosophical Path for Paracelsian Medicine: The Idea, Intellectual Context, and Influence of Petrus Severinus* (Copenhagen: Museum Tusculanum).

Sinning, Jens (1991). *Tale om Nødvendigheden af Filosofiske Studier for den Teologiske Student.* Eric Jacobsen, ed. (Copenhagen: Museum Tusculanum).

Steensgaard, Niels (1995). "The cosmography of Hans Nansen, a seventeenth-century Copenhagen merchant and politician", 426–439 in Wilfried Feldenkirchen et al., eds, *Wirtschaft, Gesellschaft, Unternehmen*, vol. 1 (Stuttgart: Franz Steiner Verlag).

Steno, Nicolaus (1969). *Steno. Geological Papers.* Gustav Scherz, ed. (Odense: Odense University Press).

– (1987). *Niels Stensens Korrespondance i Dansk Oversættelse*, 2 vols. Harriet Hansen, ed. (Copenhagen: Reitzel).

– (1997). *Chaos. Niels Stensen's Chaos-Manuscript, Copenhagen 1659.* August Ziggelaar, ed. (Copenhagen: Danish National Library of Science and Medicine).

Sunesen, Anders (1985). *Hexaëmeron.* Henrik D. Schepelern, transl. (Copenhagen: Gad).

Thoren, Victor E. (1990). *The Lord of Uraniborg: A Biography of Tycho Brahe* (Cambridge: Cambridge University Press).

Thykier, Claus, ed. (1990). *Dansk Astronomi Gennem Firehundrede År*, 3 vols (Copenhagen: Rhodos).

Van der Star, Pieter, ed. (1983). *Fahrenheit's Letters to Leibniz and Boerhaave* (Amsterdam: Rodopi).

Wear, Andrew (1981). "Galen in the Renaissance", 229–262 in Vivian Nutton, ed., *Galen: Problems and Prospects* (London: The Wellcome Institute).

Westfall, Richard (1994). "Charting the scientific community", 1–14 in Kostas Gavroglu et al., eds, *Trends in the Historiography of Science* (Dordrecht: Kluwer Academic).

Wittendorff, Alex (1994). *Tyge Brahe* (Copenhagen: Gad).

Worm, Ole (1965–1968). *Breve fra og til Ole Worm*, 3 vols. Henrik D. Schepelern, ed. and transl. (Copenhagen: Munksgaard).

Part II

Albeck, Gustav (1984). *Universitet og Folk* (Copenhagen: Gyldendal).

Amundsen, Leiv (1957). *Det Norske Videnskaps-Akademi i Oslo, 1857–1957* (Oslo: Aschehoug).

Andersen, Einar (1968). *Thomas Bugge* (Copenhagen: Geodætisk Instituts Forlag).

Andersen, Hemming (1996). *En Videnskabsmand af Rang: Adam Wilhelm Hauch, 1755–1838* (Århus: Steno Museets Venner).

Anker, Jean (1951). "Georg Christian Oeders botanische Reise in Europa um die Mitte des achtzehnten Jahrhunderts", *Centaurus 1*, 242–265.

Barton, H. Arnold (1998). *Northern Arcadia. Foreign Travelers in Scandinavia, 1765–1815* (Carbondale, IL: Southern Illinois University Press).

Biermann, Kurt-Reinhold (1979). *Briefwechsel zwischen Alexander von Humboldt und Heinrich Christian Schumacher* (Berlin: Akademie-Verlag).

Bille, Steen (1849–1851). *Corvetten Galatheas Reise omkring Jorden 1845, 46 og 47* (Copenhagen: C. A. Reitzel).

Bjerg, Hans Christian (1984). *Poul Løvenørn 1751–1826* (Copenhagen: Farvandsdirektoratet).

Bostrup, Ole (1974). "Ulrik Green. Et studie i dansk fysik- og kemiundervisnings historie", *Naturens Verden*, 146–149.

– (1996). *Dansk Kemi 1770–1807. Den Kemiske Revolution* (Copenhagen: Teknisk Forlag).

Branner, Bodil and Jesper Lützen, eds (1999). *Caspar Wessel: On the Analytical Representation of Direction* (Copenhagen: C.A. Reitzel).

Caneva, Kenneth L. (1997). "Colding, Ørsted, and the meaning of force", *Historical Studies in the Physical and Biological Sciences 28*, 1–138.

Christensen, Carl (1924–1926). *Den Danske Botaniks Historie med Tilhørende Bibliografi*, 2 vols (Copenhagen: H. Hagerup).

Christensen, Dan C. (1995). "The Ørsted–Ritter partnership and the birth of Romantic natural philosophy", *Annals of Science 52*, 153–185.

– (1996). *Det Moderne Projekt: Teknik & Kultur i Danmark-Norge 1750–(1814)–1850* (Copenhagen: Gyldendal).

Clément, Adolph, ed. (1920). *Breve til og fra J. G. Forchhammer* in *J. G. Forchhammer og Jac. Berzelius 1834–1845* (Copenhagen: Thiele).

– (1926). *Breve til og fra J. G. Forchhammer, III: J. G. Forchhammer og Charles Darwin 1849–1850* (Copenhagen: Thiele).

Crosland, Maurice P., ed. (1969). *Science in France in the Revolutionary Era* (Cambridge, MA: MIT Press).

Dahl, Per F. (1972). *Ludvig Colding and the Conservation of Energy Principle. Experimental and Philosophical Contributions* (New York: Johnson Reprint Corporation).

Dahl, Svend (1941). *Den Danske Plante- og Dyreverdens Udforskning* (Copenhagen: Udvalget for Folkeoplysnings Fremme).

Dibner, Bern (1962). *Oersted and the Discovery of Electromagnetism* (New York, NY: Blaisdell Publishing Company).

Ebert, Max (1988). *Bogtrykkerne ved Sorø Akademi* (Copenhagen: Finn Jacobsens Forlag).

Fisher, Raymond H. (1977). *Bering's Voyages: Whither and Why* (Seattle, WA: University of Washington Press).

Forchhammer, Johan G. (1869). *Almeenfattelige Afhandlinger og Foredrag* (Copenhagen: Thaarup).

Gad, Finn (1976). *Grønlands Historie, III: 1728–1808* (Copenhagen: Nyt Nordisk Forlag).

Garboe, Axel (1959–1961). *Geologiens Historie i Danmark*, 2 vols (Copenhagen: Reitzel).

– (1963). "Grønlandsmineralogen K. L. Giesecke og hans danske venner", *Tidsskriftet Grønland*, 441–451.

– (1966). "Farmaceuten, kemikeren, naturforskeren B. Levy (Lewy)", *Theriaca*, No. 11.

Giesecke, Karl L. (1910). "Bericht über einer mineralogischen Reise in Grönland", *Meddelelser om Grønland 35*, 1–478.

Gilje, Nils and Tarald Rasmussen (2002). *Norsk Idéhistorie*, vol. 2: *Tankeliv i den Lutherske Stat, 1537–1814* (Oslo: Aschehoug).

Gosch, C. C. A. (1873–1878). *Udsigt over Danmarks Zoologiske Litteratur*, 3 vols (Copenhagen: Otto Schwartz's Efterfølger).

Gouk, Penelope M. (1983). "The union of arts and sciences in the eighteenth century: Lorenz Spengler (1720–1807), artistic turner and natural scientist", *Annals of Science 40*, 411–436.

Harding, M. C. (1924). *Selskabet for Naturlærens Udbredelse* (Copenhagen: Gjellerup).

Haue, Harry (2003). *Almendannelse som Ledestjerne: En Undersøgelse af Almendannelsens Funktion i Dansk Gymnasieundervisning 1775–2000* (Odense: Syddansk Universitetsforlag).

Helveg Jespersen, P. (1946). "J. C. Fabricius as an evolutionist", *Svenska Linné-Sällskapets Årsskrift 29*, 35–56.

Holten, Birgitte and Michael Sterll (2000). "The Danish Naturalist Peter Wilhelm Lund (1801–80) – Research on Early Man in Minas Gerais", *Luso, Brazilian Review 37*, 33–45.

Hultberg, Helge (1973). *Den Unge Henrik Steffens 1773–1811* (Copenhagen: Bianco Lund).

Jacobsen, Anja Skaar (2000). "A. W. Hauch's role in the introduction of antiphlogistic chemistry into Denmark", *Ambix 47*, 71–95.

– (2001). "Spirit and unity: Ørsted's fascination by Winterl's chemistry", *Centaurus 43*, 184–218.

Jacobsen, Anja Skaar and Svend Larsen, eds (2002). *H. C. Ørsteds Selvbiografi* (Århus: Steno Museets Venner).

Johansen, Hans Christian (1979). *Dansk Socialhistorie*, vol. 4: *En Samfundsorganisation i Opbrud* (Copenhagen: Gyldendal).

Johnston, James F. W. (1830). *Scientific Men and Institutions in Copenhagen* (Edinburgh).

Kjølsen, Frits H. (1965). *Capitain F. L. Norden og hans Rejse til Ægypten 1737–38* (Copenhagen: Gad).

Koch, Carl Henrik (2004). *Den Danske Idealisme* (Copenhagen: Gyldendal).

Kornerup, Bjørn et al. (1923). "Biskop Dr. Theol. Otto Fabricius. Et mindeskrift i hundredaaret for hans død", *Meddelelser om Grønland* 62, 215–400.

Kragemo, Helge (1960). "Pater Hells Vardøhusekspedition", pp. 92–125 in G. I. Willoch, ed., *Vardøhus Festning, 650 År* (Oslo: Generalinspektøren for Kystartilleriet).

Kragh, Helge (1995). "From curiosity to industry: The early history of cryolite soda manufacture", *Annals of Science* 52, 285–301.

– (2003). "Volta's apostle: C. H. Pfaff, champion of the contact theory", *Nuova Voltiana* 5, 69–82.

Kragh, Helge and Hans Jørgen Styhr Petersen (1995). *En Nyttig Videnskab. Episoder fra den Tekniske Kemis Historie i Danmark* (Copenhagen: Gyldendal).

Larsen, Børge Riis (1991). *Naturvidenskab og Dannelse. Studier i Fysik- og Kemiundervisningens Historie i den Højere Skole indtil Midten af 1800-Tallet* (Espergærde: Dansk Selskab for Historisk Kemi).

Lind, Natasha O. and Peter Ulf Møller (2003). *Under Bering's Command: New Perspectives on the Russian Kamchatka Expeditions* (Århus: Aarhus University Press).

Lomholt, Asger (1942–1973). *Det Kongelige Danske Videnskabernes Selskab 1742–1942. Samlinger til Selskabets Historie*, 5 vols (Copenhagen: Munksgaard).

Lundbye, J. T. (1929). *Den Polytekniske Læreanstalt 1829–1929* (Copenhagen: Gad).

Lützen, Jesper, ed. (2000). *Around Caspar Wessel and the Geometric Representation of Complex Numbers* (Copenhagen: C. A. Reitzel).

Meyer, Kirstine, ed. (1920). *H. C. Ørsted. Naturvidenskabelige Skrifter*, 3 vols (Copenhagen: Høst & Søn).

Midbøe, Hans (1960). *Det Kongelige Norske Videnskabers Selskabs Historie 1760–1960*, 2 vols (Trondheim: Aktietrykkeriet).

Mortensen, Peder et al. (1996). *Den Arabiske Rejse. Danske Forbindelser med den Islamiske Verden gennem 1000 År* (Århus: Forhistorisk Museum Moesgaard).

Nicolson, Malcolm (1996). "Humboldtian plant geography after Humboldt: The link to ecology", *British Journal for the History of Science* 29, 289–310.

Niebuhr, Carsten (2003–2004). *Rejsebeskrivelse fra Arabien og andre Omkringliggende Lande*, 2 vols (Copenhagen: Vandkunsten).

Nielsen, Aksel V. (1950). "Observatoriet i Altona paa Schumachers tid", *Nordisk Astronomisk Tidsskrift*, 91–95.

– (1957). "Pater Hell og venuspassagen 1769", *Nordisk Astronomisk Tidsskrift*, 77–97.

Norvin, William (1937–1940). *Københavns Universitet i Reformationens og Orthodoxiens Tidsalder*, 2 vols (Copenhagen: Gyldendal).

Olwig, Kenneth (1980). "Historical geography and the society/nature 'problematic': The perspective of J. F. Schouw, G. P. Marsh and E. Reclus", *Journal of Historical Geography* 6, 29–45.

– (1986). *Hedens Natur. Om Natursyn og Naturanvendelse gennem Tiderne* (Copenhagen: Teknisk Forlag).

Pedersen, Kurt Møller, ed. (1997). *Thomas Bugge. Journal of a Voyage through Holland and England, 1777* (Århus: History of Science Department).

– (2001). "Thomas Bugge's journal of a voyage through Germany, Holland and England, 1777", pp. 29–46 in Lützen 2001.

Peters, Christian A. F., ed. (1860–1865). *Briefwechsel zwischen C. F. Gauss und H. C. Schumacher* (Altona: Gustav Esch).

Petersen, Julius (1893). *Den Danske Lægevidenskab 1700–1750* (Copenhagen: Gyldendal).

Pihl, Mogens (1972). *Betydningsfulde Danske Bidrag til den Klassiske Fysik.* Københavns Universitets Festskrift, pp. 1–82 (Copenhagen: University of Copenhagen).

Rasmussen, Stig T., ed. (1997). *Den Arabiske Rejse 1761–1767. En Dansk Ekspedition Set i Videnskabshistorisk Perspektiv* (Copenhagen: Munksgaard, Rosinante).

Ringler, Dick (2002). *Bard of Iceland: Jónas Hallgrímsson, Poet and Scientist* (Madison, WI: University of Wisconsin Press).

Rostrup, Haagen, ed. (1985). *Francisco de Mirandas Danske Rejsedagbog 1787–1788* (Copenhagen: Rhodos).

Rømer, Ole (2001). *Ole Rømer. Korrespondance og Afhandlinger samt et Udvalg af Dokumenter.* Edited by P. Friedrichsen and C. G. Tortzen (Copenhagen: Reitzel).

Skytte Christiansen, M. (1973). *Historien om Flora Danica: To Bogværker og et Porcelænsstel* (Copenhagen: Dansk Esso).

Slottved, Ejvind (1978). *Lærestole og Lærere ved Københavns Universitet 1537–1977* (Copenhagen: Samfundet for Dansk Genealogi og Personalhistorie).

Smith, Fritze (1950). *Doktordisputatsens Historie ved Københavns Universitet* (Copenhagen: Munksgaard).

Snorrason, Egil (1974). *C. G. Kratzenstein and his Studies on Electricity during the Eighteenth Century* (Odense: Odense University Press).

Spang-Hanssen, Ebbe (1965). *Erasmus Montanus og Naturvidenskaben* (Copenhagen: Gad).

Splinter, Susan (2007). *Zwischen Nützlichkeit und Nachahmung: Eine Biografie des gelehrten Christian Gottlieb Kratzenstein* (Frankfurt am Main: Peter Lang).

Steller, Georg W. (1993). *Journal of a Voyage with Bering, 1741–1742* (Cambridge: Cambridge University Press).

Sweet, Jessie (1972). "Morten Wormskiold: Botanist", *Annals of Science 28*, 293–305.

Sylvest, Søren B. (1972). *Bidrag til Zeises Biografi* (Århus: Unpublished MSc dissertation, University of Aarhus).

Teuber, Jan, ed. (2002). *Højdepunkter i Dansk Naturvidenskab* (Copenhagen: Gad).

Thalbitzer, William (1932). "Fra Grønlandsforskningens første dage", *Københavns Universitets Festskrift* (Copenhagen: University of Copenhagen).

Thuborg, Anders (1951). *Den Kantiske Periode i Dansk Filosofi 1790–1800* (Copenhagen: Gyldendal).

Wagner, Michael F. (1999). *Det Polytekniske Gennembrud: Romantikkens Teknologiske Konstruktion 1780–1850* (Århus: Aarhus University Press).

Wentrup, Curt (2001). "Bunsen i Danmark", *Dansk Kemi 83*: 3, 35–37

Wilson, Leonard G. (1972). *Charles Lyell. The Years to 1841: The Revolution in Geology* (New Haven,

CT: Yale University Press).

Winter, Frank H. (1965). "Landmarks on the road to space travel – Danish rocketry", *Militært Tidsskrift 94*, 55–66.

– (1990). *The First Golden Age of Rocketry: Congreve and Hale Rockets of the Nineteenth Century* (Washington, DC: Smithsonian Institution Press).

Woolf, Harry (1959). *The Transits of Venus: A Study of Eighteenth-Century Science* (Princeton, NJ: Princeton University Press).

Ørsted, Hans Christian (1820). "Experiments on the effect of a current of electricity on the magnetic needle", *Annals of Philosophy 16*, 273–276.

– (1966). *The Soul in Nature* (London: Dawsons of Pall Mall).

– (2003). *H. C. Ørsted's Theory of Force. An Unpublished Textbook in Dynamical Chemistry*. Edited by Anja S. Jacobsen et al. (Copenhagen: C. A. Reitzel).

Part III

Anon.

– (1859). "Til Læserne", *Illustreret Tidende* (2 October 1859, p. 1).

– (1890). "Det botaniske Laboratorium", *Illustreret Tidende* (2 February 1890, p. 211).

– (1900). "Professor Segelcke", *Illustreret Tidende* (4 February 1900).

– (1902). "Statens Serum-Institut", *Illustreret Tidende* (14 September 1902, p. 786).

– (1902). "Meteorologisk Institut og den vaade Sommer", *Illustreret Tidende* (10 August 1902, p. 718).

– (1904). "Niels R. Finsen", *Illustreret Tidende* (2 October 1904).

– (1907). "Direktør Adam Poulsen", *Illustreret Tidende* (10 January 1907).

– (1909). "Julius Thomsen", *Illustreret Tidende* (21 February 1909).

– (1909). "Ugen og Dagen", *Illustreret Tidende* (7 November 1909, p. 72).

– (1915). "Fra Rigsdagen", *Kvinden og Samfundet* (5 June 1915, p. 169).

Abildgaard, Hanne (1994). *Tidlig modernisme, Ny Dansk Kunsthistorie*, vol. 6 (Copenhagen: Forlaget Palle Fogtdal).

Andersen, Casper and Hans Henrik Hjermitslev (2006). "Videnskab på landet", pp. 251–260 in Peter C. Kjærgaard 2006a.

Andersen, Hans Christian (1861). "Det ny Aarhundredes Muse" in *Nye Eventyr of Historier* (Copenhagen: Reitzel).

Atkinson, G. M. (1872). "On a Kitchen-midden in Cork Habour". *The Journal of the Anthropological Institute of Great Britain and Ireland 1*, 213–214.

Been, C. (1910). "Den frie Udstilling", *Illustreret Tidende* (15 May 1910, p. 408).

Bensaude-Vincent, Bernadette (2003). "Chemistry", pp. 196–220 in Cahan 2003.

Brandes, Georg (1903 [1872]). "Forklaring og Forsvar" in *Samlede Skrifter*, vol. 13 (Copenhagen:

Gyldendalske Boghandels Forlag), 387–433.

– (1910). "Det gamle Kjøbenhavn, Voldene" in *Samlede Skrifter*, vol. 18 (Copenhagen: Gyldendalske Boghandel).

Bravo, Michael and Sverker Sörlin, eds (2002). *Narrating the Arctic: A Cultural History of Nordic Scientific Practices* (Canton MA: Science History Publications).

Broberg Nielsen, Jørgen and Ejvind Slottved (1983), "Fakultetets almene historie", pp. 1–112 in Mogens Pihl, ed., *Det matematisk-naturvidenskabelige Fakultet, 1. del, Københavns Universitet 1479–1979*, vol. 12 (Copenhagen: Gads Forlag).

Browne, Janet (2005). "Presidential Address: Commemorating Darwin", *British Journal for the History of Science 38*, 251–274.

Buchwald, Jed Z. and Sungook Hong (2003). "Physics", pp. 163–195 in Cahan 2003.

Buhl, Hans (2005). *Buesenderen: Valdemar Poulsens radiosystem* (Aarhus: Aarhus University Press).

Burton, Richard F. (1866). "On a Kjokkenmodding at Santos, Brazil". *Journal of the Anthropological Society of London 4*, cxciii–cxciv.

Cahan, David, ed. (2003). *From Natural Philosophy to the Sciences: Writing the History of Nineteenth-Century Science* (Chicago, IL: The University of Chicago Press).

Cantor, Geoffrey (1995). "The scientist as hero: public images of Michael Faraday" in Michael Shortland and Michael Yeo, eds, *Telling Lives in Science: Essays on Scientific Biography* (Cambridge: Cambridge University Press).

Chambers, Charles H. (1864). "The Danish Kitchen Middens". *Anthropological Review 2*, 60–61.

Christensen, Christian Villads (1904/1905). "Et besøg paa Maanen", *Det ny Aarhundrede*, 2nd year, vol. 1 (October 1904 – March 1905), 27–33, 99–107.

– (1912). *København i Kristian den Ottendes og Frederik den Syvendes Tid 1840 til 1857* (Copenhagen: Gads Forlag).

Claussen, Sophus (1983). "Ny Aand" in *Lyrik*, vol. 6 (Copenhagen: Det danske Sprog- og Litteraturselskab).

Clark, Mark and Henry Nielsen (1995), "Crossed Wires and Missing Connections: Valdemar Poulsen, The American Telephone Company, and the Failure to Commercialize Magnetic Recording". *Business History Review* 69 (1), 1-41.

Deans, James (1876). "Eating and Cooking Implements Found on the Shores of the San Fransico Bay". *The Journal of the Anthropological Institute of Great Britain and Ireland 5*, 489–490.

Dorph-Petersen, Jes (1961). *Danmarks Naturvidenskabelige Samfund 1911–1961* (Copenhagen: Berlingske).

Drachmann, Holger (1987). *En Overkomplet* (Copenhagen: Borgen).

Dreyer, Carl T. (1910). "Om Flyveindustrien og dens Betingelser i Danmark", *Illustreret Tidende* (29 May 1910, p. 432).

Ellinger, H. O. G. (1897–1898). *Naturen og dens Kræfter – Populær Fysik* (Copenhagen: Frem).

– (1919). "Den kgl. Veterinær- og Landbohøjskole", *Illustreret Tidende* (12 May 1919, p. 280).

Fischer, Anders and Kristian Kristiansen, eds (2002). *The Neolithisation of Denmark. 150 years of debate* (Sheffield: J. R. Collis Publications).

Flammarion, Camille (1910). *Verdens Undergang* (Copenhagen: Gyldendalske Boghandel – Nordisk Forlag).

Flower, J. G. (1869). "Notices of a Kjökken-Mödding in the Island of Herm". *Journal of the Anthropological Society of London 7*, cxv–cxix.

Fraenkel, A. (1897). "Gamle Carlsberg", *Illustreret Tidende* (7 November 1897, p. 97).

Frey, S. L. (1878). "Relic Hunting on the Mohawk". *The American Naturalist 12* (12), 777–785.

Gershenson, S. M. (1990). "Difficult Years in Soviet Genetics". *Quarterly Review of Biology 65*, 447—456.

Funder, Heidi (2003). "En historie om foranderlighed. Evolutionsteorien i Danmark 1860–80", *Historisk Tidsskrift 102* (2), 306–335.

Glamann, Kristof (1990). *Bryggeren. J. C. Jacobsen på Carlsberg* (Copenhagen: Gyldendal).

Glamann, K. and K. Glamann (2004). *Nordens Pasteur. Fortællingen om Emil Chr. Hansen.* (Copenhagen: Gyldendal).

Gooch, W. D. (1882). "The Stone Age in South Africa". *The Journal of the Anthropological Institute of Great Britain and Ireland 11*, 124–183.

Greibe, Valdemar. (1907). "Steins Laboratorium 1857–1907", *Illustreret Tidende* (23 June 1907, p. 500).

Hansen, H. C. (1985). *Poul la Cour – grundtvigianer, opfinder og folkeoplyser* (Askov: Askov Højskoles Forlag).

Hansen, H. L. (1887). "Den danske Presse og Spiritualismen", pp. 165–179 in Alfred Russel Wallace, *Spiritualismen for Videnskabens Domstol* (Copenhagen: V. F. Levinsons Forlag).

Hansen, B. S. and K. R. Olsen (1980). *Darwinism: en tekstsamling til belysning af sammenhængen mellem naturvidenskab og samfundsopfattelse* (Copenhagen: Gyldendal 1980).

Hagner, Michael (2003), "Scientific Medicine", pp. 49–87 in Cahan 2003.

Harding, M. C. (1912). "Den Polytekniske Læreanstalt – under G. A. Hagemanns Styrelse 1902–1912", *Illustreret Tidende* (12 May 1912, p. 400).

Hauch, Carsten (1929). *Udvalgte Skrifter*, vol. 3 (Copenhagen: J. Jørgensen og Co. Bogtrykkeri).

Haue, Harry, Jørgen Olsen and Jørn Aarup-Kristensen (1985). *Det ny Danmark 1890–1985 – Udviklingslinjer og tendens* (Copenhagen: Munksgaard).

Heegaard, Poul (1902). *Populær Astronomi* (Copenhagen: Det Nordiske Forlag).

Heilbron, John and Thomas Kuhn (1969). "The genesis of the Bohr atom", *Historical Studies of the Physical Sciences 1*, 211–290.

Hessenbruch, Arne and Flemming Petersen (2001). "Niels R. Finsen (1903): Banishing Darkness and Disease" in Henry Nielsen and Keld Nielsen, eds, *Neighbouring Nobel: The History of Thirteen Danish Nobel Prizes* (Aarhus: Aarhus University Press).

Hjermitslev, Hans Henrik, Casper Andersen and Peter C. Kjærgaard (2006). "Populærvidenskab og folkeoplysning", pp. 345–374 in Peter C. Kjærgaard 2006a.

Holter, H. and K. Max Møller (1976). *The Carlsberg Laboratory 1876–1976* (Copenhagen: Rhodos).

Humboldt, Alexander (1877). "The Physical Conditions and the Distinctive Characteristics of the Lapplanders, and the Races Inhabiting the North Coast of Europe". *The Journal of the Anthropological Institute of Great Britain and Ireland 6*, 316–323.

Hvidt, Kristian (1990). *Det folkelige gennembrud og dets mænd, Gyldendal og Politikens Danmarks-historie*, vol. 11 (Copenhagen: Gyldendal og Politikens Forlag).

Hyldtoft, Ole (2003). "Den anden industrielle revolution – i forskningen og i Danmark", *Den Jyske Historiker 102–103*, 18–46.

Ibsen, Eugen (1866). "Professor E. A. Scharling", *Illustreret Tidende* (23 September 1866, pp. 427–428).

Jacobsen, J. P. (1918). *Samlede Skrifter*, vol. 1 (Copenhagen: Gyldendalske Boghandel).

Jacobsen, Lis, ed. (1925). *Kvindelige Akademikere 1875–1925* (Copenhagen: Gyldendalske Boghandel, Nordisk Forlag).

Johannsen W. (1903a). *Om arvelighed i samfund og i rene linier*. Oversigt over det Kongelige Danske Videnskabernes Selskabs forhandlinger, No. 3: 247–270.

– (1903b). "Om Darwinismen set fra Arvelighedslærens Standpunkt", *Tilskueren*, 525–541.

– (1925). "Udtalelse", pp. 75–80 in Lis Jacobsen, ed., *Kvindelige Akademikere 1875–1925* (Copenhagen: Gyldendalske Boghandel, Nordisk Forlag).

– (1976). "Johan Kjeldahl", pp. 50–62 in H. Holter and K. Max Møller, *The Carlsberg Laboratory 1876–1976* (Copenhagen: Rhodos).

Kildebæk Nielsen, Anita (2000). "The Chemists. Danish Chemical Communities and Networks, 1900–1940" (Aarhus: Unpublished PhD thesis, History of Science Department, University of Aarhus).

– (2001). "En disciplins demarkation og selvforståelse: Dansk kemisk tidsskriftlitteratur i 1800-tallet", *Rotunden 16*, 25–53.

– (2006). "Kemi", pp. 119–140 in Peter C. Kjærgaard 2006a.

Kjærgaard, Peter C. (2000). "Defending Science – 'Genuine Scientific Men' and the Limits of Natural Knowledge" (Aarhus: Unpublished PhD thesis, Department of the History of Ideas, University of Aarhus).

– (2002). "Competing Allies: Professionalisation and the Hierarchy of Science in Victorian Britain". *Centaurus 44*, 248–288.

– (2004). " 'Within the Bounds of Science': Redirecting Controversies to *Nature*" in Louise Henson, Geoffrey Cantor, Gowan Dawson, Richard Noakes, Sally Shuttleworth and Jonathan Topham, eds, *Culture and Science in the Nineteenth-Century Media* (Aldershot: Ashgate).

– ed. (2006a). *Lys over Landet: Dansk Naturvidenskabs Historie*, vol. 3 (Aarhus: Aarhus University Press).

– (2006b). "Kundskab er magt", pp. 11–40 in Peter C. Kjærgaard 2006a.

– (2006c). "Naturvidenskaben styrkes", pp. 43–59 in Peter C. Kjærgaard 2006a.

– (2006d). "Den videnskabelige helt", pp. 263–266 in Peter C. Kjærgaard 2006a.

– (2006e). "Videnskabens kvinder", pp. 331–340 in Peter C. Kjærgaard 2006a.

– (2006f). "Den nye offentlighed", pp. 342–344 in Peter C. Kjærgaard 2006a.
– (2006g). "Videnskabens baggård", pp. 463–472 in Peter C. Kjærgaard 2006a.
– (2006h). " 'To the Benefit of Mankind, to Honour the Nation' – National Identity in an International World of Science", pp. 265–284 in Siegmund-Schulze, Reinhard and Henrik Kragh Sørensen, eds., *Perspectives on Scandinavian Science in the Early Twentieth Century* (Oslo: Novus Forlag).
– (2009, forthcoming). "Electric Wonderland: Visions of a Brighter Future in the Danish Periodical Press 1870–1917" in Graeme Gooday et al., ed., *Electrifying Cultures.*

Kjærgaard, Peter C. and Niels Henrik Gregersen (2006), "Darwinism comes to Denmark: The early Danish reception of Charles Darwin's *Origin of Species*", *Ideas in History 1*, 151–175.

Kjærgaard, Peter C., Niels Henrik Gregersen and Hans Henrik Hjermitslev (2008), "Darwinizing the Danes 1859–1909" in Eve-Marie Engels and Thomas Glick, eds., *The Reception of Charles Darwin in Europe* (London: Continuum).

Kjærgaard, Peter C. and Jens Erik Kristensen (2003), "Universitetets idéhistorie", pp. 31–144 in Hans Fink, Peter C. Kjærgaard, Helge Kragh and Jens Erik Kristensen, *Universitet og videnskab – Universitetets idéhistorie, videnskabsteori og etik* (Copenhagen: Hans Reitzels Forlag).

Klöcker, Albert (1976). "Emil Christian Hansen 1842–1909", pp. 168–189 in H. Holter and K. Max Møller, *The Carlsberg Laboratory 1876–1976* (Copenhagen: Rhodos).

Knudsen, Henrik (2005). "Konsensus og konflikt: Organiseringen af den tekniske forskning i Danmark 1900–1960" (Aarhus: PhD thesis, Steno Institute, University of Aarhus).

Knudsen, Henrik and Peter C. Kjærgaard (2006). "Industriel og privat forskning", pp. 239–250 in Peter C. Kjærgaard 2006a.

Knudsen, Susanne (1984). "Dansk Kvindesamfund, et rids" in Anne Margrete Berg, Lis Frost and Anne Olsen, eds, *Kvindfolk 2 – En danmarkshistorie fra 1600 til 1980* (Copenhagen: Gyldendal).

Koch, Carl Henrik (2006). "Filosofi og teologi", pp. 443–462 in Peter C. Kjærgaard 2006a.

Kohlstedt, Sally Gregory (1987). "International exchange and national style: A view of natural history museums in the United States, 1850–1900", pp. 167–190 in Nathan Reingold and Marc Rothenberg, eds, *Scientific colonialism: A cross-cultural comparison* (Washington, DC: Smithsonian Institution Press).

Korsgaard, Ove (1997). *Kampen om lyset – Dansk voksenoplysning gennem 500 år* (Copenhagen: Gyldendal).

Kragh, Helge (2006). "Fysik og astronomi", pp. 95–118 in Peter C. Kjærgaard 2006a.

Kragh, Helge and Hans Jørgen Styhr Petersen (1995). *En nyttig videnskab. Episoder fra den tekniske kemis historie i Danmark* (Copenhagen: Gyldendal).

Krebs, Johanne (1885). "Om Arbejdsfordelingen mellem Mænd og Kvinder", *Kvinden og Samfundet* (September, p. 114).

Larsen, C. C. (1899). *En lille letfattelig Husholdningskemi: Med grafisk Fortegnelse over Fødemidlernes Næringsindhold: Hvad enhver Husmoder bør vide* (Aarhus).

Larsen, Susanne and Birte Maj Tønsberg (1984). "Lærerinde eller sygeplejerske?" in Anne Margrete

Berg, Lis Frost and Anne Olsen, eds, *Kvindfolk 1 – En danmarkshistorie fra 1600 til 1980* (Copenhagen: Gyldendal).

Laws, Edward (1878). "On a 'Kitchen Midden' found in a Cave Near Tenby, Pembrokeshire and Explored by Wilmot Powers". *The Journal of the Anthropological Institute of Great Britain and Ireland 7*, 84–89.

Lehmann, Alfred (1920). *Overtro og Trolddom*, vol. 1–2 (Copenhagen: J. Frimodts Forlag).

Linderstrøm-Lang, K. (1976). "S. P. L. Sørensen 1868–1939", pp. 63–81 in H. Holter and K. Max Møller, *The Carlsberg Laboratory 1876–1976* (Copenhagen: Rhodos).

Lundbye, J. T. (1929). *Den Polytekniske Læreanstalt 1829–1929* (Copenhagen: Gads Forlag).

Lynning, Kristine Hays (2006). "Den lærde skole", pp. 399–414 in Peter C. Kjærgaard 2006a.

Lütken, Christian (1863). "Darwin's Theorie om Arternes Oprindelse", *Tidsskrift for Populære Fremstillinger af Naturvidenskaben*, 1–33, 131–162, 217–243, 243.

Meyer, Kirstine (1925). "Udtalelse", pp. 81–85 in Lis Jacobsen, red., *Kvindelige Akademikere 1875–1925* (Copenhagen: Gyldendalske Boghandel, Nordisk Forlag).

Milne, John (1888). "The Stone Age in Japan; with Notes on Recent Geological Changes Which Have Taken Place". *The Journal of the Anthropological Institute of Great Britain and Ireland 10*, 389–423.

Mygdal-Meyer, Toni (2002). *Da danskerne fik vinger* (Copenhagen: Gyldendal).

Møller, Jes Fabricius (2000), "Teologiske reaktioner på darwinismen i Danmark 1860–1900", *Historisk Tidsskrift 100* (1), 69–91.

– (2006). "Den videnskabelige by", pp. 415–430 in Peter C. Kjærgaard 2006a.

Mørch, Søren (1982). *Den ny danmarkshistorie 1880–1960* (Copenhagen: Gyldendal).

Nielsen, Henry and Birgitte Wistoft (1996). *Industriens mænd: Et Krøyer-maleris tilblivelse og industrihistoriske betydning* (Aarhus: Klim).

Oldroyd, David R. (2003), "The Earth Sciences", pp. 88–128 in Cahan 2003.

Oxenløwe, Ragna Heyn (2006). "Bygninger, politik og penge", pp. 61–94 in Peter C. Kjærgaard 2006a.

Pedersen, Johannes (1942). *Carlsbergfondet* (Copenhagen: Munksgaard).

Pedersen, Kenneth (1998). "Is-interferenser: København som verdenshovedstad for den etnografiske eskimoforskning i perioden 1900–1940" in Thomas Söderqvist, Jan Faye, Helge Kragh and Frank Allan Rasmussen, eds, *Videnskabernes København* (Roskilde: Roskilde Universitetsforlag).

Petersen, Jesper Krogh (2001). "Er der nogen? Danske forestillinger om liv uden for Jorden fra det nittende århundrede til første verdenskrig" (Aarhus: Unpublished MSc dissertation, History of Science Department, Univeristy of Aarhus).

Petersen, Niels (1993). "Københavns Universitet 1848–1902" in Leif Grane and Kai Hørby, eds, *Københavns Universitet 1479–1979*, vol. 2 (Copenhagen: Gads Forlag).

Reingold, Nathan (1991). "The peculiarities of the Americans, or, Are there national styles in the sciences?", *Science in Context 4*, 347–366.

Richards, Robert J. (2003). "Biology", pp. 16–48 in Cahan 2003.

Ries, Christopher J. (2003). *Retten, magten og æren. Lauge Koch Sagen: en strid om Grønlands geologiske udforskning* (Copenhagen: Lindhardt og Ringhof).

– (2006). "Polarforskeren", pp. 295–318 in Peter C. Kjærgaard 2006a.

Robson, Mike (1982). "Darwinismen i Danmark: debatten i offentlig sammenhæng", *Naturens Verden*, 224–236.

Rosenbeck, Bente (1994). "Kønnets grænser" in Marianne Alenius, Nanna Damsholt and Bente Rosenbeck, eds, *Clios døtre gennem hundrede år – I anledning af Anna Hudes disputats 1893* (Copenhagen: Museum Tusculanums Forlag).

Rubin, Marcus (1899). *Statistisk Bureaus Historie – Et Omrids udarbejdet i Anledning af Bureauets 50-aarige Bestaaen* (Copenhagen: Gyldendal).

Salomonsen, Carl Julius (1917). "Bemærkninger om 'naturvidenskabelig dannelse' " in *Smaa-Arbejder* (Copenhagen).

Sand-Jensen, K. (2003). *Den sidste naturhistoriker. Naturforkæmperen og videnskabsmanden Carl Wesenberg-Lund* (Copenhagen: Gads Forlag).

– (2006). "Naturhistorie", pp. 141–192 in Peter C. Kjærgaard 2006a.

Schmidt Kjærgaard, Rikke (2006). "Videnskabens offentlige rum", pp. 375–398 in Peter C. Kjærgaard 2006a.

Schmidt-Nielsen, B. (1997). *August og Marie Krogh. Et fælles liv for videnskaben* (Copenhagen: Gyldendal).

Sestoft, Ingolf (1972). "Dansk Meteorologi gennem tiderne", in *Meteorologisk Institut 1872–1972* (Copenhagen: Det Danske Meteorologiske Institut).

Skydsgaard, Morten A. (2006). "Medicin", pp. 217–238 in Peter C. Kjærgaard 2006a.

Slottved, Ejvind (2000). "Det naturvidenskabelige Fakultet fylder 150 år!", pp. 3–21 in Jesper Lützen, ed., *Fakultære højdepunkter. Episoder fra Det naturvidenskabelige Fakultets 150-årige historie* (Copenhagen: The Faculty of Science, University of Copenhagen).

Spärck, R. (1932). "Japetus Steenstrup, 1812–1897", pp. 115–119 in V. Meisen, ed., *Prominent Danish Scientists Through the Ages with Facsimiles from their Works* (Copenhagen: Levin & Munksgaard).

Steenstrup, Japetus (1848). "Nogle Iagttagelser Angaaende Tiden, da visse hævede Lag af Östers- og Muslingeskaller vare dannede". *Oversigt over Videnskabernes Selskabs Forhandlinger* 1848, 1–12.

– *Les* kjøkkenmøddings *de l'âge de la pierre et sur la faune et la flore préhistorique de Danmark* (Copenhagen: Thiele).

Steenstrup, Johannes (1909). "Darwins Brevveksling med Professor Japetus Steenstrup", *Tilskueren*, 217–223.

Stuckenberg, Viggo (1910). *I Gennembrud*, Samlede Værker, vol. 1 (Copenhagen: Gyldendal).

Sørensen, Henrik Kragh (2006). "Matematik og statistik", pp. 193–216 in Peter C. Kjærgaard 2006a.

Thomsen, Julius (1856). Vandringer paa Naturvidenskabens Gebeet (Copenhagen: C. A. Reitzel).

– (1884). "Om Molekuler og Atomer", *Indbydelsesskrift til Kjøbenhavns Universitets Aarsfest til Erindring om Kirkens Reformation* (Copenhagen: J. H. Schultz).

– (1887). "Om Materiens Enhed", *Indbydelsesskrift til Kjøbenhavns Universitets Aarsfest i Anledning af Hans Majestæt Kongens Fødselsdag den 8de April 1887* (Copenhagen: J. H. Schultz).

Thomsen, Niels (1998). *Hovedstrømninger 1870–1914 – Idélandskabet under dansk kultur, politik og hverdagsliv* (Odense: Odense Universitetsforlag).

Thomson, J. J. (1936). *Recollections and Reflections* (London: G. Bell and Sons, Ltd.).

Tregear, Edward (1888). "The Maori and the Mao", *The Journal of the Anthropological Institute of Great Britain and Ireland 17*, 292–305.

Videbech, Poul (1909). "Finsens Lysinstitut", *Illustreret Tidende* (18 August 1901, p. 720).

Wagner, Michael (2006). "Ingeniøren", pp. 269–294 in Peter C. Kjærgaard 2006a.

Wallace, Alfred Russell (1887). *Spiritualismen for Videnskabens Domstol* (Copenhagen: V. F. Levinsons Forlag).

Wengenroth, Ulrich (2003). "Science, Technology, and Industry", pp. 212–253 in David Cahan 2003a.

Wesenberg-Lund, Carl (1898). "Über dänischen Rotiferen und über die Ferpflanzungsverhältnisse der Rotiferen", *Zoologischer Anzeiger 21*, 200–211.

Wistoft, Birgitte, Flemming Petersen and Harriet M. Hansen (1991). *Elektricitetens Aarhundrede – Dansk elforsyningshistorie*, vol. 1, 1891–1940 (Danske Elværkers Forening).

Part IV

Aaserud, Finn (1990). *Redirecting Science. Niels Bohr, Philanthropy, and the Rise of Nuclear Physics* (Cambridge: Cambridge University Press).

Aaserud, Finn (1998). "Videnskabernes København i 1920'erne belyst af amerikansk filantropi", pp. 201–221 in Thomas Söderqvist et al., eds, *Videnskabernes København* (Copenhagen: Roskilde Universitetsforlag).

Aaserud, Finn and Henry Nielsen (2006). "Niels Bohrs verdenscenter for teoretisk fysik", pp. 41–66 in Nielsen and Nielsen 2006.

Agrell, Wilhelm (2002). *Svenska förintelsesvapen. Utveckling af kemiska och nukleära stridsmedel 1928–70* (Falun: Historiska Media).

Andersen, Torkild (2004). "Niels Bjerrums rolle i Carlsbergfondet", pp. 139–154 in Anita Kildebæk Nielsen, ed., *Niels Bjerrum (1879–1958). Liv og værk* (Copenhagen: Dansk Selskab for Historisk Kemi).

Andreasen, Alfred H. M., ed. (1954). *Den polytekniske Læreanstalt* (Copenhagen: Gjellerups Forlag).

Ballhausen, Carl J. (1982). "Carlsbergfondets Biologiske Institut under Albert Fischer 1931–1956", *Carlsbergfondet, Årsskrift 1982*, pp. 44–45.

Barker Jørgensen, C. (1979). "Dyrefysiologi og gymnastikteori", pp. 447–488 in *Københavns*

Universitet 1479–1979, vol. 13 (Copenhagen: G. E. C. Gads Forlag).

Berg Hansen, Inge, ed. (1979). *DTH Polyteknisk undervisning og forskning i det 20. århundrede* (Copenhagen: Polyteknisk Forlag).

Bernal, John Desmond (1939). *The Social Function of Science* (London: Routledge).

Blegvad, Mogens (1992). *Det Kongelige Danske Videnskabernes Selskab 1942–1992* (Copenhagen: Munksgaard).

Bliss, Michael (2000). *The Discovery of Insulin* (Toronto: University of Toronto Press).

Blædel, Niels (1954). "Den videnskabelige medarbejder har brug for selv at være uvidende", pp. 121–127 in E. Hansen and H. L. Monberg, eds, *Journalisten og avisen. 21 journalister fortæller om deres arbejde* (Copenhagen: Journalistforbundet).

Bohr, Aage and Ben Mottelson (1969 & 1975). *Nuclear Structure*, vols 1–2 (New York, NY: W. A. Benjamin 1969 and 1975).

Bohr, Niels (1921). "Tale ved Indvielsen af Universitetets Institut for teoretisk Fysik (3. Marts 1921)", manuscript reprinted in *Niels Bohr Collected Works*, vol. 3, 284–293.

Bomholt, Julius (1942). "Forskellige oplysningsbestræbelser", pp. 260–277 in C. Dahl, ed., *Danmarks Kultur ved Aar 1940*, vol. 6 (Copenhagen: Det danske Selskab).

Borgen, Mikkel (1945). "Danske Videnskabsmænds Indsats i Danmarks Frihedskamp", *Illustreret Familie-Journal*, No. 25, 19 June 1945, p. 2.

Bruun, Anton F., Svend Greve, Hakon Mielche, and Ragnar Spärck, eds (1956). *The Galathea Deep Sea Expedition 1950–1952. Described by Members of the Expedition* (London: George Allen and Unwin).

Bush, Vannevar (1945). *Science – The Endless Frontier. A Report to the President* (Washington, DC: United States Government Printing Office).

Bøggild, Jørgen (1969). "Europas Store Maskine", *Politiken* 27 May; "Det lille land og grundvidenskaberne", *Politiken* 28 May.

Carson, Rachel (1963). *Silent Spring* (Boston: Houghton).

Christiansen, Niels Finn (1993). *Hartvig Frisch. Mennesket og politikken. En biografi* (Copenhagen: Chr. Ejlers' Forlag).

Crawford, Elisabeth (1987). *The Beginnings of the Nobel Institution. The Science Prizes, 1901–1915* (Cambridge: Cambridge University Press).

Danish Broadcasting Corporation, Press Service (1966). Various press releases about the programme *Spørg Århus*, issued on 27 September 1966, 10 October 1966, 15 November 1966 and 14 December 1966. Statens Mediesamling, the State Library, Aarhus.

Danish Broadcasting Corporation, Press Service (1967). Press releases about the programme *Spørg Århus*, issued on 16 October 1967. Statens Mediesamling, the State Library, Aarhus.

Dansgaard, Willi (1969). "One Thousand Centuries of Climatic Record from Camp Century on the Greenland Ice Cap", *Science 166*, pp. 377–381.

Dansgaard, Willi (2000). *Grønland – i istid og nutid* (Copenhagen: Rhodos).

Dansgaard, Willi (2004). *Frozen Annals – Greenland Ice Sheet Research* (Copenhagen: University of

Copenhagen).

Dansk Naturvidenskabs Historie, vols 1–4 (2005–2006). Helge Kragh, Peter C. Kjærgaard, Henry Nielsen and Kristian H. Nielsen, eds (Aarhus: Aarhus University Press).

Deckert, Torsten (1998). *Dr. med. H. C. Hagedorn og det danske insulin-eventyr* (Herning: Poul Kristensens Forlag).

Estrup, Knud (1923). *Om Atomer og Molekyler* (Copenhagen: V. Pios Boghandel).

Fog, Mogens (1977). *Efterskrift. 1946 – og resten* (Copenhagen: Gyldendal).

Forland, Astrid (1987). *På leiting etter uran. Institutt for Atomenergi og internasjonalt samarbeid 1945–51* (Oslo: Institutt for Forsvarsstudier).

Forland, Astrid (1988). *Atomer for krig eller fred? Etablering af Institutt for Atomenergi 1945–48* (Oslo: Institutt for Forsvarsstudier).

Forman, Paul (1971). "Weimar Culture, Causality, and Quantum Theory, 1918–1927: Adaption by German Physicists and Mathematicians to a Hostile Intellectual Environment", *Historical Studies in the Physical Sciences*, iii (1971), 1–115.

Freeman, Chris (1999). "The Social Function of Science", pp. 101–131 in B. Swann and F. Aprahamian, eds, *J. D. Bernal. A Life in Science and Politics* (London: Verso).

Frisch, Otto Robert (1964). "The Interest is Focussing on the Atomic Nucleus", pp. 137–148 in Stefan Rozental, ed., *Niels Bohr: His Life and Work as Seen by his Friends and Colleagues* (Amsterdam: North Holland Publishing Company).

Frisch, Otto Robert (1979). *What Little I Remember* (Cambridge: Cambridge University Press).

Gade Rasmussen, H. (1975). "Da hele universitetet demonstrerede", pp. 11–18 in *Om et håb eller to blev brudt… Festskrift til moderate studenter 1970–1975* (Copenhagen: Forlaget ST).

Galison, Peter (1999). "The Many Faces of Big Science" pp. 1–20 in P. Galison and B. Hevly, eds, *Big Science. The Growth of Large-Scale Research* (Stanford, CA: Stanford University Press).

Galison, Peter and Bruce Hevly, eds (1999). *Big Science. The Growth of Large-Scale Research* (Stanford, CA: Stanford University Press).

Gasgoigne, John (1998). *Science in the Service of Empire: Joseph Banks, the British State, and the Uses of Science in the Age of Revolution* (Cambridge: Cambridge University Press).

Glamann, Kristof (1988). *Louis Pasteur and Carlsberg. A contribution to the history of pure yeast cultivation* (Copenhagen: The Carlsberg Foundation).

Glamann, Kristof (1991). *Jacobsen of Carlsberg: Brewer and Philantropist* (Copenhagen: Gyldendal).

Glamann, Kristof (2003). *The Carlsberg Foundation – The Early Years* (Copenhagen: The Carlsberg Foundation).

Graae, F. (1941). *Dansk videnskab og udlandet* (Copenhagen: Gads Forlag).

Gregory, Jane and Steven Miller (1998). *Science in Public. Communication, Culture and Credibility* (Cambridge, MA: Perseus Publishing).

Grønbæk, David (2001). *Mellem politik og videnskab. Reform af det danske forskningsrådssystem 1994–1997* (Copenhagen: PhD thesis, Department of Political Science, University of Copenhagen).

Grønborg, Per (1996). *Teknisk-videnskabelig forskning i 50 år* (Copenhagen: Statens Teknisk-videnskabelige Forskningsråd).

Guggenheim, E. A. (1960). "The Niels Bjerrum Memorial Lecture", *Proceedings of the Chemical Society of London,* 104–114.

Hansen, Peder G. (1996). "The SC: Isolde and Nuclear Structure", pp. 327–413 in John Krige, ed., *History of CERN,* vol. III.

Heiberg, J. L. (1919). "Videnskabens udsigter efter krigen", *Tilskueren,* vol. 34, 21–27.

Heisenberg, Werner (1930a). *Die physikalische Prinzipien der Quantentheorie* (Leipzig: S. Hirzel).

Heisenberg, Werner (1930b). *The Physical Principles of the Quantum Theory* (New York, NY: Dover).

Heisenberg, Werner (1971). *Del og helhed* (Copenhagen: Thanning og Appel).

Hermann, Armin, John Krige, Dominique Pestre and Ulrike Mersits (1987). *History of CERN,* vol. I (Amsterdam: North Holland).

Hermann, Armin, John Krige, Dominique Pestre and Ulrike Mersits (1990). *History of CERN,* vol. II (Amsterdam: North Holland).

Hermansen, Niels K. and Max A. Lobedanz (1958). *Den kongelige veterinær- og landbohøjskole 1858–1958* (Copenhagen: Kandrup & Wunsch).

Holmgaard, Jørgen (2000). "Universitetet", pp. 658–694 in Tim Knudsen, ed., *Dansk Forvaltningshistorie, 1901–1953. Folkestyret og forvaltningen* (Copenhagen: Jurist- og Økonomforbundets Forlag).

Holter, Heinz (1965). "Videnskab og Samfund", feature article in *Politiken* (23 March 1965).

Holter, Heinz and K. Max Møller, eds (1976). *The Carlsberg Laboratory 1876–1976* (Copenhagen: Rhodos).

Iliopoulos, John (1996). "Physics in the CERN Theory Division", pp. 277–326 in Armin Hermann et al., *History of CERN,* vol. III (Amsterdam: North Holland).

Johnsen, Sigfus J. et al. (1976). "Climatic Oscillations 1200–2000 AD", *Nature* 227, pp. 482–483.

Jungnickel, Christa and Russell McCormmach (1986). *Intellectual Mastery of Nature: Theoretical Physics from Ohm to Einstein,* vol. 2: *The Now Mighty Theoretical Physics 1870–1925* (Chicago, IL: Chicago University Press).

Kaijser, Arne (1992). "Redirecting Nuclear Power: Swedish Nuclear Policies in Historical Perspective", *Annual Review of Energy and Environment* 1992, 437–462.

Kampmann, Viggo (1957a). *Samfundet og fremtidens tekniske udvikling* (Copenhagen: Socialdemokratisk Forbund).

Kampmann, Viggo (1957b). "Videnskaben og Samfundet", feature article in *Dagens Nyheder* (25 November 1957).

Kevles, Daniel J. (1997). "Big science and big politics in the United States: Reflections on the death of the SCC and the life of the human genome project", *Journal for the History of Physical and Biological Sciences 22* (2), 269–297.

Kevles, Daniel J. (1995). *The Physicists – The History of a Scientific Community in Modern America*

(New York, NY: Harvard University Press).
Kildebæk Nielsen, Anita and Helge Kragh (1997). "An Institute for Dollars: Physical Chemistry in Copenhagen Between the Wars", *Centaurus 39*, 311–331.
Kildebæk Nielsen, Anita (1997a). "Johannes Nicolaus Brønsted", pp. 11–29 in Riis Larsen, ed., *J.N. Brønsted. En dansk kemiker* (Copenhagen: Dansk Selskab for Historisk Kemi).
Kildebæk Nielsen, Anita (1998). "Kemien i København – netværk og niveau", pp. 180–200 in Thomas Söderqvist et al., eds, *Videnskabernes København* (Copenhagen: Roskilde Universitetsforlag).
Kildebæk Nielsen, Anita (2000). *The Chemists. Danish Chemical Communities and Networks, 1900–1940* (Aarhus: Unpublished PhD thesis, History of Science Department, University of Aarhus).
Kildebæk Nielsen, Anita (2001). "August Krogh 1920: Scientist Explains the Blushing of Girls", pp. 430–460 in Nielsen and Nielsen 2001.
Kildebæk Nielsen, Anita (2004). "Niels Bjerrum – Et liv i kemien", pp. 101–128 in Anita Kildebæk Nielsen, ed., *Niels Bjerrum (1879–1958). Liv og værk* (Copenhagen: Dansk Selskab for Historisk Kemi).
Kildebæk Nielsen, Anita and Torkild Andersen (2006). "Dansk kemi fra 1920 til 1960'erne", pp. 67–84 in Nielsen and Nielsen 2006.
Kildebæk Nielsen, Anita and Thomas Vorup-Jensen (2006). "Fysiologisk forskning i verdensklasse", pp. 165–182 in Nielsen and Nielsen 2006.
Knudsen, Henrik (2000). *Det teknisk-videnskabelige Forskningsråd – en beretning om småkævl og teknisk forskning.* (Aarhus: Unpublished paper, History of Science Department, University of Aarhus).
Knudsen, Henrik (2005). *Konsensus og konflikt: Organiseringen af den tekniske forskning i Danmark 1900–1960* (Aarhus: PhD thesis, Steno institute, University of Aarhus).
Knudsen, Henrik (2006). "Politik, penge og forskningsvilkår", pp. 323–346 in Nielsen and Nielsen 2006.
Knudsen, Henrik and Kristian H. Nielsen (2006). "Forskning uden for universiteterne", pp. 259–282 in Nielsen and Nielsen 2006.
Koch, Hans H. (1967). "Science and Administration", pp. 310–314 in Stefan Rozental, ed., *Niels Bohr. His life and work as seen by his friends and colleagues* (Amsterdam: North Holland Publishing Company).
Koch, Lene (1996). *Racehygiejne i Danmark 1920–56* (Copenhagen: Gyldendal).
Kohler, Robert E. (1991). *Partners in Science: Foundations and Natural Scientists 1900–1945* (Chicago, IL and London: University of Chicago Press).
Kohler, Robert E. (2002). *Landscapes and Labscapes. Exploring the Lab-Field Border in Biology* (Chicago, IL and London: University of Chicago Press).
Kragh, Helge and Hans Jørgen Styhr Petersen (1995). *En nyttig Videnskab. Episoder fra den tekniske kemis historie i Danmark* (Copenhagen: Gyldendal).

Kragh, Helge (1999). *Quantum Generations. A History of Physics in the Twentieth Century* (Princeton, NJ: Princeton University Press).

Kragh, Helge, Peter C. Kjærgaard, Henry Nielsen and Kristian H. Nielsen, eds (2005–2006). *Dansk Naturvidenskabs Historie*, vols 1–4 (Aarhus: Aarhus University Press).

Krige, John, ed. (1996). History of CERN, vol. III (Amsterdam: North Holland).

Krige, John (1999). "The Ford Foundation, European Physics, and the Cold War", *Historical Studies in the Physical and Biological Sciences*, 333–361.

Krige, John (2006). *American Hegemony and the Postwar Reconstruction of Science in Europe* (Cambridge, MA: MIT Press).

Krogh, August (1924). *Insulin: en Opdagelse og dens Betydning* (Copenhagen: Gad).

Krogh, August (1930). "Institute of Physiology, University of Copenhagen", *Methods and Problems of Medical Education 18*, 173–188.

Kryger Larsen, Hans (1993). "Københavns Universitet 1902–1936", pp. 455–604 in *Københavns Universitet 1479–1979*, vol. 2 (Copenhagen: G. E. C. Gads Forlag).

Kuklick, Henrika and Robert E. Kohler, eds (1996a). *Science in the Field, Osiris 11*.

Kuklick, Henrika and Robert E. Kohler (1996). "Introduction", pp. 1–14 in H. Kuklick and R. E. Kohler, eds, *Science in the Field, Osiris 11*.

Källen, Gunnar (1964). "Nordiske Institut for elementarpartikelfysik", *Nordiske Betænkninger 1964* (3).

Københavns Universitet 1479–1979, vols 1–14 (1979–2005). Svend Ellehøj and Leif Grane, eds (Copenhagen: Gads Forlag).

Lagaaij, A. and G. Verbong (1998). *Kerntechniek in Nederland 1945–1974* (Eindhoven: KiVI).

Larsen, Erik H. (2002). "Hans H. Ussing – Scientific work: contemporary significance and perspectives", *Biochimica et biophysica acta 1566*, 2–15.

Larsson, Ulf (2001). *Cultures of Creativity: The Centennial Exhibition of the Nobel Prize* (Canton, MA: Science History Publications).

Lindqvist, Svante (1993). "Introductory Essay: Harry Martinson and the Periphery of the Atom", pp. xi–lv in Svante Lindqvist et al, eds, *Center on the Periphery. Historical Aspects of 20th-Century Physics* (Canton, MA: Science History Publications).

Linderström-Lang, K.U. (1952). "Proteins and Enzymes", *Lane Medical Lectures,* Stanford University Publications, University Series, Medical Science, vol. 6, Stanford University Press.

Lindström, Stefan (1991). *Hela nationens tacksamhet: Svensk forskningspolitik på atomenergiområdet 1945:56* (Stockholm: Statsvetenskapliga Institutionen).

Lolck, Maiken (2004). *Klima, kold krig og iskerner: En historie om baggrunden for dansk iskerneforskning og den første internationale dybdeboring i Grønland* (Aarhus: Unpublished MSc dissertation, Steno Institute, University of Aarhus).

Lolck, Maiken (2007). *Klima, kold krig og iskerner* (Aarhus: Aarhus University Press).

Lundager Jensen, Bjarne (1996). "Dansk forskningspolitik – fra finkultur til national strategi", *Økonomi & Politik 69* (4), 30–39.

MacLeod, Roy, ed. (2000). "Nature and Empire", *Osiris 15*, 1–323.

Madsen, Thorvald (1940). *Statens Seruminstitut 1902–1940* (Copenhagen: Bianco Lunos Bogtrykkeri).

Marner, Joakim (1998). "Niels Bohr og Københavnerskolen i fysik", pp. 159–179 in Thomas Söderqvist et al., eds, *Videnskabernes København* (Copenhagen: Roskilde Universitetsforlag).

Max Møller, Knud (1982). "Albert Fischer – forskeren bag oprettelsen af Carlsbergfondets Biologiske Institut", *Carlsbergfondet, Årsskrift 1982*, pp. 38–43.

Merton, Robert K. (1968). "The Matthew Effect in Science", *Science 159*, 56–63.

Meyer-Thurow, G. (1982). "The Industrialization of Invention: A Case Study from the German Chemical Industry", *Isis 73*, 363–381.

Michelsen, Børge (1940a). "Takseret til 3 Aars Fængsel", *Illustreret Familie-Journal 64* (3), 5–6.

Michelsen, Børge (1940b). "Min ven, Betjenten på Hjørnet", *Illustreret Familie-Journal 64* (20), 14–15.

Michelsen, Børge (1940c). "De farlige Vinduesfag", *Illustreret Familie-Journal 64* (34), 5–7.

Michelsen, Børge (1941). *Vore unge Videnskabsmænd arbejder*, vol. 1–3 (Copenhagen: Chr. Erichsen).

Michelsen, Børge (1943). "Dansk Videnskab paa Vej mod Proletarisering", *Politiken* 3 February 1943, 8–9.

Michelsen, Børge (1946). "Ekstrabladets Ærespris uddelt i Dag: Skaberen af Gravimeteret faar i Dag de 10.000 Kroner", *Ekstrabladet* 12 February 1946, 1.

Michelsen, Børge (1947). "Populærvidenskaben og Verden i Dag", *Ekstrabladet* 7 November 1947, 7–8.

Michelsen, Børge (1954). *Rejsende i populærvidenskab* (Copenhagen: Jespersen og Pios Forlag).

Mielche, Hakon (1951). *Galathea lægger ud. Rejsen rundt om Afrika* (Copenhagen: Steen Hasselbalchs Forlag).

Mielche, Hakon (1952). "1000 spalter forsidestof," *Journalisten 48*, 7–10.

Mielche, Hakon (1979). *Mennesker jeg mødte* (Copenhagen: Thaning og Appel).

Møller, E. Schelde (1949). "Ny journalistsammenslutning", *Videnskabsmanden 3* (2), 11.

Nielsen, Henry, Keld Nielsen, Flemming Petersen and Hans Siggaard Jensen (1998). *Til samfundets tarv – Forskningscenter Risøs historie* (Roskilde: Forskningscenter Risø).

Nielsen, Henry and Keld Nielsen, eds (2001). *Neighbouring Nobel: The History of Thirteen Danish Nobel Prizes* (Aarhus: Aarhus University Press).

Nielsen, Henry (2006). "Opbygningen af Det naturvidenskabelige Fakultet ved Aarhus Universitet", pp. 219–238 in Nielsen and Nielsen 2006.

Nielsen, Henry (2006a). "Risøs oprettelse og kamp for at finde en egnet forskningsstrategi", pp. 239–258 in Nielsen and Nielsen 2006.

Nielsen, Henry and Kristian Hvidtfelt Nielsen, eds (2006). *Viden uden grænser 1920–1970. Dansk Naturvidenskabs Historie*, vol. 4 (Aarhus: Aarhus Universitetsforlag).

Nielsen, Kristian H., Christopher J. Ries and Kristian B. Østergaard (2006). "Danske videnskabelige ekspeditioner i perioden", pp. 283–302 in Nielsen and Nielsen 2006.

Njølstad, Olav (1998). *Fra atomskip til oljeboringsplatformer. Institutt for Energiteknikk, 1948–1998* (Oslo: Tano Aschehough).

Noble, David F. (1977). *America by Design. Science, Technology, and the Rise of Corporate Capitalism* (New York, NY: Alfred A. Knopf).

Noe-Nygaard, Arne (1979). "Geologi", pp. 261–376 in *Københavns Universitet 1479–1979*, vol. 13, part 2 (Copenhagen: Gads Forlag).

Nørregaard Rasmussen, Poul (1968). "De nye forskningsråd", *Uddannelse 1*, 187–189.

OECD (1963). *Science, Economic Growth and Government Policy* (Paris: OECD Publications).

OECD (2002). *Frascati Manual* (Paris: OECD Publications).

Rasmussen, Klaus (2002). *Det danske engagement i CERN 1950–70* (Aarhus: Unpublished MSc dissertation, History of Science Department, University of Aarhus, accessible at http://www.ivh.au.dk/hosta/hosta010.pdf).

Rasmussen, Klaus, Simon O. Rebsdorf and Henry Nielsen (2006). "Det lille land og big science", pp. 303–322 in Nielsen and Nielsen 2006.

Reich, Leonard (1985). *The Making of American Industrial Research. Science and Business at the GE and Bell, 1876–1926* (New York, NY: Cambridge University Press).

Richter-Friis, Helge (1991). *Livet på Novo* (Copenhagen: Gyldendal).

Rozental, Stefan (1994). *Niels Bohr – Erindringer om et samarbejde* (Copenhagen: Christian Ejlerts' Forlag).

Robertson, Peter (1979). *The Early Years. The Niels Bohr Institute 1921–1930* (Copenhagen: Akademisk Forlag).

Pais, Abraham (1991). *Niels Bohr's Times in Physics, Philosophy, and Polity* (Oxford: Oxford University Press).

Pais, Abraham (1994). *Niels Bohr og hans tid i fysik, filosofi og samfundet* (Copenhagen: Spektrum).

Paulsen, Ove (1919). "International Videnskab efter Krigen", *Naturens Verden 3*, 145–149.

Pestre, Dominique (1997). "Science, Political Power and the State", pp. 61–76 in John Krige and Dominique Pestre, eds, *Science in the Twentieth Century* (Amsterdam: Harwood Academic Publishers).

Price, Derek J. de Solla (1963). *Little Science, Big Science* (New York, NY: Columbia University Press).

Produktions- og Raastofkommissionen (1942). *Betænkning angaaende Teknisk-videnskabelig Forskning* (Copenhagen: Schultz)

Pyenson, Lewis (1985). *Cultural Imperialism and the Exact Sciences: German Expansion Overseas, 1900–1930* (New York, NY: P. Lang).

Reingold, Nathan and Marc Rothenberg, eds (1986). *Scientific Colonialism. A Cross-Cultural Comparison* (Washington, DC: Smithsonian Institution Press).

Ries, Christopher Jacob (2003). *Retten, magten og æren. Lauge Koch Sagen – en strid om Grønlands udforskning* (Copenhagen: Lindhardt og Ringhof).

Robinson, Joseph D. (1997). *Moving questions – a history of membrane transport and bioenergetics* (Oxford: Oxford University Press).

Schellman, J.A. and C.G. Schellman (1997). "Kaj Ulrik Linderström-Lang (1896–1959)". *Protein Science* 6 (5), 1092–1100.

Schiebinger, Londa (2004). *Plants and Empire: Colonial Bioprospecting in the Atlantic World* (Cambridge, MA: Harvard University Press).

Schiebinger, Londa (2005). "Forum Introduction: The European Colonial Science Complex", *Isis* *96*, 52–55.

Schmidt-Nielsen, Bodil (1995). *August and Marie Krogh. Lives in Science* (New York, NY: Oxford)

Schmidt-Nielsen, Bodil (1997). *August og Marie Krogh – Et fælles liv for videnskaben* (Copenhagen: Gyldendal).

Shapin, Steven (1988). "The house of experiment", *Isis 79*, 373–404.

Shapin, Steven (1995). "Truth, honesty, and the authority of science", pp. 388–408 in R. E. Bulger et al., eds, *Society's Choices: Social and Ethical Decision Making in Biomedicine* (Washington, DC: National Academy Press).

Skovmand, Roar (1949). *Lys over Landet* (Copenhagen: Arbejdernes Oplysningsforbund).

Spengler, Oswald (1918 and 1922). *Untergang des Abendlandes* (Munich: C. H. Beck Verlag). This book appeared in Danish in 1962, published by Aschehoug as *Vesterlandets undergang: omrids af en verdenshistoriens morfologi.*

Stang, Frederik (1917). *Norden som centralsted for internationalt videnskabelig arbeide. Foredrag ved det nordiske interparlamentariske delegeretmøte i Kristiania 29de juni 1917* (Kristiania: Centraltrykkeriet).

Steeman Nielsen, Ejnar (1956). "Measuring the Productivity of the Sea", pp. 53–64 in A. F. Bruun et al., eds, *The Galathea Deep Sea Expedition 1950–1952. Described by Members of the Expedition* (London: George Allen and Unwin).

Stokes, Donald E. (1997). *Pasteur's Quadrant. Basic Science and Technological Innovation* (Washington, DC: Brookings Institution Press).

Swan, Claudia and Londa Schiebinger, eds (2005). *Colonial Botany: Science, Commerce, and Politics* (Philadelphia, PA: University of Philadelphia Press).

Söderqvist, Thomas et al., eds (1998). *Videnskabernes København* (Copenhagen: Roskilde Universitetsforlag).

Tegart, Greg, Francois Jacq, Josef Rembser, Kathriné E. Barker, and Albert H. Teich (1995). *Megascience Policy Issues* (Paris: OECD Publications).

The Rockefeller Foundation (1932). *Annual Report 1932* (New York, NY: The Rockefeller Foundation).

The Rockefeller Foundation (1972). *Directory of Fellowships and Scholarships 1917–1970* (New York, NY: The Rockefeller Foundation).

Thomsen, Niels (1972). *Dagbladskonkurrencen 1870–1970. Politik, journalistik og økonomi i dansk dagspresses strukturudvikling* (Copenhagen: Universitetsforlaget i København).

Thomsen, Niels (1986). "Københavns Universitet 1936–1966", pp. 1–287 in *Københavns Universitet 1479–1979*, vol. 3 (Copenhagen: G. E. C. Gads Forlag).

Thorsen, Svend (1947). *Den danske dagspresse I. Avisernes Udvikling indtil 1940. De førende Mænd* (Copenhagen: Det danske Selskab).

Tovborg Jensen, Aksel (1958), "Dansk kemiker af verdensry død", *Politiken* 1 October 1958, 5.

Ussing, Hans (1980). "Life with tracers", *Annual Review of Physiology 42*, 1–16.

Ussing, Hans and Karl Zerahn (1951). "Active transport of sodium as the source of electric current in the short-circuited isolated frog skin", *Acta Physiologica Scandinavica 23*, 110–127.

van Berkel, Klaas (1999). "The Interbellum Period, 1914–1940", pp. 170–209 in Klaas van Berkel et al., eds, *A History of Science in the Netherlands. Survey, Themes and Reference* (Leiden: Brill).

von Ungern-Sternberg, Jürgen and Wolfgang von Ungern-Sternberg (1996). *Der Aufruf 'An die Kulturwelt!'* (Stuttgart: Franz Steiner Verlag).

Vorup-Jensen, Thomas (2001). "Jens Christian Skou (1997) – Research on the Periphery", in Nielsen and Nielsen 2001.

Weinberg, Alvin M. (1961). "Impact of Large-Scale Science on the United States", *Science 134*, 161–164.

Weis-Fogh, Torkel (1964). "Staten må investere i talentet", feature article in *B.T.* 22 February 1964.

Wise, George (1985). "Science and Technology", *Osiris 1*, 229–246.

Wolff, Torben (1967). *Danske ekspeditioner på verdenshavene. Dansk havforskning gennem 200 år* (Copenhagen: Rhodos).

Wolff, Torben (1979). "Zoologi", pp. in 1–162 in *Københavns Universitet 1479–1979*, vol. 13, part 1 (Copenhagen: Gads Forlag).

Worboys, Michael (1976). "The Emergence of Tropical Medicine: A Study in the Establishment of Scientific Speciality", pp. 76–98 in G. Lemaine et al., eds, *Perspectives on the Emergence of Scientific Disciplines* (The Hague: Mouton).

Index

E

K

L

T

W

X

Z

Bergen
Oslo
NORWAY
Stavanger
Stockholm
Kristianssand
SWEDEN
Skagerak
Gothenburg
Aalborg
HALLAND
Kattegat
BLEKINGE
NORTH SEA
DENMARK
Aarhus
Elsinore
Lund
JUTLAND
SCANIA
Copenhagen
ZEALAND
Esbjerg
FUNEN
Flensburg
BALTIC SEA
Kiel
GERMANY